G. Hamm • G. Burk

Tables for the Automotive Trade

Wiley Eastern Limited
New Delhi Bangalore Bombay Calcutta Guwahati
Hyderabad Lucknow Madras Pune

Special edition for the

Deutsche Gesellschaft für Technische Zusammenarbeit
(GTZ) GmbH, Eschborn (Federal Republic of Germany)

Special edition within the scope of Technical Cooperation in the field of Vocational Training

Original Title: Tabellenbuch Kraftfahrtechnik

Licence edition:
Translated by: M. V. Ranganathan

Third Wiley Eastern Reprint, June 1992

ISBN 0 85226 350 3

Published by order of the
Deutsche Gesellschaft für Techniscne Zusammenarbeit
(GTZ) GmbH, Eschborn, Federal Republic of Germany

Printed in India by Wiley Eastern Limited
at A.P. Offset, Delhi 110 032.

Table of Contents

Preface

This book of tables of automotive engineering consists of four sections:

- Technical mathematics
- Theory of automotive engineering
- Properties of materials
- Technical drawing with symbols used in automotive engineering

References to DIN standards correspond to the German standardization. The lettering/abbreviations and indices within the drawings and formulas are derived from the German expressions. They are explained by the technical terms where needed.

Frequently used abbreviations are the following:

A = Outlet
AK = Outlet port
Aö = Outlet valve opened
As = Outlet valve closed
AV = Outlet valve
E = Inlet
Eb = Begin of injection
Ee = End of injection
EK = Inlet port
Eö = Inlet valve opened
Es = Inlet valve closed
EV = Inlet valve
OT = Top dead center
Ü = Valve overlap
ÜK = Transfer port
UT = Bottom dead center
Ve = End of combustion
Z = Ignition

Calculation table

Examples of use

1. Given : Square-section steel, length of side 36 mm.
   To find : Cross-sectional area in mm².

   $n=d \quad n^2$
   $36{,}0 \longrightarrow 1296$

   Cross-sectional area = 36.0 mm. 36.0 mm = 1296 mm²

2. Given : Cube, length of side 52 mm.
   To find : Volume in mm³

   $n=d \quad n^3$
   $52{,}0 \longrightarrow 140603$

   Volume = 52 mm. 52 mm. 52 mm = 140608 mm³

3 Given : Square-section steel, cross-sectional area 81 mm².
   To find : length of side in mm.

   a) $n=d \quad \sqrt{n}$
   $81{,}0 \longrightarrow 9{,}0$

   b) $n=d \quad n^2$
   $9{,}0 \longleftarrow 81$

   Length of side $=\sqrt{81\text{ mm}^2}=9{,}0$ mm

4. Given : Cube, Volume 64 cm³
   To find : Length of side in cm

   a) $n=d \quad \sqrt[3]{n}$
   $64{,}0 \longrightarrow 4{,}0$

   b) $n=d \quad n^3$
   $4{,}0 \longleftarrow 64$

   Length of side $=\sqrt[3]{64\text{ cm}^3}=4{,}0$ cm

5. Given : Flywheel, diameter 270 mm.
   To find : Circumference in cm.

   $n=d \quad \pi\cdot d$
   $27{,}0 \longrightarrow 84{,}82$

   Circumference $=\pi\cdot 27{,}0$ cm $= 84{,}82$ cm

6. Given : Rim, unwound circumference 1.885 m
   To find : Rim diameter in cm

   $n=d \quad \pi\cdot d$
   $60{,}0 \longleftarrow 188{,}5$

   Diameter = 188,5 cm : π = 60,0 cm

7. Given : Round steel, cross-sectional area 56.75 cm²
   To find : Diameter in mm

   $n=d \quad \frac{\pi\cdot d^2}{4}$
   $8{,}5 \longleftarrow 56{,}75$

   $$\text{Diameter} = \sqrt{\frac{4\cdot 56{,}75\text{ mm}^2}{\pi}} = 8{,}5\text{ mm}$$

Shift of decimal point

$n=d$	n^2	n^3	$\sqrt{n}$	$\sqrt[3]{n}$	$\pi\cdot d$	$\frac{\pi\cdot d^2}{4}$
1	2	3	–	–	1	2
2	4	6	1	–	2	4
3	6	9	–	1	3	6

$n=d$	n^2	n^3	$\sqrt{n}$	$\sqrt[3]{n}$	$\pi\cdot d$	$\frac{\pi\cdot d^2}{4}$
0,1	0,01	0,001	0,316	0,464	0,3142	0,00785
0,2	0,04	0,008	0,447	0,585	0,6283	0,03142
0,3	0,09	0,027	0,548	0,669	0,9425	0,07069
0,4	0,16	0,064	0,632	0,737	1,2566	0,12566
0,5	0,25	0,125	0,707	0,794	1,5708	0,19635
0,6	0,36	0,216	0,775	0,843	1,8850	0,28274
0,7	0,49	0,343	0,837	0,888	2,1991	0,38485
0,8	0,64	0,512	0,894	0,928	2,5133	0,50266
0,9	0,81	0,729	0,949	0,965	2,8274	0,63617
1,0	1,00	1,000	1,000	1,000	3,142	0,7854
1,1	1,21	1,331	1,049	1,032	3,456	0,9503
1,2	1,44	1,728	1,095	1,063	3,770	1,1310
1,3	1,69	2,197	1,140	1,091	4,084	1,3273
1,4	1,96	2,744	1,183	1,119	4,398	1,5394
1,5	2,25	3,375	1,225	1,145	4,712	1,7671
1,6	2,56	4,096	1,265	1,170	5,027	2,0106
1,7	2,89	4,913	1,304	1,194	5,341	2,2697
1,8	3,24	5,832	1,342	1,216	5,655	2,5447
1,9	3,61	6,859	1,378	1,239	5,969	2,8353
2,0	4,00	8,000	1,414	1,260	6,283	3,1416
2,1	4,41	9,261	1,449	1,281	6,597	3,4636
2,2	4,84	10,648	1,483	1,301	6,912	3,8013
2,3	5,29	12,167	1,517	1,320	7,226	4,1548
2,4	5,76	13,824	1,549	1,339	7,540	4,5239
2,5	6,25	15,625	1,581	1,357	7,854	4,9087
2,6	6,76	17,576	1,613	1,375	8,168	5,3093
2,7	7,29	19,683	1,643	1,393	8,482	5,7256
2,8	7,84	21,952	1,673	1,410	8,797	6,1575
2,9	8,41	24,389	1,703	1,426	9,111	6,6052
3,0	9,00	27,000	1,732	1,442	9,425	7,0686
3,1	9,61	29,791	1,761	1,458	9,739	7,5477
3,2	10,24	32,768	1,789	1,474	10,053	8,0425
3,3	10,89	35 937	1,817	1,489	10,367	8,5530
3,4	11,56	39,304	1,844	1,504	10,681	9,0792
3,5	12,25	42,875	1,871	1,518	10,996	9,6211
3,6	12,96	46,656	1,897	1,533	11,310	10,1788
3,7	13,69	50,653	1,924	1,547	11,624	10,7521
3,8	14,44	54,872	1,949	1,561	11,938	11,3411
3,9	15,21	59,319	1,975	1,574	12,252	11,9459
4,0	16,00	64,000	2,000	1,587	12,566	12,5664
4,1	16,81	68,921	2,025	1,601	12,881	13,2025
4,2	17,64	74,088	2,049	1,613	13,195	13,8544
4,3	18,49	79,507	2,074	1,626	13,509	14,5220
4,4	19,36	85,184	2,098	1,639	13,823	15,2053
4,5	20,25	91,125	2,121	1,651	14,137	15,9043
4,6	21,16	97,336	2,145	1,663	14,451	16,6190
4,7	22,09	103,823	2,168	1,675	14,765	17,3494
4,8	23,04	110,592	2,191	1,687	15,080	18,0956
4,9	24,01	117,649	2,214	1,699	15,394	18,8574
5,0	25,00	125,000	2,236	1,710	15,708	19,6350
5,1	26,01	132,65	2,258	1,721	16,022	20,4282
5,2	27,04	140,61	2,280	1,733	16,336	21,2372
5,3	28,09	148,88	2,302	1,744	16,650	22,0618
5,4	29,16	157,46	2,324	1,754	16,965	22,9022
5,5	30,25	166,38	2,345	1,765	17,279	23,7583
5,6	31,36	175,62	2,366	1,776	17,593	24,6301
5,7	32,49	185,19	2,387	1,786	17,907	25,5176
5,8	33,64	195,11	2,408	1,797	18,221	26,4208
5,9	34,81	205,38	2,429	1,807	18,535	27,3397
6,0	36,00	216,00	2,450	1,817	18,850	28,2743

$n = d$	n^2	n^3	$\sqrt{n}$	$\sqrt[3]{n}$	$\pi \cdot d$	$\frac{\pi \cdot d^2}{4}$	$n = d$	n^2	n^3	$\sqrt{n}$	$\sqrt[3]{n}$	$\pi \cdot d$	$\frac{\pi \cdot d^2}{4}$
6,1	37,21	226,98	2,470	1,827	19,164	29,22	12,6	158,8	2000	3,550	2,327	39,58	124,7
6,2	38,44	238,33	2,490	1,837	19,478	30,19	12,7	161,3	2048	3,564	2,333	39,90	126,7
6,3	39,69	250,05	2,510	1,847	19,792	31,17	12,8	163,8	2097	3,578	2,339	40,21	128,7
6,4	40,96	262,14	2,530	1,857	20,106	32,17	12,9	166,4	2147	3,592	2,345	40,53	130,7
6,5	42,25	274,63	2,550	1,866	20,420	33,18	**13,0**	169,0	2197	3,606	2,351	40,84	132,7
6,6	43,56	287,50	2,569	1,876	20,735	34,21	13,1	171,6	2248	3,619	2,357	41,16	134,8
6,7	44,89	300,76	2,588	1,885	21,049	35,26	13,2	174,2	2300	3,633	2,363	41,47	136,8
6,8	46,24	314,43	2,608	1,894	21,363	36,32	13,3	176,9	2353	3,647	2,369	41,78	138,9
6,9	47,61	328,51	2,627	1,904	21,677	37,39	13,4	179,6	2406	3,661	2,375	42,10	141,0
7,0	49,00	343,00	2,646	1,913	21,991	38,48	**13,5**	182,3	2460	3,674	2,381	42,41	143,1
7,1	50,41	357,91	2,665	1,922	22,305	39,59	13,6	185,0	2515	3,688	2,387	42,73	145,3
7,2	51,84	373,25	2,683	1,931	22,619	40,72	13,7	187,7	2571	3,701	2,393	43,04	147,4
7,3	53,29	389,02	2,702	1,940	22,934	41,85	13,8	190,4	2628	3,715	2,399	43,35	149,6
7,4	54,76	405,22	2,720	1,949	23,248	43,01	13,9	193,2	2686	3,728	2,404	43,67	151,7
7,5	56,25	421,88	2,739	1,957	23,562	44,18	**14,0**	196,0	2744	3,742	2,410	43,98	153,9
7,6	57,76	438,98	2,757	1,966	23,876	45,36	14,1	198,8	2803	3,755	2,416	44,30	156,1
7,7	59,29	456,53	2,775	1,975	24,190	46,57	14,2	201,6	2863	3,768	2,422	44,61	158,4
7,8	60,84	474,55	2,793	1,983	24,504	47,78	14,3	204,5	2924	3,782	2,427	44,93	160,6
7,9	62,41	493,04	2,811	1,992	24,819	49,02	14,4	207,4	2986	3,795	2,433	45,24	162,9
8,0	64,00	512,00	2,828	2,000	25,133	50,27	**14,5**	210,3	3049	3,808	2,438	45,55	165,1
8,1	65,61	531,44	2,846	2,008	25,447	51,53	14,6	213,2	3112	3,821	2,444	45,87	167,4
8,2	67,24	551,37	2,864	2,017	25,761	52,81	14,7	216,1	3177	3,834	2,450	46,18	169,7
8,3	68,89	571,79	2,881	2,025	26,075	54,11	14,8	219,0	3242	3,847	2,455	46,50	172,0
8,4	70,56	592,70	2,898	2,033	26,389	55,42	14,9	222,0	3308	3,860	2,461	46,81	174,4
8,5	72,25	614,13	2,915	2,041	26,704	56,75	**15,0**	225,0	3375	3,873	2,466	47,12	176,7
8,6	73,96	636,06	2,933	2,049	27,018	58,09	15,1	228,0	3443	3,886	2,472	47,44	179,1
8,7	75,69	658,50	2,950	2,057	27,332	59,45	15,2	231,0	3512	3,899	2,477	47,75	181,5
8,8	77,44	681,47	2,966	2,065	27,646	60,82	15,3	234,1	3582	3,912	2,483	48,07	183,9
8,9	79,21	704,97	2,983	2,072	27,960	62,21	15,4	237,2	3652	3,924	2,488	48,38	186,3
9,0	81,00	729,00	3,000	2,080	28,274	63,62	**15,5**	240,3	3724	3,937	2,493	48,69	188,7
9,1	82,81	753,57	3,017	2,088	28,588	65,04	15,6	243,4	3796	3,950	2,499	49,01	191,1
9,2	84,64	778,69	3,033	2,095	28,903	66,48	15,7	246,5	3870	3,962	2,504	49,32	193,6
9,3	86,49	804,36	3,050	2,103	29,217	67,93	15,8	249,6	3944	3,975	2,509	49,64	196,1
9,4	88,36	830,58	3,066	2,111	29,531	69,40	15,9	252,8	4020	3,987	2,515	49,95	198,6
9,5	90,25	857,38	3,082	2,118	29,845	70,88	**16,0**	256,0	4096	4,000	2,520	50,27	201,1
9,6	92,16	884,74	3,098	2,125	30,159	72,38	16,1	259,2	4173	4,012	2,525	50,58	203,6
9,7	94,09	912,67	3,115	2,133	30,473	73,90	16,2	262,4	4252	4,025	2,530	50,89	206,1
9,8	96,04	941,19	3,131	2,140	30,788	75,43	16,3	265,7	4331	4,037	2,535	51,21	208,7
9,9	98,01	970,30	3,146	2,147	31,102	76,98	16,4	269,0	4411	4,050	2,541	51,52	211,2
10,0	100,00	1 000,00	3,162	2,154	31,416	78,54	**16,5**	272,3	4492	4,062	2,546	51,84	213,8
10,1	102,01	1 030,30	3,178	2,162	31,730	80,12	16,6	275,6	4574	4,074	2,551	52,15	216,4
10,2	104,04	1 061,21	3,194	2,169	32,044	81,71	16,7	278,9	4657	4,087	2,556	52,47	219,0
10,3	106,09	1 092,73	3,209	2,176	32,358	83,32	16,8	282,2	4742	4,099	2,561	52,78	221,7
10,4	108,16	1 124,86	3,225	2,183	32,673	84,95	16,9	285,6	4827	4,111	2,566	53,09	224,3
10,5	110,25	1 157,63	3,240	2,190	32,987	86,59	**17,0**	289,0	4913	4,123	2,571	53,41	227,0
10,6	112,4	1 191	3,256	2,197	33,30	88,25	17,1	292,4	5000	4,135	2,576	53,72	229,7
10,7	114,5	1 225	3,271	2,204	33,62	89,92	17,2	295,8	5088	4,147	2,581	54,04	232,4
10,8	116,6	1 260	3,286	2,210	33,93	91,61	17,3	299,3	5178	4,159	2,586	54,35	235,1
10,9	118,8	1 295	3,302	2,217	34,24	93,31	17,4	302,8	5268	4,171	2,591	54,66	237,8
11,0	121,0	1 331	3,317	2,224	34,56	95,03	**17,5**	306,3	5359	4,183	2,596	54,98	240,5
11,1	123,2	1 368	3,332	2,231	34,87	96,77	17,6	309,8	5452	4,195	2,601	55,29	243,3
11,2	125,4	1 405	3,347	2,237	35,19	98,52	17,7	313,3	5545	4,207	2,606	55,61	246,1
11,3	127,7	1 443	3,362	2,244	35,50	100,3	17,8	316,8	5640	4,219	2,611	55,92	248,8
11,4	130,0	1 482	3,376	2,251	35,81	102,1	17,9	320,4	5735	4,231	2,616	56,24	251,6
11,5	132,3	1 521	3,391	2,257	36,13	103,9	**18,0**	324,0	5832	4,243	2,621	56,55	254,5
11,6	134,6	1 561	3,406	2,264	36,44	105,7	18,1	327,6	5930	4,254	2,626	56,86	257,3
11,7	136,9	1 602	3,421	2,270	36,76	107,5	18,2	331,2	6029	4,266	2,630	57,18	260,2
11,8	139,2	1 643	3,435	2,277	37,07	109,4	18,3	334,9	6128	4,278	2,635	57,49	263,0
11,9	141,6	1 685	3,450	2,283	37,38	111,2	18,4	338,6	6230	4,290	2,640	57,81	265,9
12,0	144,0	1 728	3,464	2,289	37,70	113,1	**18,5**	342,3	6332	4,301	2,645	58,12	268,8
12,1	146,4	1 772	3,479	2,296	38,01	115,0	18,6	346,0	6435	4,313	2,650	58,43	271,7
12,2	148,8	1 816	3,493	2,302	38,33	116,9	18,7	349,7	6539	4,324	2,654	58,75	274,6
12,3	151,3	1 861	3,507	2,308	38,64	118,8	18,8	353,4	6645	4,336	2,659	59,06	277,6
12,4	153,8	1 907	3,521	2,315	38,96	120,8	18,9	357,2	6751	4,347	2,664	59,38	280,6
12,5	156,3	1 953	3,536	2,321	39,27	122,7	**19,0**	361,0	6859	4,359	2,668	59,69	283,5

Calculation table

$n = d$	n^2	n^3	$\sqrt{n}$	$\sqrt[3]{n}$	$\pi \cdot d$	$\frac{\pi \cdot d^2}{4}$	$n = d$	n^2	n^3	$\sqrt{n}$	$\sqrt[3]{n}$	$\pi \cdot d$	$\frac{\pi \cdot d^2}{4}$
19,1	364,8	6968	4,370	2,673	60,00	286,5	25,6	655,4	16777	5,060	2,947	80,42	514,7
19,2	368,6	7078	4,382	2,678	60,32	289,5	25,7	660,5	16975	5,070	2,951	80,74	518,7
19,3	372,5	7189	4,393	2,682	60,63	292,6	25,8	665,6	17174	5,079	2,955	81,05	522,8
19,4	376,4	7301	4,405	2,687	60,95	295,6	25,9	670,8	17374	5,089	2,959	81,37	526,9
19,5	380,3	7415	4,416	2,692	61,26	298,6	**26,0**	676,0	17576	5,099	2,962	81,68	530,9
19,6	384,2	7530	4,427	2,696	61,58	301,7	26,1	681,2	17780	5,109	2,966	82,00	535,0
19,7	388,1	7645	4,438	2,701	61,89	304,8	26,2	686,4	17985	5,119	2,970	82,31	539,1
19,8	392,0	7762	4,450	2,705	62,20	307,9	26,3	691,7	18191	5,128	2,974	82,62	543,3
19,9	396,0	7881	4,461	2,710	62,52	311,0	26,4	697,0	18400	5,138	2,978	82,94	547,4
20,0	400,0	8000	4,472	2,714	62,83	314,2	**26,5**	702,3	18610	5,148	2,981	83,25	551,5
20,1	404,0	8121	4,483	2,719	63,15	317,3	26,6	707,6	18821	5,158	2,985	83,57	555,7
20,2	408,0	8242	4,494	2,723	63,46	320,5	26,7	712,9	19034	5,167	2,989	83,88	559,9
20,3	412,1	8365	4,506	2,728	63,77	323,7	26,8	718,2	19249	5,177	2,993	84,19	564,1
20,4	416,2	8490	4,517	2,732	64,09	326,9	26,9	723,6	19465	5,187	2,996	84,51	568,3
20,5	420,3	8615	4,528	2,737	64,40	330,1	**27,0**	729,0	19683	5,196	3,000	84,82	572,6
20,6	424,4	8742	4,539	2,741	64,72	333,3	27,1	734,4	19903	5,206	3,004	85,14	576,8
20,7	428,5	8870	4,550	2,746	65,03	336,5	27,2	739,8	20124	5,215	3,007	85,45	581,1
20,8	432,6	8999	4,561	2,750	65,35	339,8	27,3	745,3	20346	5,225	3,011	85,77	585,3
20,9	436,8	9129	4,572	2,755	65,66	343,1	27,4	750,8	20571	5,235	3,015	86,08	589,6
21,0	441,0	9261	4,583	2,759	65,97	346,4	**27,5**	756,3	20797	5,244	3,018	86,39	594,0
21,1	445,2	9394	4,593	2,763	66,29	349,7	27,6	761,8	21025	5,254	3,022	86,71	598,3
21,2	449,4	9528	4,604	2,768	66,60	353,0	27,7	767,3	21254	5,263	3,026	87,02	602,6
21,3	453,7	9664	4,615	2,772	66,92	356,3	27,8	772,8	21485	5,273	3,029	87,34	607,0
21,4	458,0	9800	4,626	2,776	67,23	359,7	27,9	778,4	21718	5,282	3,033	87,65	611,4
21,5	462,3	9938	4,637	2,781	67,54	363,1	**28,0**	784,0	21952	5,292	3,037	87,96	615,8
21,6	466,6	10078	4,648	2,785	67,86	366,4	28,1	789,6	22188	5,301	3,040	88,28	620,2
21,7	470,9	10218	4,658	2,789	68,17	369,8	28,2	795,2	22426	5,310	3,044	88,59	624,6
21,8	475,2	10360	4,669	2,794	68,49	373,3	28,3	800,9	22665	5,320	3,047	88,91	629,0
21,9	479,6	10503	4,680	2,798	68,80	376,7	28,4	806,6	22906	5,329	3,051	89,22	633,5
22,0	484,0	10648	4,690	2,802	69,12	380,1	**28,5**	812,3	23149	5,339	3,055	89,54	637,9
22,1	488,4	10794	4,701	2,806	69,43	383,6	28,6	818,0	23394	5,348	3,058	89,85	642,4
22,2	492,8	10941	4,712	2,811	69,74	387,1	28,7	823,7	23640	5,357	3,062	90,16	646,9
22,3	497,3	11090	4,722	2,815	70,06	390,6	28,8	829,4	23888	5,367	3,065	90,48	651,4
22,4	501,8	11239	4,733	2,819	70,37	394,1	28,9	835,2	24138	5,376	3,069	90,79	656,0
22,5	506,3	11391	4,743	2,823	70,69	397,6	**29,0**	841,0	24389	5,385	3,072	91,11	660,5
22,6	510,8	11543	4,754	2,827	71,00	401,1	29,1	846,8	24642	5,394	3,076	91,42	665,1
22,7	515,3	11697	4,764	2,831	71,31	404,7	29,2	852,6	24897	5,404	3,079	91,73	669,7
22,8	519,8	11852	4,775	2,836	71,63	408,3	29,3	858,5	25154	5,413	3,083	92,05	674,3
22,9	524,4	12009	4,785	2,840	71,94	411,9	29,4	864,4	25412	5,422	3,086	92,36	678,9
23,0	529,0	12167	4,796	2,844	72,26	415,5	**29,5**	870,3	25672	5,431	3,090	92,68	683,5
23,1	533,6	12326	4,806	2,848	72,57	419,1	29,6	876,2	25934	5,441	3,093	92,99	688,1
23,2	538,2	12487	4,817	2,852	72,88	422,7	29,7	882,1	26198	5,450	3,097	93,31	692,8
23,3	542,9	12649	4,827	2,856	73,20	426,4	29,8	888,0	26464	5,459	3,100	93,62	697,5
23,4	547,6	12813	4,837	2,860	73,51	430,1	29,9	894,0	26731	5,468	3,104	93,93	702,2
23,5	552,3	12978	4,848	2,864	73,83	433,7	**30,0**	900,0	27000	5,477	3,107	94,25	706,9
23,6	557,0	13144	4,858	2,868	74,14	437,4	30,1	906,0	27271	5,486	3,111	94,56	711,6
23,7	561,7	13312	4,868	2,872	74,46	441,2	30,2	912,0	27544	5,495	3,114	94,88	716,3
23,8	566,4	13481	4,879	2,876	74,77	444,9	30,3	918,1	27818	5,505	3,117	95,19	721,1
23,9	571,2	13652	4,889	2,880	75,08	448,6	30,4	924,2	28094	5,514	3,121	95,50	725,8
24,0	576,0	13824	4,899	2,884	75,40	452,4	**30,5**	930,3	28373	5,523	3,124	95,82	730,6
24,1	580,8	13998	4,909	2,888	75,71	456,2	30,6	936,4	28653	5,532	3,128	96,13	735,4
24,2	585,6	14172	4,919	2,892	76,03	460,0	30,7	942,5	28934	5,541	3,131	96,45	740,2
24,3	590,5	14349	4,930	2,896	76,34	463,8	30,8	948,6	29218	5,550	3,135	96,76	745,1
24,4	595,4	14527	4,940	2,900	76,65	467,6	30,9	954,8	29504	5,559	3,138	97,08	749,9
24,5	600,3	14706	4,950	2,904	76,97	471,4	**31,0**	961,0	29791	5,568	3,141	97,39	754,8
24,6	605,2	14887	4,960	2,908	77,28	475,3	31,1	967,2	30080	5,577	3,145	97,70	759,6
24,7	610,1	15069	4,970	2,912	77,60	479,2	31,2	973,4	30371	5,586	3,148	98,02	764,5
24,8	615,0	15253	4,980	2,916	77,91	483,1	31,3	979,7	30664	5,595	3,151	98,33	769,4
24,9	620,0	15438	4,990	2,920	78,23	487,0	31,4	986,0	30959	5,604	3,155	98,65	774,4
25,0	625,0	15625	5,000	2,924	78,54	490,9	**31,5**	992,3	31256	5,612	3,158	98,96	779,3
25,1	630,0	15813	5,010	2,928	78,85	494,8	31,6	998,6	31554	5,621	3,162	99,27	784,3
25,2	635,0	16003	5,020	2,932	79,17	498,8	31,7	1005	31855	5,630	3,165	99,59	789,2
25,3	640,1	16194	5,030	2,936	79,48	502,7	31,8	1011	32157	5,639	3,168	99,90	794,2
25,4	645,2	16387	5,040	2,940	79,80	506,7	31,9	1018	32462	5,648	3,171	100,2	799,2
25,5	650,3	16581	5,050	2,943	80,11	510,7	**32,0**	1024	32768	5,657	3,175	100,5	804,2

$n = d$	n^2	n^3	$\sqrt{n}$	$\sqrt[3]{n}$	$\pi \cdot d$	$\frac{\pi \cdot d^2}{4}$	$n = d$	n^2	n^3	$\sqrt{n}$	$\sqrt[3]{n}$	$\pi \cdot d$	$\frac{\pi \cdot d^2}{4}$
32,1	1030	33076	5,666	3,178	100,8	809,3	38,6	1490	57512	6,213	3,380	121,3	1170
32,2	1037	33386	5,675	3,181	101,2	814,3	38,7	1498	57961	6,221	3,382	121,6	1176
32,3	1043	33698	5,683	3,185	101,5	819,4	38,8	1505	58411	6,229	3,385	121,9	1182
32,4	1050	34012	5,692	3,188	101,8	824,5	38,9	1513	58864	6,237	3,388	122,2	1188
32,5	1056	34328	5,701	3,191	102,1	829,6	**39,0**	1521	59319	6,245	3,391	122,5	1195
32,6	1063	34646	5,710	3,195	102,4	834,7	39,1	1529	59776	6,253	3,394	122,8	1201
32,7	1069	34966	5,718	3,198	102,7	839,8	39,2	1537	60236	6,261	3,397	123,2	1207
32,8	1076	35288	5,727	3,201	103,0	845,0	39,3	1544	60698	6,269	3,400	123,5	1213
32,9	1082	35611	5,736	3,204	103,4	850,1	39,4	1552	61163	6,277	3,403	123,8	1219
33,0	1089	35937	5,745	3,208	103,7	855,3	**39,5**	1560	61630	6,285	3,406	124,1	1225
33,1	1096	36265	5,753	3,211	104,0	860,5	39,6	1568	62099	6,293	3,409	124,4	1232
33,2	1102	36594	5,762	3,214	104,3	865,7	39,7	1576	62571	6,301	3,411	124,7	1238
33,3	1109	36926	5,771	3,217	104,6	870,9	39,8	1584	63045	6,309	3,414	125,0	1244
33,4	1116	37260	5,779	3,220	104,9	876,2	39,9	1592	63521	6,317	3,417	125,3	1250
33,5	1122	37595	5,788	3,224	105,2	881,4	**40,0**	1600	64000	6,325	3,420	125,7	1257
33,6	1129	37933	5,797	3,227	105,6	886,7	40,1	1608	64481	6,332	3,423	126,0	1263
33,7	1136	38273	5,805	3,230	105,9	892,0	40,2	1616	64965	6,340	3,426	126,3	1269
33,8	1142	38614	5,814	3,233	106,2	897,3	40,3	1624	65451	6,348	3,428	126,6	1276
33,9	1149	38958	5,822	3,236	106,5	902,6	40,4	1632	65939	6,356	3,431	126,9	1282
34,0	1156	39304	5,831	3,240	106,8	907,9	**40,5**	1640	66430	6,364	3,434	127,2	1288
34,1	1163	39652	5,840	3,243	107,1	913,3	40,6	1648	66923	6,372	3,437	127,5	1295
34,2	1170	40002	5,848	3,246	107,4	918,6	40,7	1656	67419	6,380	3,440	127,9	1301
34,3	1176	40354	5,857	3,249	107,8	924,0	40,8	1665	67917	6,387	3,443	128,2	1307
34,4	1183	40708	5,865	3,252	108,1	929,4	40,9	1673	68418	6,395	3,445	128,5	1314
34,5	1190	41064	5,874	3,255	108,4	934,8	**41,0**	1681	68921	6,403	3,448	128,8	1320
34,6	1197	41422	5,882	3,259	108,7	940,2	41,1	1689	69427	6,411	3,451	129,1	1327
34,7	1204	41782	5,891	3,262	109,0	945,7	41,2	1697	69935	6,419	3,454	129,4	1333
34,8	1211	42144	5,899	3,265	109,3	951,1	41,3	1706	70445	6,427	3,457	129,7	1340
34,9	1218	42509	5,908	3,268	109,6	956,6	41,4	1714	70958	6,434	3,459	130,1	1346
35,0	1225	42875	5,916	3,271	110,0	962,1	**41,5**	1722	71473	6,442	3,462	130,4	1353
35,1	1232	43244	5,925	3,274	110,3	967,6	41,6	1731	71991	6,450	3,465	130,7	1359
35,2	1239	43614	5,933	3,277	110,6	973,1	41,7	1739	72512	6,458	3,468	131,0	1366
35,3	1246	43987	5,941	3,280	110,9	978,7	41,8	1747	73035	6,465	3,471	131,3	1372
35,4	1253	44362	5,950	3,283	111,2	984,2	41,9	1756	73560	6,473	3,473	131,6	1379
35,5	1260	44739	5,958	3,287	111,5	989,8	**42,0**	1764	74088	6,481	3,476	131,9	1385
35,6	1267	45118	5,967	3,290	111,8	995,4	42,1	1772	74618	6,488	3,479	132,3	1392
35,7	1274	45499	5,975	3,293	112,2	1001	42,2	1781	75151	6,496	3,482	132,6	1399
35,8	1282	45883	5,983	3,296	112,5	1007	42,3	1789	75687	6,504	3,484	132,9	1405
35,9	1289	46268	5,992	3,299	112,8	1012	42,4	1798	76225	6,512	3,487	133,2	1412
36,0	1296	46656	6,000	3,302	113,1	1018	**42,5**	1806	76766	6,519	3,490	133,5	1419
36,1	1303	47046	6,008	3,305	113,4	1024	42,6	1815	77309	6,527	3,493	133,8	1425
36,2	1310	47438	6,017	3,308	113,7	1029	42,7	1823	77854	6,535	3,495	134,1	1432
36,3	1318	47832	6,025	3,311	114,0	1035	42,8	1832	78403	6,542	3,498	134,5	1439
36,4	1325	48229	6,033	3,314	114,4	1041	42,9	1840	78954	6,550	3,501	134,8	1445
36,5	1332	48627	6,042	3,317	114,7	1046	**43,0**	1849	79507	6,557	3,503	135,1	1452
36,6	1340	49028	6,050	3,320	115,0	1052	43,1	1858	80063	6,565	3,506	135,4	1459
36,7	1347	49431	6,058	3,323	115,3	1058	43,2	1866	80622	6,573	3,509	135,7	1466
36,8	1354	49836	6,066	3,326	115,6	1064	43,3	1875	81183	6,580	3,512	136,0	1473
36,9	1362	50243	6,075	3,329	115,9	1069	43,4	1884	81747	6,588	3,514	136,3	1479
37,0	1369	50653	6,083	3,332	116,2	1075	**43,5**	1892	82313	6,595	3,517	136,7	1486
37,1	1376	51065	6,091	3,335	116,6	1081	43,6	1901	82882	6,603	3,520	137,0	1493
37,2	1384	51479	6,099	3,338	116,9	1087	43,7	1910	83453	6,611	3,522	137,3	1500
37,3	1391	51895	6,107	3,341	117,2	1093	43,8	1918	84028	6,618	3,525	137,6	1507
37,4	1399	52314	6,116	3,344	117,5	1099	43,9	1927	84605	6,626	3,528	137,9	1514
37,5	1406	52734	6,124	3,347	117,8	1104	**44,0**	1936	85184	6,633	3,530	138,2	1521
37,6	1414	53157	6,132	3,350	118,1	1110	44,1	1945	85766	6,641	3,533	138,5	1527
37,7	1421	53583	6,140	3,353	118,4	1116	44,2	1954	86351	6,648	3,536	138,9	1534
37,8	1429	54010	6,148	3,356	118,8	1122	44,3	1962	86938	6,656	3,538	139,2	1541
37,9	1436	54440	6,156	3,359	119,1	1128	44,4	1971	87528	6,663	3,541	139,5	1548
38,0	1444	54872	6,164	3,362	119,4	1134	**44,5**	1980	88121	6,671	3,544	139,8	1555
38,1	1452	55306	6,173	3,365	119,7	1140	44,6	1989	88717	6,678	3,546	140,1	1562
38,2	1459	55743	6,181	3,368	120,0	1146	44,7	1998	89315	6,686	3,549	140,4	1569
38,3	1467	56182	6,189	3,371	120,3	1152	44,8	2007	89915	6,693	3,552	140,7	1576
38,4	1475	56623	6,197	3,374	120,6	1158	44,9	2016	90519	6,701	3,554	141,1	1583
38,5	1482	57067	6,205	3,377	121,0	1164	**45,0**	2025	91125	6,708	3,557	141,4	1590

$n = d$	n^2	n^3	$\sqrt{n}$	$\sqrt[3]{n}$	$\pi \cdot d$	$\frac{\pi \cdot d^2}{4}$
45,1	2034	91734	6,716	3,560	141,7	1598
45,2	2043	92345	6,723	3,562	142,0	1605
45,3	2052	92960	6,731	3,565	142,3	1612
45,4	2061	93577	6,738	3,567	142,6	1619
45,5	2070	94196	6,745	3,570	142,9	1626
45,6	2079	94819	6,753	3,573	143,3	1633
45,7	2088	95444	6,760	3,575	143,6	1640
45,8	2098	96072	6,768	3,578	143,9	1647
45,9	2107	96703	6,775	3,580	144,2	1655
46,0	2116	97336	6,782	3,583	144,5	1662
46,1	2125	97972	6,790	3,586	144,8	1669
46,2	2134	98611	6,797	3,588	145,1	1676
46,3	2144	99253	6,804	3,591	145,5	1684
46,4	2153	99897	6,812	3,593	145,8	1691
46,5	2162	100545	6,819	3,596	146,1	1698
46,6	2172	101195	6,826	3,599	146,4	1706
46,7	2181	101848	6,834	3,601	146,7	1713
46,8	2190	102503	6,841	3,604	147,0	1720
46,9	2200	103162	6,848	3,606	147,3	1728
47,0	2209	103823	6,856	3,609	147,7	1735
47,1	2218	104487	6,863	3,611	148,0	1742
47,2	2228	105154	6,870	3,614	148,3	1750
47,3	2237	105824	6,877	3,616	148,6	1757
47,4	2247	106496	6,885	3,619	148,9	1765
47,5	2256	107172	6,892	3,622	149,2	1772
47,6	2266	107850	6,899	3,624	149,5	1780
47,7	2275	108531	6,907	3,627	149,9	1787
47,8	2285	109215	6,914	3,629	150,2	1795
47,9	2294	109902	6,921	3,632	150,5	1802
48,0	2304	110592	6,928	3,634	150,8	1810
48,1	2314	111285	6,935	3,637	151,1	1817
48,2	2323	111980	6,943	3,639	151,4	1825
48,3	2333	112679	6,950	3,642	151,7	1832
48,4	2343	113380	6,957	3,644	152,1	1840
48,5	2352	114084	6,964	3,647	152,4	1847
48,6	2362	114791	6,971	3,649	152,7	1855
48,7	2372	115501	6,979	3,652	153,0	1863
48,8	2381	116214	6,986	3,654	153,3	1870
48,9	2391	116930	6,993	3,657	153,6	1878
49,0	2401	117649	7,000	3,659	153,9	1886
49,1	2411	118371	7,007	3,662	154,3	1893
49,2	2421	119095	7,014	3,664	154,6	1901
49,3	2430	119823	7,021	3,667	154,9	1909
49,4	2440	120554	7,029	3,669	155,2	1917
49,5	2450	121287	7,036	3,672	155,5	1924
49,6	2460	122024	7,043	3,674	155,8	1932
49,7	2470	122763	7,050	3,677	156,1	1940
49,8	2480	123506	7,057	3,679	156,5	1948
49,9	2490	124251	7,064	3,682	156,8	1956
50,0	2500	125000	7,071	3,684	157,1	1964
50,1	2510	125752	7,078	3,686	157,4	1971
50,2	2520	126506	7,085	3,689	157,7	1979
50,3	2530	127264	7,092	3,691	158,0	1987
50,4	2540	128024	7,099	3,694	158,3	1995
50,5	2550	128788	7,106	3,696	158,7	2003
50,6	2560	129554	7,113	3,699	159,0	2011
50,7	2570	130324	7,120	3,701	159,3	2019
50,8	2581	131097	7,127	3,704	159,6	2027
50,9	2591	131872	7,134	3,706	159,9	2035
51,0	2601	132651	7,141	3,708	160,2	2043
51,1	2611	133433	7,148	3,711	160,5	2051
51,2	2621	134218	7,155	3,713	160,8	2059
51,3	2632	135006	7,162	3,716	161,2	2067
51,4	2642	135797	7,169	3,718	161,5	2075
51,5	2652	136591	7,176	3,721	161,8	2083
51,6	2663	137388	7,183	3,723	162,1	2091
51,7	2673	138188	7,190	3,725	162,4	2099
51,8	2683	138992	7,197	3,728	162,7	2107
51,9	2694	139798	7,204	3,730	163,0	2116
52,0	2704	140608	7,211	3,733	163,4	2124
52,1	2714	141421	7,218	3,735	163,7	2132
52,2	2725	142237	7,225	3,737	164,0	2140
52,3	2735	143056	7,232	3,740	164,3	2148
52,4	2746	143878	7,239	3,742	164,6	2157
52,5	2756	144703	7,246	3,744	164,9	2165
52,6	2767	145532	7,253	3,747	165,2	2173
52,7	2777	146363	7,259	3,749	165,6	2181
52,8	2788	147198	7,266	3,752	165,9	2190
52,9	2798	148036	7,273	3,754	166,2	2198
53,0	2809	148877	7,280	3,756	166,5	2206
53,1	2820	149721	7,287	3,759	166,8	2215
53,2	2830	150569	7,294	3,761	167,1	2223
53,3	2841	151419	7,301	3,763	167,4	2231
53,4	2852	152273	7,308	3,766	167,8	2240
53,5	2862	153130	7,314	3,768	168,1	2248
53,6	2873	153991	7,321	3,770	168,4	2256
53,7	2884	154854	7,328	3,773	168,7	2265
53,8	2894	155721	7,335	3,775	169,0	2273
53,9	2905	156591	7,342	3,777	169,3	2282
54,0	2916	157464	7,348	3,780	169,7	2290
54,1	2927	158340	7,355	3,782	170,0	2299
54,2	2938	159220	7,362	3,784	170,3	2307
54,3	2948	160103	7,369	3,787	170,6	2316
54,4	2959	160989	7,376	3,789	170,9	2324
54,5	2970	161879	7,382	3,791	171,2	2333
54,6	2981	162771	7,389	3,794	171,5	2341
54,7	2992	163667	7,396	3,796	171,8	2350
54,8	3003	164567	7,403	3,798	172,2	2359
54,9	3014	165469	7,409	3,801	172,5	2367
55,0	3025	166375	7,416	3,803	172,8	2376
55,1	3036	167284	7,423	3,805	173,1	2384
55,2	3047	168197	7,430	3,808	173,4	2393
55,3	3058	169112	7,436	3,810	173,7	2402
55,4	3069	170031	7,443	3,812	174,0	2411
55,5	3080	170954	7,450	3,814	174,4	2419
55,6	3091	171880	7,457	3,817	174,7	2428
55,7	3102	172809	7,463	3,819	175,0	2437
55,8	3114	173741	7,470	3,821	175,3	2445
55,9	3125	174677	7,477	3,824	175,6	2454
56,0	3136	175616	7,483	3,826	175,9	2463
56,1	3147	176558	7,490	3,828	176,2	2472
56,2	3158	177504	7,497	3,830	176,6	2481
56,3	3170	178454	7,503	3,833	176,9	2489
56,4	3181	179406	7,510	3,835	177,2	2498
56,5	3192	180362	7,517	3,837	177,5	2507
56,6	3204	181321	7,523	3,839	177,8	2516
56,7	3215	182281	7,530	3,842	178,1	2525
56,8	3226	183250	7,537	3,844	178,4	2534
56,9	3238	184220	7,543	3,846	178,8	2543
57,0	3249	185193	7,550	3,849	179,1	2552
57,1	3260	186169	7,556	3,851	179,4	2561
57,2	3272	187149	7,563	3,853	179,7	2570
57,3	3283	188133	7,570	3,855	180,0	2579
57,4	3295	189119	7,576	3,857	180,3	2588
57,5	3306	190109	7,583	3,860	180,6	2597
57,6	3318	191103	7,589	3,862	181,0	2606
57,7	3329	192100	7,596	3,864	181,3	2615
57,8	3341	193101	7,603	3,866	181,6	2624
57,9	3352	194105	7,609	3,869	181,9	2633
58,0	3364	195112	7,616	3,871	182,2	2642

n = d	n^2	n^3	$\sqrt{n}$	$\sqrt[3]{n}$	$\pi \cdot d$	$\frac{\pi \cdot d^2}{4}$
58,1	3376	196123	7,622	3,873	182,5	2651
58,2	3387	197137	7,629	3,875	182,8	2660
58,3	3399	198155	7,635	3,878	183,2	2669
58,4	3411	199177	7,642	3,880	183,5	2679
58,5	3422	200202	7,649	3,882	183,8	2688
58,6	3434	201230	7,655	3,884	184,1	2697
58,7	3446	202262	7,662	3,886	184,4	2706
58,8	3457	203297	7,668	3,889	184,7	2715
58,9	3469	204336	7,675	3,891	185,0	2725
59,0	3481	205379	7,681	3,893	185,4	2734
59,1	3493	206425	7,688	3,895	185,7	2743
59,2	3505	207475	7,694	3,897	186,0	2753
59,3	3516	208528	7,701	3,900	186,3	2762
59,4	3528	209585	7,707	3,902	186,6	2771
59,5	3540	210645	7,714	3,904	186,9	2781
59,6	3552	211709	7,720	3,906	187,2	2790
59,7	3564	212776	7,727	3,908	187,6	2799
59,8	3576	213847	7,733	3,911	187,9	2809
59,9	3588	214922	7,740	3,913	188,2	2818
60,0	3600	216000	7,746	3,915	188,5	2827
60,1	3612	217082	7,752	3,917	188,8	2837
60,2	3624	218167	7,759	3,919	189,1	2846
60,3	3636	219256	7,765	3,921	189,4	2856
60,4	3648	220349	7,772	3,924	189,8	2865
60,5	3660	221445	7,778	3,926	190,1	2875
60,6	3672	222545	7,785	3,928	190,4	2884
60,7	3684	223649	7,791	3,930	190,7	2894
60,8	3697	224756	7,797	3,932	191,0	2903
60,9	3709	225867	7,804	3,934	191,3	2913
61,0	3721	226981	7,810	3,936	191,6	2922
61,1	3733	228099	7,817	3,939	192,0	2932
61,2	3745	229221	7,823	3,941	192,3	2942
61,3	3758	230346	7,829	3,943	192,6	2951
61,4	3770	231476	7,836	3,945	192,9	2961
61,5	3782	232608	7,842	3,947	193,2	2971
61,6	3795	233745	7,849	3,949	193,5	2980
61,7	3807	234885	7,855	3,951	193,8	2990
61,8	3819	236029	7,861	3,954	194,2	3000
61,9	3832	237177	7,868	3,956	194,5	3009
62,0	3844	238328	7,874	3,958	194,8	3019
62,1	3856	239483	7,880	3,960	195,1	3029
62,2	3869	240642	7,887	3,962	195,4	3039
62,3	3881	241804	7,893	3,964	195,7	3048
62,4	3894	242971	7,899	3,966	196,0	3058
62,5	3906	244141	7,906	3,968	196,3	3068
62,6	3919	245314	7,912	3,971	196,7	3078
62,7	3931	246492	7,918	3,973	197,0	3088
62,8	3944	247673	7,925	3,975	197,3	3097
62,9	3956	248858	7,931	3,977	197,6	3107
63,0	3969	250047	7,937	3,979	197,9	3117
63,1	3982	251240	7,944	3,981	198,2	3127
63,2	3994	252436	7,950	3,983	198,5	3137
63,3	4007	253636	7,956	3,985	198,9	3147
63,4	4020	254840	7,962	3,987	199,2	3157
63,5	4032	256048	7,969	3,990	199,5	3167
63,6	4045	257259	7,975	3,992	199,8	3177
63,7	4058	258475	7,981	3,994	200,1	3187
63,8	4070	259694	7,987	3,996	200,4	3197
63,9	4083	260917	7,994	3,998	200,7	3207
64,0	4096	262144	8,000	4,000	201,1	3217
64,1	4109	263375	8,006	4,002	201,4	3227
64,2	4122	264609	8,012	4,004	201,7	3237
64,3	4134	265848	8,019	4,006	202,0	3247
64,4	4147	267090	8,025	4,008	202,3	3257
64,5	4160	268336	8,031	4,010	202,6	3267

n = d	n^2	n^3	$\sqrt{n}$	$\sqrt[3]{n}$	$\pi \cdot d$	$\frac{\pi \cdot d^2}{4}$
64,6	4173	269586	8,037	4,013	202,9	3278
64,7	4186	270840	8,044	4,015	203,3	3288
64,8	4199	272098	8,050	4,017	203,6	3298
64,9	4212	273359	8,056	4,019	203,9	3308
65,0	4225	274625	8,062	4,021	204,2	3318
65,1	4238	275894	8,068	4,023	204,5	3329
65,2	4251	277168	8,075	4,025	204,8	3339
65,3	4264	278445	8,081	4,027	205,1	3349
65,4	4277	279726	8,087	4,029	205,5	3359
65,5	4290	281011	8,093	4,031	205,8	3370
65,6	4303	282300	8,099	4,033	206,1	3380
65,7	4316	283593	8,106	4,035	206,4	3390
65,8	4330	284890	8,112	4,037	206,7	3400
65,9	4343	286191	8,118	4,039	207,0	3411
66,0	4356	287496	8,124	4,041	207,3	3421
66,1	4369	288805	8,130	4,043	207,7	3432
66,2	4382	290118	8,136	4,045	208,0	3442
66,3	4396	291434	8,142	4,047	208,3	3452
66,4	4409	292755	8,149	4,049	208,6	3463
66,5	4422	294080	8,155	4,051	208,9	3473
66,6	4436	295408	8,161	4,053	209,2	3484
66,7	4449	296741	8,167	4,055	209,5	3494
66,8	4462	298078	8,173	4,058	209,9	3505
66,9	4476	299418	8,179	4,060	210,2	3515
67,0	4489	300763	8,185	4,062	210,5	3526
67,1	4502	302112	8,191	4,064	210,8	3536
67,2	4516	303464	8,198	4,066	211,1	3547
67,3	4529	304821	8,204	4,068	211,4	3557
67,4	4543	306182	8,210	4,070	211,7	3568
67,5	4556	307547	8,216	4,072	212,1	3578
67,6	4570	308916	8,222	4,074	212,4	3589
67,7	4583	310289	8,228	4,076	212,7	3600
67,8	4597	311666	8,234	4,078	213,0	3610
67,9	4610	313047	8,240	4,080	213,3	3621
68,0	4624	314432	8,246	4,082	213,6	3632
68,1	4638	315821	8,252	4,084	213,9	3642
68,2	4651	317215	8,258	4,086	214,3	3653
68,3	4665	318612	8,264	4,088	214,6	3664
68,4	4679	320014	8,270	4,090	214,9	3675
68,5	4692	321419	8,276	4,092	215,2	3685
68,6	4706	322829	8,283	4,094	215,5	3696
68,7	4720	324243	8,289	4,096	215,8	3707
68,8	4733	325661	8,295	4,098	216,1	3718
68,9	4747	327083	8,301	4,100	216,5	3728
69,0	4761	328509	8,307	4,102	216,8	3739
69,1	4775	329939	8,313	4,104	217,1	3750
69,2	4789	331374	8,319	4,106	217,4	3761
69,3	4802	332813	8,325	4,108	217,7	3772
69,4	4816	334255	8,331	4,109	218,0	3783
69,5	4830	335702	8,337	4,111	218,3	3794
69,6	4844	337154	8,343	4,113	218,7	3805
69,7	4858	338609	8,349	4,115	219,0	3816
69,8	4872	340068	8,355	4,117	219,3	3826
69,9	4886	341532	8,361	4,119	219,6	3837
70,0	4900	343000	8,367	4,121	219,9	3848
70,1	4914	344472	8,373	4,123	220,2	3859
70,2	4928	345948	8,379	4,125	220,5	3870
70,3	4942	347429	8,385	4,127	220,9	3882
70,4	4956	348914	8,390	4,129	221,2	3893
70,5	4970	350403	8,396	4,131	221,5	3904
70,6	4984	351896	8,402	4,133	221,8	3915
70,7	4998	353393	8,408	4,135	222,1	3926
70,8	5013	354895	8,414	4,137	222,4	3937
70,9	5027	356401	8,420	4,139	222,7	3948
71,0	5041	357911	8,426	4,141	223,1	3959

$n = d$	n^2	n^3	$\sqrt{n}$	$\sqrt[3]{n}$	$\pi \cdot d$	$\frac{\pi \cdot d^2}{4}$	$n = d$	n^2	n^3	$\sqrt{n}$	$\sqrt[3]{n}$	$\pi \cdot d$	$\frac{\pi \cdot d^2}{4}$
71,1	5055	359425	8,432	4,143	223,4	3970	77,6	6022	467289	8,809	4,265	243,8	4729
71,2	5069	360944	8,438	4,145	223,7	3982	77,7	6037	469097	8,815	4,267	244,1	4742
71,3	5084	362467	8,444	4,147	224,0	3993	77,8	6053	470911	8,820	4,269	244,4	4754
71,4	5098	363994	8,450	4,149	224,3	4004	77,9	6068	472729	8,826	4,271	244,7	4766
71,5	5112	365526	8,456	4,151	224,6	4015	**78,0**	6084	474552	8,832	4,273	245,0	4778
71,6	5127	367062	8,462	4,152	224,9	4026	78,1	6100	476380	8,837	4,274	245,4	4791
71,7	5141	368602	8,468	4,154	225,3	4038	78,2	6115	478212	8,843	4,276	245,7	4803
71,8	5155	370146	8,473	4,156	225,6	4049	78,3	6131	480049	8,849	4,278	246,0	4815
71,9	5170	371695	8,479	4,158	225,9	4060	78,4	6147	481890	8,854	4,280	246,3	4827
72,0	5184	373248	8,485	4,160	226,2	4072	**78,5**	6162	483737	8,860	4,282	246,6	4840
72,1	5198	374805	8,491	4,162	226,5	4083	78,6	6178	485588	8,866	4,284	246,9	4852
72,2	5213	376367	8,497	4,164	226,8	4094	78,7	6194	487443	8,871	4,285	247,2	4865
72,3	5227	377933	8,503	4,166	227,1	4106	78,8	6209	489304	8,877	4,287	247,6	4877
72,4	5242	379503	8,509	4,168	227,5	4117	78,9	6225	491169	8,883	4,289	247,9	4889
72,5	5256	381078	8,515	4,170	227,8	4128	**79,0**	6241	493039	8,888	4,291	248,2	4902
72,6	5271	382657	8,521	4,172	228,1	4140	79,1	6257	494914	8,894	4,293	248,5	4914
72,7	5285	384241	8,526	4,174	228,4	4151	79,2	6273	496793	8,899	4,294	248,8	4927
72,8	5300	385828	8,532	4,176	228,7	4162	79,3	6288	498677	8,905	4,296	249,1	4939
72,9	5314	387420	8,538	4,177	229,0	4174	79,4	6304	500566	8,911	4,298	249,4	4951
73,0	5329	389017	8,544	4,179	229,3	4185	**79,5**	6320	502460	8,916	4,300	249,8	4964
73,1	5344	390618	8,550	4,181	229,7	4197	79,6	6336	504358	8,922	4,302	250,1	4976
73,2	5358	392223	8,556	4,183	230,0	4208	79,7	6352	506262	8,927	4,303	250,4	4989
73,3	5373	393833	8,562	4,185	230,3	4220	79,8	6368	508170	8,933	4,305	250,7	5001
73,4	5388	395447	8,567	4,187	230,6	4231	79,9	6384	510082	8,939	4,307	251,0	5014
73,5	5402	397065	8,573	4,189	230,9	4243	**80,0**	6400	512000	8,944	4,309	251,3	5027
73,6	5417	398688	8,579	4,191	231,2	4254	80,1	6416	513922	8,950	4,311	251,6	5039
73,7	5432	400316	8,585	4,193	231,5	4266	80,2	6432	515850	8,955	4,312	252,0	5052
73,8	5446	401947	8,591	4,195	231,8	4278	80,3	6448	517782	8,961	4,314	252,3	5064
73,9	5461	403583	8,597	4,196	232,2	4289	80,4	6464	519718	8,967	4,316	252,6	5077
74,0	5476	405224	8,602	4,198	232,5	4301	**80,5**	6480	521660	8,972	4,318	252,9	5090
74,1	5491	406869	8,608	4,200	232,8	4312	80,6	6496	523607	8,978	4,320	253,2	5102
74,2	5506	408518	8,614	4,202	233,1	4324	80,7	6512	525558	8,983	4,321	253,5	5115
74,3	5520	410172	8,620	4,204	233,4	4336	80,8	6529	527514	8,989	4,323	253,8	5128
74,4	5535	411831	8,626	4,206	233,7	4347	80,9	6545	529475	8,994	4,325	254,2	5140
74,5	5550	413494	8,631	4,208	234,0	4359	**81,0**	6561	531441	9,000	4,327	254,5	5153
74,6	5565	415161	8,637	4,210	234,4	4371	81,1	6577	533412	9,006	4,329	254,8	5166
74,7	5580	416833	8,643	4,212	234,7	4383	81,2	6593	535387	9,011	4,330	255,1	5178
74,8	5595	418509	8,649	4,213	235,0	4394	81,3	6610	537368	9,017	4,332	255,4	5191
74,9	5610	420190	8,654	4,215	235,3	4406	81,4	6626	539353	9,022	4,334	255,7	5204
75,0	5625	421875	8,660	4,217	235,6	4418	**81,5**	6642	541343	9,028	4,336	256,0	5217
75,1	5640	423565	8,666	4,219	235,9	4430	81,6	6659	543338	9,033	4,337	256,4	5230
75,2	5655	425259	8,672	4,221	236,2	4441	81,7	6675	545339	9,039	4,339	256,7	5242
75,3	5670	426958	8,678	4,223	236,6	4453	81,8	6691	547343	9,044	4,341	257,0	5255
75,4	5685	428661	8,683	4,225	236,9	4465	81,9	6708	549353	9,050	4,343	257,3	5268
75,5	5700	430369	8,689	4,227	237,2	4477	**82,0**	6724	551368	9,055	4,344	257,6	5281
75,6	5715	432081	8,695	4,228	237,5	4489	82,1	6740	553388	9,061	4,346	257,9	5294
75,7	5730	433798	8,701	4,230	237,8	4501	82,2	6757	555412	9,066	4,348	258,2	5307
75,8	5746	435520	8,706	4,232	238,1	4513	82,3	6773	557442	9,072	4,350	258,6	5320
75,9	5761	437245	8,712	4,234	238,4	4525	82,4	6790	559476	9,077	4,352	258,9	5333
76,0	5776	438976	8,718	4,236	238,8	4536	**82,5**	6806	561516	9,083	4,353	259,2	5346
76,1	5791	440711	8,724	4,238	239,1	4548	82,6	6823	563560	9,088	4,355	259,5	5359
76,2	5806	442451	8,729	4,240	239,4	4560	82,7	6839	565609	9,094	4,357	259,8	5372
76,3	5822	444195	8,735	4,241	239,7	4572	82,8	6856	567664	9,099	4,359	260,1	5385
76,4	5837	445944	8,741	4,243	240,0	4584	82,9	6872	569723	9,105	4,360	260,4	5398
76,5	5852	447697	8,746	4,245	240,3	4596	**83,0**	6889	571787	9,110	4,362	260,8	5411
76,6	5868	449455	8,752	4,247	240,6	4608	83,1	6906	573856	9,116	4,364	261,1	5424
76,7	5883	451218	8,758	4,249	241,0	4620	83,2	6922	575930	9,121	4,366	261,4	5437
76,8	5898	452985	8,764	4,251	241,3	4632	83,3	6939	578010	9,127	4,367	261,7	5450
76,9	5914	454757	8,769	4,252	241,6	4645	83,4	6956	580094	9,132	4,369	262,0	5463
77,0	5929	456533	8,775	4,254	241,9	4657	**83,5**	6972	582183	9,138	4,371	262,3	5476
77,1	5944	458314	8,781	4,256	242,2	4669	83,6	6989	584277	9,143	4,373	262,6	5489
77,2	5960	460100	8,786	4,258	242,5	4681	83,7	7006	586376	9,149	4,374	263,0	5502
77,3	5975	461890	8,792	4,260	242,8	4693	83,8	7022	588480	9,154	4,376	263,3	5515
77,4	5991	463685	8,798	4,262	243,2	4705	83,9	7039	590590	9,160	4,378	263,6	5529
77,5	6006	465484	8,803	4,264	243,5	4717	**84,0**	7056	592704	9,165	4,380	263,9	5542

$n = d$	n^2	n^3	$\sqrt{n}$	$\sqrt[3]{n}$	$\pi \cdot d$	$\frac{\pi \cdot d^2}{4}$	$n = d$	n^2	n^3	$\sqrt{n}$	$\sqrt[3]{n}$	$\pi \cdot d$	$\frac{\pi \cdot d^2}{4}$
84,1	7073	594823	9,171	4,381	264,2	5555	90,6	8208	743677	9,518	4,491	284,6	6447
84,2	7090	596948	9,176	4,383	264,5	5568	90,7	8226	746143	9,524	4,493	284,9	6461
84,3	7106	599077	9,182	4,385	264,8	5581	90,8	8245	748613	9,529	4,495	285,3	6475
84,4	7123	601212	9,187	4,386	265,2	5595	90,9	8263	751089	9,534	4,496	285,6	6490
84,5	7140	603351	9,192	4,388	265,5	5608	**91,0**	8281	753571	9,539	4,498	285,9	6504
84,6	7157	605496	9,198	4,390	265,8	5621	91,1	8299	756058	9,545	4,500	286,2	6518
84,7	7174	607645	9,203	4,392	266,1	5635	91,2	8317	758551	9,550	4,501	286,5	6533
84,8	7191	609800	9,209	4,393	266,4	5648	91,3	8336	761048	9,555	4,503	286,8	6547
84,9	7208	611960	9,214	4,395	266,7	5661	91,4	8354	763552	9,560	4,505	287,1	6561
85,0	7225	614125	9,220	4,397	267,0	5675	**91,5**	8372	766061	9,566	4,506	287,5	6576
85,1	7242	616295	9,225	4,399	267,3	5688	91,6	8391	768575	9,571	4,508	287,8	6590
85,2	7259	618470	9,230	4,400	267,7	5701	91,7	8409	771095	9,576	4,509	288,1	6604
85,3	7276	620650	9,236	4,402	268,0	5715	91,8	8427	773621	9,581	4,511	288,4	6619
85,4	7293	622836	9,241	4,404	268,3	5728	91,9	8446	776152	9,586	4,513	288,7	6633
85,5	7310	625026	9,247	4,405	268,6	5741	**92,0**	8464	778688	9,592	4,514	289,0	6648
85,6	7327	627222	9,252	4,407	268,9	5755	92,1	8482	781230	9,597	4,516	289,3	6662
85,7	7344	629423	9,257	4,409	269,2	5768	92,2	8501	783777	9,602	4,518	289,7	6677
85,8	7362	631629	9,263	4,411	269,5	5782	92,3	8519	786330	9,606	4,519	290,0	6691
85,9	7379	633840	9,268	4,412	269,9	5795	92,4	8538	788889	9,612	4,521	290,3	6706
86,0	7396	636056	9,274	4,414	270,2	5809	**92,5**	8556	791453	9,618	4,523	290,6	6720
86,1	7413	638277	9,279	4,416	270,5	5822	92,6	8575	794023	9,623	4,524	290,9	6735
86,2	7430	640504	9,284	4,417	270,8	5836	92,7	8593	796598	9,628	4,526	291,2	6749
86,3	7448	642736	9,290	4,419	271,1	5849	92,8	8612	799179	9,633	4,527	291,5	6764
86,4	7465	644973	9,295	4,421	271,4	5863	92,9	8630	801765	9,638	4,529	291,9	6778
86,5	7482	647215	9,301	4,423	271,7	5877	**93,0**	8649	804357	9,644	4,531	292,2	6793
86,6	7500	649462	9,306	4,424	272,1	5890	93,1	8668	806954	9,649	4,532	292,5	6808
86,7	7517	651714	9,311	4,426	272,4	5904	93,2	8686	809558	9,654	4,534	292,8	6822
86,8	7534	653972	9,317	4,428	272,7	5917	93,3	8705	812166	9,659	4,536	293,1	6837
86,9	7552	656235	9,322	4,429	273,0	5931	93,4	8724	814781	9,664	4,537	293,4	6851
87,0	7569	658503	9,327	4,431	273,3	5945	**93,5**	8742	817400	9,670	4,539	293,7	6866
87,1	7586	660776	9,333	4,433	273,6	5958	93,6	8761	820026	9,675	4,540	294,1	6881
87,2	7604	663055	9,338	4,434	273,9	5972	93,7	8780	822657	9,680	4,542	294,4	6896
87,3	7621	665339	9,343	4,436	274,3	5986	93,8	8798	825294	9,685	4,544	294,7	6910
87,4	7639	667628	9,349	4,438	274,6	5999	93,9	8817	827936	9,690	4,545	295,0	6925
87,5	7656	669922	9,354	4,440	274,9	6013	**94,0**	8836	830584	9,695	4,547	295,3	6940
87,6	7674	672221	9,359	4,441	275,2	6027	94,1	8855	833238	9,701	4,548	295,6	6955
87,7	7691	674526	9,365	4,443	275,5	6041	94,2	8874	835897	9,706	4,550	295,9	6969
87,8	7709	676636	9,370	4,445	275,8	6055	94,3	8892	838662	9,711	4,552	266,3	6984
87,9	7726	679151	9,375	4,446	276,1	6068	94,4	8911	841232	9,716	4,553	296,6	6999
88,0	7744	681472	9,381	4,448	276,5	6082	**94,5**	8930	843909	9,721	4,555	296,9	7014
88,1	7762	683798	9,386	4,450	276,8	6096	94,6	8949	846591	9,726	4,556	297,2	7029
88,2	7779	686129	9,391	4,451	277,1	6110	94,7	8968	849278	9,731	4,558	297,5	7044
88,3	7797	688465	9,397	4,453	277,4	6124	94,8	8987	851971	9,737	4,560	297,8	7058
88,4	7815	690807	9,402	4,455	277,7	6138	94,9	9006	854670	9,742	4,561	298,1	7073
88,5	7832	693154	9,407	4,456	278,0	6151	**95,0**	9025	857375	9,747	4,563	298,5	7088
88,6	7850	695506	9,413	4,458	278,3	6165	95,1	9044	860085	9,752	4,565	298,8	7103
88,7	7868	697864	9,418	4,460	278,7	6179	95,2	9063	862801	9,757	4,566	299,1	7118
88,8	7885	700227	9,423	4,461	279,0	6193	95,3	9082	865523	9,762	4,568	299,4	7133
88,9	7903	702595	9,429	4,463	279,3	6207	95,4	9101	868251	9,767	4,569	299,7	7148
89,0	7921	704969	9,434	4,465	279,6	6221	**95,5**	9120	870984	9,772	4,571	300,0	7163
89,1	7939	707348	9,439	4,466	279,9	6235	95,6	9139	873723	9,778	4,572	300,3	7178
89,2	7957	709732	9,445	4,468	280,2	6249	95,7	9158	876467	9,783	4,574	300,7	7193
89,3	7974	712122	9,450	4,470	280,5	6263	95,8	9178	879218	9,788	4,576	301,0	7208
89,4	7992	714517	9,455	4,471	280,9	6277	95,9	9197	881974	9,793	4,577	301,3	7223
89,5	8010	716917	9,460	4,473	281,2	6291	**96,0**	9216	884736	9,798	4,579	301,6	7238
89,6	8028	719323	9,466	4,475	281,5	6305	96,1	9235	887504	9,803	4,580	301,9	7253
89,7	8046	721734	9,471	4,476	281,8	6319	96,2	9254	890277	9,808	4,582	302,2	7268
89,8	8064	724151	9,476	4,478	282,1	6333	96,3	9274	893056	9,813	4,584	302,5	7284
89,9	8082	726573	9,482	4,480	282,4	6348	96,4	9293	895841	9,818	4,585	302,8	7299
90,0	8100	729000	9,487	4,481	282,7	6362	**96,5**	9312	898632	9,823	4,587	303,2	7314
90,1	8118	731433	9,492	4,483	283,1	6376	96,6	9332	901429	9,829	4,588	303,5	7329
90,2	8136	733871	9,497	4,485	283,4	6390	96,7	9351	904231	9,834	4,590	303,8	7344
90,3	8154	736314	9,503	4,486	283,7	6404	96,8	9370	907039	9,839	4,592	304,1	7359
90,4	8172	738763	9,508	4,488	284,0	6418	96,9	9390	909853	9,844	4,593	304,4	7375
90,5	8190	741218	9,513	4,490	284,3	6433	**97,0**	9409	912673	9,849	4,595	304,7	7390

$n = d$	n^2	n^3	$\sqrt{n}$	$\sqrt[3]{n}$	$\pi \cdot d$	$\frac{\pi \cdot d^2}{4}$	$n = d$	n^2	n^3	$\sqrt{n}$	$\sqrt[3]{n}$	$\pi \cdot d$	$\frac{\pi \cdot d^2}{4}$
97,1	9428	915499	9,854	4,596	305,0	7405	107,2	11491	1231925	10,354	4,751	336,8	9025
97,2	9448	918330	9,859	4,598	305,4	7420	107,4	11534	1238833	10,363	4,753	337,4	9059
97,3	9467	921167	9,864	4,599	305,7	7436	107,6	11577	1245766	10,373	4,756	338,0	9092
97,4	9487	924010	9,869	4,601	306,0	7451	107,8	11620	1252726	10,382	4,759	338,7	9127
97,5	9506	926859	9,874	4,603	306,3	7466	**108,0**	11664	1259712	10,392	4,762	339,3	9160
97,6	9526	929714	9,879	4,604	306,6	7482	108,2	11707	1266723	10,401	4,765	339,9	9194
97,7	9545	932575	9,884	4,606	306,9	7497	108,4	11750	1273760	10,411	4,768	340,6	9228
97,8	9565	935441	9,889	4,607	307,2	7512	108,6	11793	1280824	10,420	4,771	341,2	9263
97,9	9584	938314	9,894	4,609	307,6	7528	108,8	11837	1287913	10,430	4,774	341,8	9297
98,0	9604	941192	9,899	4,610	307,9	7543	**109,0**	11881	1295029	10,440	4,777	342,4	9331
98,1	9624	944076	9,905	4,612	308,2	7558	109,2	11924	1302170	10,449	4,780	343,1	9365
98,2	9643	946966	9,910	4,614	308,5	7574	109,4	11968	1309338	10,459	4,783	343,7	9399
98,3	9663	949862	9,915	4,615	308,8	7589	109,6	12012	1316532	10,468	4,785	344,3	9434
98,4	9683	952764	9,920	4,617	309,1	7605	109,8	12056	1323753	10,478	4,788	344,9	9468
98,5	9702	955672	9,925	4,618	309,4	7620	**110,0**	12100	1331000	10,488	4,791	345,6	9503
98,6	9722	958585	9,930	4,620	309,8	7636	110,2	12144	1338273	10,497	4,794	346,2	9537
98,7	9742	961505	9,935	4,621	310,1	7651	110,4	12188	1345572	10,507	4,797	346,8	9572
98,8	9761	964430	9,940	4,623	310,4	7667	110,6	12232	1352899	10,516	4,800	347,5	9607
98,9	9781	967362	9,945	4,625	310,7	7682	110,8	12276	1360251	10,526	4,803	348,1	9642
99,0	9801	970299	9,950	4,626	311,0	7698	**111,0**	12321	1367631	10,536	4,806	348,7	9676
99,1	9821	973242	9,955	4,628	311,3	7713	111,2	12365	1375036	10,545	4,809	349,4	9711
99,2	9841	976191	9,960	4,629	311,6	7729	111,4	12409	1382469	10,554	4,812	349,9	9746
99,3	9860	979147	9,965	4,631	312,0	7744	111,6	12454	1389928	10,564	4,814	350,6	9781
99,4	9880	982108	9,970	4,632	312,3	7760	111,8	12499	1397415	10,573	4,817	351,2	9816
99,5	9900	985075	9,975	4,634	312,6	7776	**112,0**	12544	1404928	10,583	4,820	351,9	9852
99,6	9920	988048	9,980	4,635	312,9	7791	112,2	12588	1412467	10,592	4,823	352,5	9887
99,7	9940	991027	9,985	4,637	313,2	7807	112,4	12633	1420034	10,602	4,826	353,1	9922
99,8	9960	994012	9,990	4,638	313,5	7823	112,6	12678	1427628	10,611	4,829	353,7	9957
99,9	9980	997003	9,995	4,640	313,8	7838	112,8	12723	1435249	10,621	4,832	354,4	9993
100,0	10000	1000000	10,000	4,642	314,2	7854	**113,0**	12769	1442897	10,630	4,835	355,0	10028
100,2	10040	1006008	10,010	4,645	314,8	7885	113,2	12814	1450571	10,639	4,838	355,6	10064
100,4	10080	1012032	10,020	4,648	315,4	7917	113,4	12859	1458274	10,649	4,841	356,3	10099
100,6	10120	1018100	10,030	4,650	316,0	7949	113,6	12904	1466003	10,658	4,843	356,9	10135
100,8	10160	1024200	10,040	4,653	316,6	7980	113,8	12950	1473760	10,667	4,846	357,5	10171
101,0	10200	1030300	10,050	4,657	317,3	8012	**114,0**	12996	1481544	10,677	4,849	358,1	10207
101,2	10241	1036433	10,060	4,660	317,9	8043	114,2	13041	1489355	10,686	4,852	358,7	10242
101,4	10281	1042590	10,070	4,663	318,6	8075	114,4	13087	1497193	10,696	4,855	359,4	10278
101,6	10322	1048772	10,080	4,666	319,2	8107	114,6	13133	1505060	10,705	4,857	360,0	10314
101,8	10363	1054977	10,090	4,669	319,8	8139	114,8	13179	1512953	10,714	4,860	360,6	10350
102,0	10404	1061208	10,099	4,672	320,4	8171	**115,0**	13225	1520875	10,724	4,863	361,3	10386
102,2	10444	1067462	10,109	4,675	321,1	8203	115,2	13271	1528823	10,733	4,866	361,9	10423
102,4	10485	1073741	10,119	4,678	321,7	8235	115,4	13317	1536800	10,742	4,869	362,5	10477
102,6	10526	1080045	10,129	4,681	322,3	8267	115,6	13363	1544804	10,751	4,871	363,2	10495
102,8	10567	1086373	10,139	4,684	323,0	8299	115,8	13409	1552836	10,761	4,874	363,8	10531
103,0	10609	1092727	10,149	4,687	323,6	8332	**116,0**	13456	1560896	10,770	4,877	364,4	10568
103,2	10650	1099104	10,159	4,691	324,2	8364	116,2	13502	1568983	10,779	4,880	365,1	10604
103,4	10691	1105507	10,169	4,694	324,8	8397	116,4	13548	1577098	10,789	4,883	365,7	10641
103,6	10732	1111934	10,178	4,697	325,5	8429	116,6	13595	1585242	10,798	4,885	366,3	10677
103,8	10774	1118386	10,188	4,700	326,1	8462	116,8	13642	1593413	10,808	4,888	366,9	10714
104,0	10816	1124864	10,198	4,703	326,7	8494	**117,0**	13689	1601613	10,817	4,891	367,6	10751
104,2	10857	1131366	10,208	4,706	327,4	8527	117,2	13735	1609840	10,826	4,894	368,2	10788
104,4	10899	1137893	10,217	4,709	328,0	8560	117,4	13782	1618096	10,835	4,897	368,8	10825
104,6	10941	1144445	10,226	4,712	328,6	8593	117,6	13829	1626379	10,845	4,899	369,5	10861
104,8	10983	1151022	10,235	4,715	329,2	8626	117,8	13876	1634691	10,854	4,902	370,1	10898
105,0	11025	1157625	10,247	4,718	329,8	8659	**118,0**	13924	1643032	10,863	4,905	370,7	10935
105,2	11067	1164252	10,257	4,721	330,5	8692	118,2	13971	1651400	10,872	4,908	371,3	10973
105,4	11109	1170905	10,267	4,724	331,1	8725	118,4	14018	1659797	10,881	4,911	372,0	11010
105,6	11151	1177583	10,277	4,727	331,7	8758	118,6	14065	1668222	10,891	4,913	372,6	11047
105,8	11193	1184287	10,287	4,730	332,4	8791	118,8	14113	1676676	10,900	4,916	373,2	11084
106,0	11236	1191016	10,296	4,733	333,0	8824	**119,0**	14161	1685159	10,909	4,919	373,9	11122
106,2	11278	1197770	10,306	4,736	333,6	8858	119,2	14208	1693669	10,918	4,921	374,5	11159
106,4	11320	1204550	10,315	4,739	334,3	8891	119,4	14256	1702209	10,927	4,924	375,1	11196
106,6	11363	1211355	10,325	4,742	334,9	8924	119,6	14304	1710777	10,937	4,927	375,7	11234
106,8	11406	1218186	10,334	4,745	335,5	8958	119,8	14352	1719374	10,946	4,929	376,4	11272
107,0	11449	1225043	10,344	4,748	336,2	8992	**120,0**	14400	1728000	10,955	4,932	377,0	11309

▼ **Sine 0...45°**

Degree	0′	10′	20′	30′	40′	50′	60′	
0	0,000	0,003	0,006	0,009	0,012	0,015	0,017	89
1	0,017	0,020	0,023	0,026	0,029	0,032	0,035	88
2	0,035	0,038	0,041	0,044	0,047	0,049	0,052	87
3	0,052	0,055	0,058	0,061	0,064	0,067	0,070	86
4	0,070	0,073	0,075	0,078	0,081	0,084	0,087	**85**
5	0,087	0,090	0,093	0,096	0,099	0,102	0,105	84
6	0,105	0,107	0,110	0,113	0,116	0,119	0,122	83
7	0,122	0,125	0,128	0,131	0,133	0,136	0,139	82
8	0,139	0,142	0,145	0,148	0,151	0,154	0,156	81
9	0,156	0,159	0,162	0,165	0,168	0,171	0,174	**80**
10	0,174	0,177	0,179	0,182	0,185	0,188	0,191	79
11	0,191	0,194	0,197	0,199	0,202	0,205	0,208	78
12	0,208	0,211	0,214	0,216	0,219	0,222	0,225	77
13	0,225	0 228	0,231	0,233	0,236	0,239	0,242	76
14	0,242	0,245	0,248	0,250	0,253	0,256	0,259	**75**
15	0,259	0,262	0,264	0,267	0,270	0,273	0,276	74
16	0,276	0,278	0,281	0,284	0,287	0,290	0,292	73
17	0,292	0,295	0,298	0,301	0,303	0,306	0,309	72
18	0,309	0,312	0,315	0,317	0,320	0,323	0,326	71
19	0,326	0,328	0,331	0,334	0,337	0,339	0,342	**70**
20	0,342	0,345	0,347	0,350	0,353	0,356	0,358	69
21	0,358	0,361	0,364	0,367	0,369	0,372	0,375	68
22	0,375	0,377	0,380	0,383	0,385	0,388	0,391	67
23	0,391	0,393	0,396	0,399	0,401	0,404	0,407	66
24	0,407	0,409	0,412	0,415	0,417	0,420	0,423	**65**
25	0,423	0,425	0,428	0,431	0,433	0,436	0,438	64
26	0,438	0,441	0,444	0,446	0,449	0,451	0,454	63
27	0,454	0,457	0,459	0,462	0,464	0,467	0,469	62
28	0,469	0,472	0,475	0,477	0,480	0,482	0,485	61
29	0,485	0,487	0,490	0,492	0,495	0,497	0,500	**60**
30	0,500	0,503	0,505	0,508	0,510	0,513	0,515	59
31	0,515	0,518	0,520	0,522	0,525	0,527	0,530	58
32	0,530	0,532	0,535	0,537	0,540	0,542	0,545	57
33	0,545	0,547	0,550	0,552	0,554	0,557	0,559	56
34	0,559	0,562	0,564	0,566	0,569	0,571	0,574	**55**
35	0,574	0,576	0,578	0,581	0,583	0,585	0,588	54
36	0,588	0,590	0,592	0,595	0,597	0,599	0,602	53
37	0,602	0,604	0,606	0,609	0,611	0,613	0,616	52
38	0,616	0,618	0,620	0,623	0,625	0,627	0,629	51
39	0,629	0,632	0,634	0,636	0,638	0,641	0,643	**50**
40	0,643	0,645	0,647	0,649	0,652	0,654	0,656	49
41	0,656	0,658	0,660	0,663	0,665	0,667	0,669	48
42	0,669	0,671	0,673	0,676	0,678	0,680	0,682	47
43	0,682	0,684	0,686	0,688	0,690	0,693	0,695	46
44	0,695	0,697	0,699	0,701	0,703	0.705	0,707	**45**
	60′	50′	40′	30′	20′	10′	0′	Degree

Cosine 45...90° ▲

▼ **Sine 45...90°**

Degree	0′	10′	20′	30′	40′	50′	60′	
45	0,707	0,709	0,711	0,713	0,715	0,717	0,719	44
46	0,719	0,721	0,723	0,725	0,727	0,729	0,731	43
47	0,731	0,733	0,735	0,737	0,739	0,741	0,743	42
48	0,743	0,745	0,747	0,749	0,751	0,753	0,755	41
49	0,755	0,757	0,759	0,760	0,762	0,764	0,766	**40**
50	0,766	0,768	0,770	0,772	0,773	0,775	0,777	39
51	0,777	0,779	0,781	0,783	0,784	0,786	0,788	38
52	0,788	0,790	0,792	0,793	0,795	0,797	0,799	37
53	0,799	0,800	0,802	0,804	0,806	0,807	0,809	36
54	0,809	0,811	0,812	0,814	0,816	0,817	0,819	**35**
55	0,819	0,821	0,822	9,824	0,826	0,827	0,829	34
56	0,829	0,831	0,832	0,834	0,835	0,837	0,839	33
57	0,839	0,840	0,842	0,843	0,845	0,847	0,848	32
58	0,848	0,850	0,851	0,853	0,854	0,856	0,857	31
59	0,857	0,859	0,860	0,862	0,863	0,865	0,866	**30**
60	0,866	0,867	0 869	0,870	0,872	0,873	0,875	29
61	0,875	0,876	0,877	0,879	0,880	0,882	0,883	28
62	0,883	0,884	0,886	0,887	0,888	0,890	0,891	27
63	0,891	0,892	0,894	0,895	0,896	0,898	0,899	26
64	0,899	0,900	0,901	0,903	0,904	0,905	0,906	**25**
65	0,906	0,908	0,909	0,910	0,911	0,912	0,914	24
66	0,914	0,915	0,916	0,917	0,918	0,919	0,921	23
67	0,921	0,922	0,923	0,924	0,925	0,926	0,927	22
68	0,927	0,928	0,929	0,930	0,931	0,933	0,934	21
69	0,934	0,935	0,936	0,937	0,938	0,939	0,940	**20**
70	0,940	0,941	0,942	0,943	0,944	0,945	0,946	19
71	0,946	0,946	0,947	0,948	0,949	0,950	0,951	18
72	0,951	0,952	0,953	0,954	0,955	0,955	0,956	17
73	0,956	0,957	0,958	0,959	0,960	0,960	0,961	16
74	0,961	0,962	0,963	0,964	0,964	0,965	0,966	**15**
75	0,966	0,967	0,967	0,968	0,969	0,970	0,970	14
76	0,970	0,971	0,972	0,972	0,973	0,974	0,974	13
77	0,974	0,975	0,976	0,976	0,977	0,978	0,978	12
78	0,978	0,979	0,979	0,980	0,981	0,981	0,982	11
79	0,982	0,982	0,983	0,983	0,984	0,984	0,985	**10**
80	0,985	0,985	0,986	0,986	0,987	0,987	0,988	9
81	0,988	0,988	0,989	0,989	0,989	0,990	0,990	8
82	0,990	0,991	0,991	0,991	0,992	0,992	0,993	7
83	0,993	0,993	0,993	0,994	0,994	0,994	0,995	6
84	0,995	0,995	0,995	0,995	0,996	0,996	0,996	**5**
85	0,996	0,996	0,997	0,997	0,997	0,997	0,998	4
86	0,998	0,998	0,998	0,998	0,998	0,998	0,999	3
87	0,999	0,999	0,999	0,999	0,999	0,999	0,999	2
88	0,999	0,999	1.000	1,000	1,000	1,000	1,000	1
09	1,000	1,000	1,000	1,000	1,000	1,000	1,000	0
	60′	50′	40′	30′	20′	10′	0′	Degree

Cosine 0...45° ▲

Example 1

Given : Angle α To find : Sin α	$=24°16'$
sin 24°20′ sin 24°10′	$=0,412$ $=0,409$
Difference for 10′	$=0,003$
Value for 1′ Value for 6′	$=0,0003$ $=0,0018$
sin 24°10′ +sin 6′	$=0,409$ $=0,002$
sin 24°16′	$=0,411$

Example 2

Given : Angle β To find : Cos β	$=40°14'$
cos 40°10′ cos 40°20′	$=0,764$ $=0,762$
Difference for 10′	$=0,002$
Value for 1′ Value for 4′	$=0,0002$ $=0,0008$
cos 40°10′ − cos 4′	$=0,764$ $=0,001$
cos 40°14′	$=0,763$

▼ **Tangent 0...45°**

Degree	0′	10′	20′	30′	40′	50′	60′	
0	0,000	0,003	0,006	0,009	0,012	0,015	0,017	**89**
1	0,017	0,020	0,023	0,026	0,029	0,032	0,035	88
2	0,035	0,038	0,041	0,044	0,047	0,049	0,052	87
3	0,052	0,055	0,058	0,061	0,064	0,067	0,070	86
4	0,070	0,073	0,076	0,079	0,082	0,085	0,087	**85**
5	0,087	0,090	0,093	0,096	0,099	0,102	0,105	84
6	0,105	0,108	0,111	0,114	0,117	0,120	0,123	83
7	0,123	0,126	0,129	0,132	0,135	0,138	0,141	82
8	0,141	0,144	0,146	0,149	0,152	0,155	0,158	81
9	0,158	0,161	0,164	0,167	0,170	0,173	0,176	**80**
10	0,176	0,179	0,182	0,185	0,188	0,191	0,194	79
11	0,194	0,197	0,200	0,203	0,206	0,210	0,213	78
12	0,213	0,216	0,219	0,222	0,225	0,228	0,231	77
13	0,231	0,234	0,237	0,240	0,243	0,246	0,249	76
14	0,249	0,252	0,256	0,259	0,262	0,265	0,268	**75**
15	0,268	0,271	0,274	0,277	0,280	0,284	0,287	74
16	0,287	0,290	0,293	0,296	0,299	0,303	0,306	73
17	0,306	0,309	0,312	0,315	0,318	0,322	0,325	72
18	0,325	0,328	0,331	0,335	0,338	0,341	0,344	71
19	0,344	0,348	0,351	0,354	0,357	0,361	0,364	**70**
20	0,364	0,367	0,371	0,374	0,377	0,381	0,384	69
21	0,384	0,387	0,391	0,394	0,397	0,401	0,404	68
22	0,404	0,407	0,411	0,414	0,418	0,421	0,424	67
23	0,424	0,428	0,431	0,435	0,438	0,442	0,445	66
24	0,445	0,449	0,452	0,456	0,459	0,463	0,466	**65**
25	0,466	0,470	0,473	0,477	0,481	0,484	0,488	64
26	0,488	0,491	0,495	0,499	0,502	0,506	0,510	63
27	0,510	0,513	0,517	0,521	0,524	0,528	0,532	62
28	0,532	0,535	0,539	0,543	0,547	0,551	0,554	61
29	0,554	0,558	0,562	0,566	0,570	0,573	0,577	**60**
30	0,577	0,581	0,585	0,589	0,593	0,597	0,601	59
31	0,601	0,605	0,609	0,613	0,617	0,621	0,625	58
32	0,625	0,629	0,633	0,637	0,641	0,645	0,649	57
33	0,649	0,654	0,658	0,662	0,666	0,670	0,675	56
34	0,675	0,679	0,683	0,687	0,692	0,696	0,700	**55**
35	0,700	0,705	0,709	0,713	0,718	0,722	0,727	54
36	0,727	0,731	0,735	0,740	0,744	0,749	0,754	53
37	0,754	0,758	0,763	0,767	0,772	0,777	0,781	52
38	0,781	0,786	0,791	0,795	0,800	0,805	0,810	51
39	0,810	0,815	0,819	0,824	0,829	0,834	0,839	**50**
40	0,839	0,844	0,849	0,854	0,859	0,864	0,869	49
41	0,869	0,874	0,880	0,885	0,890	0,895	0,900	48
42	0,900	0,906	0,911	0,916	0,922	0,927	0,933	47
43	0,933	0,938	0,943	0,949	0,955	0,960	0,966	46
44	0,966	0,971	0,977	0,983	0,988	0,994	1,000	**45**
	60′	50′	40′	30′	20′	10′	0′	Degree

Cotangent 45...90° ▲

▼ **Tangent 45...90**

Degree	0′	10′	20′	30′	40′	50′	60′	
45	1,000	1,006	1,012	1,018	1,024	1,030	1,036	44
46	1,036	1,042	1,048	1,054	1,060	1,066	1,072	43
47	1,072	1,079	1,085	1,091	1,098	1,104	1,111	42
48	1,111	1,117	1,124	1,130	1,137	1,144	1,150	41
49	1,150	1,157	1,164	1,171	1,178	1,185	1,192	**40**
50	1,192	1,199	1,206	1,213	1,220	1,228	1,235	39
51	1,235	1,242	1,250	1,257	1,265	1,272	1,280	38
52	1,280	1,288	1,295	1,303	1,311	1,319	1,327	37
53	1,327	1,335	1,343	1,351	1,360	1,368	1,376	36
54	1,376	1,385	1,393	1,402	1,411	1,419	1,428	**35**
55	1,428	1,437	1,446	1,455	1,464	1,473	1,483	34
56	1,483	1,492	1,501	1,511	1,520	1,530	1,540	33
57	1,540	1,550	1,560	1,570	1,580	1,590	1,600	32
58	1,600	1,611	1,621	1,632	1,643	1,653	1,664	31
59	1,664	1,675	1,686	1,698	1,709	1,720	1,732	**30**
60	1,732	1,744	1,756	1,767	1,780	1,792	1,804	29
61	1,804	1,816	1,829	1,842	1,855	1,868	1,881	28
62	1,881	1,894	1,907	1,921	1,935	1,949	1,963	27
63	1,963	1,977	1,991	2,006	2,020	2,035	2,050	26
64	2,050	2,066	2,081	2,097	2,112	2,128	2,145	**25**
65	2,145	2,161	2,177	2,194	2,211	2,229	2,246	24
66	2,246	2,264	2,282	2,300	2,318	2,337	2,356	23
67	2,356	2,375	2,394	2,414	2,434	2,455	2,475	22
68	2,475	2,496	2,517	2,539	2,560	2,583	2,605	21
69	2,605	2,628	2,651	2,675	2,699	2,723	2,747	**20**
70	2,747	2,773	2,798	2,824	2,850	2,877	2,904	19
71	2,904	2,932	2,960	2,989	3,018	3,047	3,078	18
72	3,078	3,108	3,140	3,172	3,204	3,237	3,271	17
73	3,271	3,305	3,340	3,376	3,412	3,450	3,487	16
74	3,487	3,526	3,566	3,606	3,647	3,689	3,732	**15**
75	3,732	3,776	3,821	3,867	3,914	3,962	4,011	14
76	4,011	4,061	4,113	4,165	4,219	4,275	4,331	13
77	4,331	4,390	4,449	4,511	4,574	4,638	4,705	12
78	4,705	4,773	4,843	4,915	4,989	5,066	5,145	11
79	5,145	5,226	5,309	5,396	5,485	5,576	5,671	**10**
80	5,671	5,769	5,871	5,976	6,084	6,197	6,314	9
81	6,314	6,435	6,561	6,691	6,827	6,968	7,115	8
82	7,115	7,296	7,429	7,596	7,770	7,953	8,144	7
83	8,144	8,345	8,556	8,777	9,010	9,255	9,514	6
84	9,514	9,788	10,08	10,39	10,71	11,06	11,43	**5**
85	11,43	11,83	12,25	12,71	13,20	13,73	14,30	4
86	14,30	14,92	15,60	16,35	17,17	18,07	19,08	3
87	19,08	20,21	21,47	22,90	24,54	26,43	28,64	2
88	28,64	31,24	34,37	38,19	42,96	49,10	57,29	1
89	57,29	68,75	85,94	114,6	171,9	343,8	[illegible]	0
	60′	50′	40′	30′	20′	10′	0′	Degree

Cotangent 0...45° ▲

Example 3

Given : Tan α = 0,856
To find : Angle α in degrees

tan 40°40′	=0,859
tan 40°30′	=0,854
Difference for 10′	=0,005
Value for 1′	=0,0005
Given value	=0,856
− tan 40°30′	=0,854
Difference	=0,002
0,002:0,0005	=4′
α=40°30+4′	=40°34′

Example 4

Given : Cot β = 1,416
To find : Angle β in degrees

cot 35°10′	=1,419
cot 35°20′	=1,411
Difference for 10′	=0,008
Value for 1′	=0,0008
Given value	=1,416
− cot 35°20′	=1,411
Difference	=0,005
0,005:0,0008	=6,25′
β=35°20′−6′	=35°14′

Units in Measuring Practice

Legal Units

The units of the International System of units (Systeme International = SI) are legally binding for official and business transactions in the Federal Republic of Germany by the "Law on Units in Measuring Practice" of 2 July 1969. Legal units are: 1. the base units of the International System of Units: metre, kilogram, second, electric current, kelvin and candela, 2. the units derived from the base units and 3. the decimal multiples and submultiples of base units and derived units.

SI—Base Units

Base quantity		Base unit		Definition
Length	l	metre	m	1 metre is 1 650 763.73 times the wave length of the orange-red radiation from krypton atom.
Mass	m	kilogram	kg	1 kilogram is the mass of the international kilogram prototype which is kept as platinum-iridium cylinder in Sevres near Paris.
Time	t	second	s	1 second is the period of 9 192 631 770 vibrations of the caesium atom in the atomic clock.
Electric Current	I	ampere	A	1 ampere is the strength of a current which would bring about an electroydynamic force of $0.2.10^{-6}$N per 1m length of conductor between two parallel conductors placed at a distance of 1m.
Temperature	T	kelvin	K	1 kelvin is the 273.16th part of the thermodynamic temperature of the triple point of water.
Light intensity	I_v	candela	cd	1 candela is the light intensity with which 1/600 000 m^2 of the surface of a black radiator illuminates perpendicular to its surface at a temperature of platinum solidifying at a pressure of 101 325 N/m^2

Units in Measuring Practice

Length	l	metre	m	1 metre is 1 650 763.73 times the wave length of the orange-red radiation from krypton atom.
Area	A, S	square metre	m^2	1 square metre is equal to the area of a square of length of side 1 m.
Volume	V	cubic metre	m^3	1 cubic metre is equal to the volume of a cube of edge length 1 m.
Angle	α, β, γ	radian	rad	1 radian is equal to the angle at the centre which cuts off an arc of length 1 m from a circle of radius 1 m. 1 rad = 1 m/m.
Mass	m	kilogram	kg	1 kilogram is the mass of the international kilogram prototype which is kept as platinumiridium cylinder in Sevres near Paris.
Linear mass	m'	kilogram per metre	$\frac{kg}{m}$	1 kilogram per metre is the mass of 1 kg on 1m length of a homogeneous body of constant crosssectional area.
Area mass	m''	kilogram per square metre	$\frac{kg}{m^2}$	1 kilogram per square metre is the mass of 1 kg on 1 m^2 area of a homogeneous body of constant thickness.
Density	∂	kilogram per cubic metre	$\frac{kg}{m^3}$	1 kilogram per cubic metre is equal to the density of a homogeneous body which has a volume of 1 m^3 for 1 kg mass.
Time	t	second	s	1 second is the period of 9 192 631 770 vibrations of caesium atom in the atomic clock.
Frequency, Periodic frequency	f	hertz	Hz	1 hertz is equal to the frequency of a periodic occurance of period 1s. In unit equations 1 Hz is put equal to 1/s
Rotational speed, Rotational frequency	n	one per second	$\frac{1}{s}$	1 per second (reciprocal second) is equal to the rotational movement of one rotation in the duration of 1s.
Speed Velocity	v	metre per second	$\frac{m}{s}$	1 metre per second squared is equal to the acceleration of a body moving in straight line whose velocity uniformly changes by 1m/s during 1s.
Acceleration Retardation	a	metre per second squared	$\frac{m}{s^2}$	1 metre per second squared is equal to the acceleration of a body moving in straight line whose velocity uniformly changes by 1m/s during 1s.
Angular velocity	ω	radian per second	$\frac{rad}{s}$	1 radian per second is the velocity of a uniformly rotating body which rotates by the angle of 1 radian during the time of 1s about the axis of rotation.
Flow	V, Q	cubic metre per second	$\frac{m^3}{s}$	1 cubic metre per second is the flow of 1m^3 of a homogeneous fluid through a flow cross-sectional area during the time of 1s.

Additional units

Length		m	dm	cm	mm
kilometre	km	1 000	10 000	100 000	1 000 000
metre	m	1	10	100	1 000
decimetre	dm	0,1	1	10	100
centimetre	cm	0,01	0,1	1	10
millimetre	mm	0,001	0,01	0,1	1
micrometre	µm	0,000001	0,00001	0,0001	0,001
nautical mile	sm	1 852	—	—	

Area	*A*	m²	dm²	cm²	mm²
square kilometre	km²	1 000 000	—	—	—
hectare	ha	10 000	—	—	—
are	a	100	—	—	—
square metre	m²	1	100	10 000	1 000 000
square decimetre	dm²	0,01	1	100	10 000
square centimetre	cm²	0,0001	0,01	1	100
square millimetre	mm²	0,000001	0,0001	0,01	1

Volume	*V*	m³	dm³	cm³	mm³
cubic metre	m³	1	1 000	1 000 000	—
cubic decimetre, litre	dm³, l	0,001	1	1 000	1 000 000
cubic centimetre	cm³	0,000001	0,001	1	1 000
cubic millimetre	mm³	—	0,000001	0,001	1

Angle	α, β, γ ...	rad	°	′	″
full angle	pla	$2 \cdot \pi$	360	21 600	1 296 000
right angle	∟	$\pi/2$	90	5 400	324 000
radian	rad	1	57,296	3 438	206 265
grad	°	$\pi/180$	1	60	3 600
gon	gon	$\pi/200$	0,9	54	3 240
minute	′	—	0,01667	1	60
second	″	—	0,00028	0,01667	1

Mass	*m*	t	kg	g	mg
ton	t	1	1 000	1 000 000	—
kilogram	kg	0,001	1	1 000	1 000 000
gram	g	0,000001	0,001	1	1 000
milligram	mg	—	0,000001	0,001	1

Density	ϱ	kg/m³	kg/dm³	g/cm³
kilogram per cubic metre	kg/m³	1	0,001	0,001
kilogram per cubic decimetre	kg/dm³	1 000	1	1
gram per cubic centimetre	g/cm³	1 000	1	1

Time	*t*	d	h	min	s
year (in the energy economics)	a	365	8 760	—	—
day	d	1	24	1 440	86 400
hour	h	0,0417	1	60	3 600
minute	min	—	0,0167	1	60
second	s	—	0,00028	0,0167	1

length of time: 3h, point of time: 3^h = 3 o'clock

Velocity	*v*	m/s	km/h	m/min	kn
metre per second	m/s	1	3,6	60	1,944
kilometre per hour	km/h	0,2778	1	16,667	0,540
metre per minute	m/min	0,0167	0,06	1	—
nautical mile per hour (knot = kn)	kn	0,5144	1,852	30,866	1

Units in measuring practice

Quantity		Unit		Explanation
Force	F	newton	N	1 newton is equal to the force which imports an acceleration of $1 m/s^2$ to a body of mass 1 kg. $1N = 1 kg\ m/s^2$.
Pressure	P	newton per square metre pascal	$\frac{N}{m^2}$ Pa	1 newton per square metre (1 pascal) is equal to the pressure with which the force of 1 N is exercised perpendicular to the area of $1m^2$ 1 Pa = 1 N/m^2. 1 Bar (bar) is the special name for 100 000 Pa.
Normal stress, tensile or compressive stress Shear stress		newton per square metre	$\frac{N}{m^2}$	1 newton per square metre (1 pascal) is equal to the mechanical stress with which the force of 1 N is exercised on the area of $1m^2$. In many branches of engineering the mechanical stress and strength are specified in N/m^2: 1 N/m^2 = 1000 000 Pa = 1 MPa.
Heat Energy Quantity of heat	W	joule	J	1 joule is equal to the work that is done when the point of application of the force of 1 N is shifted by 1m in the direction of the force. 1 J = 1 Nm = 1 Ws = 1 kgm^2/s^2, 3600 000 J = 1 kWh
Moment of a force, torque	M	newton metre, joule	Nm J	1 newton is equal to the moment of a force which results from the product of the force of 1 N and the lever arm of 1m. 1 Nm = 1 J = 1 Ws = 1 kgm^2/s^2
Power Energy flow Heat flow	P ϕ	watt	W	1 watt is equal to the power with which the energy of 1 J is converted during the time of 1s. The unit watt is also called volt ampere in the specification of apparent electric power. 1 W = 1 J/s = 1 Nm/s = 1 VA
Specific heat value	Hu	joule per kilogram	$\frac{J}{kg}$	1 joule per kilogram is equal to the quantity of heat which on complete burning of the mass of 1 kg releases the energy of 1 J.
Fuel consumption	b	gram per kilowatt-hour	$\frac{g}{kWh}$	1 gram per kilowatt-hour is equal to the fuel consumption of the mass of 1g for the work of 1 kWh.
Dynamic viscosity	μ	pascal second	Pa.s	1 pascal second is equal to the dynamic viscosity of a flowing fluid in which the shear stress of 1 Pa prevails between two parallel layers at a distance of 1m and a velocity difference of 1m/s.
Kinematic velocity	ϑ	square metre per second	$\frac{m^2}{s}$	1 square metre per second is equal to the kinematic viscosity of a fluid of dynamic viscosity 1 Pa.s and density 1 kg/m^3
Temperature	T	kelvin	K	1 kelvin is 273.16th part of the thermodynamic temperature of the triple point of water.
Electric Current	I	ampere	A	1 ampere is the strength of a current which would bring about an electrodynamic force of $0.2.10^{-6}$ N per 1m length between two parallel conductors placed at a distance of 1m.
Electric voltage	U	volt	V	1 volt is equal to the electric voltage between two points of a metallic conductor in which a power of 1 W is expended for a current of 1 A strength.
Electric resistance	R	ohm	Ω	1 ohm is equal to the electric resistance between two points of a metallic conductor in which an electric current of 1 A flows at a voltage of 1 V.
Electric conductance	G	siemens	S	1 siemens is equal to the electric conductance of a conductor of electric resistance of 1 Ω
Quantity of electricity	Q	coulomb ampere-second	C As	1 coulomb is equal to the quantity of electricity which flows through the conductor cross-section during the time of 1s at an electric current of 1A.

Prefixes for decimal multiples and submultiples

Prefix		Value		Use		
Mega	M	1 000 000	10^6	1 Megapascal	= 1 MPa	= 1 000 000 Pa
Kilo	k	1 000	10^3	1 Kilowatt	= 1 kW	= 1 000 W
Hecto	h	100	10^2	1 Hectolitre	= 1 hl	= 100 l
Deca	da	10	10^1	1 Decanewton	= 1 daN	= 10 N
Deci	d	0,1	10^{-1}	1 Decimetre	= 1 dm	= 0,1 m
Centi	c	0,01	10^{-2}	1 Centimetre	= 1 cm	= 0,01 m
Milli	m	0,001	10^{-3}	1 Millimetre	= 1 mm	= 0,001 m
Micro	μ	0,000001	10^{-6}	1 Micrometre	= 1 μm	= 0,000001 m

Additional units

Force	*F*	kN	daN	N	mN
Meganewton	MN	1 000	100 000	1 000 000	—
Kilonewton	kN	1	100	1 000	1 000 000
Decanewton	daN	0,01	1	10	10 000
Newton	N	0,001	0,1	1	1 000
Millinewton	mN	0,000001	0,0001	0,001	1

Pressure	*p*	MPa N/mm²	bar daN/cm²	mbar	Pa N/m²
Mega pascal (newton/square millimetre)	MPa	1	10	10 000	1 000 000
Bar (decanewton/square centimetre)	bar	0,1	1	1 000	100 000
Millibar	mbar	0,0001	0,001	1	100
Pascal (newton/square metre)	Pa	0,000001	0,00001	0,01	1

Mechanical stress	*p*, *σ*, *τ*	daN/mm²	MPa N/mm²	daN/cm² bar	Pa N/m²
Decanewton per square millimetre	daN/mm²	1	10	100	10 000 000
Megapascal (newton/square millimetre)	MPa	0,1	1	10	1 000 000
Decanewton per square centimetre (bar)	daN/cm²	0,01	0,1	1	100 000
Pascal (newton/square metre)	Pa	—	0,000001	0,00001	1

Work, energy, quantity of heat	*W*, *Q*	kWh	MJ	kJ	J
Kilowatt-hour	kWh	1	3,6	3 600	3 600 000
Megajoule	MJ	0,2778	1	1 000	1 000 000
Kilojoule	kJ	0,000278	0,001	1	1 000
Joule (newton metre, watt second)	J	—	0,000001	0,001	1

Power, energy flow, heat flow	*P*	MW	kW	W	mW
Megawatt	MW	1	1 000	1 000 000	—
Kilowatt	kW	0,001	1	1 000	1 000 000
Watt (joule/second, newton metre/second)	W	0,000001	0,001	1	1 000
Milliwatt	mW	—	0,000001	0,001	1

Temperature		Point of temperature		Temperature difference	
	T	K	°C	K	°C
Kelvin	K	0	− 273	1	1
Degree Celsius	°C	273	0	1	1

Electric current	*I*	kA	A	mA
Kiloampere	kA	1	1 000	1 000 000
Ampere	A	0,001	1	1 000
Milliampere	mA	0,000001	0,001	1

Electric voltage	*U*	kV	V	mV
Kilovolt	kV	1	1 000	1 000 000
Volt (watt/ampere)	V	0,001	1	1 000
Millivolt	mV	0,000001	0,001	1

Electric resistance	*R*	kΩ	Ω	mΩ
Kiloohm	kΩ	1	1 000	1 000 000
Ohm	Ω	0,001	1	1 000
Milliohm	mΩ	0,000001	0,001	1

Quantity of electricity	*Q*	Ah	C (As)
Ampere hour	Ah	1	3 600
Coulomb (ampere second=As)	C	0,000278	1

Conversion factors for units up to 31.12.77

Quantity		Unit used till now		New unit		Conversion factor	
Angle	α	New degree New minute New second	g c cc	Gon Centigon Milligon	gon cgon mgon	1 g = 1 gon 1 c = 1 cgon 1 cc = 1/10 mgon 1 mgon = 10 cc	Up to 31.12.1974
Force	*F*	Kilopond	kp	Newton	N	1 kp = 9,80665 N 1 N = 0,10197 kp	≈ 10 N ≈ 0,1 kp
Pressure	*p*	Kilopond/Centimetre^2	kp/cm^2	Pascal N/m^2	Pa	1 kp/cm^2 = 98 066,5 Pa 1 MPa = 10,197 kp/cm^2	≈ 100 000 Pa ≈ 10 kp/cm^2
		Kilopond/Centimetre^2	kp/cm^2	Decanewton/Centimetre^2	daN/cm^2	1 kp/cm^2 = 0,980665 daN/cm^2 1 daN/cm^2 = 1,0197 kp/cm^2	≈ 1 daN/cm^2 ≈ 1 kp/cm^2
		Technical atmosphere	at	Bar	bar	1 at = 0,980665 bar 1 bar = 1,0197 at	≈ 1 bar ≈ 1 at
		Physical atmosphere	atm	Bar	bar	1 atm = 1,01325 bar 1 bar = 0,98692 atm	≈ 1 bar ≈ 1 atm
		m-water column	m WS	Millibar	mbar	1 m WS = 98,0665 mbar 1 mbar = 10,197 mm WS	≈ 100 mbar ≈ 10 mm WS
		mm-mercury column	mm Hg	Millibar	mbar	1 mm Hg = 1,33322 mbar 1 mbar = 0,75006 mm Hg	
Mechanical stress, strength	*p* *σ*	Kilopond/Centimetre^2	kp/cm^2	Pascal N/m^2	Pa	1 kp/cm^2 = 98066,5 Pa 1 MPa = 10,197 kp/cm^2	≈ 100000 Pa ≈ 10 kp/cm^2
		Kilopond/Centimetre^2	kp/cm^2	Decanewton/Centimetre^2	daN/cm^2	1 kp/cm^2 = 0,980665 daN/cm^2 1 daN/cm^2 = 1,0197 kp/cm^2	≈ 1 daN/cm^2 ≈ 1 kp/cm^2
		Kilopond/Millimetre^2	kp/mm^2	Newton/Millimetre^2	N/mm^2	1 kp/mm^2 = 9,80665 N/mm^2 1 N/mm^2 = 0,10197 kp/mm^2	≈ 10 N/mm^2 ≈ 0,1 kp/mm^2
Moment of a force	*M*	Kilopond metre	kpm	Newton metre	Nm	1 kpm = 9,80665 Nm 1 Nm = 0,10197 kpm	≈ 10 Nm ≈ 0,1 kpm
Work	*W*	Kilopond metre	kpm	Joule 1 J = 1 Nm = 1 Ws	J	1 kpm = 9,80665 J 1 J = 0,10197 kpm	≈ 10 J ≈ 0,1 kpm
Quantity of heat	*Q*	Kilocaloric	kcal	Kilojoule	kJ	1 kcal = 4,1868 kJ 1 kJ = 0,2388 kcal	≈ 4,2 kJ ≈ 0,24 kcal
Power	*P*	Horse power	PS	Kilowatt 1 W = 1 Nm/s = 1 J/s	kW	1 PS = 0,7355 kW 1 kW = 1,3596 PS	≈ 0,736 kW ≈ 1,36 PS
Specific heat value	H_u	Kilocalorie/Kilogram	kcal/kg	Kilojoule/Kilogram	kJ/kg	1 kcal/kg = 4,1868 kJ/kg 1 kJ/kg = 0,2388 kcal/kg	≈ 4,2 kJ/kg ≈ 0,24 kcal/kg
		Kilocalorie/Cubic metre	kcal/m^3	Kilojoule/Cubic metre	kJ/m^3	1 kcal/m^3 = 4,1868 kJ/m^3 1 kJ/m^3 = 0,2388 kcal/m^3	≈ 4,2 kJ/m^3 ≈ 0,24 kcal/m^3
Fuel consumption	*b*	Gram/Horse power-hour	g/PSh	Gram/Kilowatt-hour	g/kWh	1 g/PSh = 1,3596 g/kWh 1 g/kWh = 0,7355 g/PSh	≈ 1,36 g/kWh ≈ 0,736 g/PSh
Temperature	*T*	Degree Kelvin Degree	°K grd	Kelvin Kelvin	K K	1 °K = 1 K 1 grd = 1 K	Upto 5.7.1975 Upto 31.12.1974
Length	*l*	Inch	″	Millimetre	mm	1 ″ = 25,4 mm 1 mm = 0,03937	(No more usable in standards)

Degree ▶ Radian (rad)

$1 \text{ Degree} = \frac{\pi}{180} \text{ rad}$

Degree	0	1	2	3	4	5	6	7	8	9
0	0,0000	0,0175	0,0349	0,0524	0,0698	0,0873	0,1047	0,1222	0,1396	0,1571
10	0,1745	0,1920	0,2094	0,2269	0,2443	0,2618	0,2793	0,2967	0,3142	0,3316
20	0,3491	0,3665	0,3840	0,4014	0,4189	0,4363	0,4538	0,4712	0,4887	0,5061
30	0,5236	0,5411	0,5585	0,5760	0,5934	0,6109	0,6283	0,6458	0,6632	0,6807
40	0,6981	0,7156	0,7330	0,7505	0,7679	0,7854	0,8029	0,8203	0,8378	0,8552
50	0,8727	0,8901	0,9076	0,9250	0,9425	0,9599	0,9774	0,9948	1,0123	1,0297
60	1,0472	1,0647	1,0821	1,0996	1,1170	1,1345	1,1519	1,1694	1,1868	1,2043
70	1,2217	1,2392	1,2566	1,2741	1,2915	1,3090	1,3265	1,3439	1,3614	1,3788
80	1,3963	1,4137	1,4312	1,4486	1,4661	1,4835	1,5010	1,5184	1,5359	1,5533
90	1,5708	1,5882	1,6057	1,6232	1,6406	1,6580	1,6755	1,6930	1,7104	1,7279
100	1,7453	1,7628	1,7802	1,7977	1,8151	1,8326	1,8500	1,8675	1,8850	1,9024
110	1,9199	1,9373	1,9548	1,9722	1,9897	2,0071	2,0246	2,0420	2,0595	2,0769
120	2,0944	2,1118	2,1293	2,1468	2,1642	2,1817	2,1991	2,2166	2,2340	2,2515
130	2,2689	2,2864	2,3038	2,3213	2,3387	2,3562	2,3736	2,3911	2,4086	2,4260
140	2,4435	2,4609	2,4784	2,4958	2,5133	2,5307	2,5482	2,5656	2,5831	2,6005
150	2,6180	2,6354	2,6529	2,6704	2,6878	2,7053	2,7227	2,7402	2,7576	2,7751
160	2,7925	2,8100	2,8274	2,8449	2,8623	2,8798	2,8972	2,9147	2,9322	2,9496
170	2,9671	2,9845	3,0020	3,0194	3,0369	3,0543	3,0718	3,0892	3,1067	3,1241
180	3,1416	3,1590	3,1764	3,1939	3,2114	3,2288	3,2463	3,2637	3,2812	3,2986
190	3,3161	3,3335	3,3510	3,3684	3,3859	3,4033	3,4208	3,4382	3,4557	3,4731
200	3,4906	3,5081	3,5255	3,5430	3,5604	3,5779	3,5953	3,6128	3,6302	3,6477
210	3,6651	3,6826	3,7000	3,7175	3,7349	3,7524	3,7698	3,7873	3,8048	3,8222
220	3,8397	3,8571	3,8746	3,8920	3,9095	3,9269	3,9444	3,9618	3,9793	3,9967
230	4,0142	4,0316	4,0491	4,0665	4,0840	4,1015	4,1189	4,1364	4,1538	4,1713
240	4,1887	4,2062	4,2236	4,2411	4,2585	4,2760	4,2934	4,3109	4,3283	4,3458
250	4,3632	4,3807	4,3982	4,4156	4,4331	4,4505	4,4680	4,4852	4,5029	4,5203
260	4,5378	4,5552	4,5727	4,5901	4,6076	4,6250	4,6425	4,6600	4,6774	4,6549
270	4,7123	4,7298	4,7472	4,7647	4,7821	4,7996	4,8170	4,8345	4,8519	4,4694
280	4,8868	4,9043	4,9217	4,9392	4,9567	4,9741	4,9916	5,0090	5,0265	5,0439
290	5,0614	5,0788	5,0963	5,1137	5,1312	5,1486	5,1661	5,1835	5,2010	5,2184
300	5,2359	5,2534	5,2708	5,2883	5,3057	5,3232	5,3406	5,3581	5,3755	5,3930
310	5,4104	5,4279	5,4453	5,4628	5,4802	5,4977	5,5151	5,5326	5,5501	5,5675
320	5,5850	5,6024	5,6199	5,6373	5,6548	5,6722	5,6897	5,7071	5,7246	5,7420
330	5,7595	5,7769	5,7944	5,8118	5,8293	5,8468	5,8642	5,8817	5,8991	5,9166
340	5,9340	5,9515	5,9689	5,9864	6,0038	6,0213	6,0387	6,0562	6,0736	6,0911
350	6,1086	6,1260	6,1435	6,1609	6,1784	6,1958	6,2133	6,2307	6,2482	6,2656
360	6,2831									

Degree ▶ Gon

1 Degree=1,111 gon

Degree	0	1	2	3	4	5	6	7	8	9
0	0,00	1,11	2,22	3,33	4,44	5,56	6,67	7,78	8,89	10,00
10	11,11	12,22	13,33	14,44	15,56	16,67	17,78	18,89	20,00	21,11
20	22,22	23,33	24,44	25,56	26,67	27,78	28,89	30,00	31,11	32,22
30	33,33	34,44	35,56	36,67	37,78	38,89	40,00	41,11	42,22	43,33
40	44,44	45,56	46,67	47,78	48,89	50,00	51,11	52,22	53,33	54,44
50	55,56	56,67	57,78	58,89	60,00	61,11	62,22	63,33	64,44	65,56
60	66,67	67,78	68,89	70,00	71,11	72,22	73,33	74,44	75,65	76,67
70	77,78	78,89	80,00	81,11	82,22	83,33	84,44	85,56	86,67	87,78
80	88,89	90,00	91,11	92,22	93,33	94,44	95,56	96,67	97,78	98,89
90	100,00									

Decimal ports of the hour, of the degree ▶ Minutes and Seconds

1 h=60 min

Hours (Degree)	0,00	0,01	0,02	0,03	0,04	0,05	0,06	0,07	0,08	0,09
0,00	0.00	0.36	1.12	1.48	2.24	3.00	3.36	4.12	4.48	5.24
0,10	6.00	6.36	7.12	7.48	8.24	9.00	9.36	10.12	10.48	11.24
0,20	12.00	12.36	13.12	13.48	14.24	15.00	15.36	16.12	16.48	17.24
0,30	18.00	18.36	19.12	19.48	20.24	21.00	21.36	22.12	22.48	23.24
0,40	24.00	24.36	25.12	25.48	26.24	27.00	27.36	28.12	28.48	29.24
0,50	30.00	30.36	31.12	31.48	32.24	33.00	33.36	34.12	34.48	35.24
0,60	36.00	36.36	37.12	37.48	38.24	39.00	39.36	40.12	40.48	41.24
0,70	42.00	42.36	43.12	43.48	44.24	45.00	45.36	46.12	46.48	47.24
0,80	48.00	48.36	49.12	49.48	50.24	51.00	51.36	52.12	52.48	53.24
0.90	54.00	54.36	55.12	55.48	56.24	57.00	57.36	58.12	58.48	59.24

Conversion tables

Minutes ▶ Decimal parts of the hour (of the degree)

$1\ \text{min} = \frac{1}{60}\ \text{h}$

Minutes	0	1	2	3	4	5	6	7	8	9
0	0,000	0,017	0,033	0,050	0,067	0,083	0,100	0,117	0,133	0,150
10	0,167	0,183	0,200	0,217	0,233	0,250	0,267	0,283	0,300	0,317
20	0,333	0,350	0,367	0,383	0,400	0,417	0,433	0 450	0,467	0,483
30	0,500	0,517	0,533	0,550	0,567	0,583	0,600	0,617	0,633	0,650
40	0,667	0,683	0,700	0,717	0,733	0,750	0,767	0,783	0,800	0,817
50	0,833	0,850	0,867	0,883	0,900	0,917	0,933	0,950	0,967	0,983

Seconds ▶ Decimal parts of the hour (of the degree)

$1\ \text{s} = \frac{1}{3600}\ \text{h}$

Second	0	1	2	3	4	5	6	7	8	9
0	0,00000	0,00028	0,00056	0,00083	0,00111	0,00139	0,00167	0,00194	0,00222	0,00250
10	0,00278	0,00306	0,00333	0,00361	0,00389	0,00417	0,00444	0,00472	0,00500	0,00528
20	0,00556	0,00583	0,00611	0,00639	0,00667	0,00694	0,00722	0,00750	0,00779	0,00806
30	0,00833	0,00861	0,00889	0,00917	0,00944	0,00972	0,01000	0,01028	0,01056	0,01083
40	0,01111	0,01139	0,01167	0,01194	0,01222	0,01250	0,01278	0,01306	0,01333	0,01361
50	0,01389	0,01417	0,01444	0,01472	0,01500	0,01528	0,01556	0,01583	0,01611	0,01639
60	0,01667									

Metre per second ▶ Kilometre per hour

1 m/s = 3,6 km/h

m/s	0	1	2	3	4	5	6	7	8	9
0	0,00	3,60	7,20	10,80	14,40	18,00	21,60	25,20	28,80	32,40
10	36,00	39,60	43,20	46,80	50,40	54,00	57,60	61,20	64,80	68,40
20	72,00	75,60	79,20	82,80	86,40	90,00	93,60	97,20	100,80	104,40
30	108,00	111,60	115,20	118,80	122,40	126,00	129,60	133,20	136,80	140,40
40	144,00	147,60	151,20	154,80	158,40	162,00	165,60	169,20	172,80	176,40
50	180,00	183,60	187,20	190,80	194,40	198,00	201,60	205,20	208,80	212,40
60	216,00	219,60	223,20	226,80	230,40	234,00	237,60	241,20	244,80	248,40
70	252,00	255,60	259,20	262,80	266,40	270,00	273,60	277,20	280,80	284,40
80	288,00	291,60	295,20	298,80	302,40	306,00	309,60	313,20	316,80	320,40
90	324,00	327,60	331,20	334,80	338,40	342,00	345,60	349,20	352,80	356,40

Kilometre per hour ▶ Metre per second

$1\ \text{km/h} = \frac{1}{3,6}\ \text{m/s}$

km/h	0	1	2	3	4	5	6	7	8	9
0	0,000	0,278	0,556	0,833	1,111	1,389	1,667	1 944	2,222	2,500
10	2,778	3,056	3,333	3,611	3,889	4,167	4,444	4,722	5,000	5,278
20	5,556	5,833	6,111	6,389	6,667	6,944	7,222	7,500	7,778	8,056
30	8,333	8,611	8,889	9,167	9,444	9,722	10,000	10,278	10,556	10,833
40	11,111	11,389	11,667	11,944	12,222	12,500	12,778	13,056	13,333	13,611
50	13,889	14,167	14,444	14,722	15,000	15,278	15,556	15,833	16,111	16,389
60	16,667	16,944	17,222	17,500	17,778	18,056	18,333	18,611	18,889	19,167
70	19,444	19,722	20,000	20,278	20,555	20,833	21,111	21,389	21,667	21,944
80	22,222	22,500	22,778	23,055	23,333	23,611	23,889	24,167	24,444	24,722
90	25,000	25,278	25,555	25,833	26,111	26,389	26,667	26,944	27,222	27,500
100	27,778	28,055	28,333	28,611	28,889	29,167	29,444	29,722	30,000	30,278
110	30,555	30,833	31,111	31,389	31,667	31,944	32,222	32,500	32,778	33,055
120	33,333	33,611	33,889	34,167	34,444	34,722	35,000	35,278	35,555	35,833
130	36,111	36,389	36,667	36,944	37,222	37,500	37,778	38,055	38,333	38,611
140	38,889	39,167	39,444	39,722	40,000	40,278	40,555	40,833	41,111	41,389
150	41,667	41,944	42,222	42,500	42,778	43,055	43,333	43,611	43,889	44,167
160	44,444	44,722	45,000	45,278	45,555	45,833	46,111	46,389	46,667	46,944
170	47,222	47,500	47,778	48,055	48,333	48,611	48,889	49,167	49,444	49,722
180	50,000	50,278	50,555	50,833	51,111	51,389	51,667	51,944	52,222	52,500
190	52,778	53,055	53,333	53,611	53,889	54,167	54,444	54,722	55,000	55,278
200	55,555	55,833	56,111	56,389	56,667	56,944	57,222	57,500	57,778	58,055
210	58,333	58,611	58,889	59,167	59,444	59,722	60,000	60,278	60,555	60,833
220	61,111	61,389	61,667	61,944	62,222	62,500	62,778	63,056	63,333	63,611
230	63,889	64,167	64,444	64,722	65,000	65,278	65,556	65,833	66,111	66,389
240	66,667	66,944	67,222	67,500	67,778	68,056	68,333	68,611	68,889	69,167

Braking distance in m

$$s = \frac{v^2}{2 \cdot a}$$

Retardation ▶	4,0	4,2	4,4	4,6	4,8	5,0	5,2	5,4	5,6	5,8 m/s²
v in km/h										
30	8,7	8,3	7,9	7,5	7,2	6,9	6,7	6,4	6,2	6,0
40	15,4	14,7	14,0	13,4	12,9	12,3	11,9	11,4	11,0	10,6
50	24,1	23,0	21,9	21,0	20,1	19,3	18,5	17,9	17,2	16,6
60	34,7	33,1	31,6	30,2	28,9	27,8	26,7	25,7	24,8	23,9
70	47,3	45,0	42,9	41,1	39,4	37,8	36,4	35,0	33,8	32,6
80	61,7	58,8	56,1	53,7	51,5	49,4	47,5	45,7	44,1	42,6
90	78,1	74,4	71,0	67,9	65,1	62,5	60,1	57,9	55,8	53,9
100	96,4	91,8	87,6	83,9	80,4	77,2	74,2	71,4	68,9	66,5
110	116,7	111,1	106,1	101,5	97,3	93,4	89,8	86,5	83,4	80,5
120	138,9	132,2	126,2	120,8	115,8	111,1	106,8	102,9	99,2	95,8

Track in mm

$$c = \frac{2 \cdot \pi \cdot d_F \cdot \beta}{360°}$$

Rimϕ ▶ Track	10	12	13	14	15	16	17	18	20	22 Inch
0° 05′	0,41	0,49	0,52	0,56	0,62	0,65	0,70	0,73	0,80	0,90
0° 10′	0,83	0,97	1,05	1,12	1,25	1,30	1,40	1,45	1,60	1,80
0° 15′	1,24	1,46	1,51	1,68	1,87	1,95	2,10	2,18	2,40	2,70
0° 20′	1,65	1,95	2,10	2,25	2,50	2,60	2,80	2,91	3,20	3,60
0° 25′	2,06	2,43	2,65	2,82	3,12	3,25	3,45	3,63	4,00	4,55
0° 30′	2,48	2,92	3,15	3,40	3,75	3,90	4,10	4,36	4,80	5,50
0° 35′	2,89	3,41	3,67	3,97	4,37	4,55	4,80	5,09	5,60	6,40
0° 40′	3,30	3,89	4,20	4,55	5,00	5,20	5,50	5,82	6,40	7,30
0° 45′	3,71	4,38	4,72	5,12	5,60	5,85	6,20	6,54	7,20	8,20
0° 50′	4,12	4,87	5,25	5,70	6,20	6,50	6,90	7,27	8,00	9,10
0° 55′	4,54	5,35	5,77	6,25	6,85	7,15	7,60	8,00	8,80	10,05
1° 00′	4,95	5,84	6,30	6,80	7,50	7,80	8,30	8,72	9,60	11,00
1° 05′	5,36	6,33	6,82	7,37	8,10	8,45	9,00	9,45	10,40	11,90
1° 10′	5,78	6,81	7,35	7,95	8,70	9,10	9,70	10,17	11,20	12,80
1° 15′	6,19	7,30	7,87	8,52	9,30	9,75	10,35	10,90	12,00	13,70
1° 20′	6,60	7,79	8,40	9,10	9,90	10,40	11,00	11,63	12,80	14,60
1° 25′	7,01	8,27	8,92	9,65	10,55	11,05	11,70	12,36	13,60	15,50
1° 30′	7,42	8,76	9,45	10,20	11,20	11,70	12,70	13,10	14,40	16,40
1° 35′	7,84	9,25	9,97	10,80	11,80	12,35	13,10	13,83	15,20	17,30
1° 40′	8,25	9,73	10,50	11,40	12,40	13,00	13,80	14,55	16,00	18,20
1° 45′	8,66	10,22	11,02	11,95	13,00	13,65	14,50	15,27	16,80	19,10
1° 50′	9,07	10,71	11,55	12,50	13,60	14,30	15,20	16,00	17,60	20,00
1° 55′	9,49	11,19	12,07	13,05	14,20	14,95	15,90	16,72	18,40	20,90
2° 00′	9,90	11,68	12,60	13,60	14,80	15,60	16,60	17,45	19,20	21,80
2° 05′	10,31	12,16	13,13	14,16	15,62	16,25	17,29	18,16	20,00	22,71

Torque in Nm

$$M = \frac{9550 \cdot P_{eff}}{n}$$

n ▶	2000	2400	2800	3200	3600	4000	4400	4800	5200	5600/min
P in kW										
30	143,2	119,4	102,3	89,5	79,6	71,6	65,1	59,7	55,1	51,2
35	167,1	139,3	119,4	104,4	92,8	83,6	76,0	69,6	64,3	59,7
40	191,0	159,2	136,4	119,4	106,1	95,5	86,8	79,6	73,5	68,2
45	214,9	179,1	153,5	134,3	119,4	107,4	97,7	89,5	82,6	76,7
50	238,8	199,0	170,5	149,2	132,6	119,4	108,5	99,5	91,8	85,3
55	262,6	218,9	187,6	164,1	145,9	131,3	119,4	109,4	101,0	93,8
60	286,5	238,8	204,6	179,1	159,2	143,2	130,2	119,4	110,2	102,3
65	310,4	258,6	221,7	194,0	172,4	155,2	141,1	129,3	119,4	110,8
70	334,2	278,5	238,8	208,9	185,7	167,1	151,9	139,3	128,6	119,4
75	358,1	298,4	255,8	223,8	199,0	179,1	162,8	149,2	137,7	127,9
80	382,0	318,3	272,9	238,8	212,2	191,0	173,6	159,2	146,9	136,4
85	405,9	338,2	289,9	253,7	225,5	202,9	184,5	169,1	156,1	145,0
90	429,8	358,1	307,0	268,6	238,8	214,9	195,3	179,1	165,3	153,5
95	453,6	378,0	324,0	283,5	252,0	226,8	206,2	189,0	174,5	162,0
100	477,5	397,9	341,1	298,4	265,3	238,8	217,0	199,0	183,7	170,5
105	501,4	417,8	358,1	313,4	278,5	250,7	227,9	208,9	192,8	179,1
110	525,2	437,7	375,2	328,3	291,8	262,6	238,8	218,9	202,0	187,6
115	549,1	457,6	392,2	343,2	305,1	274,6	249,6	228,8	211,2	196,1
120	573,0	477,5	409,3	358,1	318,3	286,5	260,5	238,8	220,4	204,6
125	596,9	497,4	426,3	373,0	331,6	298,4	271,3	248,7	229,6	213,2

Travelling speed in km/h

$$v = \frac{60 \cdot \pi \cdot d \cdot n}{1000}$$

Tyre ⌀ ▶ / *n* tyre in 1/min	550	560	570	580	590	600	610	620	630	640 mm
100	10,4	10,6	10,7	10,9	11,1	11,3	11,5	11,7	11,9	12,1
200	20,7	21,1	21,5	21,9	22,2	22,6	23,0	23,4	23,8	24,1
300	31,1	31,7	32,2	32,8	33,4	33,9	34,5	35,1	35,6	36,2
400	41,5	42,2	43,0	43,7	44,5	45,2	46,0	46,7	47,5	48,3
500	51,8	52,8	53,7	54,7	55,6	56,6	57,5	58,4	59,4	60,3
600	62,2	63,3	64,5	65,6	66,7	67,9	69,0	70,1	71,3	72,4
700	72,6	73,9	75,2	76,5	77,8	79,2	80,5	81,8	83,1	84,4
800	82,9	84,4	86,0	87,5	89,0	90,5	92,0	93,5	95,0	96,5
900	93,3	95,0	96,7	98,4	100,1	101,8	103,5	105,2	106,9	108,6
1 000	103,7	105,6	107,4	109,3	111,2	113,1	115,0	116,9	118,8	120,6
1 100	114,0	116,1	118,2	120,3	122,3	124,4	126,5	128,6	130,6	132,7
1 200	124,4	126,7	128,9	131,2	133,5	135,7	138,0	140,2	142,5	144,8
1 300	134,8	137,2	139,7	142,1	144,6	147,0	149,5	151,9	154,4	156,8
1 400	145,1	147,8	150,4	153,1	155,7	158,3	161,0	163,6	166,3	168,9
1 500	155,5	158,3	161,2	164,0	166,8	169,6	172,5	175,3	178,1	181,0

Piston velocity in m/s

$$v_m = \frac{s \cdot n}{30}$$

Stroke ▶ / *n* engine in 1/min	70	72	74	76	78	80	82	84	86	88 mm
3 200	7,47	7,68	7,89	8,11	8,32	8,53	8,75	8,96	9,17	9,39
3 400	7,93	8,16	8,39	8,61	8,84	9,07	9,29	9,52	9,75	9,97
3 600	8,40	8,64	8,88	9,12	9,36	9,60	9,84	10,08	10,32	10,56
3 800	8,87	9,12	9,37	9,63	9,88	10,13	10,39	10,64	10,89	11,15
4 000	9,33	9,60	9,87	10,13	10,40	10,67	10,93	11,20	11,47	11,73
4 200	9,80	10,08	10,36	10,64	10,92	11,20	11,48	11,76	12,04	12,32
4 400	10,27	10,56	10,85	11,15	11,44	11,73	12,03	12,32	12,61	12,91
4 600	10,73	11,04	11,35	11,65	11,96	12,27	12,57	12,88	13,19	13,49
4 800	11,20	11,52	11,84	12,16	12,48	12,80	13,12	13,44	13,76	14,08
5 000	11,67	12,00	12,33	12,67	13,00	13,33	13,67	14,00	14,33	14,67
5 200	12,13	12,48	12,83	13,17	13,52	13,87	14,21	14,56	14,91	15,25
5 400	12,60	12,96	13,32	13,68	14,04	14,40	14,76	15,12	15,48	15,84
5 600	13,07	13,44	13,81	14,19	14,56	14,93	15,31	15,68	16,05	16,43
5 800	13,53	13,92	14,31	14,69	15,08	15,47	15,85	16,24	16,63	17,01
6 000	14,00	14,40	14,80	15,20	15,60	16,00	16,40	16,80	17,20	17,60

Cylinder stroke volume in cm³

$$V_h = \frac{\pi \cdot d^2}{4} \cdot s$$

Stroke ▶ / Bore in mm	70	72	74	76	78	80	82	84	86	88 mm
75	309,3	318,1	326,9	335,8	344,6	353,4	362,3	371,1	379,9	388,8
76	317,5	326,6	335,7	344,7	353,8	362,9	372,0	381,0	390,1	399,2
77	326,0	335,3	344,6	353,9	363,2	372,6	381,9	391,2	400,5	409,8
78	334,5	344,0	353,6	363,1	372,7	382,2	391,8	401,4	410,9	420,5
79	343,1	352,9	362,7	372,6	382,4	392,2	402,0	411,8	421,6	431,4
80	351,9	361,9	372,0	382,1	392,1	402,2	412,2	422,3	432,3	442,4
81	360,7	371,0	381,3	391,6	401,9	412,2	422,5	432,9	443,2	453,5
82	369,7	380,2	390,8	401,4	411,9	422,5	433,0	443,6	454,2	464,7
83	378,8	389,6	400,4	411,2	422,1	432,9	443,7	454,5	465,3	476,2
84	387,9	399,0	410,1	421,2	432,3	443,4	454,4	465,5	476,6	487,7
85	397,2	408,6	420,0	431,3	442,6	454,0	465,4	476,7	488,0	499,4
86	406,6	418,2	429,9	441,5	453,1	464,7	476,3	488,0	499,6	511,2
87	416,2	428,0	439,9	451,8	463,7	475,6	487,5	499,4	511,3	523,2
88	425,7	437,9	450,1	462,2	474,4	486,6	498,7	510,9	523,0	535,2
89	435,5	447,9	460,4	472,8	485,2	497,7	510,1	522,6	535,0	547,4
90	445,3	458,1	470,8	483,5	496,2	509,0	521,7	534,4	547,1	559,9
92	465,4	478,7	492,0	505,2	518,5	531,8	545,1	558,4	571,7	585,0
94	485,8	499,7	513,6	527,4	541,3	555,2	569,1	583,0	596,8	610,7
96	506,7	521,1	535,6	550,1	564,6	579,0	593,6	608,0	622,4	636,9
98	528,0	543,1	558,2	573,3	588,4	603,4	618,5	633,6	648,7	663,8

Correction factor k for Otto engines in 1/1000

$P_0 = k \cdot P_{eff}$ 1 000 mbar, 25 °C

Air pressure in mbar ▶		973	976	979	981	984	987	989	992	995	997	1 000	1 003	1 005	1 008	1 011	1 013
in mm Hg ▶		730	732	734	736	738	740	742	744	746	748	750	752	754	756	758	760
Temperature °C	K																
10	283	1 002	998	995	993	990	987	985	982	979	977	975	972	970	967	964	962
11	284	1 003	1 000	997	995	992	989	987	984	981	979	976	973	971	968	966	964
12	285	1 005	1 002	999	997	994	991	989	986	983	981	978	975	973	970	967	965
13	286	1 007	1 004	1 001	999	996	993	991	988	985	983	980	977	975	972	969	967
14	287	1 009	1 006	1 002	1 000	997	994	992	989	986	984	981	978	976	974	971	969
15	288	1 010	1 007	1 004	1 002	999	996	994	991	988	986	983	980	978	975	972	970
16	289	1 012	1 009	1 006	1 004	1 001	998	996	993	990	988	985	982	980	977	974	972
17	290	1 014	1 011	1 008	1 006	1 003	999	997	994	991	989	986	984	982	979	976	974
18	291	1 016	1 012	1 009	1 007	1 004	1 001	999	996	993	991	988	985	983	980	977	976
19	292	1 017	1 014	1 011	1 009	1 006	1 003	1 001	998	995	993	990	987	985	982	979	977
20	293	1 019	1 016	1 013	1 011	1 008	1 005	1 003	1 000	997	995	992	989	987	984	981	979
21	294	1 021	1 018	1 015	1 013	1 009	1 006	1 004	1 001	998	996	993	990	988	985	982	981
22	295	1 023	1 019	1 016	1 014	1 011	1 008	1 006	1 003	1 000	998	995	992	990	987	984	982
23	296	1 024	1 021	1 018	1 016	1 013	1 010	1 008	1 005	1 002	1 000	997	994	992	989	986	984
24	297	1 026	1 023	1 020	1 018	1 015	1 011	1 009	1 006	1 003	1 001	998	995	993	990	987	986
25	298	1 028	1 025	1 021	1 019	1 016	1 013	1 011	1 008	1 005	1 003	1 000	997	995	992	989	987
26	299	1 029	1 026	1 023	1 021	1 018	1 015	1 013	1 010	1 007	1 005	1 002	999	997	994	991	989
27	300	1 031	1 028	1 025	1 023	1 020	1 017	1 015	1 011	1 008	1 006	1 003	1 000	998	995	992	990
28	301	1 033	1 030	1 027	1 024	1 021	1 018	1 016	1 013	1 010	1 008	1 005	1 002	1 000	997	994	992
29	302	1 035	1 031	1 028	1 026	1 023	1 020	1 018	1 015	1 012	1 010	1 007	1 004	1 002	999	996	994
30	303	1 036	1 033	1 030	1 028	1 025	1 022	1 020	1 016	1 013	1 011	1 008	1 005	1 003	1 000	997	995
31	304	1 038	1 035	1 032	1 030	1 026	1 023	1 021	1 018	1 015	1 013	1 010	1 007	1 005	1 002	999	997
32	305	1 040	1 037	1 033	1 031	1 028	1 025	1 023	1 020	1 017	1 015	1 012	1 009	1 007	1 004	1 001	999
33	306	1 041	1 038	1 035	1 033	1 030	1 027	1 025	1 022	1 018	1 016	1 013	1 010	1 008	1 005	1 002	1 000
34	307	1 043	1 040	1 037	1 035	1 031	1 028	1 026	1 023	1 020	1 018	1 015	1 012	1 010	1 007	1 004	1 002
35	308	1 045	1 042	1 038	1 036	1 033	1 030	1 028	1 025	1 022	1 020	1 017	1 014	1 012	1 009	1 006	1 004

The influence of the relative humidity on the value of the correction factor can be neglected within the temperature range from 10°C to 35°C.

Braking in %

$$z = \frac{F_B}{m \cdot g}$$

Braking force ▶ / Mass kg	5500	5750	6000	6250	6500	6750	7000	7250	7500	7750	8000	8250	8500	8750	9000	9250 N
800	70,1	73,3	76,5	79,6	82,8	86,0	89,2	92,4	95,6	98,7	101,9	105,1	108,3	111,5	114,7	117,9
825	68,0	71,0	74,1	77,2	80,3	83,4	86,5	89,6	92,7	95,8	98,8	101,9	105,0	108,1	111,2	114,3
850	66,0	69,0	72,0	75,0	78,0	80,9	83,9	86,9	89,9	92,9	95,9	98,9	101,9	104,9	107,9	110,9
875	64,1	67,0	69,9	72,8	75,7	78,6	81,5	84,5	87,4	90,3	93,2	96,1	99,0	101,9	104,8	107,8
900	62,2	65,1	68,0	70,8	73,6	76,5	79,3	82,1	84,9	87,8	90,6	93,4	96,3	99,1	101,9	104,8
925	60,6	63,4	66,1	68,9	71,6	74,4	77,1	79,9	82,6	85,4	88,2	90,9	93,7	96,4	99,2	101,9
950	59,1	61,7	64,4	67,1	69,7	72,4	75,1	77,8	80,5	83,2	85,8	88,5	91,2	93,9	96,6	99,3
975	57,5	60,1	62,7	65,3	67,9	70,6	73,2	75,8	78,4	81,0	83,6	86,3	88,9	91,5	94,1	96,7
1 000	56,1	58,6	61,2	63,7	66,3	68,8	71,4	73,9	76,5	79,0	81,5	84,1	86,6	89,2	91,7	94,3
1 025	54,7	57,2	59,7	62,2	64,6	67,1	69,6	72,1	74,6	77,1	79,6	82,0	84,5	87,0	89,5	92,0
1 050	53,4	55,8	58,2	60,7	63,1	65,5	67,9	70,4	72,8	75,2	77,7	80,1	82,5	84,9	87,4	89,8
1 075	52,2	54,5	56,9	59,3	61,6	64,0	66,4	68,7	71,1	73,5	75,9	78,2	80,6	82,9	85,3	87,7
1 100	51,0	53,3	55,6	57,9	60,2	62,5	64,9	67,2	69,5	71,8	74,1	76,5	78,8	81,1	83,4	85,7
1 125	49,8	52,1	54,4	56,6	58,9	61,2	63,4	65,7	68,0	70,2	72,5	74,8	77,0	79,3	81,5	83,8
1 150	48,8	51,0	53,2	55,4	57,6	59,8	62,0	64,3	66,5	68,7	70,9	73,1	75,3	77,6	79,8	82,0
1 175	47,7	49,9	52,1	54,2	56,4	58,6	60,7	62,9	65,1	67,2	69,4	71,6	73,7	75,9	78,1	80,2
1 200	46,7	48,8	51,0	53,1	55,2	57,3	59,5	61,6	63,7	65,8	68,0	70,1	72,2	74,3	76,5	78,6
1 125	45,8	47,8	49,9	52,0	54,1	56,2	58,2	60,3	62,4	64,5	66,6	68,6	70,7	72,8	74,9	77,0
1 250	44,8	46,9	48,9	51,0	53,0	55,0	57,1	59,1	61,1	63,2	65,2	67,3	69,3	71,3	73,4	75,4
1 275	44,0	46,0	48,0	50,0	52,0	54,0	56,0	58,0	60,0	62,0	64,0	66,0	68,0	70,0	72,0	74,0
1 300	43,1	45,1	47,0	49,0	51,0	52,9	54,9	56,8	58,8	60,8	62,7	64,7	66,6	68,6	70,6	72,5
1 325	42,3	44,2	46,2	48,1	50,0	51,9	53,9	55,8	57,7	59,6	61,5	63,5	65,4	67,3	69,2	71,2
1 350	41,5	43,4	45,3	47,2	49,1	51,0	52,9	54,7	56,6	58,5	60,4	62,3	64,2	66,1	68,0	69,8
1 375	40,8	42,6	44,5	46,3	48,2	50,0	51,9	53,7	55,6	57,5	59,3	61,2	63,0	64,9	66,7	68,6
1 400	40,0	41,9	43,7	45,5	47,3	49,1	51,0	52,8	54,6	56,4	58,2	60,1	61,9	63,7	65,5	67,4
1 425	39,3	41,1	42,9	44,7	46,5	48,3	50,1	51,9	53,7	55,4	57,2	59,0	60,8	62,6	64,4	66,2
1 450	38,7	40,4	42,2	43,9	45,7	47,5	49,2	51,0	52,7	54,5	56,2	58,0	59,8	61,5	63,3	65,0
1 475	38,0	39,7	41,5	43,2	44,9	46,6	48,4	50,1	51,8	53,6	55,3	57,0	58,7	60,5	62,2	63,9
1 500	37,4	39,1	40,8	42,5	44,2	45,9	47,6	49,3	51,0	52,7	54,4	56,1	57,8	59,5	61,2	62,9

General calculation

Rule of Three

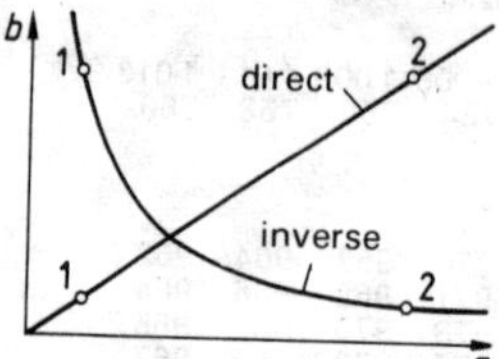

Direct proportion

20 l oil cost 210. — DM (Statement)
12 l oil cost ? DM (Question)

$$\text{Price} = \frac{210,-\ \text{DM} \cdot 12\ l}{20\ l} = 126,-\ \text{DM}$$

Inverse proportion

20 Workers need 210 h (Statement)
12 Workers need ? h (Question)

$$\text{Time} = \frac{210\ \text{h}.20\ \text{workers}}{12\ \text{workers}}$$

Given : Wage for 40 h = 600 DM.
To find : Wage for 176 h.

$$\text{Wage} = \frac{600\ \text{DM} \cdot 176\ \text{h}}{40\ \text{h}} = 2640\ \text{DM}$$

Given : 3 Workers need 16 h.
To find : Working time for 8 workers.

3 workers need 16 hours.
8 workers need ? hours.

$$\text{Time} = \frac{16\ \text{h}.3\ \text{workers}}{8\ \text{workers}} = 6\ \text{hours}$$

Mixture calculation

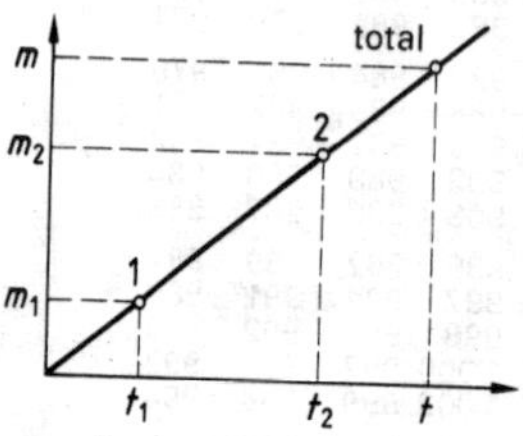

$$m_1 = \frac{m \cdot t_1}{t} \qquad m_1 = \frac{m \cdot p_1}{100}$$

$$m = \frac{m_1 \cdot t}{t_1} \qquad m = \frac{100 \cdot m_1}{p_1}$$

$$t_1 = \frac{m_1 \cdot t}{m} \qquad p_1 = \frac{100 \cdot m_1}{m}$$

$$t = t_1 + t_2 + \ldots \qquad m = m_1 + m_2 + \ldots$$

m Total quantity
m_1 Part quantity
p_1 Percent parts
t Total parts
t_1 Components, Part number

Given : Coolant mixture m = 8 l, Non-freeze t_1 = 4, Water t_2 = 6.
To find : Non-freeze and Water quantities

$$m_1 = \frac{8\ l \cdot 4\ \text{parts}}{(4+6)\ \text{parts}} = 3{,}20\ l$$

$$m_2 = 8{,}0 - 3{,}2 = 4{,}80\ l$$

Given : Two-stroke mixture, m = 39 l
Mixture ratio 1:25.
To find : Oil and petrol volumes in l

$$m_{oil} = \frac{39\ l \cdot 1\ \text{Part}}{(1+25)\ \text{Parts}} = 1{,}50\ l$$

$$m_{petrol} = 39{,}0\ l - 1{,}5\ l = 37{,}5\ l$$

Distribution calculation

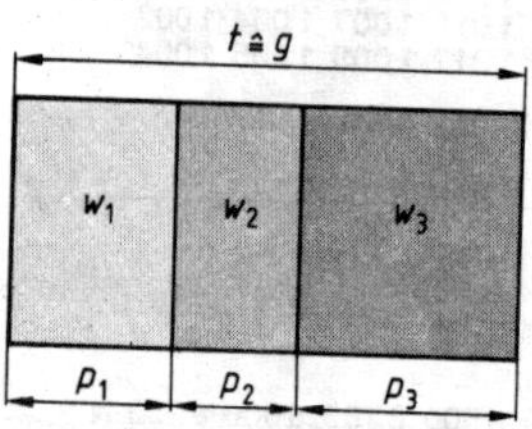

$$w_1 = \frac{g \cdot t_1}{t} \qquad w_1 = \frac{g \cdot p_1}{100\%}$$

$$g = \frac{w_1 \cdot t}{t_1} \qquad g = \frac{100\% \cdot w_1}{p_1}$$

$$t_1 = \frac{w_1 \cdot t}{g} \qquad p_1 = \frac{100\% \cdot w_1}{g}$$

$$t = t_1 + t_2 + \ldots \qquad g = w_1 + w_2 + \ldots$$

g Total value
p_1 Percentage
t Total parts
t_1 Components, Component number
w_1 Part value

Given: Bonus distribution g = 2112 DM.
Filter A t_1 165 hours
Filter B t_2 187 hours
Filter C t_3 154 hours
Filter D t_4 198 hours
To find: Bonus value $w_1 \ldots w_4$ in DM.

$$w_1 = \frac{2112,-\ \text{DM} \cdot 165\ \text{h}}{704\ \text{h}} = 495,-\ \text{DM}$$

$$w_2 = \frac{2112,-\ \text{DM} \cdot 187\ \text{h}}{704\ \text{h}} = 561,-\ \text{DM}$$

$$w_3 = \frac{2112,-\ \text{DM} \cdot 154\ \text{h}}{704\ \text{h}} = 462,-\ \text{DM}$$

$$w_4 = \frac{2112,-\ \text{DM} \cdot 198\ \text{h}}{704\ \text{h}} = 594,-\ \text{DM}$$

Percentage calculation

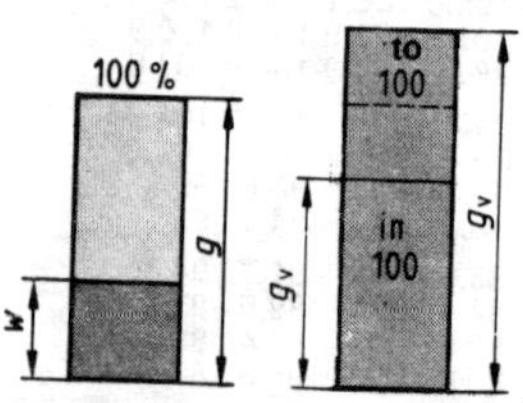

$$w = \frac{g \cdot p}{100\%} \qquad w = \frac{g_v \cdot p}{100\% \pm p}$$

$$g = \frac{100\% \cdot w}{p} \qquad g_v = \frac{w \cdot (100\% \pm p)}{p}$$

$$p = \frac{100\% \cdot w}{g} \qquad p = \frac{100\% \cdot w}{g_v \mp w}$$

g Basic value
g_v Increased value
Reduced value
p Percentage
w Percentage value

Given : Bill amount g = 600 DM.
Discount p = 3%
To find : Discount and Cash payment in DM.

$$\text{Discount } w = \frac{600\ \text{DM} \cdot 3\%}{100\%} = 18\ \text{DM}$$

Cash payment = 600 DM − 18 DM = 582 DM

Given : Bill amount with surplus expense
g_v = 2 260 DM, Surplus p = 13%.
To find : Surplus expenses w = in DM.

$$\text{Surplus } w = \frac{2\,260\ \text{DM} \cdot 13\%}{100\% + 13\%} = 260\ \text{DM}$$

Interest calculation

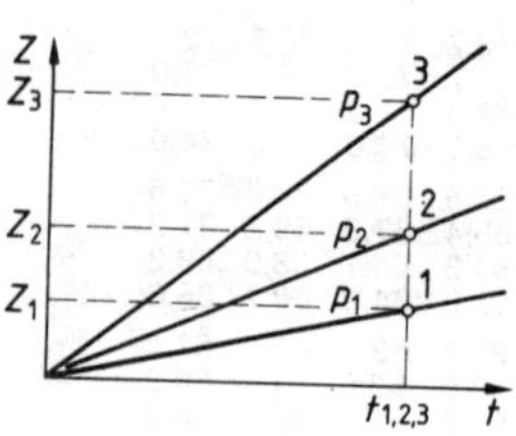

$$Z = \frac{K \cdot p \cdot t}{36\,000} \qquad Z = \frac{\text{Interest number}}{\text{Interest divider}}$$

$$K = \frac{36\,000 \cdot Z}{p \cdot t} \qquad \text{Interest number} = \frac{K \cdot t}{100}$$

$$p = \frac{36\,000 \cdot Z}{K \cdot t} \qquad \text{Interest divider} = \frac{360}{p}$$

$$t = \frac{36\,000 \cdot Z}{K \cdot p}$$

Use in case of several amounts with different periods but uniform Rate of interest in %

K Principal in DM
p Rate of interest
z Interest in %
Time in days
1 Month 30 days
1 Year 360 days

Given: Principal K = 8000—DM
Rate of interest p = 8% Time = 315 days
To find: a) Interest Z in DM.
b) End principal in DM.

$$Z = \frac{8\,000\ \text{DM} \cdot 8 \cdot 315}{36\,000} = 560,-\ \text{DM}$$

or

$$\text{Interest number} = \frac{8\,000 \cdot 315}{100} = 25\,200$$

Interest divider = 360 : 8 = 45

$$Z = \frac{25\,200}{45} = 560,-\ \text{DM}$$

End principal = 8 000 + 560 = 8560 DM

General	Letters are place holders for definite numbers. Same quantities are always designated with same letters. Numbers for the letters specify how many times the variable is to be taken. Calculation quantities belonging together are put in brackets.	$+a+a+a = 3$ times $(+a) = +3a$ $-a-a-a = 3$ times $(-a) = -3a$ $3a$ number 3 Coefficient a Variable, place holder $5\cdot(3a+5b)$ $7\cdot[4a-2\cdot(5a+6b-3c)]$
Pre-signs, Calculation signs General	Numbers without signs are always positive. Numbers with minus sign are negative.	$5a = +5a$ $(-5a) = -5a$
Adding and subtracting of numbers	If the calculation and pre-signs are same, the number is added; if the calculation and pre-signs are different, the number is subtracted.	$+(+a) = +a$ $-(-a) = +a$ $+(-a) = -a$ $-(+a) = -a$
Adding and subtracting of sums	When solving a plus bracket, the calculation sign remains unchanged; when solving a minus bracket, the calculation signs change. Inner brackets are solved first.	$a+(b+c-d) = a+b+c-d$ $a-(b+c-d) = a-b-c+d$ $-[a-(b+c-d)] = -[a-b-c+d]$ $= -a+b+c-d$
Multiplying of numbers	The product of two numbers with same sign is always positive; the product of two numbers with different signs is always negative.	$(+a)\cdot(+b) = +(ab)$ $(-a)\cdot(-b) = +(ab)$ $(+a)\cdot(-b) = -(ab)$ $(-a)\cdot(+b) = -(ab)$
Addition Subtraction	Coefficients of the same alphabets are added or subtracted according to calculation symbol. only same letters are added.	$+18b+5b+7b = 30b$ $-25a-8a-5a = -38a$ $+15c-7c+3c = 11c$ $+12b+4a-7b = 4a+5b$
Multiplication Numbers with numbers	Coefficients multiplied by one another and all alphabets without coefficients are arranged after the product of the coefficients.	$2\cdot 4a = 2\cdot 4\cdot a = 8a$ $3a\cdot 5b\cdot 4 = 3\cdot 5\cdot 4\cdot a\cdot b = 60ab$ $5a\cdot(-7b) = 5\cdot(-7)\cdot a\cdot b = -35ab$
Numbers with sums	Every term of the sum in the brackets is multiplied by the number before or after the brackets.	$4\cdot(a+b) = 4a+4b$ $3a\cdot(4b-7) = 12ab-21a$ $(-2a)\cdot(5b-3c) = -10ab+6ac$
Sums with sums	Every term of a sum (summand) in one set of brackets is multiplied by every term of a sum in the other set of brackets.	$(a+b)\cdot(a+b) = a^2+2ab+b^2$ $(a+b)\cdot(a-b) = a^2-b^2$
Division Number by number	Coefficients are divided by each other and same letters simplified with each other.	$10a : 2 = (10:2)\cdot a = 5a$ $28a : (-4a) = [28:(-4)]\cdot(a:a) = -7$
Sums by numbers	Every term of the sum within the brackets (dividend) is divided by the number after the brackets (divisor).	$(9a+6b):3 = 3a+2b$ $9a$ $- \quad +6b$ $6b$ $-$
Sums by sums	Every term of the sum within one set of brackets (dividend) is divided by every term of the sum within the other set of brackets (divisor)	$(a^2+2ab+b^2):(a+b) = a+b$ a^2+ab $- \quad ab+b^2$ $ab+b^2$ $- \quad -$
Resolution into factors Formation of a set of brackets	A common factor of a sum is placed before a set of brackets and the remaining factors are combined within the brackets.	$8a+8b+8c = 8\cdot(a+b+c)$ $4ab-4ac+4ae = 4a\cdot(b-c+e)$ $3bx-9ax+15cx = 3x\cdot(b-3a+5c)$
Solution of brackets	a) Calculation carried out within the brackets. b) Every term within the brackets multiplied with the number before the brackets. c) Plus brackets removed without change; in case of minus brackets, calculation signs changed.	$(5a+2a-3a)+8a = 4a+8a = 12a$ $(5a+2b-3c)\cdot 8 = 40a+16b-24c$ $(-8)\cdot(5a+2b-3c) = -40a-16b+24c$ $5a+(4b-3c+2d) = 5a+4b-3c+2d$ $5a-(4b-3c+2d) = 5a-4b+3c-2d$

Calculation of fractions

General	A fraction is a number which divides a unit—a whole unit—into a certain number of equal parts. The number below the fraction line is called denominator, it designates the size of the part pieces. The number above the fraction line is called numerator, it indicates the number of part pieces.	$\frac{2a}{2b} = \frac{\text{numerator}}{\text{denominator}}$ $\frac{3}{4}$ = proper fraction $\frac{7}{4}$ = improper fraction $1\frac{3}{4}$ mixed fraction 0.75 decimal fraction
Reduction Numbers	Numerator and denominator divided by the same number. The value of the fraction remains same.	$\frac{12ab}{9a} = \frac{12ab:3a}{9a:3a} = \frac{4b}{3}$
Sums	In case of sums in the numerator or denominator, all terms of the sum divided by the same number.	$\frac{ab+ac}{a} = \frac{(ab+ac):a}{a:a} = b+c$
Extension Numbers	Numerator and denominator multiplied by the same number. The fraction value remains same.	$\frac{3a}{4b} = \frac{3a \cdot 2c}{4b \cdot 2c} = \frac{6ac}{8bc}$
Sums	In case of sums in the numerator or denominator, all terms multiplied by the same number.	$\frac{a+b}{2} = \frac{(a+b)\cdot 2c}{2 \cdot 2c} = \frac{2ac+2bc}{4c}$
Main denominator (Lowest common Multiple)	The main denominator is the smallest common number in which all denominators are contained. It is determined from the prime numbers of the denominators. 1. Denominators resolved into prime numbers, 2. Prime numbers are so combined that all prime numbers of a denominator are contained once in the main denominator, 3. Common prime numbers multiplied with one another.	Number: 6 ab, 8b, 10 ab, 12, 15 a It is determined Prime numbers? $6ab = 2 \cdot 3 \cdot a \cdot b$ $8b = 2 \cdot 2 \cdot 2 \cdot b$ $10ab = 2 \cdot 5 \cdot a \cdot b$ $12 = 2 \cdot 2 \cdot 3$ $15a = 3 \cdot 5 \cdot a$ $2 \cdot 2 \cdot 2 \cdot 3 \cdot 5 \cdot a \cdot b = 120ab$
Addition, Subtraction Correspondent fractions	Numerators added or subtracted from one another according to calculation sign and the common denominator retained.	$\frac{2b}{3a} + \frac{5b}{3a} - \frac{4b}{3a} = \frac{2b+5b-4b}{3a}$ $= \frac{3b}{3a} = \frac{b}{a}$
Non-correspondent fractions	Denominator made correspondent first; 1. Main denominator determined, 2. Extension number M found, 3. Fractions made correspondant by extension, 4. Numerators added or subtracted according to calculation sign.	$\frac{3x}{4a} + \frac{5x}{8a} - \frac{7x}{12a} = \frac{18x+15x-14x}{24a} = \frac{19x}{24a}$ $4a = 2 \cdot 2 \cdot a$ $M = 2 \cdot 3 = 6$ $8a = 2 \cdot 2 \cdot 2 \cdot a$ $M = \cdot 3 = 3$ $12a = 2 \cdot 2 \cdot 3 \cdot a$ $M = \cdot 2 = 2$ $2 \cdot 2 \cdot 2 \cdot 3 \cdot a = 24a$
Multiplication Fraction with number	Individual number multiplied with the numerator or individual number divided by the denominator	$12b \cdot \frac{4a}{3b} = \frac{12b \cdot 4a}{3b} = \frac{48ab}{3b} = 16a$
Fraction with fraction	Numerator multiplied with numerator and denominator multiplied with denominator.	$\frac{5a}{9b} \cdot \frac{2c}{7a} = \frac{5a \cdot 2c}{9b \cdot 7a} = \frac{10ac}{63ab} = \frac{10c}{63b}$
Division Fraction by number	Numerator divided by the individual number or denominator multiplied with the individual number.	$\frac{6ac}{7b} : 3a = \frac{6ac:3a}{7b} = \frac{2c}{7b}$ $\frac{6ac}{7b} : 3a = \frac{6ac}{7b \cdot 3a} = \frac{6ac}{21ab} = \frac{2c}{7b}$
Fraction by fraction	Divisor (second fraction) reversed and the dividend (first fraction) multiplied with the inverse value of the divisor.	$\frac{21x}{8b} : \frac{7a}{4b} = \frac{21x}{8b} \cdot \frac{4b}{7a} = \frac{21x \cdot 4b}{8b \cdot 7a} = \frac{3x}{2a}$
Transformation Fraction into a decimal number	Numerator divided by the denominator or fraction extended so that the denominator is a ten.	$\frac{7}{8} = 7:8 = 0{,}875$
Decimal number into a fraction	Decimal number extended with 10 (100,1000) and the result written in the form of fraction.	$0{,}4 = \frac{0{,}4 \cdot 10}{1 \cdot 10} = \frac{4}{10}$

Powers

General	Equal factors of a product can be represented in short form as power.	$5 \cdot 5 \cdot 5 = 5^3$ (five power three) $= 125$
	The number, denoting how often the factor is multiplied by itself is written as exponent after the base number.	5^3 Power 5 Base number, base 3 Exponent 125 Value of power
Value of power	The value of power is calculated by multiplying the base number by itself as many times as prescribed by the exponent.	$4^3 = 4 \cdot 4 \cdot 4 = 64$ $a^4 = a \cdot a \cdot a \cdot a$
Sign Base number positive	The value of power with positive base number is always positive.	$(+4)^3 = (+4) \cdot (+4) \cdot (+4) = +4^3 = +64$ $(+a)^4 = (+a) \cdot (+a) \cdot (+a) \cdot (+a) = +a^4$
Base number negative, exponent even	Values of powers of negative base number and an even exponent are always positive.	$(-4)^2 = (-4) \cdot (-4) = +4^2 = +16$ $(-a)^4 = (-a) \cdot (-a) \cdot (-a) \cdot (-a) = +a^4$
Base number negative, exponent odd	Values of power of a negative base number and an odd exponent are always negative.	$(-4)^3 = (-4) \cdot (-4) \cdot (-4) = -4^3 = -64$ $(-a)^3 = (-a) \cdot (-a) \cdot (-a) = -a^3$
Addition	Preliminary number of equal powers are added and the sum of the preliminary members multiplied by the common power.	$3 \cdot 4^2 + 5 \cdot 4^2 = (3+5) \cdot 4^2 = 8 \cdot 4^2$ $5a^2 + 7a^2 + 4a^2 = (5+7+4) \cdot a^2 = 16a^2$
Subtraction	Preliminary numbers of equal powers are subtracted and the difference of the preliminary numbers is multiplied by the common exponent.	$9 \cdot 5^3 - 7 \cdot 5^3 = (9-7) \cdot 5^3 = 2 \cdot 5^3$ $15a^2 - 8a^2 - 4a^2 = (15-8-4) \cdot a^2 = 3a^2$
Multiplication Base numbers equal	Add the exponents and raise the common base number to the power of the sum of exponents.	$4^2 \cdot 4^3 \cdot 4^2 = 4^{2+3+2} = 4^7$ $2a^2 \cdot 5a^3 \cdot a^4 = 2 \cdot 5 \cdot a^{2+3+4} = 10a^9$
Exponents equal	Multiply the base numbers and raise the product of the base numbers to the power of the common exponent.	$4^2 \cdot 3^2 \cdot 5^2 = (4 \cdot 3 \cdot 5)^2 = 60^2$ $3a^2 \cdot 5b^2 = 3 \cdot 5 \cdot (a \cdot b)^2 = 15(ab)^2$
Division Base numbers equal	Subtract the exponents from each other and raise the common base number to the power of the difference of exponents.	$4^5 : 4^2 = 4^{5-2} = 4^3$ $8a^5 : 2a^3 = 8 : 2 \cdot a^{5-3} = 4a^2$
Exponents equal	Divide the base numbers by each other and raise the quotient of the base numbers to the power of the common exponent.	$10^3 : 2^3 = (10:2)^3 = 5^3$ $12a^2 : 4b^2 = (12:4) \cdot (a:b)^2 = 3 \cdot (a/b)^2$
Raising to a power	Multiply the exponents and raise the base numbers to the power of the product of exponents.	$(3^2)^3 = 3^{2 \cdot 3} = 3^6$ $(a^3)^4 = a^{3 \cdot 4} = a^{12}$
Special Cases Exponent negative	A power with negative exponent is equal to the reciprocal of the power with positive exponent.	$3^{-2} = \frac{1}{3^2} = \frac{1}{9}$ $a^{-3} = \frac{1}{a^3}$
Exponent = 1	A power with the exponent 'one' is equal to the base number.	$4^1 = 4$ $a^1 = a$
Exponent = 0	A power with the exponent 'zero' is equal to 'one'.	$4^0 = 1$ $a^0 = 1$

Root calculation

General	Root calculation is the inversion of power calculation. Exponents and values of power are known. The base numbers are found. The method of writing used for this: $\sqrt{25} = x$, say the 'second root of 25' or square root of 25'.	$\sqrt[3]{64} = 4$ — Third root of 64 or Cube root of 64 $\sqrt[3]{64}$ 3 — Index of the root 64 — Radical 4 — Root value
Addition	Preliminary numbers of roots with equal exponents and equal radicals added and the sum multiplied by the common root.	$6 \cdot \sqrt{9} + 4 \cdot \sqrt{9} = (6 + 4) \cdot \sqrt{9} = 10 \cdot \sqrt{9}$ $8 \cdot \sqrt{a} + 6 \cdot \sqrt{a} = (8 + 6) \cdot \sqrt{a} = 14 \cdot \sqrt{a}$
Subtraction	Preliminary numbers of roots with equal exponents and equal radicals subtracted, and the difference is multiplied by the common root.	$9 \cdot \sqrt{4} - 6 \cdot \sqrt{4} = (9 - 6) \cdot \sqrt{4} = 3 \cdot \sqrt{4}$ $6 \cdot \sqrt{a} - 4 \cdot \sqrt{a} = (6 - 4) \cdot \sqrt{a} = 2 \cdot \sqrt{a}$
Multiplication Root exponents equal	Radicals of roots with equal exponents multiplied and the common root of the product extracted.	$\sqrt{4} \cdot \sqrt{9} = \sqrt{4 \cdot 9} = \sqrt{36}$ $2\sqrt{a} \cdot 3\sqrt{b} = 2 \cdot 3 \cdot \sqrt{a \cdot b} = 6\sqrt{ab}$
Division Root exponents equal	Radicals of roots with equal exponents divided by each other and the common root of the product extracted.	$\sqrt{64} : \sqrt{4} = \sqrt{64 : 4} = \sqrt{16} = 4$ $\sqrt{a} : \sqrt{b} = \sqrt{a : b}$
Raising to power	Exponent of the radicals multiplied by the exponent of power, and the common root of the power extracted.	$(\sqrt{4})^3 = \sqrt{4^3} = \sqrt{64}$ $(\sqrt{4^3})^2 = \sqrt{4^{3 \cdot 2}} = \sqrt{4^6}$
Extracting the root Number	Determine the number which has to be multiplied with itself as many times as indicated by the root exponent to obtain the radical. In practice the root is found out with the help of table, a slide-rule or a pocket calculator.	$\sqrt{64} = 8$ — Check: $8 \cdot 8 = 64$ $\sqrt[3]{64} = 4$ — Check: $4 \cdot 4 \cdot 4 = 64$ $\sqrt{a^2} = a$ — Check: $a \cdot a = a^2$ $\sqrt[3]{a^3} = a$ — Check: $a \cdot a \cdot a = a^3$
Sum, Difference	Combine the terms according to the calculation sign and extract the root of the sum. Extract the root of the sum only, not of the individual terms.	$\sqrt{9 + 16} = \sqrt{25} = 5$ $\sqrt{64 - 28} = \sqrt{36} = 6$
Product	Multiply the factors by each other and extract the root of the product or extract the root of each factor and multiply the root values by each other.	$\sqrt{9 \cdot 16} = \sqrt{144} = 12$ $\sqrt{9 \cdot 16} = \sqrt{9} \cdot \sqrt{16} = 3 \cdot 4 = 12$
Quotient	Divide the dividend by the divisor and extract the root of the quotient or extract the roots of the dividend and divisor and divide the root values by each other.	$\sqrt{\frac{16}{4}} = \sqrt{4} = 2$ or $\sqrt{\frac{16}{4}} = \frac{\sqrt{16}}{\sqrt{4}} = \frac{4}{2} = 2$
Power	Determine the power value and extract the root of the power value so found or extract the root of the radical and raise the root value to the power of the power exponent.	$\sqrt{4^3} = \sqrt{4 \cdot 4 \cdot 4} = \sqrt{64} = 8$ $\sqrt{4^3} = (\sqrt{4})^3 = 2^3 = 2 \cdot 2 \cdot 2 = 8$
	Equal root exponents and power exponents cancel each other; root value equal to the base number of the radicand.	$\sqrt{a^2} = a$ $\sqrt[3]{a^3} = a$
Calculation of the square root	1. Resolve larger radicands into groups of two digits each from the decimal point towards left. 2. Extract the roots according to the formula $(a + b)^2 = a^2 + 2\,ab + b^2$. 3. In case of radicands with more than two groups apply the method many times one after another.	$\sqrt{13\,69} = 30 + 7 = 37$ $a^2 = 30^2 \quad = 9\,00$ $\quad 4\,69$ $2ab = 2 \cdot 30 \cdot 7 = 4\,20$ $\quad 49$ $b^2 = 7^2 \quad = 49$ $\quad -$ Check: $(30 + 7)^2 = 30 \cdot 30 + 2 \cdot 30 \cdot 7 + 7 \cdot 7 = 1369$

Topic and rule	Examples	Further examples
General Equations consist of two sides. Since both sides are of equal value, the equal sign is placed between them.	$3+5=8$ $x-6=12$	$24+15-18+12=33$ $4a+8a-3a+9a=18a$
Change of terms The terms of a side can be changed in any sequence as desired.	$15+7=22$ ▶ $16a+5x=37$ ▶	$7+15=22$ $5x+16a=37$
Exchange of sides The left and the right sides of an equation can be mutually exchanged.	$28-16=12$ ▶ $12b+5x=45$ ▶	$12=28-16$ $45=12b+5x$
Modification of sides Any modification of the equation must be done in an identical manner on both sides.	$6+14=20$ ▶ $x+12=28$ ▶	$6+14-5=20-5$ $x+12-12=28-12$
Reciprocal value of sides If the reciprocal value is written for one side of the equation, the other side must also be written with reciprocal values.	$\frac{9}{15}=\frac{3}{5}$ $\left(\frac{9}{15}\right)=\left(\frac{3}{5}\right)$ $\frac{75}{x}=\frac{3}{2}$ ▶	$\frac{15}{9}=\frac{5}{3}$ $\frac{x}{75}=\frac{2}{3}$
Transformation of sums and products An equation is transformed till the unknown quantity x (y or z) is on the left side with the positive sign. **Calculation sign** If a term is transferred from one side of the equation to the other, then the calculation sign of the changed term is modified as follows: from + it becomes − from − it becomes + from × it will become ÷ from ÷ it will become ×	$x+5=22$ $x=22-5$ $x-4=16$ $x=16+4$ $x\cdot 5=30$ $x=30:5$ $x:2=12$ $x=12\cdot 2$	$x+a=8a$ $x=8a-a$ $x-a=8a$ $x=8a+a$ $x\cdot a=8a$ $x=8a:a$ $x:a=8a$ $x=8a\cdot a$
Transformation of fractions If, in an equation, the unknown quantity x (y, z) is combined with fractions, then these fractions must be eliminated by transformation or expansion. In doing so: 'numerator' becomes 'denominator' 'denominator' becomes 'numerator'.	$\frac{5\cdot x}{8}=15$ $5\cdot x=15\cdot 8$ $x=\frac{15\cdot 8}{5}$ $x=24$	$\frac{3x}{8}=\frac{12}{16}$ $\frac{3x\cdot 8}{8}=\frac{12\cdot 8}{16}$ $3x=6$ $x=6:3$ $x=2$
Several non-homologous fractions If the equation consists of several non-homologous fractions, then the fractions must be brought to a main denominator by expansion before transformation.	$\frac{4x}{3}+\frac{5x}{4}-\frac{3x}{6}=50$ $\frac{16x+15x-6x}{12}=50$ $\frac{25x}{12}=50$ $25x=50\cdot 12$ $x=24$	
Unknown quantities in denominator If the unknown quantity x (y, z) occurs in the denominator of a fraction, then before transformation, both sides of the equation are written with the reciprocal value.	$\frac{3}{x+2}=\frac{6}{10}$ $\frac{x+2}{3}=\frac{10}{6}$ $x+2=\frac{3\cdot 10}{6}$ $x=5-2$ $x=3$	$\frac{5}{x}=\frac{15}{24}$ $\frac{x}{5}=\frac{24}{15}$ $x=\frac{24\cdot 5}{15}$ $x=8$
Transformation of powers If the unknown quantity x (y, z) is on one side of the equation as power, then the root of the other side is extracted during transformation: 'power' becomes 'root' 'root' becomes 'power'	$5^2=25$ $5=\sqrt{25}$ $x^2=16$ $x=\sqrt{16}$	$\sqrt{16}=4$ $16=4^2$ $\sqrt{a}=x$ $a=x^2$
Transformation of brackets If the unknown quantity x (y, z) is inside brackets, then the brackets are first solved, or all known quantities must be separated from the brackets by bringing them outside the brackets through side change.	$5\cdot(x+2)=40$ $5x+10=40$ $5x=40-10$ $x=\frac{30}{5}=6$	$5\cdot(x+2)=40$ $(x+2)=\frac{40}{5}$ $x=8-2$ $x=6$

Ratios, Proportional Equations

Proportions	Similar quantities can be set in ratios to each other. By setting opposite each other, definite indices are obtained which simply and clearly project the facts of the case. There are arithmetical and geometrical ratios. The latter appear most frequently in engineering. Ratios have at least two terms, and they are written in the division or fractional form.	$2:3=\frac{2}{3}$ $\quad a:b=\frac{a}{b}$ Speed ratio $n_1 : n_2$ Compression ratio $(V_h+V_c):V_c$ stroke ratio. $s:d$ pitch ratio. $h:l$ Cross-sectional ratio $H:B$ Trigonometrcial tunction. $\sin\alpha = a:c$
Reduction	In a ratio, both terms can be divided by the same number, without altering the value of the ratio. By reduction the ratios can be expressed with the smallest whole numbers, or relate the ratio of the quantities to number 'one'.	$\frac{24}{18}=\frac{24:3}{18:3}=\frac{8}{6}$ $\quad \frac{15}{25}=\frac{15:5}{25:5}=\frac{3}{5}$ $\frac{15}{18}=\frac{15:3}{18:3}=\frac{5}{6}$ $\quad \frac{32}{20}=\frac{32:20}{20:20}=\frac{1{,}6}{1}$ $\frac{14}{28}=\frac{14:14}{28:14}=\frac{1}{2}$ $\quad \frac{14}{28}=\frac{14:28}{28:28}=\frac{0{,}5}{1}$
Extension	In a ratio, both terms can be multiplied with the same number. This does not alter the value of the ratio. Ratios can be expressed in percentage or 1:100 form by extension.	$\frac{2}{3}=\frac{2\cdot 12}{3\cdot 12}=\frac{24}{36}$ $\quad \frac{25}{40}=\frac{25\cdot 8}{40\cdot 8}=\frac{200}{320}$ $\frac{4}{1}=\frac{4\cdot 100}{1\cdot 100}=\frac{400}{100}=400:100=400\%$
Multiplication	Several homologous ratios can be condensed into a single total ratio by mutually multiplying the numerators and denominators.	$(3:5)\cdot(4:7)=(3\cdot 4):(5\cdot 7)=12:35$ $\frac{24}{16}\cdot\frac{32}{20}\cdot\frac{10}{36}=\frac{24\cdot 32\cdot 10}{16\cdot 20\cdot 36}=\frac{7680}{11520}=\frac{2}{3}$
Proportional Equation General Direct proportion Inverse proportion	Two homologous ratios can be combined into one equation. The equation is called the proportional equation or proportion. A proportion has 4 terms: 2 outer terms and 2 inner terms. If the sequence of terms is identical on both sides, then it is a straight or direct proportion. If the sequence of both terms is inverse on both sides, then it is an inverse or indirect proportion.	$4:5=8:10$ $\quad 2:1=50:25$ $a_1:a_2=b_1:b_2$ $\quad a_1:a_2=b_2:b_1$ $a_1:a_2=b_1:b_2$ $\quad$ Hydraulic transmission: $F_1:F_2=A_1:A_2$ $a_1:a_2=b_2:b_1$ $\quad$ Gear drive $n_1:n_2=z_2:z_1$
Substitution Inner terms Outer terms Inner terms with outer terms	In a proportional equation the internal terms can be substituted with each another. In a proportional equation the outer terms can be substituted with each other. In a proportional equation the inner terms can be substituted with outer terms.	$a:b=c:d$ will become $a:c=b:d$ $2:3=4:6$ $\quad 2:4=3:6$ $a:b=c:d$ will become $d:b=c:a$ $2:3=4:6$ $\quad 6:3=4:2$ $a:b=c:d$ will become $b:a=d:c$ $2:3=4:6$ $\quad 3:2=6:4$
Transformation	If three terms of a proportional equation are given, then the fourth can be calculated by transformation.	$x:b=c:d$ $\quad x=\frac{b\cdot c}{d}$
Product equation	In a proportional equation the product of inner terms is equal to the product of outer terms. Two homologous products together constitute a product equation.	$2:3=6:9$ will become $2\cdot 9=3\cdot 6$ $a:b=c:d$ will become $a\cdot d=b\cdot c$ Belt drive: $n_1:n_2=d_2:d_1$ will become $n_1\times d_1=n_2\times d_2$

Sine 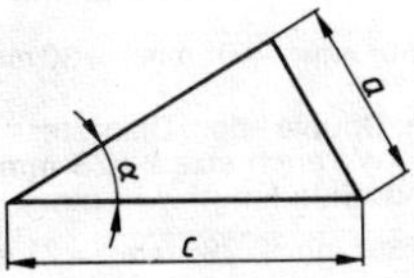 	Sine = $\frac{\text{Opposite side}}{\text{Hypotenuse}}$ $\sin\alpha = \frac{a}{c}$ $a = c \cdot \sin\alpha$ $c = \frac{a}{\sin\alpha}$ a Opposite side c Hypotenuse α Angle	Given: Right angled triangle To find: Angle α in degrees Sin $\alpha = \frac{75\text{ mm}}{150\text{ mm}} = 0.500$ $\alpha = 30°$ Given: Right angled triangle To find: Opposite side a in mm Sin 32° from Table = 0.530 $c = 120$ mm · 0,530 = 63,6 mm
Cosine 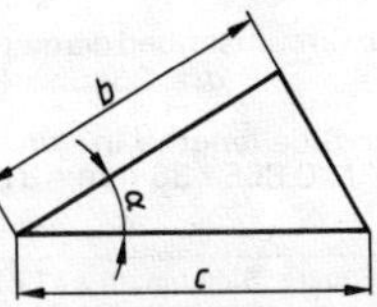 	Cosine = $\frac{\text{Adjacent side}}{\text{Hypotenuse}}$ $\cos\alpha = \frac{b}{c}$ $b = c \cdot \cos\alpha$ $c = \frac{b}{\cos\alpha}$ b = Adjacent side c = Hypotenuse α = Angle	Given: Right angled triangle $b = 130$ mm, $c = 150$ mm To find: α in degrees $\cos\alpha = \frac{130\text{ mm}}{150\text{ mm}} = 0{,}866$ $\alpha = 30°$ Given: Right angled triangle To find: Adjacent side b in mm cos 51°20′ from Table = 0.625 $b = 152$ mm × 0.625 = 95.0 mm
Tangent 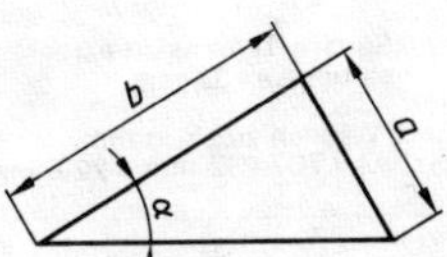 	Tangent = $\frac{\text{Opposite side}}{\text{Adjacent side}}$ $\tan\alpha = \frac{a}{b}$ $a = b \cdot \tan\alpha$ $b = \frac{a}{\tan\alpha}$ a Opposite side b Adjacent side α Angle	Given: Right angled triangle To find: Angle α in degrees $\tan\alpha = \frac{75\text{ mm}}{130\text{ mm}} = 0{,}577$ $\alpha = 30°$ Given: Right angled triangle To find: Opposite side a in mm tan 39°10′ from Table 0.815 $a = 160$ mm × 0.815 = 130.4 mm
Cotangent 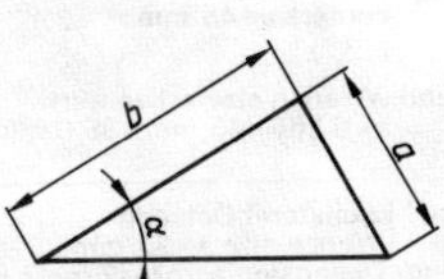 	Contangent = $\frac{\text{Adjacent side}}{\text{Opposite side}}$ $\cot\alpha = \frac{b}{a}$ $b = a \cdot \cot\alpha$ $a = \frac{b}{\cot\alpha}$ a Opposite side b Adjacent side α Angle	Given: Right angled triangle $a = 75$ mm, $b = 130$ mm To find: Angle α in degrees $\cot\alpha = \frac{130\text{ mm}}{75\text{ mm}} = 1{,}733$ $\alpha = 30°$ Given: Right angled triangle To find: Opposite side b in mm cot 39″40′ from Table = 1.250 $b = 68$ mm × 1.250 = 85.0 mm
Pythagoras Theorem 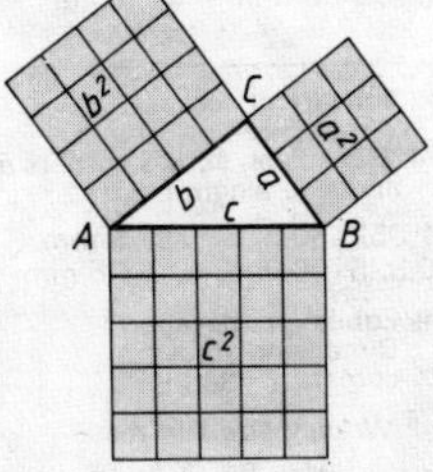 	In a right angled triangle the square of the hypotenuse is equal to the sum of the squares of the other two sides. $c^2 = a^2 + b^2$ $c = \sqrt{a^2 + b^2}$ $a^2 = c^2 - b^2$ $a = \sqrt{c^2 - b^2}$ $b^2 = c^2 - a^2$ $b = \sqrt{c^2 - a^2}$ In a triangle, if the sides are 3:4:5, then it is always right angled. a, b : Sides of the right angle c : Hypotenuse	Given: Right angled triangle To find: Hypotenuse c in mm. $c = \sqrt{60^2\text{ mm}^2 + 80^2\text{ mm}^2} = 100$ mm Given: Right angled triangle To find: Side b in mm $b = \sqrt{35^2\text{ mm}^2 - 21^2\text{ mm}^2} = 28$ mm Given: Right angled triangle To find: Side a in mm $a = \sqrt{60^2\text{ mm}^2 - 48^2\text{ mm}^2} = 36$ mm

Wrench size, Dimensions across corners

Double edge

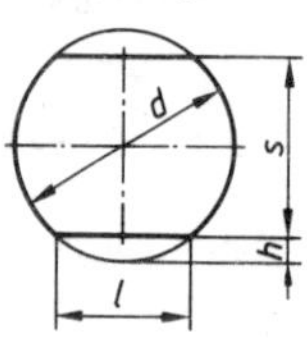

$s = \sqrt{d^2 - l^2}$ $\quad d = \sqrt{l^2 + s^2}$

$l = \sqrt{d^2 - s^2}$ $\quad h = \dfrac{d-s}{2}$

d Diameter
h Height
l Side length
s Wrench size

Designation of a Wrench size of e.g., $s = 24$ mm SW 24

Given: Double edge, $d = 50$ mm
Side length $l = 40$ mm
To find: Wrench size s in mm

$s = \sqrt{50^2\ \text{mm}^2 - 40^2\ \text{mm}^2} = 30$ mm

Given: Double edge, Diameter = 35 mm
Wrench size $s = 28$ mm
To find: Side length l in mm.

$l = \sqrt{35^2\ \text{mm}^2 - 28^2\ \text{mm}^2} = 21$ mm

Equilateral triangle

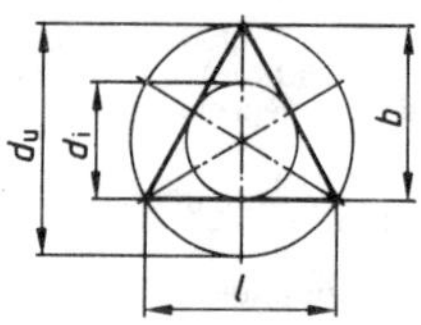

$d_u = 1{,}155 \cdot l$ $\quad d_i = 0{,}578 \cdot l$

$l = 0{,}866 \cdot d_u$ $\quad l = 1{,}730 \cdot d_i$
$b = 0{,}866 \cdot l$ $\quad l = 1{,}155 \cdot b$

b Width, Height
d_i Inscribed diameter
d_u Circumscribed diameter
l Side length

Given: Equilateral triangle
Side length $l = 24$ mm

$d_u = 1{,}155 \cdot 24$ mm $= 27{,}7$ mm

To find: Circumscribed diameter
d_u = 36mm

To find: Side length l in mm.
$l = 0.866 \times 36$ mm = 31.2 mm

Square

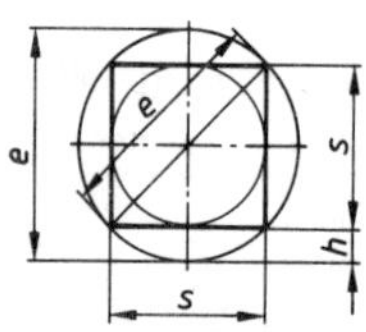

$e = 1{,}414 \cdot s$ $\quad e = \sqrt{2} \cdot s$

$s = 0{,}707 \cdot e$ $\quad s = \sqrt{2} \cdot \dfrac{e}{2}$

$h = \dfrac{e-s}{2}$ $\quad s = e - 2 \cdot h$

e Dimension across corners, Circumscribed diameter
h Height
s Wrench size, Side length

Given: Square, Side length $s = 17$ mm
To find: a) Dimension across corners e in mm
b) Height h in mm

$e = 1.414 \times 17$ mm $= 24.0$ mm

$h = \dfrac{24\ \text{mm} - 17\ \text{mm}}{2} = 3.5$ mm

Given: Square, Dimension across corners $e = 32$ mm.

To find: Wrench size s in mm.
$s = 0.707 \times 32$ mm $= 22.6$ mm

Equilateral Hexagon

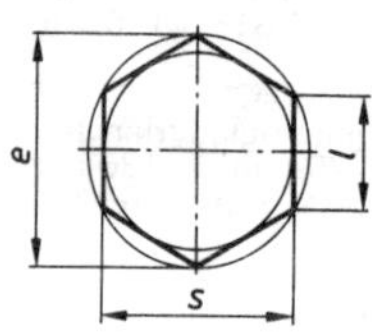

$e = 1{,}155 \cdot s$ $\quad l = 0{,}577 \cdot s$

$s = 0{,}866 \cdot e$ $\quad s = 1{,}733 \cdot l$

$e = 2{,}0 \cdot l$ $\quad l = 0{,}5 \cdot e$

e Dimension across corners, Circumscribed diameter
l Side length
s Wrench size, Inscribed diameter

Given: Equilateral hexagon
Wrench size $s = 21$ mm

To find: Dimension across corners e and side length l

Given: Equilateral hexagon
Dimension across corners $e = 45$ mm

To find: Wrench size s 1 in mm
$s = 0.866 \times 45$ mm $= 39.0$ mm

Equilateral Octagon

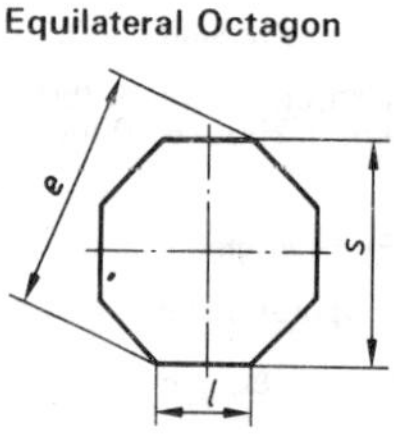

$e = 1{,}082 \cdot s$ $\quad l = 0{,}414 \cdot s$

$s = 0{,}924 \cdot e$ $\quad s = 2{,}415 \cdot l$

$e = 2{,}611 \cdot l$ $\quad l = 0{,}383 \cdot e$

e Dimension across corners, Circumscribed diameter
l Side length
s Wrench size, Inscribed diameter

Given: Equilateral Octagon
Wrench size $s = 32$ mm
To find: Dimension across corners e and side length l

$e = 1{,}082 \cdot 32$ mm $= 34{,}6$ mm
$l = 0{,}414 \cdot 32$ mm $= 13{,}25$ mm

Given: Equilateral octagon
$e = 45$ mm
To find: Wrench size s in mm

$s = 0{,}924 \cdot 45$ mm $= 41{,}6$ mm

Equilateral Decagon

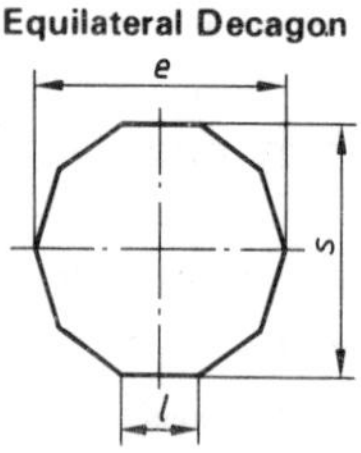

$e = 1{,}052 \cdot s$ $\quad l = 0{,}325 \cdot s$

$s = 0{,}951 \cdot e$ $\quad s = 3{,}077 \cdot l$

$e = 3{,}236 \cdot l$ $\quad l = 0{,}309 \cdot e$

e Dimension across corners, Circumscribed diameter
l Side length
s Wrench size, Inscribed diameter

Given: Equilateral decagon
Wrench size $s = 50$ mm
To find: Dimension across corners e and side length l

$e = 1{,}052 \cdot 50$ mm $= 52{,}6$ mm
$l = 0{,}325 \cdot 50$ mm $= 16{,}25$ mm

Given: Equilateral decagon
Dimension across corners $e = 36$ mm

To find: Wrench size s in mm

$s = 0{,}951 \cdot 36$ mm $= 34{,}2$ mm

Scales in graphic representation

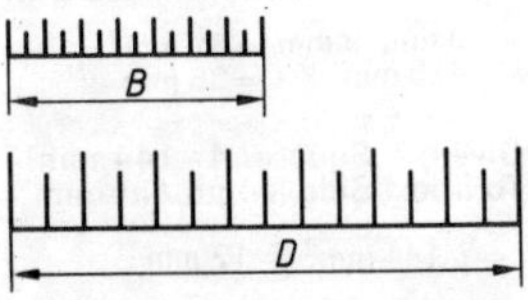

$$\mu = \frac{\text{Picture size } B}{\text{Matter size } D} \text{ e.g., } \frac{1\text{ mm}}{5\text{ cm}} \text{ or } 0.2\ \frac{\text{mm}}{\text{cm}}$$

$$\mu = \frac{B}{D} \qquad B = D \cdot \mu \qquad D = \frac{B}{\mu}$$

B = Picture size, Drawing size
D = Matter size, Natural size
μ = Scale e.g. 1 mm 5 cm

Given: Bolt, length D = 150mm, Scale μ = 1:5
To find: Picture size B in mm

$$B = 150\text{mm} \times \frac{1\text{mm}}{5\text{mm}} = 30\text{mm}$$

Given: Travel map μ = 1:40,000
Distance between two points B = 85mm
To find: Actual distance D in km.

$$D = 85\text{ mm} : \frac{1\text{mm}}{0.04\text{ km}} = 3.4\text{ km}$$

Slope rise

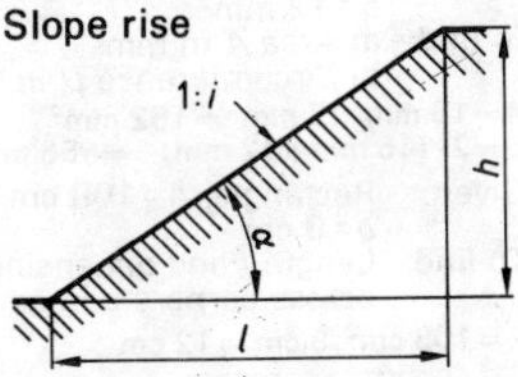

5% gradient is 5 m height difference over 100 m horizontal length

$$1{:}i = \frac{h}{l} \qquad P = \frac{100\% \cdot h}{l} \qquad \tan\alpha = \frac{h}{l}$$

$$h = \frac{l}{i} \qquad h = \frac{P \cdot l}{100\%} \qquad h = l \cdot \tan\alpha$$

$$l = h \cdot i \qquad l = \frac{100\% \cdot h}{P} \qquad l = \frac{h}{\tan\alpha}$$

h Height difference
l Horizontal length
p Gradient in percent
$1{:}i$ Gradient ratio
α Gradient angle

Given: Mountain road, horizontal length 1 = 2.5 km, height difference h = 125mm
To find: a) Gradient ratio 1 : i
b) Gradient in %
c) Gradient angle α

$$1{:}i = \frac{125\text{ m}}{2\,500\text{ m}} = 1{:}20$$

$$P = \frac{100\% \cdot 125\text{ m}}{2\,500\text{ m}} = 5{,}0\%$$

$$\tan\alpha = \frac{125\text{ m}}{2500\text{ m}} = 0{,}050 \qquad \alpha = 3°00'$$

Pitch

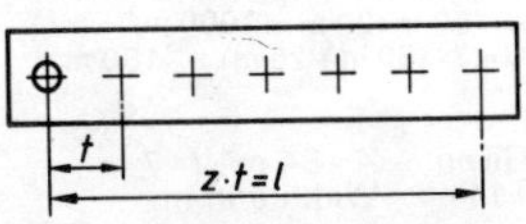

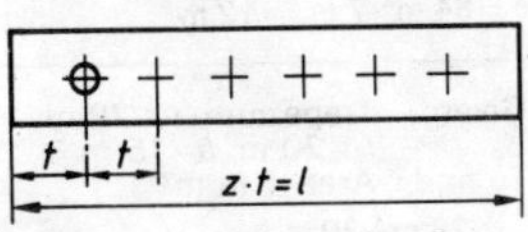

$$t = \frac{\text{Pitch length}}{\text{number of pitches}} \qquad t = \frac{l}{z}$$

$l = t \cdot z$ $\quad z = n - 1$ (without end piece)
$z = n + 1$ (with end pieces)
$z = n$ (endless, e.g. chains or hole circle)

$$z = \frac{l}{t}$$

l graduation lengths, hole circle circumference.
n graduation marks, number of holes.
t graduation, graduated part's length, distance between hole centres number of graduations,
z number of pieces.

Given: Roller chain l = 533.4 mm t = 12.7 mm
To find: Number of links z

$$z = \frac{533.4\text{ mm}}{12.7\text{ mm}} = 42 \text{ links}$$

Given: Tail piece, l = 1800 mm with end pieces, n = 5 pitch graduation
To find: Pitch t in mm

z = 5 + 1 = pitches

$$t = \frac{1800\text{ mm}}{6\text{ pitches}} = 300\text{ mm}$$

Developed length

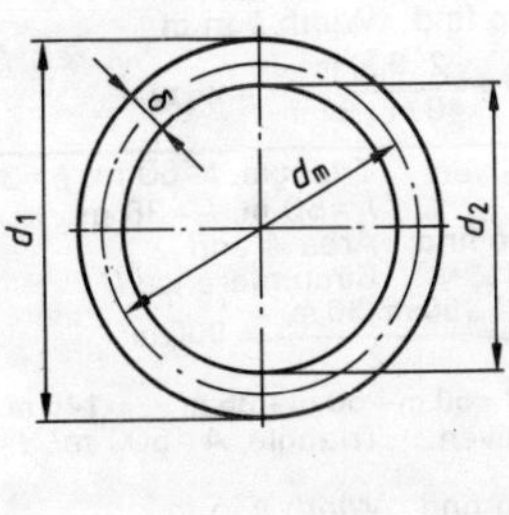

$$l = \pi \cdot d_m \qquad l = \pi \cdot \frac{d_1 + d_2}{2}$$

$$d_m = \frac{l}{\pi} \qquad d_1 = \frac{2 \cdot l}{\pi} - d_2$$

$$d_m = \frac{d_1 + d_2}{2} \qquad d_2 = \frac{2 \cdot l}{\pi} - d_1$$

$$d_m = d_2 + \delta \qquad d_m = d_1 - \delta$$

d_1 Outer diameter
d_2 Inner diameter
d_m Mean diameter
l Developed length
Length of neutral fibre
δ Material thickness (only for $\delta > 3$ mm)

Given: Steel ring, d_1 = 120 mm
Material thickness δ = 20 mm
To find: a) Mean diameter d_m
b) Developed length/in mm

d_m = 120 mm − 20 mm = 100 mm
l = π · 100 mm = 314,16 mm

Given: Steel tyre, d_2 = 600 mm
Material thickness δ = 30 mm
To find: a) Mean diameter d_m
b) Length of the neutral fibre l in mm.

d_m = 600 mm + 30 mm = 630 mm
l = π · 630 mm = 1979,2 mm

Circular arc length

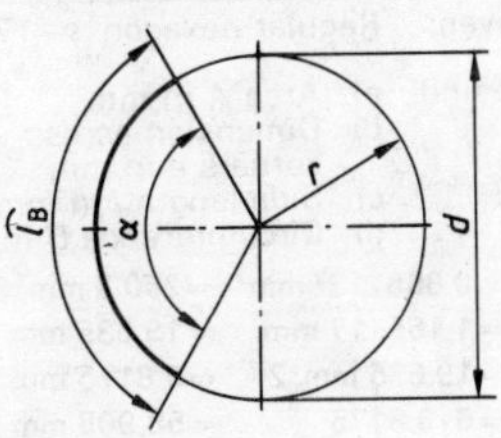

$$l_B = \frac{\pi \cdot d \cdot \alpha}{360°} \qquad l_B = \frac{2 \cdot \pi \cdot r \cdot \alpha}{360°}$$

$$d = \frac{360° \cdot l_B}{\pi \cdot \alpha} \qquad r = \frac{360° \cdot l_B}{2 \cdot \pi \cdot \alpha}$$

$$\alpha = \frac{360° \cdot l_B}{\pi \cdot d} \qquad \alpha = \frac{360° \cdot l_B}{2 \cdot \pi \cdot r}$$

d Diameter
l_B Arc length
r Radius
α Central Angle in degree

Given: Belt drive, Pulley diameter d = 85 mm,
Central angle α = 120°
To find: Arc length l_B in mm

$$l_B = \frac{\pi \cdot 85\text{ mm} \cdot 120°}{360°} = 89{,}0\text{ mm}$$

Given: Curvacious travel, Tread circle diameter d = 12 m, α = 120°
To find: Curve length l_B in m.

$$l_B = \frac{\pi \cdot 12\text{ m} \cdot 120°}{360°} = 12{,}566\text{ m}$$

Areas

Square

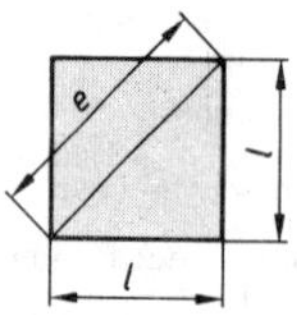

$A = l \cdot l$ $\quad U = 4 \cdot l$

$l = \sqrt{A}$ $\quad l = \frac{U}{4}$

$e = 1{,}414 \cdot l$ $\quad l = \frac{e}{1{,}414}$

A Area
e Dimension across corners
l Side length
U Circumference

Given: Square, $l = 9$ mm
To find: a) Area A in mm²
b) Circumference U in mm

$A = 9\text{ mm} \cdot 9\text{ mm} = 81\text{ mm}^2$
$U = 4 \cdot 9\text{ mm} = 36\text{ mm}$

Given: Square, $A = 144$ mm²
To find: Side length l in mm

$l = \sqrt{144\text{ mm}^2} = 12\text{ mm}$

Rectangle

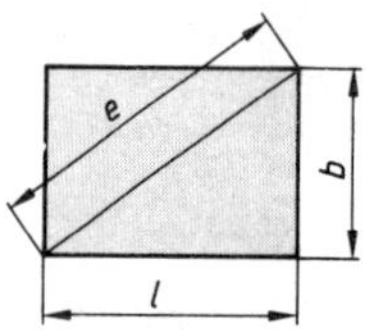

$A = l \cdot b$ $\quad U = 2 \cdot (l + b)$

$l = \frac{A}{b}$ $\quad l = \frac{U}{2} - b$

$b = \frac{A}{l}$ $\quad e = \sqrt{l^2 + b^2}$

A Area
b Width
l Length
e Dimension across corners
U Circumference

Given: Rectangle, $l = 16$ mm, $b = 12$ mm
To find: a) Area A in mm²
b) Circumference U in mm
$A = 16\text{ mm} \cdot 12\text{ mm} = 192\text{ mm}^2$
$U = 2 \cdot (16\text{ mm} + 12\text{ mm}) = 56\text{ mm}$

Given: Rectangle, $A = 108$ cm², $b = 9$ cm
To find: Length l and dimension across corners e in cm
$l = 108\text{ cm}^2 : 9\text{ cm} = 12\text{ cm}$
$e = \sqrt{12^2\text{ cm}^2 + 9^2\text{ cm}^2} = 15\text{ cm}$

Parallelogram

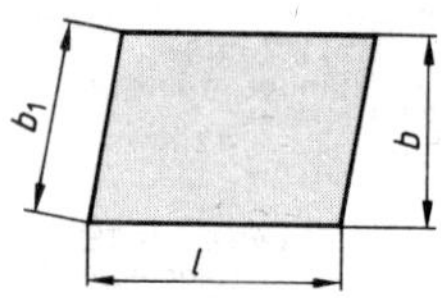

$A = l \cdot b$ $\quad U = 2 \cdot (l + b_1)$

$l = \frac{A}{b}$ $\quad l = \frac{U}{2} - b_1$

$b = \frac{A}{l}$ $\quad b_1 = \frac{U}{2} - l$

A Area
b Width
b_1 Inclined width
l Length
U Circumference

Given: Parallelogram, $l = 50$ m, $b = 20$ m, $b_1 = 25$ mm
To find: a) Area A in m²
b) Circumference U in m.
$A = 50\text{ m} \cdot 20\text{ m} = 1000\text{ m}^2$
$U = 2 \cdot (50\text{ m} + 25\text{ m}) = 150\text{ m}$

Given: $A = 84$ m², $l = 7$ m
To find: Width b in m
$b = 84\text{ m}^2 : 7\text{ m} = 12\text{ m}$

Trapezium

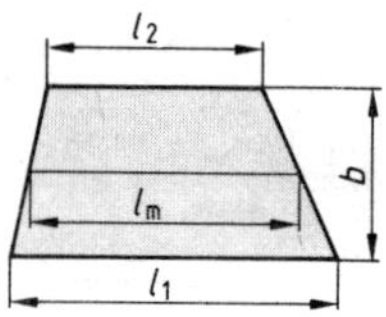

$A = l_m \cdot b$ $\quad A = \frac{l_1 + l_2}{2} \cdot b$

$l_m = \frac{l_1 + l_2}{2}$ $\quad l_1 = \frac{2 \cdot A}{b} - l_2$

$b = \frac{2 \cdot A}{l_1 + l_2}$ $\quad l_2 = \frac{2 \cdot A}{b} - l_1$

A Area
b Width
l_1 Large length
l_2 Small length
l_m Mean length
U Circumference

Given: Trapezium, $l_1 = 28$ m, $l_2 = 20$ m, $b = 15$ m
To find: Area A in m²
$A = \frac{28\text{ m} + 20\text{ m}}{2} \cdot 15\text{ m} = 360\text{ m}^2$

Given: $A = 900$ m², $l_1 = 40$ m, $l_2 = 32$ m.
To find: Width b in m
$b = \frac{2 \cdot 900\text{ m}^2}{40\text{ m} + 32\text{ m}} = 25\text{ m}$

Triangle

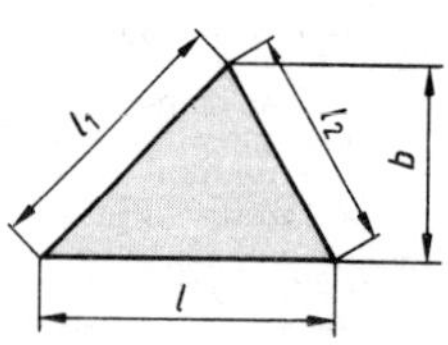

$A = \frac{l \cdot b}{2}$ $\quad$ U = Sum of sides

$l = \frac{2 \cdot A}{b}$ $\quad U = l + l_1 + l_2$

$b = \frac{2 \cdot A}{l}$

A Area
b Width
l Side length
l_1 Side length
l_2 Side length
U Circumference

Given: Triangle, $l = 60$ m, $b = 30$ m, $l_1 = 50$ m, $l_2 = 36$ m
To find: Area A and Circumference U
$A = \frac{60\text{ m} \cdot 30\text{ m}}{2} = 900\text{ m}^2$
$U = 60\text{ m} + 50\text{ m} + 36\text{ m} = 146\text{ m}$

Given: Triangle, $A = 500$ m², $l = 40$ m.
To find: Width b in m
$b = \frac{2 \cdot 500\text{ m}^2}{40\text{ m}} = 25\text{ m}$

Regular Polygon

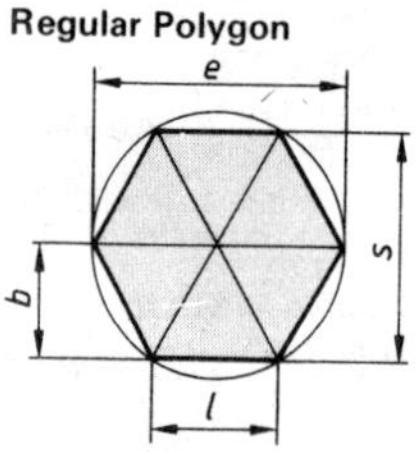

$A = \frac{l \cdot b}{2} \cdot n$ $\quad U = l \cdot n$

Hexagon
$A = 0{,}866 \cdot s^2$
$s = 0{,}866 \cdot e$
$e = 1{,}155 \cdot s$

Octagon
$A = 0{,}828 \cdot s^2$
$s = 0{,}924 \cdot e$
$e = 1{,}082 \cdot s$

A Area
b Width
e Dimension across corners
n = Number of edges
s = Wrench size
U = Circumference

Given: Regular hexagon $s = 17$ mm
To find: a) Area A in mm²
b) Dimension across corners e in mm
c) Side length l in mm
d) Circumference U in mm

$A = 0{,}866 \cdot 17^2\text{ mm}^2 = 250{,}3\text{ mm}^2$
$e = 1{,}155 \cdot 17\text{ mm} = 19{,}635\text{ mm}$
$l = 19{,}635\text{ mm} : 2 = 9{,}8175\text{ mm}$
$U = 6 \cdot 9{,}8175 = 58{,}905\text{ mm}$

Irregular Polygon

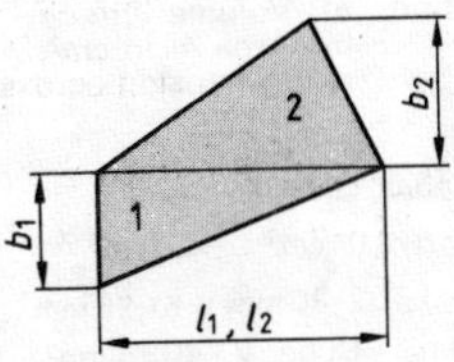

$A = A_1 + A_2 + A_3 + \ldots$ U = sum of sides

$A_1 = \frac{l_1 \cdot b_1}{2}$

$A_2 = \frac{l_2 \cdot b_2}{2}$

Polygon to be split into partial areas which can be calculated.

A Total area; A_1 Partial area; A_2 Partial area; b_1, b_2 Widths; l_1, l_2 Lengths; U Circumference

Given: Irregular polygon $l_1 = 110$ mm, $l_2 = 110$ mm, $b_1 = 100$ mm, $b_2 = 35$ mm

To find: a) Partial areas A_1 and A_2 b) Total area A in mm

$A_1 = \frac{110\ \text{mm} \cdot 100\ \text{mm}}{2} = 5500\ \text{mm}^2$

$A_2 = \frac{110\ \text{mm} \cdot 35\ \text{mm}}{2} = 1925\ \text{mm}^2$

$A = A_1 + A_2 = 7425\ \text{mm}^2$

Circle

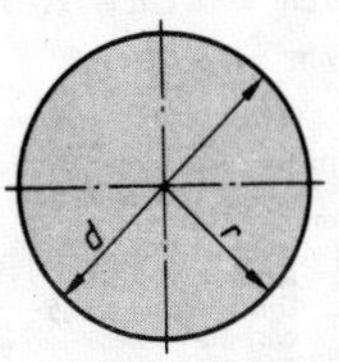

$A = \frac{\pi \cdot d^2}{4}$ $U = \pi \cdot d$

$d = \sqrt{\frac{4 \cdot A}{\pi}}$ $d = \frac{U}{\pi}$

$A = \pi \cdot r^2$ $U = 2 \cdot \pi \cdot r$

A Area; d Diameter; r Radius; U Circumference

Given: Piston of automobile engine $d = 80$ mm

To find: Piston area A in cm²

$A = \frac{\pi \cdot 8{,}0^2\ \text{cm}^2}{4} = 50{,}27\ \text{cm}^2$

Given: Diaphragm area $A = 314.159\ \text{cm}^2$

To find: Diaphragm diameter d in mm.

$d = \sqrt{\frac{4 \cdot 314{,}159\ \text{cm}^2}{\pi}} = 20\ \text{cm} = 200\ \text{mm}$

Circular ring

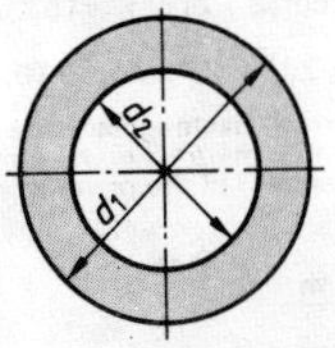

$A = \frac{\pi \cdot (d_1^2 - d_2^2)}{4}$ $A = A_1 - A_2$

$d_1 = \sqrt{d_2^2 + \frac{4 \cdot A}{\pi}}$ $A_1 = A + A_2$

$d_2 = \sqrt{d_1^2 - \frac{4 \cdot A}{\pi}}$ $A_2 = A_1 - A$

A Ring area; A_1 Total area; A_2 Inner area; d_1 Outer diameter; d_2 Inner diameter

Given: Piston pin, $d = 30$ mm, $d_2 = 24$ mm

To find: Cross sectional area A in cm²

$A = \frac{\pi \cdot (30^2 - 24^2)\ \text{mm}^2}{4} = 254{,}47\ \text{mm}^2$

Given: Steel tube, $A = 36757\ \text{mm}^2$, $d_2 = 360$ mm

To find: Outer diameter d_1 in mm.

$d_1 = \sqrt{360^2 + \frac{4 \cdot 36\,757}{\pi}} = 420\ \text{mm}$

Segment of a circle

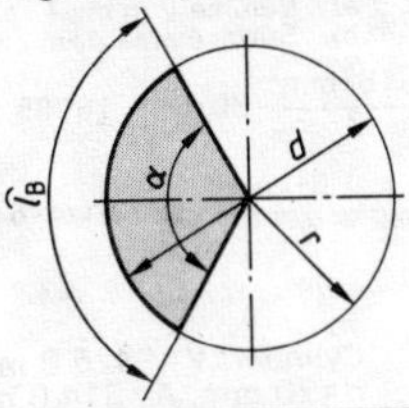

$A = \frac{\pi \cdot d^2 \cdot \alpha}{1440°}$ $A = \frac{l_B \cdot r}{2}$

$d = \sqrt{\frac{1440° \cdot A}{\pi \cdot \alpha}}$ $l_B = \frac{2 \cdot A}{r}$

$\alpha = \frac{1440° \cdot A}{\pi \cdot d^2}$

A Area; d Diameter; l_B Arc length; r Radius; α Inner angle in degrees

Given: Segment of a circle, $d = 80$ mm, $\alpha = 120°$

To find: Area A in mm²

$A = \frac{\pi \cdot 80^2\ \text{mm}^2 \cdot 120}{1440} = 1675{,}5\ \text{mm}^2$

Given: Segment of a circle $A = 884.88\ \text{cm}^2$, $d = 520$ mm

To find: Inner angle α in degrees

$\alpha = \frac{1440 \cdot 884{,}88\ \text{cm}^2}{\pi \cdot 52^2\ \text{cm}^2} = 150{,}0°$

Section of a circle

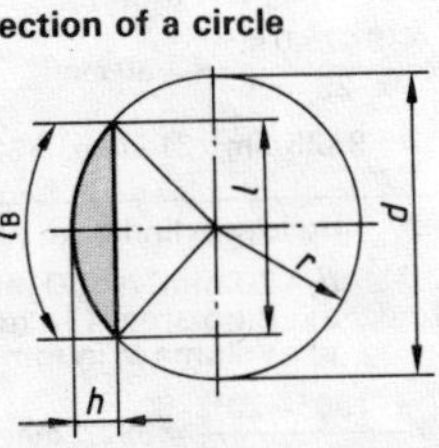

$A = \frac{l_B \cdot r - l \cdot (r - h)}{2}$ $A \approx \frac{2}{3} \cdot l \cdot h$

$h = r - \sqrt{r^2 - l^2/4}$ $l \approx \frac{3 \cdot A}{2 \cdot h}$

$l = 2 \cdot \sqrt{2 \cdot r \cdot h - h^2}$ $h \approx \frac{3 \cdot A}{2 \cdot l}$

A Area; h Height; l_a Arc length; l Chord length; r Radius

Given: Section of a circle, $b = 74.2$ mm, $r = 40$ mm, $l = 64$ mm $h = 16$ mm

To find: a) Area A in mm² (exact) b) Area A in mm² (approximate)

$A = \frac{74{,}2 \cdot 40 - 64 \cdot (40 - 16)}{2} = 716\ \text{mm}^2$

or approximately

$A \approx \frac{2}{3} \cdot 64\ \text{mm} \cdot 16\ \text{mm} = 683\ \text{mm}^2$

Ellipse

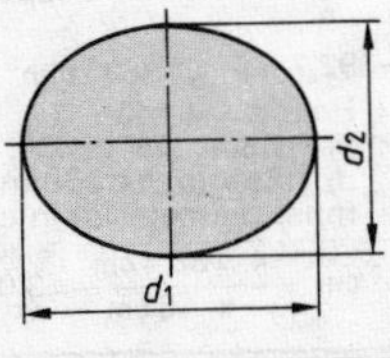

$A = \frac{\pi \cdot d_1 \cdot d_2}{4}$ $U \approx \frac{\pi}{2} \cdot (d_1 + d_2)$

$d_1 = \frac{4 \cdot A}{\pi \cdot d_2}$ $d_1 \approx \frac{2 \cdot U}{\pi} - d_2$

$d_2 = \frac{4 \cdot A}{\pi \cdot d_1}$ $d_2 \approx \frac{2 \cdot U}{\pi} - d_1$

A Area; d_1 Major axis; d_2 Minor axis; U Circumference

Given: Ellipse, $d_1 = 12.5$ mm, $d_2 = 8.0$ m

To find: Area A in m²

$A = \frac{\pi \cdot 12{,}5\ \text{m} \cdot 8\ \text{m}}{4} = 78{,}54\ \text{m}^2$

Given: Ellipse, $A = 235.619\ \text{mm}^2$, $d = 20$ mm

To find: Minor axis d_2 in mm

$d_2 = \frac{4 \cdot 235{,}619\ \text{mm}^2}{\pi \cdot 20\ \text{mm}} = 15{,}0\ \text{mm}$

Solids

Cube

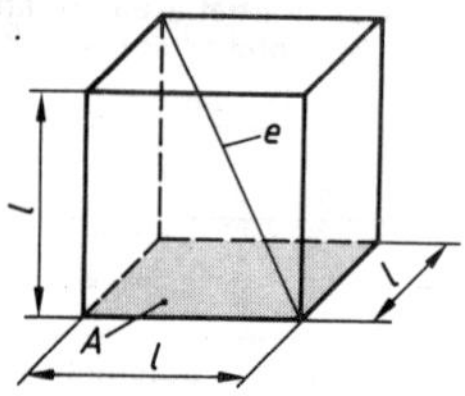

$V = l^3$ $\qquad V = A \cdot l$

$A_0 = 6 \cdot l^2$

$l = \sqrt[3]{V}$ $\qquad l = \dfrac{V}{A}$

$e = 1{,}732 \cdot l$ $\qquad A = \dfrac{V}{l}$

A Base area
A_0 Surface area
e Dimension across corners
l Side length
V volume

Given: Cube, Side length $l = 30$ mm
To find: a) Volume V in cm³
b) Area A_0 in cm²
c) Dimension across corners e in mm

$V = 3{,}0^3\ \text{cm}^3 = 27{,}0\ \text{cm}^3$

$A_0 = 6 \cdot 3{,}0^2\ \text{cm}^2 = 54.0\ \text{cm}^2$

$e = 1{,}732 \cdot 30\ \text{mm} = 51{,}96\ \text{mm}$

Given: Cube, $V = 4096$ cm³
To find: a) Side length l in mm
b) Area A_0 in cm²

$l = \sqrt[3]{4096\ \text{cm}^3} = 16\ \text{cm} = 160{,}0\ \text{mm}$

$A_0 = 6 \cdot 16^2\ \text{cm}^2 = 1536\ \text{cm}^2$

Prism

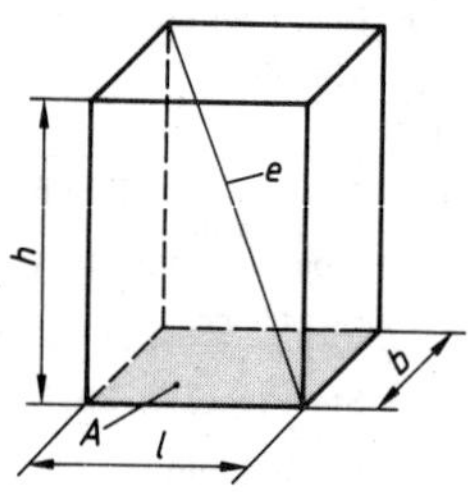

$V = l \cdot b \cdot h$ $\qquad V = A \cdot h$

$A_0 = 2 \cdot l \cdot b + 2 \cdot h \cdot (l + b)$

$h = \dfrac{V}{l \cdot b}$ $\qquad h = \dfrac{V}{A}$

$l = \dfrac{V}{b \cdot h}$ $\qquad A = \dfrac{V}{h}$

$b = \dfrac{V}{l \cdot h}$ $\qquad e = \sqrt{l^2 + b^2 + h^2}$

A Base area
A_0 Surface area
e Dimension across corners
l Side length
h Height
V Volume

Given: Prism, Rectangular base area $l = 80$ mm, $b = 20$ mm, $h = 60$ mm
To find: a) Volume V in cm³
b) Area A_0 in cm²
c) Dimension across corners e in mm

$V = 8\ \text{cm} \cdot 2\ \text{cm} \cdot 6\ \text{cm} = 96{,}0\ \text{cm}^3$

$A_0 = 2 \cdot 8\ \text{cm} \cdot 2\ \text{cm} + 2 \cdot 6\ \text{cm}\,(8 + 2)\ \text{cm} = 152{,}0\ \text{cm}^2$

$e = \sqrt{(80^2 + 20^2 + 60^2)\ \text{mm}^2} = 102\ \text{mm}$

Given: Prism, Rectangular base area $V = 72$ m³, $b = 3$ m, $h = 4$ m
To find: Length l of the prism in m.

$l = \dfrac{72\ \text{m}^3}{3\ \text{m} \cdot 4\ \text{m}} = 6{,}0\ \text{m}$

Cylinder

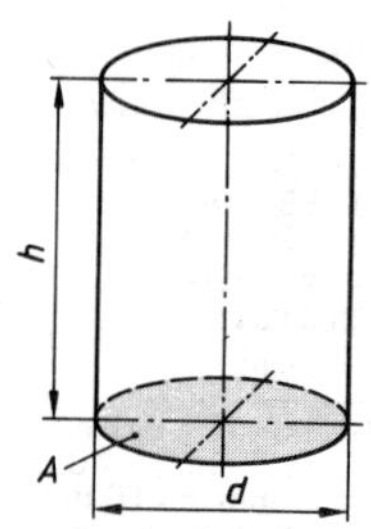

$V = \dfrac{\pi \cdot d^2}{4} \cdot h$ $\qquad V = A \cdot h$

$A_0 = 2 \cdot \dfrac{\pi \cdot d^2}{4} + \pi \cdot d \cdot h$

$h = \dfrac{4 \cdot V}{\pi \cdot d^2}$ $\qquad h = \dfrac{V}{A}$

$d = \sqrt{\dfrac{4 \cdot V}{\pi \cdot h}}$ $\qquad A = \dfrac{V}{h}$

A Base area
A_0 Surface area
d Diameter
h Height
V Volume

Given: Cylinder, $d = 16$ mm, $h = 80$ mm
To find: a) Volume V in mm³
b) Surface area A_0 in cm²

$V = \dfrac{\pi \cdot 16^2\ \text{mm}^2}{4} \cdot 80\ \text{mm} = 16085\ \text{mm}^3$

$A_0 = 2 \cdot \dfrac{\pi \cdot 1{,}6^2\ \text{cm}^2}{4} + \pi \cdot 1{,}6\ \text{cm} \cdot 8\ \text{cm} = 44{,}23\ \text{cm}^2$

Given: Cylinder, $V = 9425.0$ mm³, $d = 20$ mm, $A = 314.0$ mm²
To find: Height h in mm

$h = \dfrac{4 \cdot 9425{,}0\ \text{mm}^3}{\pi \cdot 20^2\ \text{mm}^2} = 30\ \text{mm}$

or $h = 9425\ \text{mm}^3 : 314\ \text{mm}^2 = 30\ \text{mm}$

Hollow Cylinder

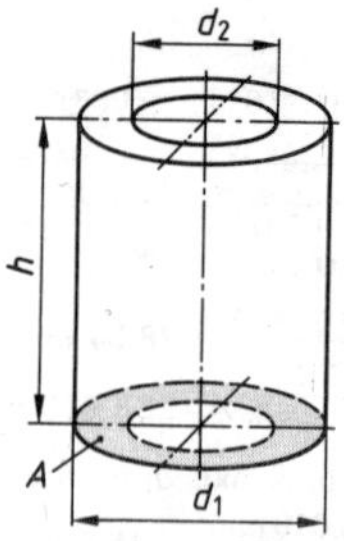

$V = \dfrac{\pi \cdot (d_1^2 - d_2^2)}{4} \cdot h$ $\qquad V = A \cdot h$

$h = \dfrac{4 \cdot V}{\pi \cdot (d_1^2 - d_2^2)}$ $\qquad h = \dfrac{V}{A}$

$d_1 = \sqrt{d_2^2 + \dfrac{4 \cdot V}{\pi \cdot h}}$ $\qquad A = \dfrac{V}{h}$

$d_2 = \sqrt{d_1^2 - \dfrac{4 \cdot V}{\pi \cdot h}}$ $\qquad A = \dfrac{\pi \cdot (d_1^2 - d_2^2)}{4}$

A Base area
d_1 Outer diameter
d_2 Inner diameter
h Height
V Volume

Given: Hollow cylinder, $d_1 = 30$ mm, $d_2 = 20$ mm, $h = 50$ mm
To find: a) Ring area A in mm²
b) Volume V in mm³

$A = \dfrac{\pi \cdot (30^2 - 20^2)}{4} = 392{,}7\ \text{mm}^2$

$V = \dfrac{\pi \cdot (30^2 - 20^2)}{4} \cdot 50 = 19635\ \text{mm}^3$

or $V = 392{,}7 \cdot 50 = 19635\ \text{mm}^3$

Given: Hollow cylinder $V = 188.4$ cm³, $d_1 = 50$ mm, $h = 150$ mm
To find: Inner diameter d_2 in cm

$d_2 = \sqrt{5^2\ \text{cm}^2 - \dfrac{4 \cdot 188{,}4\ \text{cm}^2}{\pi \cdot 15\ \text{cm}}} = 3{,}0\ \text{cm}$

Pyramid

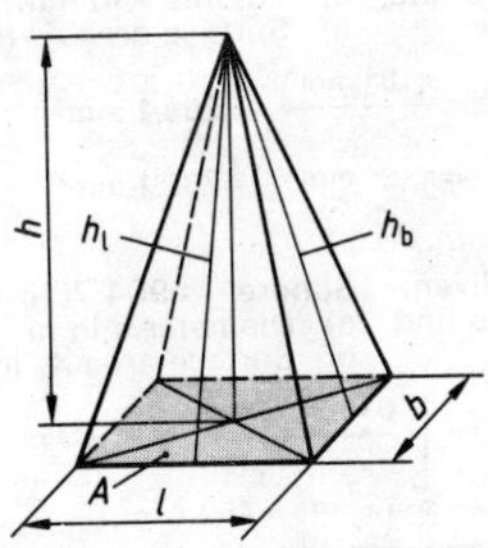

$$V = \frac{l \cdot b \cdot h}{3} \qquad V = \frac{A \cdot h}{3}$$

$$A_0 = l \cdot b + l \cdot h_l + b \cdot h_b$$

$$h = \frac{3 \cdot V}{l \cdot b} \qquad h = \frac{3 \cdot V}{A}$$

$$l = \frac{3 \cdot V}{b \cdot h} \qquad A = \frac{3 \cdot V}{h}$$

$$b = \frac{3 \cdot V}{l \cdot h} \qquad A = l \cdot b$$

A	Base area	h_f	Height over b
A_0	Surface area	h_l	Height over l
b	Width	V	Volume
h	Height	l	Length

Given: Pyramid, $l = 40$ mm, $b = 25$ mm, $h = 90$ mm, $A = 1000$ mm²

To find: Volume V in mm³

$$V = \frac{(40 \cdot 25 \cdot 90)\ \text{mm}^3}{3} = 30000\ \text{mm}^3$$

$$V = \frac{1000\ \text{mm}^2 \cdot 90\ \text{mm}}{3} = 30000\ \text{mm}^3$$

Given: Pyramid, $v = 320$ cm³, $l = 8$ cm, $b = 6$ cm, $A = 48$ cm²

To find: Height h in mm

$$h = \frac{3 \cdot 320\ \text{cm}^3}{8\ \text{cm} \cdot 6\ \text{cm}} = 20\ \text{cm} = 200\ \text{mm}$$

or

$$h = \frac{3 \cdot 320\ \text{cm}^3}{48\ \text{cm}^2} = 20\ \text{cm} = 200\ \text{mm}$$

Cone

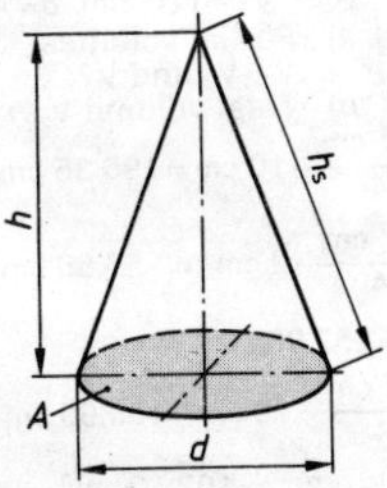

$$V = \frac{\pi \cdot d^2 \cdot h}{12} \qquad V = \frac{A \cdot h}{3}$$

$$A_0 = \frac{\pi \cdot d^2}{4} + \frac{\pi \cdot d \cdot h_s}{2}$$

$$h = \frac{12 \cdot V}{\pi \cdot d^2} \qquad h = \frac{3 \cdot V}{A}$$

$$d = \sqrt{\frac{12 \cdot V}{\pi \cdot h}} \qquad A = \frac{3 \cdot V}{h}$$

$$h_s = \sqrt{h^2 + d^2/4} \qquad A = \frac{\pi \cdot d^2}{4}$$

A	Base area	h	Height
A_0	Surface area	h_s	Generator height
d	Diameter	v	Volume

Given: Cone, $d = 200$ mm, $h = 600$ mm, $h_s = 603.8$ mm

To find: a) Volume v in cm³ b) Surface area A_0 in cm²

$$V = \frac{\pi \cdot 20^2\ \text{cm}^2 \cdot 60\ \text{cm}}{12} = 6283\ \text{cm}^3$$

$$A_0 = \frac{\pi \cdot 20^2\ \text{cm}^2}{4} + \frac{\pi \cdot 20\ \text{cm} \cdot 60{,}83\ \text{cm}}{2} = 2225\ \text{cm}^2$$

Given: Cone, $v = 78.54$ cm³, $h = 12$ cm

To find: Diameter d in mm

$$d = \sqrt{\frac{12 \cdot 78{,}54\ \text{cm}^3}{\pi \cdot 12\ \text{cm}}} = 5\ \text{cm} = 50\ \text{mm}$$

Truncated pyramid (Frustum of a pyramid)

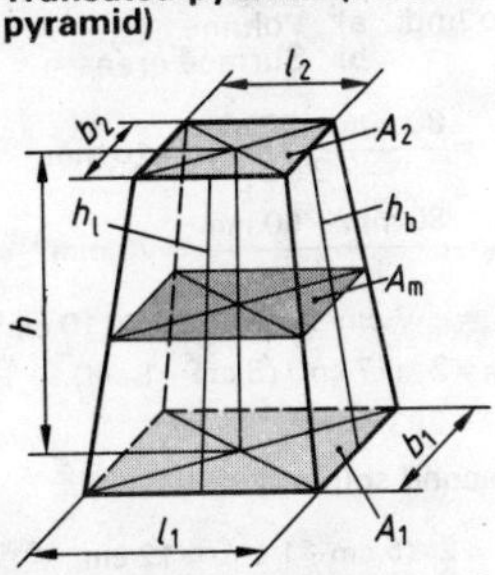

$$V \approx \frac{A_1 + A_2}{2} \cdot h \qquad V \approx A_m \cdot h$$

$$A_0 = A_1 + A_2 + h_l \cdot (l_1 + l_2) + h_b \cdot (b_1 + b_2)$$

$$h \approx \frac{2 \cdot V}{A_1 + A_2} \qquad h \approx \frac{V}{A_m}$$

$$A_1 \approx \frac{2 \cdot V}{h} - A_2 \qquad A_m \approx \frac{V}{h}$$

$$V = \frac{h}{3} \cdot (A_1 + A_2 + \sqrt{A_1 \cdot A_2})$$

A	Base area	b_1, b_2	Widths
A_2	Top area	h_b, h_d	Heights over b, l
A_m	Mean area		
A_0	Surface area	l_1, l_2	Lengths
h	Height	v	volume

Given: Truncated pyramid, $l = 40$ mm, $l_2 = 20$ mm, $b_1 = 30$ mm, $b_2 = 15$ mm, $h = 60$ mm

To find: Volume v in cm³

$A_1 = 12$ cm² $\quad A_2 = 3$ cm²

$$V \approx \frac{12\ \text{cm}^2 + 3\ \text{cm}^2}{2} \cdot 6\ \text{cm} = 45{,}0\ \text{cm}^3$$

or exact value

$$V = \frac{6\ \text{cm}}{3} \cdot (12\ \text{cm}^2 + 3\ \text{cm}^2 + \sqrt{12\ \text{cm}^2 \cdot 3\ \text{cm}^2}) = 42{,}0\ \text{cm}^3$$

Given: Truncated pyramid $v = 320$ cm³, $A_2 = 50$ cm², $A_2 = 30$ cm²

To find: Height h in mm h=

$$h \approx \frac{2 \cdot 320\ \text{cm}^3}{50\ \text{cm}^2 + 30\ \text{cm}^2} = 8\ \text{cm} = 80\ \text{mm}$$

Truncated Cone (Frustum of a cone)

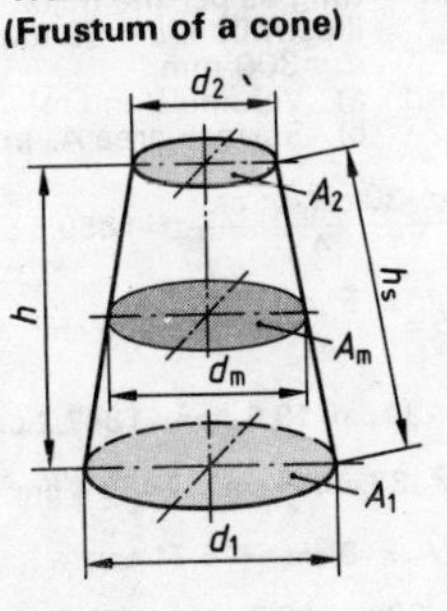

$$V \approx \frac{A_1 + A_2}{2} \cdot h \qquad V \approx A_m \cdot h$$

$$A_0 = A_1 + A_2 + \frac{\pi \cdot h_s \cdot (d_1 + d_2)}{2}$$

$$h \approx \frac{2 \cdot V}{A_1 + A_2} \qquad h \approx \frac{V}{A_m}$$

$$A_1 \approx \frac{2 \cdot V}{h} - A_2 \qquad A_m \approx \frac{V}{h}$$

$$V = \frac{\pi \cdot h}{12} \cdot (d_1^2 + d_2^2 + d_1 \cdot d_2)$$

A_1	Base area	d_1, d_2	Diameters
A_2	Top area	h	Height
A_m	Mean area	h_s	Surface height
A_0	Surface area	v	volume

Given: Truncated cone, $d_1 = 12$ cm

To find: Volume v in cm³

$A_1 = 113$ cm² $\quad A_2 = 50$ cm²

$$V \approx \frac{163\ \text{cm}^3}{2} \cdot 15\ \text{cm} = 1223\ \text{cm}^3$$

or exact value

$$V = \frac{\pi \cdot 15\ \text{cm}}{12} \cdot (12^2\ \text{cm}^2 + 8^2\ \text{cm}^2 + 12\ \text{cm} \cdot 8\ \text{cm}) = 1194\ \text{cm}^3$$

Given: Truncated cone, $v = 840$ cm³

To find: Area A_1 in cm²

$$A_1 \approx \frac{2 \cdot 840\ \text{cm}^3}{12\ \text{cm}} - 60\ \text{cm}^2 = 80{,}0\ \text{cm}^2$$

Solids

Sphere

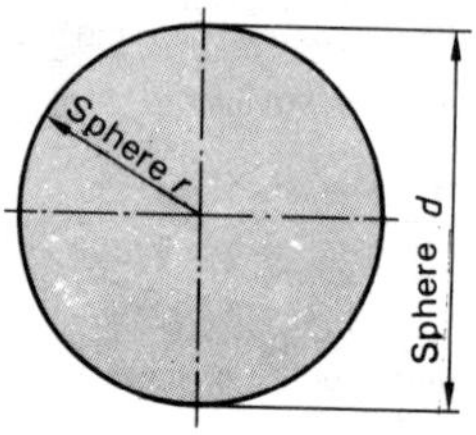

$$V = \frac{\pi \cdot d^3}{6} \qquad V = \frac{4 \cdot \pi \cdot r^3}{3}$$

$$A_0 = \pi \cdot d^2 \qquad A_0 = 4 \cdot \pi \cdot r^2$$

$$d = \sqrt[3]{\frac{6 \cdot V}{\pi}} \qquad r = \sqrt[3]{\frac{3 \cdot V}{4 \cdot \pi}}$$

$$d = \sqrt{\frac{A_0}{\pi}} \qquad r = \sqrt{\frac{A_0}{4 \cdot \pi}}$$

A_0 Surface area
d Diameter
r Radius
V Volume

Given: Sphere, $d = 8$ mm.
To find: a) Volume v in mm³
b) Surface area A_0 inmm²

$$V = \frac{\pi \cdot 8^3\ mm^3}{6} = 268{,}1\ mm^3$$

$$A_0 = \pi \cdot 8^2\ mm^2 = 201{,}1\ mm^2$$

Given: Sphere, $v = 904.78\ m^3$
To find: a) Diameter d in m
b) Surface area A_0 in m²

$$d = \sqrt[3]{\frac{6 \cdot 904{,}78\ m^3}{\pi}} = 12{,}0\ m$$

$$A_0 = \pi \cdot 12^2\ m^2 = 452{,}4\ m^2$$

Combination solids

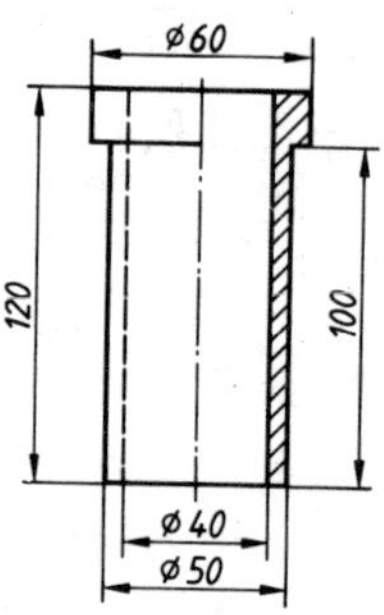

$$V = V_1 \pm V_2 \pm V_3 \pm \cdots$$

1. Split the combined solids into simple basic solids.
2. Calculate the volume of basic solids.
3. Add or subtract volumes of simple basic solids.

V Volume of combination solids
V_1 Partial volume of simple basic solids.

Given: Flange bushing as per the adjoining drawing.
Part 1 $l = 100$ mm, $d = 50$ mm
Part 2 $l = 20$ mm, $d = 60$ mm
Part 3 $l = 120$ mm, $d = 40$ mm
To find: a) Partial volumes V_1, V_2 and V_3
b) Total volume V *in cm³*

$$V_1 = \frac{\pi \cdot 5^2\ cm^2}{4} \cdot 10\ cm = 196{,}35\ cm^3$$

$$V_2 = \frac{\pi \cdot 6^2\ cm^2}{4} \cdot 2\ cm = 56{,}55\ cm^3$$

$$V_1 + V_2 = 252{,}90\ cm^3$$

$$V_3 = \frac{\pi \cdot 4^2\ cm^2}{4} \cdot 12\ cm = 150{,}80\ cm^3$$

$$V = V_1 + V_2 - V_3 = 102{,}10\ cm^3$$

Turned solid with rectangular cross section

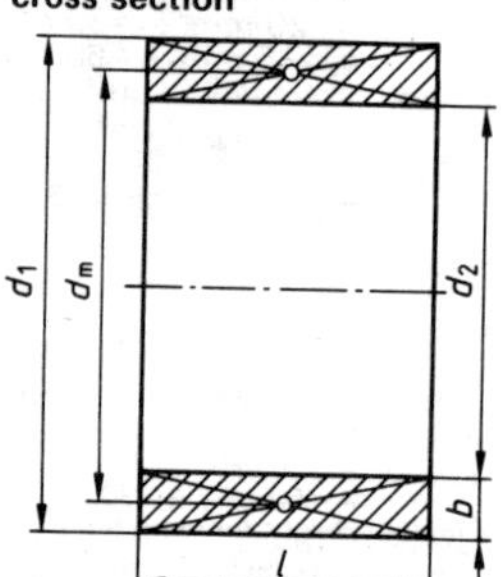

$$V = \pi \cdot d_m \cdot l \cdot b \qquad V = \pi \cdot d_m \cdot S$$

$$A_0 = 2 \cdot \pi \cdot d_m \cdot (l + b) \qquad A_0 = \pi \cdot d_m \cdot U$$

$$l = \frac{V}{\pi \cdot d_m \cdot b} \qquad S = \frac{V}{\pi \cdot d_m}$$

$$d_m = \frac{V}{\pi \cdot l \cdot b} \qquad d_m = \frac{V}{\pi \cdot S}$$

$$d_m = \frac{d_1 + d_2}{2} \qquad b = \frac{d_1 - d_2}{2}$$

A_o Surface area
l Cross section length
b Cross section width
S Cross sectional area
d_m Mean diameter
V Volume
d_1 Outer diameter
U Cross secion circumference
d_2 Inner diameter

Given: Bush as per the adjoining drawing,
To find: a) Volume
b) Surface area

$$b = \frac{80\ mm - 60\ mm}{2} = 10\ mm$$

$$d_m = \frac{80\ mm + 60\ mm}{2} = 70\ mm$$

$$V = \pi \cdot 7\ cm \cdot 5\ cm \cdot 1\ cm = 110\ cm^3$$

$$A_0 = 2 \cdot \pi \cdot 7\ cm \cdot (5\ cm + 1\ cm) = 263{,}9\ cm^2$$

Second solution

$$U = 2 \cdot (5\ cm + 1\ cm) = 12\ cm$$

$$A_0 = \pi \cdot 7\ cm \cdot 12\ cm = 263{,}9\ cm^3$$

Turned solid with cirular cross section

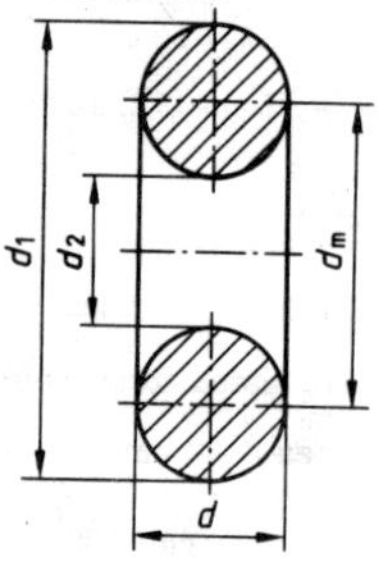

$$V = \frac{\pi^2 \cdot d_m \cdot d^2}{4} \qquad V = \pi \cdot d_m \cdot S$$

$$A_0 = \pi^2 \cdot d_m \cdot d \qquad A_0 = \pi \cdot d_m \cdot U$$

$$d = \sqrt{\frac{4 \cdot V}{\pi^2 \cdot d_m}} \qquad S = \frac{V}{\pi \cdot d_m}$$

$$d_m = \frac{4 \cdot V}{\pi^2 \cdot d^2} \qquad d_m = \frac{V}{\pi \cdot S}$$

$$d_m = \frac{d_1 + d_2}{2} \qquad S = \frac{\pi \cdot d^2}{4}$$

d Material diameter
A_o Surface area
d_m Mean diameter
S Cross sectional area
d_1 Outside diameter
U Cross section circumference
d_2 Inside diameter
V Volume

Given: Ring as per the adjoining illustration, d = 50 mm, d_m = 300 mm.
To find: a) Volume V in cm³
b) Surface area A_o in cm²

$$V = \frac{\pi^2 \cdot 30\ cm \cdot 5^2\ cm^2}{4} = 1\,850{,}6\ cm^3$$

or $S = \frac{\pi \cdot 5^2\ cm^2}{4} = 19{,}635\ cm^2$

$$V = \pi \cdot 30\ cm \cdot 19{,}6\ cm^2 = 1847{,}2\ cm^3$$

$$A_0 = \pi^2 \cdot 30\ cm \cdot 5\ cm = 1\,480{,}4\ cm^2$$

or $U = \pi \cdot 5\ cm = 15{,}71\ cm$

$$A_0 = \pi \cdot 30\ cm \cdot 15{,}7\ cm = 1479{,}7\ cm^2$$

Density

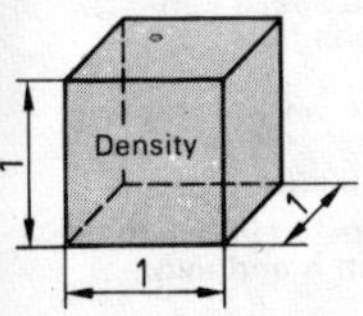

$\varrho = \frac{m}{V}$ $\varrho = \frac{G}{V \cdot g}$

g Acceleration due to gravity during a free fall all bodies experience a constant acceleration of due to the gravitational force.
G Gravitational force
m Mass
V Volume.
ϱ Density

Given: Steel shaft, $V = 1500\ cm^3$, $m = 11.79$ kg.
To find: Density ρ in kg/cm^3

$\varrho = 11{,}79\ kg : 1{,}5\ dm^3 = 7{,}86\ kg/dm^3$

Given: Oxygen, $V = 6000\ l$, $m = 8.58$ kg.
To find: Density ρ in kg/m^3

$\varrho = 8{,}58\ kg : 6{,}0\ m^3 = 1{,}43\ kg/m^3$

Mass

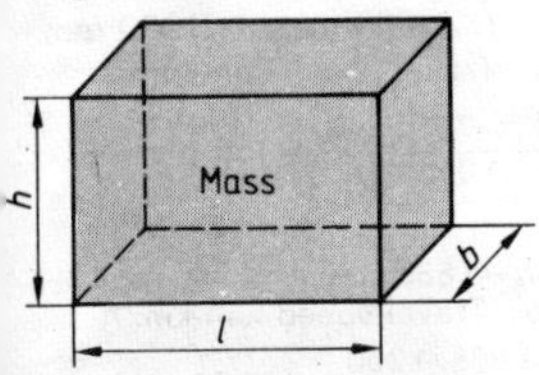

$m = V \cdot \varrho$ $m = \frac{G}{g}$

$V = \frac{m}{\varrho}$ $G = m \cdot g$

$\varrho = \frac{m}{V}$ $g = \frac{G}{m}$

g Acceleration due to gravity
G Gravitational force
m Mass
V Volume
ϱ Density

Given: Aluminium tube, $V = 8{,}0\ dm^3$, $\varrho = 2{,}7\ kg/dm^3$
To find: Mass m in kg.

$m = 8\ dm^3 \cdot 2{,}7\ kg/dm^3 = 21{,}60$ kg

Given: Steel ball, $m = 1{,}0$ kg, $\varrho = 7{,}86\ kg/dm^3$.
To find: Volume V in cm^3

$V = \frac{1\,000\ g}{7{,}86\ g/cm^3} = 127{,}22\ cm^3$

Gravitational force

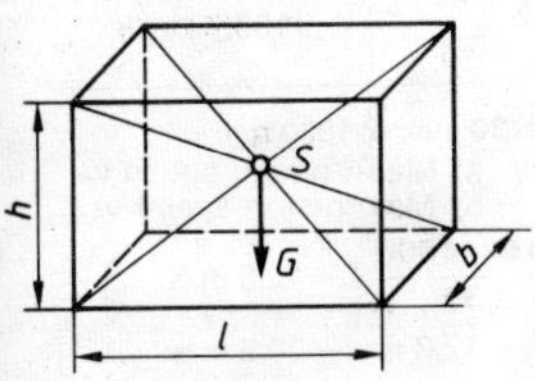

$G = m \cdot g$ $G = V \cdot \varrho \cdot g$

$m = \frac{G}{g}$ $V = \frac{G}{\varrho \cdot g}$

$g = \frac{G}{m}$ $\varrho = \frac{G}{V \cdot g}$

g Acceleration due to gravity
G Gravitational force
m Mass
V Volume
ϱ Density

Given: Passenger car, permissible total weight $m = 1200$ kg.
To find: Gravitational force G in N

$G = 1\,200\ kg \cdot 9{,}81\ m/s^2 = 11\,772$ N

Given: Light metal piston, Gravitational force
To find: Mass in g.

$m = \frac{5\,886\ mN}{9{,}81\ m/s^2} = 600{,}0$ g

Area based mass

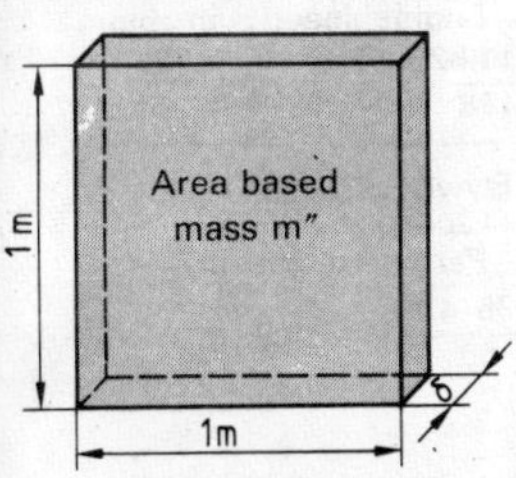

$m = m'' \cdot A$ $m = A \cdot \delta \cdot \varrho$

$A = \frac{m}{m''}$ $A = \frac{m}{\delta \cdot \varrho}$

$m'' = \frac{m}{A}$ $\delta = \frac{m}{A \cdot \varrho}$

$\varrho = \frac{m}{A \cdot \delta}$

A Area in m^2
m Mass in kg
m'' Area based mass in kg/m^2
δ Sheet thickness in mm
ϱ Density in kg/dm^3

Given: a) Steel sheet, $A = 4\ m^2$, $m = 12\ kg/m^2$.
b) Copper sheet, $A = 3\ m^2$, $\delta = 2$ mm, $\varrho = 8{,}93\ kg/dm^3$.
To find: Mass m in kg.
a) $m = 12\ kg/m^2 \cdot 4\ m^2$; $= 48{,}0$ kg
b) $m = 3 \cdot 2 \cdot 8{,}93 = 53{,}58$ kg

Given: Aluminium sheet, $m = 10.9$ kg. $m'' = 2{,}18\ kg/m^2$
To find: Area A in m^2

$A = \frac{10{,}9\ kg}{2{,}18\ kg/m^2} = 5{,}0\ m^2$

Length based mass

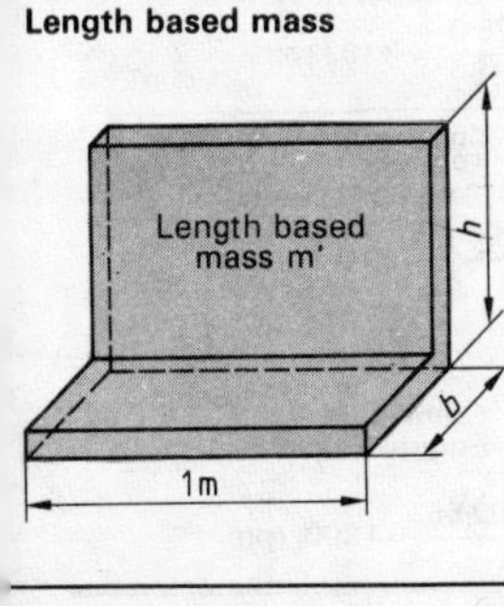

$m = m' \cdot l$ $m = \frac{S \cdot l \cdot \varrho}{1\,000}$

$l = \frac{m}{m'}$ $l = \frac{1\,000 \cdot m}{S \cdot \varrho}$

$m' = \frac{m}{l}$ $S = \frac{1\,000 \cdot m}{l \cdot \varrho}$

$\varrho = \frac{1\,000 \cdot m}{S \cdot l}$

l Length in m
m Mass in kg
m' Length based mass in kg/m
s Cross sectional area in mm^2
ϱ Density in kg/dm^3

Given: a) Steel angle, $l = 8.0$ m. $m = 2.1$ kg/m.
b) Copper tube, $S = 77.75\ mm^2$, $l = 7$ m, $\varrho = 8{,}93\ kg/dm^3$.
To find: Mass m in kg.
a) $m = 2{,}1\ kg/m \cdot 8{,}0\ m = 16{,}8$ kg
b) $m = \frac{77{,}75 \cdot 7 \cdot 8{,}93}{1\,000} = 4{,}86$ kg

Given: Steel wire, $m = 5.39$ kg. $m = 1.54$ kg/1000 m.
To find: Length l in mm

$l = \frac{5{,}39\ kg \cdot 1\,000\ m}{1{,}54\ kg} = 3\,500$ m

Speed

Average speed

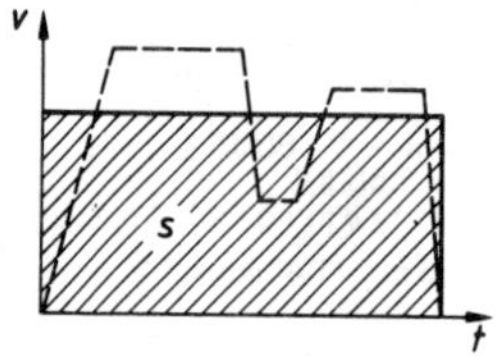

$v = \frac{s}{t}$

$t = \frac{s}{v}$ $\quad s = v \cdot t$

$1\ \text{km/h} = \frac{1}{3{,}6}\ \text{m/s}$ $\quad 1\ \text{m/s} = 3{,}6\ \text{km/h}$

$1\ \text{h} = 60\ \text{min}$ $\quad 1\ \text{min} = \frac{1}{60}\ \text{h}$

s Distance travelled
t Time, travel time
v Average speed

Given: s = 180 km, t = 2 h 30 min.
To find: Average speed v in km/h and m/s

$v = \frac{180\ \text{km}}{2{,}5\ \text{h}} = 72{,}0\ \text{km/h}$

$v = 72{,}0 : 3{,}6 = 20{,}0\ \text{m/s}$

Given: s = 72 km, v = 90 km/h
To find: Time t in h and min.

$t = \frac{72\ \text{km}}{90\ \text{km/h}} = 0{,}80\ \text{h}$

$t = 0{,}80\ \text{h} \cdot 60\ \text{min/h} = 48\ \text{min}$

Travel speed

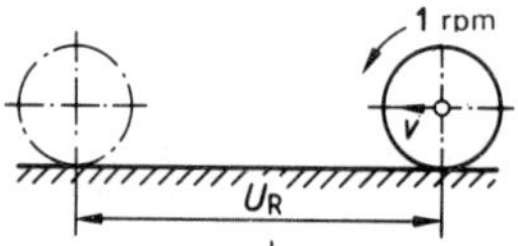

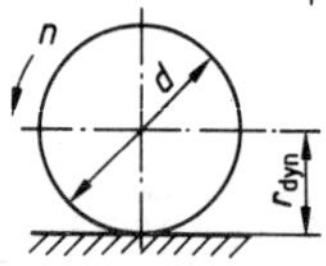

$v = \frac{120 \cdot \pi \cdot r_{dyn} \cdot n}{1\,000}$ $\quad v = \frac{60 \cdot U_R \cdot n}{1\,000}$

$n = \frac{1\,000 \cdot v}{120 \cdot \pi \cdot r_{dyn}}$ $\quad n = \frac{1\,000 \cdot v}{60 \cdot U_R}$

$r_{dyn} = \frac{1\,000 \cdot v}{120 \cdot \pi \cdot n}$ $\quad U_R = \frac{1\,000 \cdot v}{60 \cdot n}$

$v = \frac{60 \cdot \pi \cdot d \cdot n}{1\,000}$ $\quad n = \frac{1\,000 \cdot v}{60 \cdot \pi \cdot d}$

d Outside diameter of tyre in m
n Wheel speed in rpm
r_{dyn} Dynamic tyre radius in m
U_R Circumference of tyre in m
v Travel speed in km/h

Given: r_{dyn} = 300 mm, n = 1200 rpm
To find: Travel speed v in km/h

$v = \frac{120 \cdot \pi \cdot 0{,}3 \cdot 1\,200}{1\,000} = 135{,}7\ \text{km/h}$

Given: U_R = 1 885 mm, n = 1 200 rpm
To find: Travel speed v in km/h

$v = \frac{60 \cdot 1{,}885 \cdot 1\,200}{1\,000} = 135{,}7\ \text{km/h}$

Given: d = 600 mm, n = 1 200 rpm
To find: Travel speed v in km/h

$v = \frac{60 \cdot \pi \cdot 0{,}6 \cdot 1\,200}{1\,000} = 135{,}7\ \text{km/h}$

Mean Piston speed

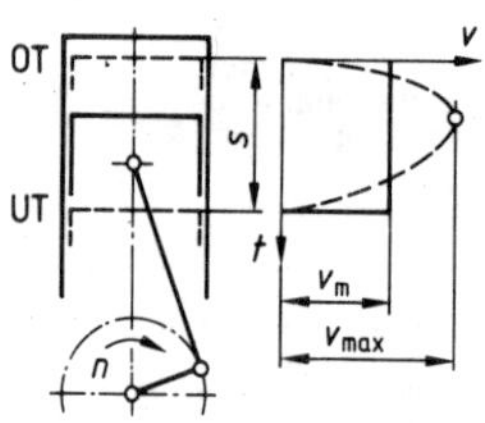

$v_m = \frac{s \cdot n}{30}$

$s = \frac{30 \cdot v_m}{n}$ $\quad n = \frac{30 \cdot v_m}{s}$

$v_{max} \approx 1{,}6 \cdot v_m$

n Engine speed in rpm
s Piston stroke in m
v_m Mean piston speed in m/s
v_{max} Max. piston speed in m/s

Given: s = 80 mm, n = 4800 rpm
To find: a) Mean piston speed v_m
b) Max. piston speed v_{max}

$v_m = \frac{0{,}08 \cdot 4800}{30} = 12{,}8\ \text{m/s}$

$v_{max} \approx 1{,}6 \cdot 12{,}8\ \text{m/s} = 20{,}5\ \text{m/s}$

Given: s = 128 mm, v_m = 11,52 m/s.
To find: Engine speed n in rpm

$n = \frac{30 \cdot 11{,}52}{0{,}128} = 2\,700\ \text{rpm}$

Peripheral speed

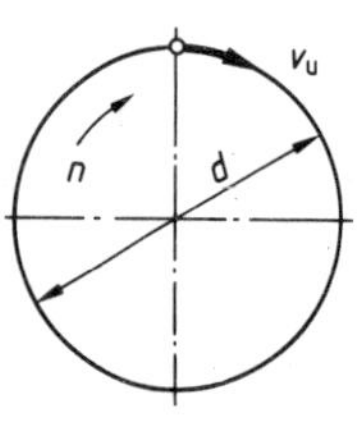

$v_u = \frac{\pi \cdot d \cdot n}{60}$

$n = \frac{60 \cdot v_u}{\pi \cdot d}$ $\quad d = \frac{60 \cdot v_u}{\pi \cdot n}$

d Diameter in m
n Speed in 1/min
v_u Peripheral speed in m/s

Given: Flywheel, d = 350 mm, n = 4 200 rpm
To find: Peripheral speed v_u

$v_u = \frac{\pi \cdot 0{,}35 \cdot 4\,200}{60} = 76{,}97\ \text{m/s}$

Given: Grinding wheel, d = 150 mm, v_u = 25 m/s.
To find: Speed n in rpm

$n = \frac{60 \cdot 25{,}0}{\pi \cdot 0{,}150} = 3\,183\ \text{rpm}$

Angular speed

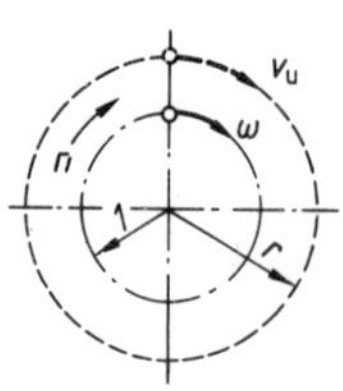

$\omega = \frac{\pi \cdot n}{30}$ $\quad n = \frac{30 \cdot \omega}{\pi}$

$\omega = \frac{v_u}{r}$ $\quad v_u = \omega \cdot r$

n Speed in rpm
r Radius in m
v_u Peripheral speed in m/s
ω Angular speed in 1/s

Given: Connecting rod, n = 4200 min, rpm
To find: Angular speed ω

$\omega = \frac{\pi \cdot 4\,200}{30} = 439{,}8\ \text{1/s}$

Given: Crankshaft, ω = 502.66 rpm
To find: Engine speed n in

$n = \frac{30 \cdot 502{,}66}{\pi} = 4800\ \text{rpm}$

Accelerating from the state of braking to stand-still

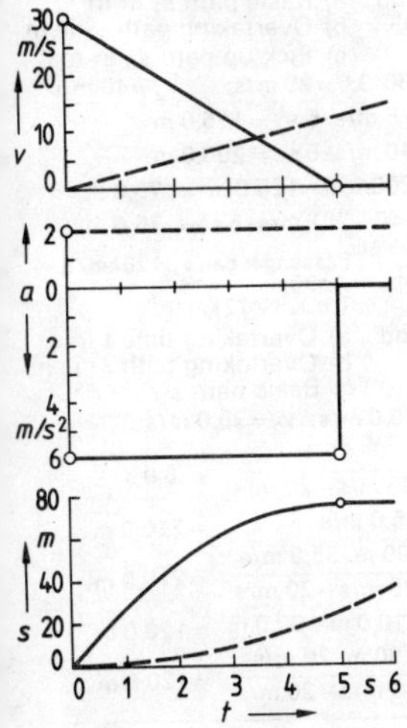

$$a = \frac{v}{t} \qquad a = \frac{v^2}{2 \cdot s} \qquad a = \frac{2 \cdot s}{t^2}$$

$$s = \frac{v \cdot t}{2} \qquad s = \frac{v^2}{2 \cdot a} \qquad s = \frac{a \cdot t^2}{2}$$

$$s_H = s_R + s \qquad s_R = v \cdot t_R \qquad s = s_H - s_R$$

$$t = \frac{v}{a} \qquad t = \frac{2 \cdot s}{v} \qquad t = \sqrt{\frac{2 \cdot s}{a}}$$

$$v = a \cdot t \qquad v = \frac{2 \cdot s}{t} \qquad v = \sqrt{2 \cdot a \cdot s}$$

- a Acceleration, deceleration in m/s²
- s Acceleration path, braking path in mm
- s_H Stopping path
- s_R Reaction path in m
- t Acceleration time, braking time in s
- t_R Reaction time in s
- v Speed after acceleration or before braking in m/s

Given: Passenger car, v=90 km/h (25 m/s). t=5 s, t_R=0,8 s.

To find: a) Deceleration a in m/s² b) Braking path s in m c) Stopping path s_H in m

$$a = 25\ \text{m/s} : 5\ \text{s} = 5{,}0\ \text{m/s}^2$$

$$s = \frac{25\ \text{m/s} \cdot 5\ \text{s}}{2} = 62{,}5\ \text{m}$$

$$s_H = 25\ \text{m/s} \cdot 0{,}8\ \text{s} + 62{,}5\ \text{m} = 82{,}5\ \text{m}$$

Given: Passenger car, s = 225 m, v = 108 km/h

To find: a) Acceleration a in m/s² b) Acceleration time t in

$$v = 108 : 3{,}6 = 30{,}0\ \text{m/s}$$

$$a = \frac{30^2\ \text{m}^2/\text{s}^2}{2 \cdot 225\ \text{m}} = 2{,}0\ \text{m/s}^2$$

$$t = \frac{2 \cdot 225\ \text{m}}{30\ \text{m/s}} = 15{,}0\ \text{s}$$

Given: Truck, t = 4 s, s = 40 m

To find: a) Brake deceleration a in m/s² b) Speed v in m/s

$$a = \frac{2 \cdot 40\ \text{m}}{4^2\ \text{s}^2} = 5{,}0\ \text{m/s}^2$$

$$v = \frac{2 \cdot 40\ \text{m}}{4\ \text{s}} = 20{,}0\ \text{m/s}$$

Accelerating and braking in movement

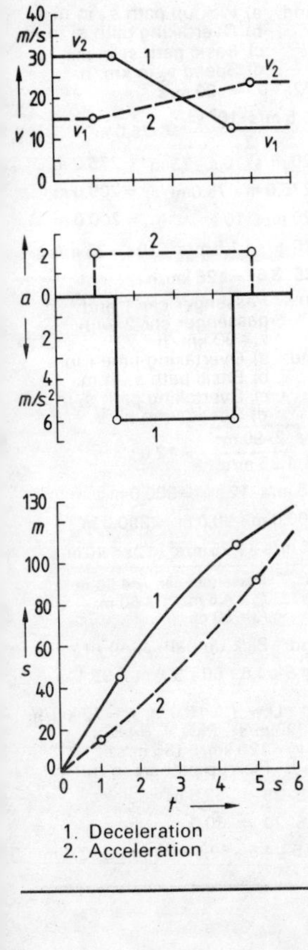

1. Deceleration
2. Acceleration

$$a = \frac{v_2 - v_1}{t} \qquad a = \frac{2 \cdot s - 2 \cdot v_1 \cdot t}{t^2}$$

$$a = \frac{v_2^2 - v_1^2}{2 \cdot s} \qquad a = \frac{2 \cdot v_2 \cdot t - 2 \cdot s}{t^2}$$

$$s = \frac{(v_1 + v_2) \cdot t}{2} \qquad s = v_1 \cdot t + \frac{a \cdot t^2}{2}$$

$$s = \frac{v_2^2 - v_1^2}{2 \cdot a} \qquad s = v_2 \cdot t - \frac{a \cdot t^2}{2}$$

$$t = \frac{v_2 - v_1}{a} \qquad t = \frac{\sqrt{v_1^2 + 2 \cdot a \cdot s} - v_1}{a}$$

$$t = \frac{2 \cdot s}{v_1 + v_2} \qquad t = \frac{v_2 - \sqrt{v_2^2 - 2 \cdot a \cdot s}}{a}$$

$$v_1 = v_2 - a \cdot t \qquad v_1 = \frac{2 \cdot s - a \cdot t^2}{2 \cdot t}$$

$$v_1 = \frac{2 \cdot s - v_2 \cdot t}{t} \qquad v_1 = \sqrt{v_2^2 - 2 \cdot a \cdot s}$$

$$v_2 = v_1 + a \cdot t \qquad v_2 = \frac{2 \cdot s + a \cdot t^2}{2 \cdot t}$$

$$v_2 = \frac{2 \cdot s - v_1 \cdot t}{t} \qquad v_2 = \sqrt{v_1^2 + 2 \cdot a \cdot s}$$

- a Acceleration, deceleration in m/s²
- s Acceleration path, braking path in m
- t Acceleration time, braking time in s
- v_1 Speed before acceleration or after braking in m/s
- v_2 Speed after acceleration or before braking in m/s

Given: Passenger car, v_1 = 54 km/h, v_2 = 90 km/h, t = 5,0 s.

To find: a) Acceleration a in m/s² b) Acceleration path s in m

$$v_1 = 54 : 3{,}6 = 15\ \text{m/s}; \quad v_2 = 25\ \text{m/s}$$

$$a = \frac{25\ \text{m/s} - 15\ \text{m/s}}{5{,}0\ \text{s}} = 2{,}0\ \text{m/s}^2$$

$$s = \frac{(15 + 25)\ \text{m/s} \cdot 5\ \text{s}}{2} = 100{,}0\ \text{m}$$

Given: Passenger car, a = 6 m/s², v_1=72 km/h, v_2=126 km/h,

To find: a) Braking path s in m. b) Braking time t in s.

$$v_1 = 72 : 3{,}6 = 20\ \text{m/s}; \quad v_2 = 35\ \text{m/s}$$

$$s = \frac{35^2\ \text{m}^2/\text{s}^2 - 20^2\ \text{m}^2/\text{s}^2}{2 \cdot 6\ \text{m/s}^2} = 68{,}75\ \text{m}$$

$$t = \frac{35\ \text{m/s} - 20\ \text{m/s}}{6\ \text{m/s}^2} = 2{,}50\ \text{s}$$

Given: Truck v_1 = 54 km/h, a = 0,7 m/s², t = 10 s.

To find: a) Speed v_2 in km/h b) Acceleration path s in m

$$v_1 = 54 : 3{,}6 = 15\ \text{m/s}$$

$$v_2 = 15\ \text{m/s} + 0{,}7\ \text{m/s}^2 \cdot 10\ \text{s} = 22\ \text{m/s}$$

$$v_2 = 22 \cdot 3{,}6 = 79{,}2\ \text{km/h}$$

$$s = 15\ \text{m/s} \cdot 10\ \text{s} + \frac{0{,}7\ \text{m/s}^2 \cdot 10^2\ \text{s}^2}{2}$$

$$= 150\ \text{m} + 35\ \text{m} = 185{,}0\ \text{m}$$

Given: Truck, v_1 = 72 km/h (20 m/s), v_2 = 18 km/h (5 m/s), s = 37,5 m.

To find: a) Braking deceleration a in m/s² b) Braking time t in s

$$a = \frac{20^2\ \text{m}^2/\text{s}^2 - 5^2\ \text{m}^2/\text{s}^2}{2 \cdot 37{,}5\ \text{m}} = 5{,}0\ \text{m/s}^2$$

$$t = \frac{2 \cdot 37{,}5\ \text{m}}{20\ \text{m/s} + 5\ \text{m/s}} = 3{,}0\ \text{s}$$

Overtaking

Overtaking with constant speed

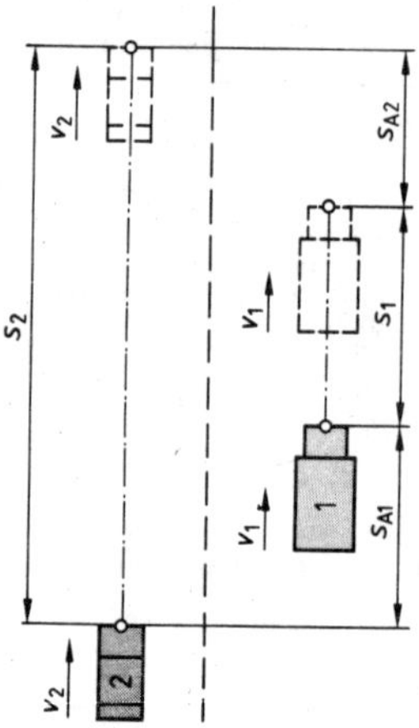

$s_2 = s_1 + s_A \qquad s_2 = v_2 \cdot t \qquad s_2 = \dfrac{s_A \cdot v_2}{v_2 - v_1}$

$s_1 = s_2 - s_A \qquad s_1 = v_1 \cdot t \qquad s_1 = \dfrac{s_A \cdot v_1}{v_2 - v_1}$

$s_A = s_2 - s_1 \qquad s_A = (v_2 - v_1) \cdot t$

$s_A = \dfrac{(v_2 - v_1) \cdot s_1}{v_1} \qquad s_A = \dfrac{(v_2 - v_1) \cdot s_2}{v_2}$

$t = \dfrac{s_1}{v_1} \qquad t = \dfrac{s_2}{v_2} \qquad t = \dfrac{s_A}{v_2 - v_1}$

- s_1 Basic path of the slower vehicle in m
- s_2 Overtaking path of the faster vehicle in m
- s_A Pick up path of the faster vehicle in m
- t Overtaking time
- v_1 Constant speed of slower vehicle in m/s
- v_2 Constant speed of the faster vehicle in m/s

Given: Truck, v_1=90 km/h, t=5 s. Passenger car v_2= 144 km/h.

To find: a) Basic path s_1 in m
b) Overtaking path s_2 in m
c) Pick up path s_A in m

$v_1 = 90 : 3{,}6 = 25$ m/s; $\quad v_2 = 40$ m/s

$s_1 = 25\ \text{m/s} \cdot 5\ \text{s} = 125{,}0$ m

$s_2 = 40\ \text{m/s} \cdot 5\ \text{s} = 200{,}0$ m

$s_A = 200{,}0\ \text{m} - 125{,}0\ \text{m} = 75{,}0$ m

$s_A = (40 - 25)\ \text{m/s} \cdot 5\ \text{s} = 75{,}0$ m

Given: Passenger car, v_2=126 km/h, s_A=90m; Truck, v_1=72 km/h.

To find: a) Overtaking time t in s
b) Overtaking path s_2 in m
c) Basic path s_1

$v_1 = 20{,}0$ m/s; $v_2 = 35{,}0$ m/s

$t = \dfrac{90\ \text{m}}{35\ \text{m/s} - 20\ \text{m/s}} = 6{,}0$ s

$s_2 = 35{,}0\ \text{m/s} \cdot 6{,}0\ \text{s} = 210{,}0$ m

$s_2 = \dfrac{90\ \text{m} \cdot 35{,}0\ \text{m/s}}{35\ \text{m/s} - 20\ \text{m/s}} = 210{,}0$ m

$s_1 = 210{,}0\ \text{m} - 90{,}0\ \text{m} = 120{,}0$ m

$s_1 = \dfrac{90\ \text{m} \cdot 20\ \text{m/s}}{35\ \text{m/s} - 20\ \text{m/s}} = 120{,}0$ m

Overtaking with constant acceleration

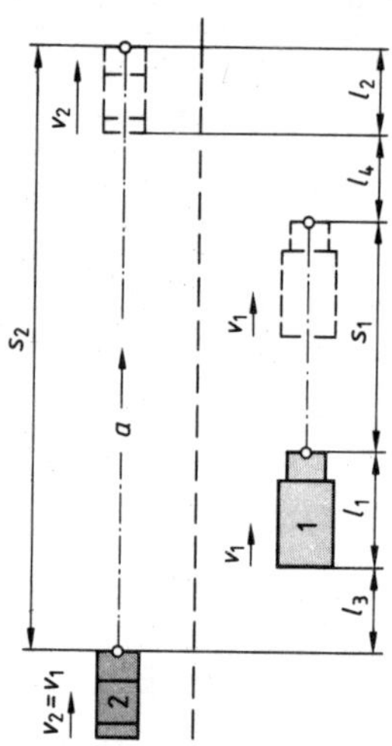

$s_2 = s_1 + s_A \qquad s_2 = v_1 \cdot t + s_A$

$s_1 = s_2 - s_A \qquad s_1 = v_1 \cdot t$

$s_A = s_2 - s_1 \qquad s_A = \dfrac{(v_2 - v_1)^2}{2 \cdot a}$

$s_A = \dfrac{a \cdot t^2}{2} \qquad s_A = \dfrac{(v_2 - v_1) \cdot t}{2}$

$t = \dfrac{v_2 - v_1}{a} \qquad t = \dfrac{2 \cdot s_A}{v_2 - v_1} \qquad t = \sqrt{\dfrac{2 \cdot s_A}{a}}$

- a Constant acceleration in m/s²
- s_1 Basic path of vehicle 1 in m
- s_2 Overtaking path of vehicle 2 in m
- s_A Pick up path of vehicle 2 in m
- t Overtaking time in s
- v_1 Constant speed of vehicle 1 or speed of vehicle 2 before overtaking in m/s
- v_2 Speed of vehicle 2 after overtaking in m/s

See page 41 for other formulas

Given: Truck and Passenger car with v_1 = 72 km/h
Passenger car a = 1.5 m/s², t = 10 s

To find: a) Pick up path s_A in m
b) Overtaking path s_2 in m
c) Basic path s_1 in m
d) Speed v_2 in km/h

$v_1 = 72 : 3{,}6 = 20$ m/s

$s_A = \dfrac{1{,}5\ \text{m/s} \cdot 10^2\ \text{s}^2}{2} = 75{,}0$ m

$s_2 = 20\ \text{m/s} \cdot 10\ \text{s} + 75\ \text{m} = 275{,}0$ m

$s_1 = 275{,}0\ \text{m} - 75{,}0\ \text{m} = 200{,}0$ m

$s_1 = 20\ \text{m/s} \cdot 10\ \text{s} = 200{,}0$ m

$v_2 = 20\ \text{m/s} + 1{,}5\ \text{m/s}^2 \cdot 10\ \text{s} = 35$ m/s

$v_2 = 35 \cdot 3{,}6 = 126$ km/h

Given: Passenger car 1 and passenger car 2 with v_1 = 90 km/h

To find: a) Overtaking time t in s
b) Basic path s_1 in m
c) Overtaking path s_2 in m
d) Speed v_2 in m/s

$t = \sqrt{\dfrac{2 \cdot 90\ \text{m}}{1{,}25\ \text{m/s}^2}} = 12{,}0$ s

$s_1 = 25\ \text{m/s} \cdot 12\ \text{s} = 300{,}0$ m

$s_2 = 300{,}0\ \text{m} + 90{,}0\ \text{m} = 390{,}0$ m

$v_2 = 25\ \text{m/s} + 1.25\ \text{m/s}^2 \cdot 12\ \text{s} = 40$ m/s

Pick up path, safety distance

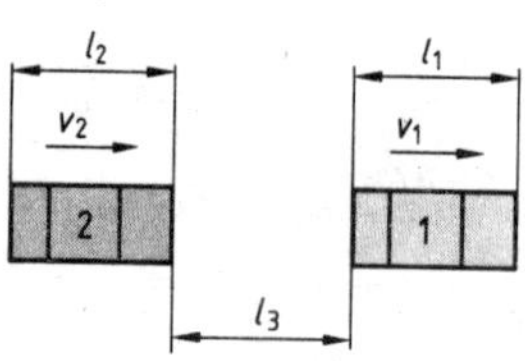

$s_A = l_1 + l_2 + l_3 + l_4$

$l_3 \approx 1{,}5 \cdot v_2 \qquad l_4 \approx 1{,}5 \cdot v_1$

- l_1 Length of vehicle 1
- l_2 Length of vehicle 2
- l_3 Safety distance for vehicle 2
- l_4 Safety distance for vehicle 1 in m
- s_A Pick up path in m
- v_1 Speed, vehicle 1 in m/s
- v_2 Speed, vehicle 2 in m/s

Given: Passenger car, l_1=4.50 m; l_2 = 4.6 m; l_3 = 60 m; l_4 = 30 m.

To find: Pick up path s_A in m

$s_A = (4{,}5 + 4{,}6 + 60 + 30)\ \text{m} = 99{,}1$ m

Given : Lkw, $l_1 = 15{,}0$ m, $v_1 = 72$ km/h (20 m/s); Pkw, $l_2 = 4{,}5$ m, $v_2 = 126$ km/h (35 m/s).

To find: Pick up path s_A in m

$l_3 \approx 1{,}5 \cdot 35 = 52{,}5$ m

$l_4 \approx 1{,}5 \cdot 20 = 30{,}0$ m

$s_A = (15 + 4{,}5 + 52{,}5 + 30)\text{m} = 102{,}0$ m

Scale of forces

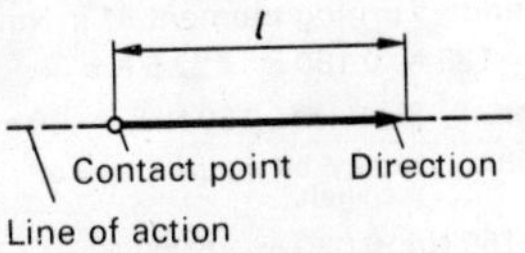

Scale = $\dfrac{\text{Picture size}}{\text{Matter size}}$ $\qquad \mu = \dfrac{l}{F}$

$l = F \cdot \mu$ $\qquad F = \dfrac{l}{\mu}$

F Magnitude of force, matter size
l Drawing length, picture size
μ Scale of forces, e.g. 1 mm ≙ 100 N
or $\mu = \dfrac{1\ \text{mm}}{100\ \text{N}}$ or $\mu = 0{,}01\ \dfrac{\text{mm}}{\text{N}}$

Given: $F = 6000$ N, $\mu = \dfrac{1\ \text{mm}}{100\ \text{N}}$
To find: Drawing length l in mm

$l = 6000\ \text{N} \cdot \dfrac{1\ \text{mm}}{100\ \text{N}} = 60\ \text{mm}$

or

$l = 6000\ \text{N} \cdot 0{,}01\ \dfrac{\text{mm}}{\text{N}} = 60\ \text{mm}$

Forces in the same line of action

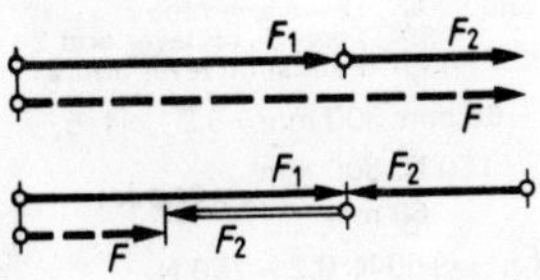

in the same direction $\qquad F = F_1 + F_2 + F_3$

in the opposite direction $\qquad F = F_1 - F_2$

F Resultant force, total force
F_1 Individual force, partial force
F_2 Individual force, partial force
F_3 Individual force, partial force

Given: $F = 2000$ N and $F_2 = 1200$ acting in the same direction
To find: Total force F in N
$F = 2000\ \text{N} + 1200\ \text{N} = 3200\ \text{N}$

Given: $F_1 = 2000$ N, $F_2 = 1200$ N acting in opposite directions
To find: Total force F in N
$F = 2000\ \text{N} - 1200\ \text{N} = 800\ \text{N}$

Forces in different lines of action

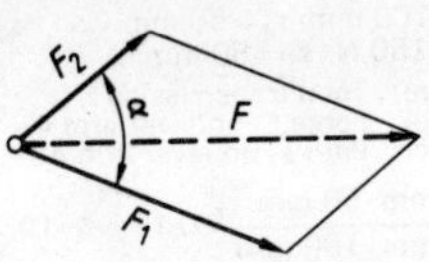

Combination of two forces

$F = \sqrt{F_1^2 + F_2^2 + 2 \cdot F_1 \cdot F_2 \cdot \cos\alpha}$

Resultant force ≙
Diagonal of the parallelogram of forces

Graphical solution:
1. Select the scale of forces
2. Draw the individual forces to scale as per magnitude and direction
3. Build a parallelogram of forces
4. The diagonal of parallelogram gives the magnitude and direction of resultant force

Given: $F_1 = 2000$ N, $F_2 = 1200$ N, $\alpha = 60°$, $\mu = 1$ mm/100 N.
To find: a) Resultant force F graphically
b) Resultant force F analytically

$l_1 = 2000\ \text{N} \cdot \dfrac{1\ \text{mm}}{100\ \text{N}} = 20\ \text{mm}$

$l_2 = 1200\ \text{N} \cdot \dfrac{1\ \text{mm}}{100\ \text{N}} = 12\ \text{mm}$

$F = 28\ \text{mm} : \dfrac{1\ \text{mm}}{100\ \text{N}} = 2800\ \text{N}$

$F = \sqrt{2^2 + 1{,}2^2 + 2 \cdot 2 \cdot 1{,}2 \cdot \cos 60°}$
$F = \sqrt{7{,}84} = 2{,}8\ \text{kN} = 2800\ \text{N}$

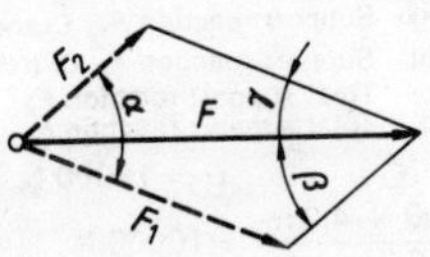

Resolution of a force

$F : F_1 : F_2 = \sin\alpha : \sin\beta : \sin\gamma$

Individual force ≙
Sides of the parallelogram of forces

Graphical solution:

1. Select the scale of forces
2. Draw the middle force to scale as per magnitude and direction
3. Draw the lines of application of individual forces through the starting point of the resultant force
1. The parallels to the lines of application through the end point of the middle force give the side forces.

Given: $F = 2800$ N, $\alpha = 60°$, $\beta = 38°\ 12'$, $\mu = 1$ mm/100 N.
To find: a) Individual forces graphically
b) Individual forces analytically

$l_1 = 2800\ \text{N} \cdot \dfrac{1\ \text{mm}}{100\ \text{N}} = 28\ \text{mm}$

$F_1 = 20\ \text{mm} : \dfrac{1\ \text{mm}}{100\ \text{N}} = 2000\ \text{N}$

$F_2 = 12\ \text{mm} : \dfrac{1\ \text{mm}}{100\ \text{N}} = 1200\ \text{N}$

$F_1 = \dfrac{2800\ \text{N} \cdot \sin 38°\ 12'}{\sin 60°} = 1996\ \text{N}$

$F_2 = \dfrac{2800\ \text{N} \cdot \sin 21°\ 48'}{\sin 60°} = 1184\ \text{N}$

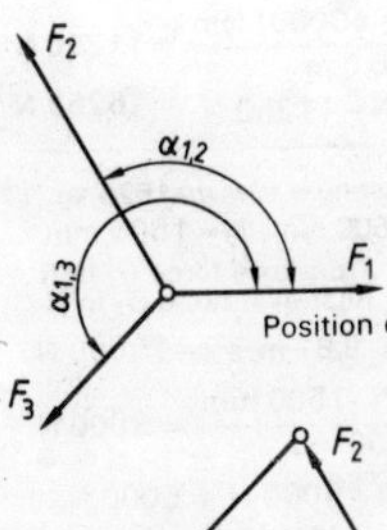

Position diagram

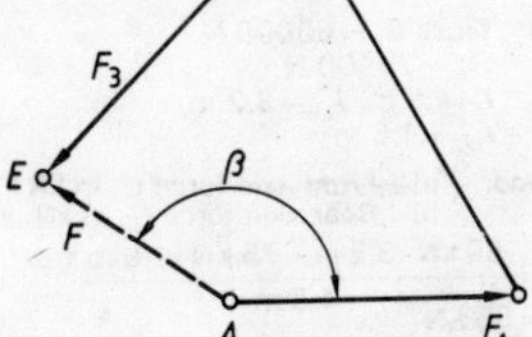

Force diagram

Combination of several forces

Resultant forces ≙
connecting line from the starting point of the first to the end point of the last force

Graphical solution:
1. Select the scale of forces
2. Arrange the individual forces to scale as per magnitude and direction on the force diagram in any desired sequence
3. The connecting line from the starting point of the first to the end point of the last force gives the resultant force

F Resultant force, Middle force
F_1, F_2, F_3 Individual forces
α, β Angles of applied forces

Given: $F_1 = 2000$ N, $F_2 = 3000$ N, $F_3 = 2500$ N, $\alpha_{1,2} = 120°$, $\alpha_{1,3} = 225°$, $\mu = 1$ mm/100 N.
To find: a) Magnitude of resultant force F
b) Angles β of resultant force

$l_1 = 2000\ \text{N} \cdot \dfrac{1\ \text{mm}}{100\ \text{N}} = 20\ \text{mm}$

$l_2 = 3000\ \text{N} \cdot \dfrac{1\ \text{mm}}{100\ \text{N}} = 30\ \text{mm}$

$l_3 = 2500\ \text{N} \cdot \dfrac{1\ \text{mm}}{100\ \text{N}} = 25\ \text{mm}$

$F = 15{,}25\ \text{mm} : \dfrac{1\ \text{mm}}{100\ \text{N}} = 1525\ \text{N}$

Angle $\beta = 147°$

Turning moment, Lever, Support reactions, Axle forces

Turning moment

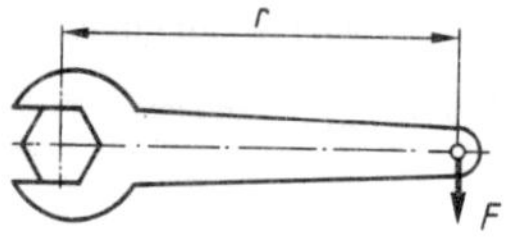

$M = F \cdot r$

$F = \frac{M}{r}$ $\quad r = \frac{M}{F}$

F Force on lever arm
M Turning moment, moment of a force
r Radius, lever arm

Given: Spanner, $r = 180$ mm, $F = 125$ N.
To find: Turning moment M in Nm.
$M = 125 \text{ N} \cdot 0{,}180 \text{ m} = 22{,}5 \text{ Nm}$

Given: Shaft, $M = 150$ Nm, $r = 30$ mm
To find: Force F on the periphery of the shaft.
$F = 150 \text{ Nm} : 0{,}030 \text{ m} = 5000 \text{ N}$

Lever

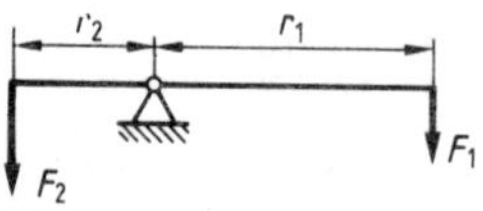

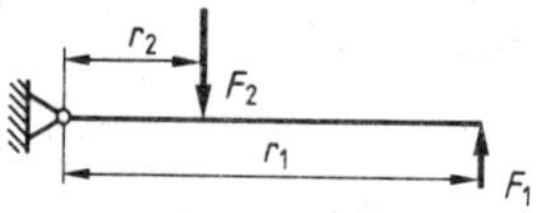

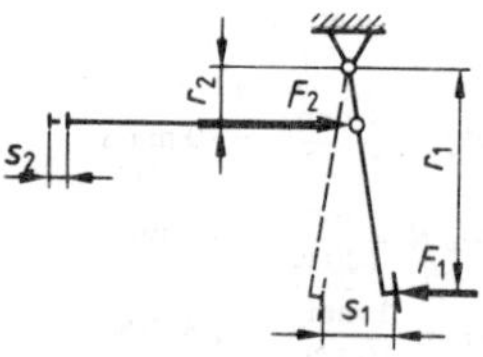

$F_1 \cdot r_1 = F_2 \cdot r_2$ $\quad i = \frac{F_1}{F_2} = \frac{r_2}{r_1} = \frac{s_2}{s_1}$

$F_1 = \frac{F_2 \cdot r_2}{r_1}$ $\quad F_1 = F_2 \cdot i$

$F_2 = \frac{F_1 \cdot r_1}{r_2}$ $\quad F_2 = \frac{F_1}{i}$

$r_1 = \frac{F_2 \cdot r_2}{F_1}$ $\quad s_1 = \frac{s_2}{i}$

$r_2 = \frac{F_1 \cdot r_1}{F_2}$ $\quad s_2 = s_1 \cdot i$

$i = i_1 \cdot i_2 \cdot i_3 \dots$ $\quad i = \frac{r_2 \cdot r_4 \cdot r_6}{r_1 \cdot r_3 \cdot r_5}$

F_1, F_2 Forces on lever
r_1, r_2 Lever arms
s_1, s_2 Force paths
i Force or path transmission lever arm ratio
i_1, i_2 Individual transmissions

Given: Lever, $r_1 = 300$ mm, $r_2 = 60$ mm, $F_1 = 150$ N, $s_1 = 50$ mm.
To find: a) Lever arm ratio i
b) Force F_2 on lever arm 2
c) Path s_2 on lever arm 2

$i = 60 \text{ mm} : 300 \text{ mm} = 0{,}2 = 1:5$

$F_2 = \frac{150 \text{ N} \cdot 300 \text{ mm}}{60 \text{ mm}} = 750 \text{ N}$

or $F_2 = 150 \text{ N} : 0{,}2 = 750 \text{ N}$

$s_2 = 50 \text{ mm} \cdot 0{,}2 = 10 \text{ mm}$

Given: Double lever transmission $r_1 = 300$ mm, $r_2 = 60$ mm, $r_3 = 100$ mm, $r_4 = 50$ mm, $F_1 = 150$ N, $s_1 = 50$ mm.
To find: a) Total transmission i
b) Force F_2 on lever arm 4
c) Path s_2 on lever arm 4

$i = \frac{60 \text{ mm} \cdot 50 \text{ mm}}{300 \text{ mm} \cdot 100 \text{ mm}} = 0{,}1 = 1:10$

$F_2 = 150 \text{ N} : 0{,}1 = 1500 \text{ N}$

$s_2 = 50 \text{ mm} \cdot 0{,}1 = 5{,}0 \text{ mm}$

Support reactions

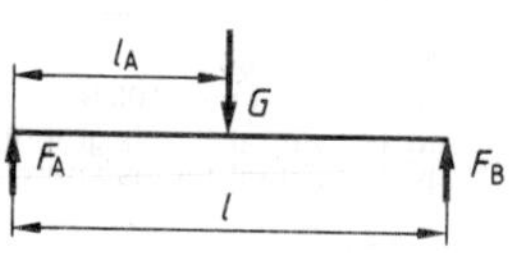

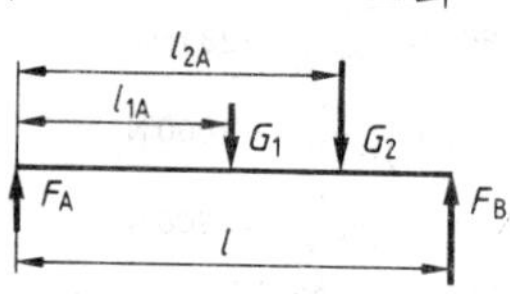

$F_A = \frac{G \cdot l_B}{l}$ $\quad F_B = G - F_A$

$F_B = \frac{G \cdot l_A}{l}$ $\quad F_A = G - F_B$

$F_A = \frac{G_1 \cdot l_{1B} + G_2 \cdot l_{2B}}{l}$ $\quad F_B = G_1 + G_2 - F_A$

$F_B = \frac{G_1 \cdot l_{1A} + G_2 \cdot l_{2A}}{l}$ $\quad F_A = G_1 + G_2 - F_B$

F_A, F_B Support reactions
G, G_1, G_2 Gravitational forces
l Distance between supports
l_A, l_B Distances between centres of gravity

Given: Carrier with crane, $l = 6{,}0$ m; Carrier $m_1 = 254{,}9$ kg, Crane $m_2 = 1529$ kg, $l_{1B} = 3{,}0$ m, $l_{2B} = 4{,}0$ m.
To find: a) Support reaction F_{A2} Crane.
b) Support reaction F_{B2} Crane.
c) Total support reaction F_A.
d) Total support reaction F_B

$G_1 = 2500 \text{ N}$ $\quad G_2 = 15000 \text{ N}$

$F_{A2} = \frac{15000 \text{ N} \cdot 4{,}0 \text{ m}}{6{,}0 \text{ m}} = 10000 \text{ N}$

$F_{B2} = 15000 \text{ N} - 10000 \text{ N} = 5000 \text{ N}$

$F_A = \frac{(7500 + 60000) \text{ Nm}}{6{,}0 \text{ m}} = 11250 \text{ N}$

$F_B = 17500 \text{ N} - 11250 \text{ N} = 6250 \text{ N}$

Axle Forces

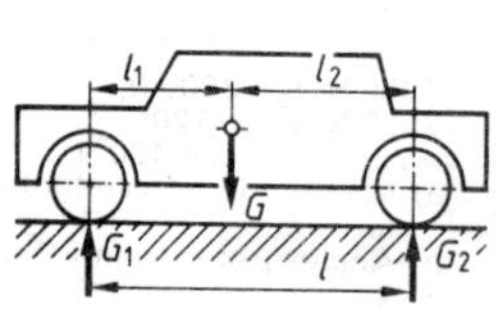

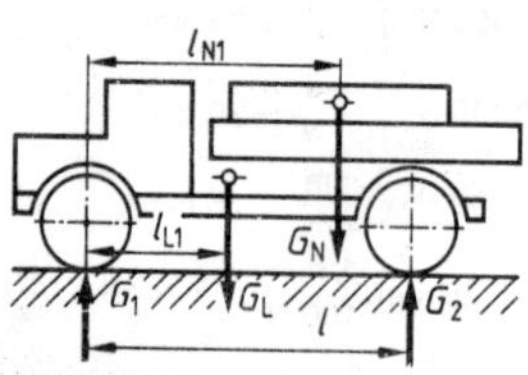

$G_1 = \frac{G \cdot l_2}{l}$ $\quad G_2 = G - G_1$

$G_2 = \frac{G \cdot l_1}{l}$ $\quad G_1 = G - G_2$

$G_1 = \frac{G_L \cdot l_{L2} + G_N \cdot l_{N2}}{l}$ $\quad G_2 = G_L + G_N - G_1$

$G_2 = \frac{G_L \cdot l_{L1} + G_N \cdot l_{N1}}{l}$ $\quad G_1 = G_L + G_N - G_2$

G Gravitational force of total weight
G_L Gravitational force of dead weight
G_N Gravitational force of pay load
G_1 Front axle force
G_2 Rear axle force
l Wheel base of vehicle
l_1 Centre of gravity distance of front axle
l_2 Centre of gravity distance of rear axle

Given: Passenger car, $m = 1529$ kg, $l = 2500$ mm, $l_2 = 1500$ mm.
To find: a) Front axle force G_1 in N
b) Rear axle force G_2 in N

$G = 1529 \text{ kg} \cdot 9{,}81 \text{ m/s}^2 = 15000 \text{ N}$

$G_1 = \frac{15000 \text{ N} \cdot 1500 \text{ mm}}{2500 \text{ mm}} = 9000 \text{ N}$

$G_2 = 15000 \text{ N} - 9000 \text{ N} = 6000 \text{ N}$

Given: Truck $G_L = 60000$ N, $G_N = 75000$ N, $l = 4{,}8$ m, $l_{L2} = 3{,}2$ m, $l_{N2} = 1{,}6$ m.
To find: a) Front axle force G_1 in kN
b) Rear axle force G_2 in kN

$G_1 = \frac{60 \text{ kN} \cdot 3{,}2 \text{ m} + 75 \text{ kN} \cdot 1{,}6 \text{ m}}{4{,}8 \text{ m}} = 65 \text{ kN}$

$G_2 = (60 + 75 - 65) \text{ kN} = 70 \text{ kN}$

Belt drive

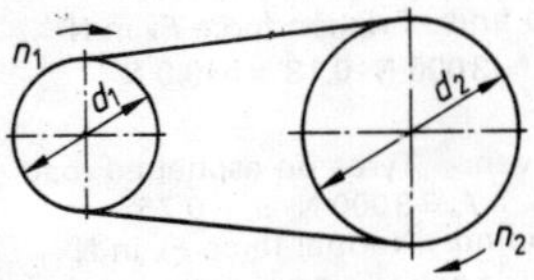

$d_1 \cdot n_1 = d_2 \cdot n_2$ $\qquad i = \frac{n_1}{n_2} = \frac{d_2}{d_1}$

$M_1 = \frac{M_2}{i \cdot \eta}$ $\qquad M_2 = M_1 \cdot i \cdot \eta$

$d_1 = \frac{d_2 \cdot n_2}{n_1}$ $\qquad d_2 = \frac{d_1 \cdot n_1}{n_2}$

$n_1 = \frac{d_2 \cdot n_2}{d_1}$ $\qquad n_2 = \frac{d_1 \cdot n_1}{d_2}$

$n_1 = n_2 \cdot i$ $\qquad n_2 = \frac{n_1}{i}$

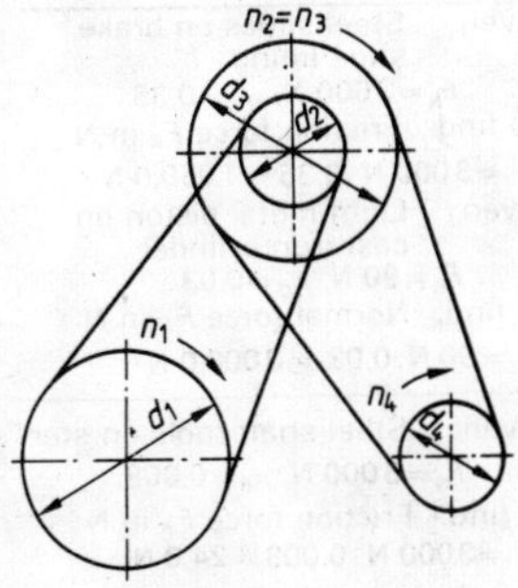

Compound transmission

$i = i_1 \cdot i_2 \ldots$ $\qquad i = \frac{d_2 \cdot d_4 \ldots}{d_1 \cdot d_3 \ldots}$

$i = \frac{n_1}{n_4}$ $\qquad n_1 = n_4 \cdot i$ $\qquad n_4 = \frac{n_1}{i}$

d_1, d_3	Diameters, driving pulleys
d_2, d_4	Diameters, driven pulleys
i	Total transmission
i_1, i_2	Individual transmissions
M_1	Torque, driving pulley
M_2	Torque, driven pulley
n_1, n_3	Speed, driving pullyes
n_2, n_4	Speeds, driven pulleys
η	Efficiency

Given: Simple belt drive $\eta = 0{,}98$, $d_1 = 60$ mm, $d_2 = 150$ mm, $n_2 = 400$/min. $M_2 = 29.4$ Nm.

To find: a) Transmission i
b) Speed n_1
c) Torque M

$i = \frac{150 \text{ mm}}{60 \text{ mm}} = 2{,}5 = 2{,}5:1$

$n_1 = \frac{150 \text{ mm} \cdot 400/\text{min}}{60 \text{ mm}} = 1\,000/\text{min}$

or $n_1 = 400/\text{min} \cdot 2{,}5 = 1\,000/\text{min}$

$M_1 = \frac{29{,}4 \text{ Nm}}{2{,}5 \cdot 0{,}98} = 12{,}0 \text{ Nm}$

Given: Double belt drive $\eta = 0{,}97$, $d_1 = 240$ mm, $d_2 = 120$ mm, $d_3 = 225$ mm, $d_4 = 90$ mm, $n_1 = 250$/min, $M_1 = 40$ Nm.

To find: a) Total transmission i
b) Speed n_4
c) Torque M_2 pulley 4

$i = \frac{120 \text{ mm} \cdot 90 \text{ mm}}{240 \text{ mm} \cdot 225 \text{ mm}} = 0{,}2 = 1:5$

$n_4 = 250/\text{min} : 0{,}20 = 1\,250/\text{min}$

$M_2 = 40 \text{ Nm} \cdot 0{,}20 \cdot 0{,}97 = 7{,}76 \text{ Nm}$

Gear drive

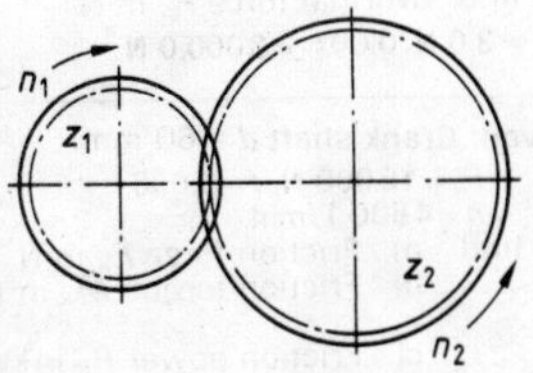

$z_1 \cdot n_1 = z_2 \cdot n_2$ $\qquad i = \frac{n_1}{n_2} = \frac{z_2}{z_1}$

$M_1 = \frac{M_2}{i \cdot \eta}$ $\qquad M_2 = M_1 \cdot i \cdot \eta$

$z_1 = \frac{z_2 \cdot n_2}{n_1}$ $\qquad z_2 = \frac{z_1 \cdot n_1}{n_2}$

$n_1 = \frac{z_2 \cdot n_2}{z_1}$ $\qquad n_2 = \frac{z_1 \cdot n_1}{z_2}$

$n_1 = n_2 \cdot i$ $\qquad n_2 = \frac{n_1}{i}$

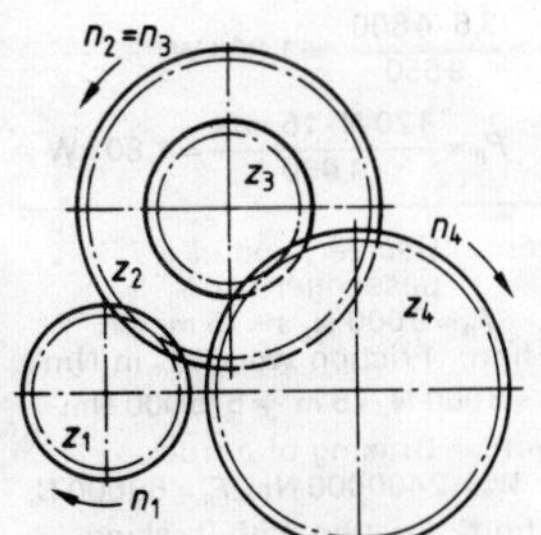

Compound transmission

$i = i_1 \cdot i_2 \ldots$ $\qquad i = \frac{z_2 \cdot z_4 \ldots}{z_1 \cdot z_3 \ldots}$

$i = \frac{n_1}{n_4}$ $\qquad n_1 = n_4 \cdot i$ $\qquad n_4 = \frac{n_1}{i}$

i	Total transmission
i_1, i_2	Individual transmissions
M_1	Torque, driving gear wheel
M_2	Torque, driven gear wheel
n_1, n_3	Speed, driving gear or speed of worm
n_2, n_4	Speed, driven gears
Z_1, Z_3	Numbers of teeth, driving gears or Z_1 number of starts of worm
Z_2, Z_4	Number of teeth, driven gears
η	Efficiency

Given: Simple gear drive, $z_1 = 24$, $z_2 = 36$, $n_1 = 975$/min, $M_1 = 120$ Nm, $\eta = 0{,}98$.

To find: a) Transmission i
b) Speed n_2
c) Torque M_2

$i = \frac{36 \text{ teeth}}{24 \text{ teeth}} = 1{,}5 = 1{,}5:1$

$n_2 = \frac{24 \text{ teeth} \cdot 975/\text{min}}{36 \text{ teeth}} = 650/\text{min}$

or $n_2 = 975/\text{min} : 1{,}5 = 650/\text{min}$

$M_2 = 120 \text{ Nm} \cdot 1{,}5 \cdot 0.98 = 176{,}4 \text{ Nm}$

Given: Double belt drive $z_1 = 18$, $z_2 = 55$, $z_3 = 25$, $z_4 = 45$, $n_4 = 800$/min, $M_2 = 627$ Nm $\eta = 0{,}95$.

To find: a) Total transmission i
b) Speed n_2
c) Torque M_1

$i = \frac{55 \text{ teeth} \cdot 45 \text{ teeth}}{18 \text{ teeth} \cdot 25 \text{ teeth}} = 5{,}5$

$n_1 = 800/\text{min} \cdot 5{,}5 = 4\,400/\text{min}$

$M_1 = \frac{627 \text{ Nm}}{5{,}5 \cdot 0{,}95} = 120{,}0 \text{ Nm}$

Planetary gear mechanism

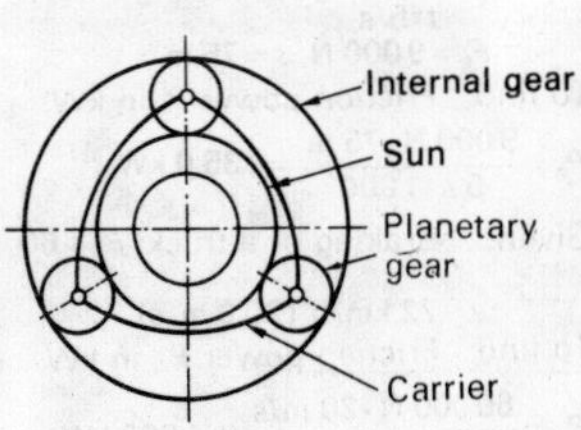

Carrier driven $\qquad i = \frac{z_1 + z_0}{z_1}$

Carrier driving $\qquad i = \frac{z_2}{z_2 + z_0}$

Carrier fixed $\qquad i = \frac{z_2}{z_1}$

i	Transmission
z_1	Number of teeth of driving gear
z_2	Number of teeth of driven gear
z_0	Number of teeth of stationary gear

Given: Simple planetary gear mechanism
Carrier driven, $\eta = 0.96$
Sun driving, $z_1 = 30$
Internal gear fixed $z_0 = 60$
$n_1 = 4200$/min, $M_1 = 150$ Nm

To find: a) Transmission i
b) Speed n_2
c) Torque M_2

$i = \frac{30 \text{ teeth} + 60 \text{ teeth}}{30 \text{ teeth}} = 3{,}0$

$n_2 = 4\,200/\text{min} : 3{,}0 = 1\,400/\text{min}$

$M_2 = 150 \text{ Nm} \cdot 3{,}0 \cdot 0{,}96 = 432{,}0 \text{ Nm}$

Friction

Static friction

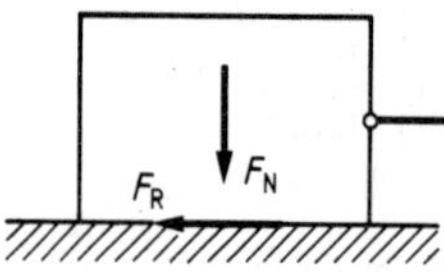

$F_R = F_N \cdot \mu_H$

$F_N = \frac{F_R}{\mu_H}$ $\qquad \mu_H = \frac{F_R}{F_N}$

F_N Normal force
F_R Friction force
μ_H Coefficient of static friction

Given: Steel on cast iron at rest
$F_N = 3000$ N, $\mu_H = 0{,}18$.
To find: Friction force F_R in N
$F_R = 3000 \text{ N} \cdot 0{,}18 = 540{,}0$ N

Given: Tyres on asphalted road
$F_R = 3000$ N, $\mu_H = 0{,}75$.
To find: Normal force F_N in N
$F_N = 3000 \text{ N} : 0{,}75 = 4000{,}0$ N

Sliding friction

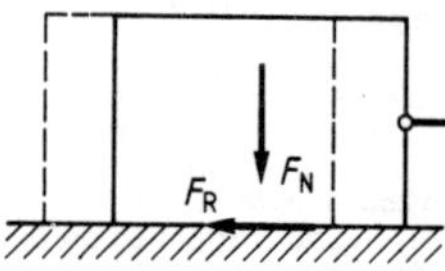

$F_R = F_N \cdot \mu_G$

$F_N = \frac{F_R}{\mu_G}$ $\qquad \mu_G = \frac{F_R}{F_N}$

F_N Normal force
F_R Friction force
μ_G Coefficient of sliding friction

Given: Steel slides on brake shoe lining
$F_N = 3000$ N, $\mu_G = 0{,}35$.
To find: Friction force F_R in N
$F_R = 3000 \text{ N} \cdot 0{,}35 = 1050{,}0$ N
Given: Light metal piston on cast iron cylinder
$F_R = 90$ N, $\mu_G = 0{,}03$.
To find: Normal force F_N in N
$F_N = 90 \text{ N} : 0{,}03 = 3000{,}0$ N

Rolling friction

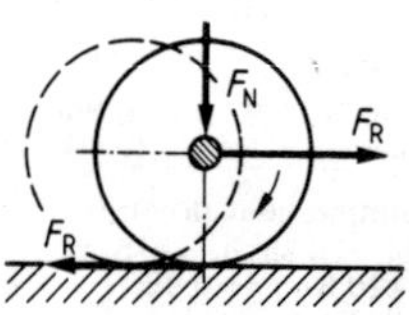

$F_R = F_N \cdot \mu_R$

$F_N = \frac{F_R}{\mu_R}$ $\qquad \mu_R = \frac{F_R}{F_N}$

F_N Normal force
F_R Friction force
μ_G Coefficient of rolling friction

Given: Steel shaft rools on steel
$F_N = 3000$ N, $\mu_R = 0{,}008$.
To find: Friction force F_R in N
$F_R = 3000 \text{ N} \cdot 0{,}008 = 24{,}0$ N

Given: Antifriction bearing with grease lubrication
$F_R = 3$ N, $\mu_R = 0{,}001$.
To find: Normal force F_N in N
$F_N = 3{,}0 \text{ N} : 0{,}001 = 3000{,}0$ N

Journal friction

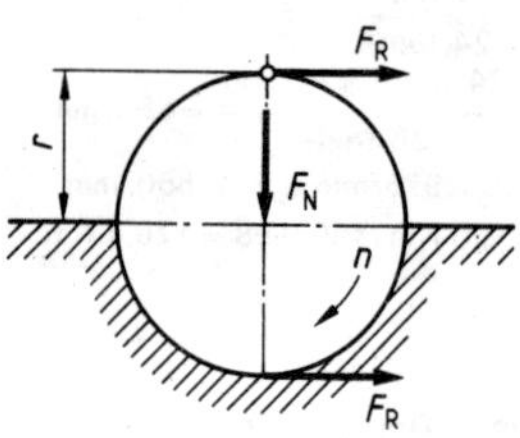

$F_R = F_N \cdot f$ $\qquad M_R = F_R \cdot r$

$P_R = \frac{F_R \cdot v_u}{1000}$ $\qquad P_R \approx \frac{M_R \cdot n}{9550}$

F_N Normal force in N
F_R Friction force in N
f Journal friction factor
M_R Friction torque in Nm
n Speed in 1/min
P_R Friction power in kW
r Radius in m
v_u Peripheral speed in m/s

Given: Crank shaft d = 60 mm.
$F_N = 15000$ N, $f = 0{,}008$,
$n = 4800$ 1/min.
To find: a) Friction force F_R in N
b) Friction torque M_R in Nm
c) Friction power P_R in kW
$F_R = 15000 \text{ N} \cdot 0{,}008 = 120{,}0$ N
$M_R = 120{,}0 \text{ N} \cdot 0{,}030 \text{ m} = 3{,}60$ Nm
$P_R \approx \frac{3{,}6 \cdot 4800}{9550} = 1{,}80$ kW
or $P_R = \frac{120 \text{ N} \cdot 15 \text{ m/s}}{1000} = 1{,}80$ kW

Friction work

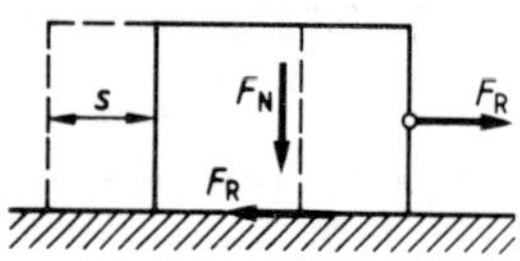

$W_R = F_R \cdot s$ $\qquad Q_R = W_R$

$F_R = \frac{W_R}{s}$ $\qquad s = \frac{W_R}{F_R}$

F_R Friction force
Q_R Friction heat
s Friction path
W_R Friction work

Given: Deceleration of a passenger car
$F_R = 9000$ N, $s = 75$ m.
To find: Friction work W_R in Nm
$W_R = 9000 \text{ N} \cdot 75 \text{ m} = 675000$ Nm
Given: Braking of a truck
$W_R = 2400000$ Nm, $F_R = 60000$ N.
To find: Friction path (braking path) s in m.
$s = 2400000 \text{ Nm} : 60000 \text{ N} = 40$ m

Friction power

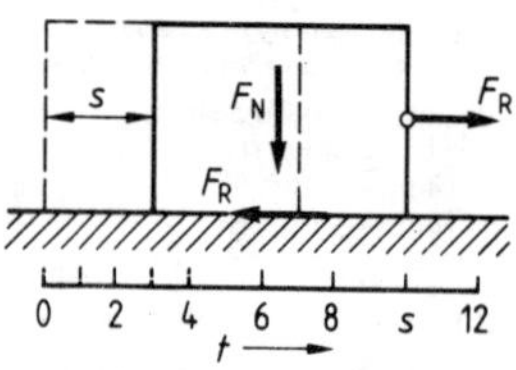

$P_R = \frac{F_R \cdot s}{t}$ $\qquad P_R = \frac{W_R}{t}$

$P_R = F_R \cdot v$ $\qquad W_R = P_R \cdot t$

F_R Friction force
P_R Friction power
s Friction path
t Friction time
v Speed
W_R Frictional work

Given: Braking a passenger car,
t=5 s,
$F_R = 9000$ N, $s = 75$ m.
To find: Friction power P_r in kW
$P_R = \frac{9000 \text{ N} \cdot 75 \text{ m}}{5 \text{ s} \cdot 1000} = 135{,}0$ kW
Given: Braking of a truck, $F_R = 60$ kN
$v = 72$ km/h (20,0 m/s).
To find: Friction power P_R in kW
$P_R = \frac{60000 \text{ N} \cdot 20 \text{ m/s}}{1000} = 1200$ kW

Surface pressure

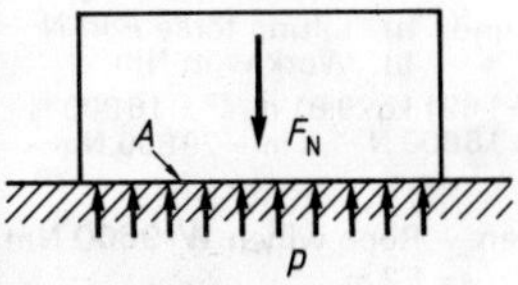

$$p = \frac{F_N}{A}$$

$$F_N = p \cdot A \qquad A = \frac{F_N}{p}$$

A Area, Bearing area
F_N Normal force
P Surface pressure

Given: Support workshop surface $F_N = 350\,000$ N, $A = 5\,000$ mm².
To find: Surface pressure p in N mm².
$p = 350\,000\text{ N} : 5000\text{ mm}^2 = 70\text{ N/mm}^2$

Given: Belt tyre for automobile $F_N = 3800$ N, $p = 2{,}0$ bar
To find: Resisting area A in cm²
$A = 3800\text{ N} : 20\text{ N/cm}^2 = 190\text{ cm}^2$

Tension, compression

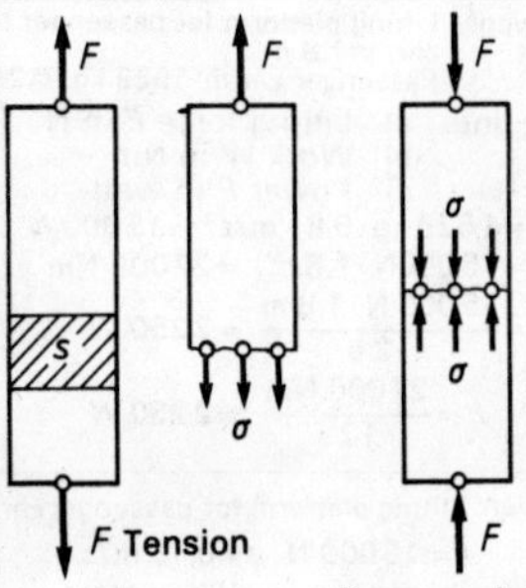

$$\sigma = \frac{F}{S} \qquad \sigma_{zul} = \frac{\sigma_B}{\nu}$$

$$F = \sigma \cdot S \qquad \sigma_B = \sigma_{zul} \cdot \nu$$

$$S = \frac{F}{\sigma} \qquad \nu = \frac{\sigma_B}{\sigma_{zul}}$$

Formulas are valid also for
σ_{zul} and F_{zul} or σ_B and F_B

F Tensile force, compressive force
F_B Tensile or compressive force at fracture
F_{zul} Permissible tensile or compressive force
S Cross-sectional area
σ Tensile stress, Compressive stress
σ_B Tensile strength (R_m) compressive strength
σ_{zul} Permissible tensile or compressive stress
ν Factor of safety, Minimum safety factor
RF_m ISO symbol for tensile strength

Given: Steel flat $S = 100$ mm² $F = 4\,250$ N, $\sigma_B = 340$ N/mm².
To find: a) Tensile stress σ in N mm²
b) Factor of safety ν
$\sigma = 4250\text{ N} : 100\text{ mm}^2 = 42{,}5\text{ N/mm}^2$
$\nu = 340\text{ N/mm}^2 : 42{,}5\text{ N/mm}^2 = 8{,}0$

Given: Towing column $S = 25$ mm²
To find: a) Permissible tensile stress σ_{zul}
b) Permissible tensile force F_{zul}
c) Tensile force F_B at fracture
$\sigma_{zul} = 1\,200\text{ N/mm}^2 . 4 = 300\text{ N/mm}^2$
$F_{zul} = 300\text{ N/mm}^2 \cdot 25\text{ mm}^2 = 7500{,}0\text{ N}$
$F_B = 1\,200\text{ N/mm}^2 \cdot 25\text{ mm}^2 = 30\,000{,}0\text{ N}$

Shear

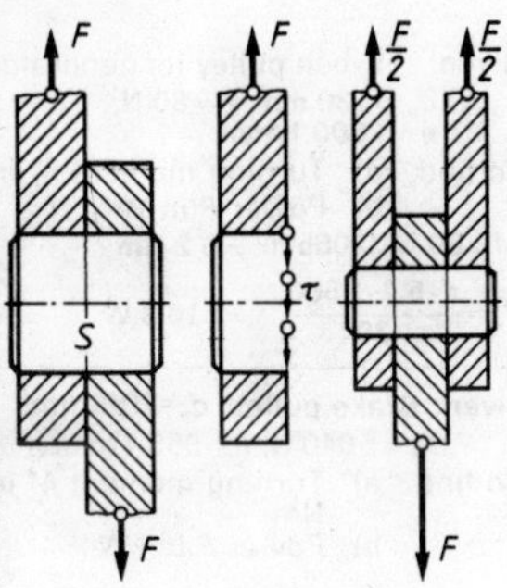

$$\tau = \frac{F}{S} \qquad \tau_{zul} = \frac{\tau_B}{\nu}$$

$$F = \tau \cdot S \qquad \tau_B = \tau_{zul} \cdot \nu$$

$$S = \frac{F}{\tau} \qquad \nu = \frac{\tau_B}{\tau_{zul}}$$

Formulas are valid also for
τ_{zul} and F_{zul} or τ_B and F_B

F Shear force
F_B Shear force
F_{zul} Permissible shear force
S Cross-sectional area
σ Shear stress
σ_B Shear stress
σ_{zul} Permissible shear stress
ν Factor of safety, Minimum safety factor
Shear stress $\tau_B \approx 0{,}8 \cdot \sigma_B$

Given: Spring bolts, double section $F = 15\,000$ N, $S = 100$ mm², $\tau_B = 480$ N/mm².
To find: a) Shear stress τ in N mm²
b) Factor of safety ν

$$\tau = \frac{15\,000\text{ N}}{2 \cdot 100\text{ mm}^2} = 75{,}0\text{ N/mm}^2$$

$\nu = 480\text{ N/mm}^2 : 75\text{ N/mm}^2 = 6{,}4$

Given: Coupling pin, double section $\tau_B = 400$ N/mm², $\nu = 5{,}0$, $F = 3160$ N.
To find: a) Permissible shear stress τ_{zul}
b) Cross section S in mm²
c) Diameter d in mm
$\tau_{zul} = 400\text{ N/mm}^2 : 5 = 80{,}0\text{ N/mm}^2$

$$S = \frac{3160\text{ N}}{2 \cdot 80\text{ N/mm}^2} = 19{,}75\text{ mm}^2$$

d according to Table $= 5{,}0$ mm

Bending

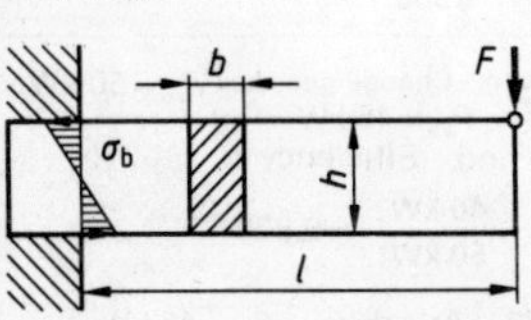

$$\sigma_b = \frac{M_b}{W_b} \qquad M_b = \sigma_b \cdot W_b \qquad W_b = \frac{M_b}{\sigma_b}$$

$$M_b = F \cdot l \qquad W_b = \frac{b \cdot h^2}{6}$$

F Force — l Lever arm
M_b Bending moment — b Width
W_b Resisting moment — h Height
$F\sigma_b$ Bending stress

Given: Steel plate clamped at one end $b = 20$ mm, $h = 42$ mm, $l = 500$ mm, $F = 1176$ N.
To find: a) Bending moment
b) Resisting moment
c) Bending stress
$M_b = 1176\text{ N} \cdot 500\text{ mm} = 588\,000\text{ Nmm}$

$$W_b = \frac{20\text{ mm} \cdot 42\text{ mm}}{6} = 5\,880\text{ mm}^2$$

$$\sigma_b = \frac{588\,000\text{ Nm}}{5\,880\text{ mm}} = 100\text{ N/mm}^2$$

Torsion

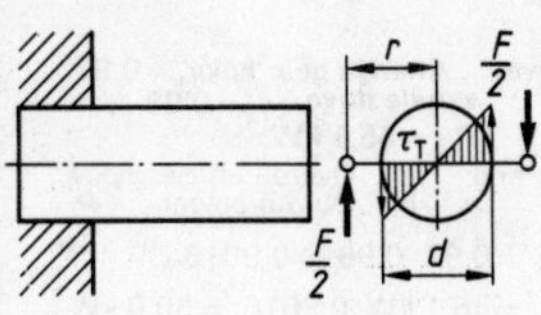

$$\tau_t = \frac{T}{W_t} \qquad T = \tau_t \cdot W_t \qquad W_t = \frac{T}{\tau_t}$$

$$T = F \cdot r \qquad W_t \approx \frac{d^3}{5}$$

F Force — d Diameter
T Torsional moment — r Lever arm
W_t Resisting torque
$F\sigma_t$ Torsional stress

Given: Torsion bar $d = 30$ mm, $r = 200$ mm, $F = 4050$ N.
To find: a) Torsional moment T in Nmm
b) Resisting torque
c) Torsional stress τ_t in N/mm²
$T = 4050\text{ N} \cdot 200\text{ mm} = 810\,000\text{ Nmm}$

$$W_t \approx \frac{30^3\text{ mm}^3}{5} = 5400\text{ mm}^3$$

$$\tau_t = \frac{810\,000\text{ Nmm}}{5400\text{ mm}^3} = 150\text{ N/mm}^2$$

Work, Power, Efficiency

Work

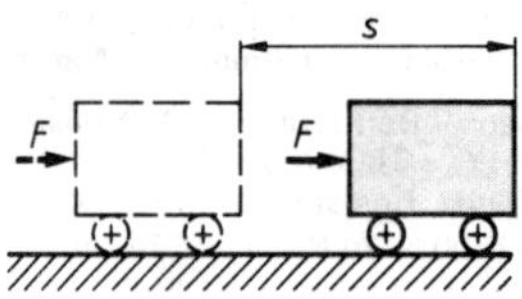

$$W = F \cdot s \qquad W = P \cdot t$$

$$F = \frac{W}{s} \qquad P = \frac{W}{t}$$

$$s = \frac{W}{F} \qquad t = \frac{W}{P}$$

F Force
P Power
s Displacement
t Time
W Work

Given: Lifting platform for passenger car, s=1.8 m.
Passenger car m=1631 kg.
To find: a) Lifting force F in N
b) Work w in Nm

$F = 1\,631 \text{ kg} \cdot 9{,}81 \text{ m/s}^2 = 16\,000 \text{ N}$
$W = 16\,000 \text{ N} \cdot 1{,}8 \text{ m} = 28\,800 \text{ Nm}$

Given: Rope winch **W=9600 Nm**
s=1,2 m.
To find: Force F on the rope
$F = 9\,600 \text{ Nm} : 1{,}2 \text{ m} = 8\,000 \text{ N}$

Power

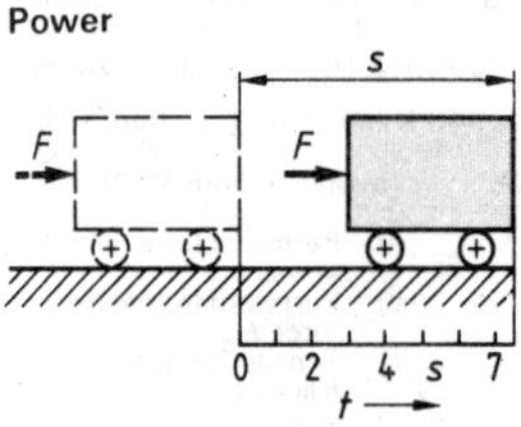

From work $P = \frac{F \cdot s}{t}$

$$P = \frac{W}{t} \qquad F = \frac{P \cdot t}{s}$$

$$W = P \cdot t \qquad s = \frac{P \cdot t}{F}$$

$$t = \frac{W}{P} \qquad t = \frac{F \cdot s}{P}$$

Given: Lifting platform for passenger car, s=1.8 m.
Passenger car m=1529 kg. t12 s
To find: a) Lifting force F in N
b) **Work W in Nm**
c) Power P in watt

$F = 1\,529 \text{ kg} \cdot 9{,}81 \text{ m/s}^2 = 15\,000 \text{ N}$
$W = 15\,000 \text{N} \cdot 1{,}8 \text{ m} = 27\,000 \text{ Nm}$

$$P = \frac{15\,000 \text{ N} \cdot 1{,}8 \text{ m}}{12 \text{ s}} = 2\,250 \text{ W}$$

or $$P = \frac{27\,000 \text{ Nm}}{12 \text{ s}} = 2\,250 \text{ W}$$

From Velocity $P = \frac{\pi \cdot F \cdot r \cdot n}{30}$

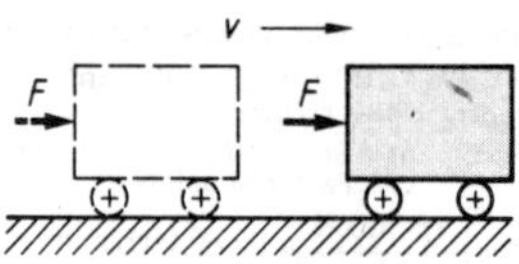

$$P = F \cdot v \qquad F = \frac{30 \cdot P}{\pi \cdot n \cdot r}$$

$$v = \frac{P}{F} \qquad F = \frac{P}{v} \qquad n = \frac{30 \cdot P}{\pi \cdot F \cdot r}$$

F Force in N
M Turning moment in Nm
n Speed in 1/min
P Power in W
s Displacement in m
t Time in s
v Velocity in m/s
W Work in Nm or J

Given: Lifting platform for passenger car
F=15 000 N, v=0,15 m/s.
To find: Power in W and kW
$P = 15\,000 \text{ N} \cdot 0{,}15 \text{ m/s} = 2\,250 \text{ W}$
$P = 2\,250 \text{ W} : 1000 = 2{,}25 \text{ kW}$

Given: V-belt pulley for generator
d_w=130 mm, F=80 N,
n=1 500 1/min.
To find: a) Turning moment M in
b) Power P in W
$M = 80 \text{ N} \cdot 0{,}065 \text{ m} = 5{,}2 \text{ Nm}$

$$P = \frac{\pi \cdot 5{,}2 \cdot 1\,500}{30} = 816{,}8 \text{ W}$$

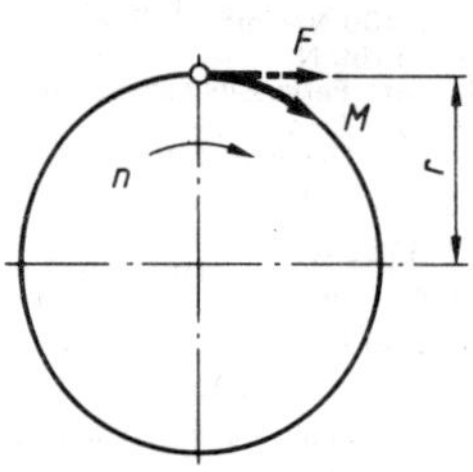

From turning moment $P \approx \frac{M \cdot n}{9550}$

$$M \approx \frac{9\,550 \cdot P}{n} \qquad n \approx \frac{9\,550 \cdot P}{M}$$

M Turning moment in Nm
n Speed in 1/min
P Power in kW

Given: Brake pulley, d_w=200 mm.
F_R=5 040 N, n=955 1/min.
To find: a) Turning moment M in Nm
b) Power P in kW

$M = 5\,040 \text{ N} \cdot 0{,}1 \text{ m} = 504 \text{ Nm}$

$$P \approx \frac{504 \cdot 955}{9\,550} = 50{,}4 \text{ kW}$$

Efficiency

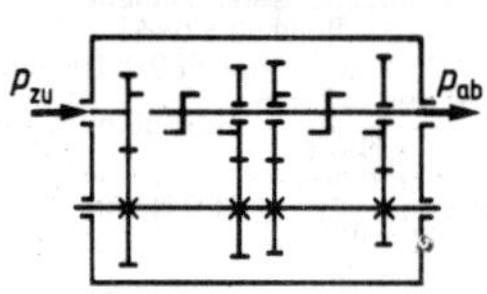

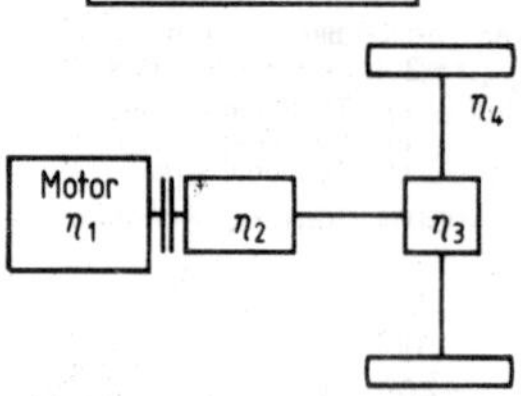

$$\eta = \frac{P_{ab}}{P_{zu}} \qquad \eta = \frac{W_{ab}}{W_{zu}}$$

$$P_{ab} = P_{zu} \cdot \eta \qquad W_{ab} = W_{zu} \cdot \eta$$

$$P_{zu} = \frac{P_{ab}}{\eta} \qquad W_{zu} = \frac{W_{ab}}{\eta}$$

$$\eta = \eta_1 \cdot \eta_2 \cdot \eta_3 \cdot \ldots$$

P_{ab} Output power
P_{zu} Input power
W_{ab} Output work
W_{zu} Input work
η Overall efficiency
η_1, η_2 Individual efficiencies

Given: Change gear box P_{zu}=50 kW,
P_{ab}=46 kW.
To find: Efficiency η.

$$\eta = \frac{46 \text{ kW}}{50 \text{ kW}} = 0{,}920$$

Given: Axle drive P_{zu}=46 kW,
η=0,98.
To find: Power at the axle driving shaft P_{ab}
$P_{ab} = 46 \text{ kW} \cdot 0{,}98 = 45{,}1 \text{ kW}$

Given: Change gear box η_1=0,92;
Axle drive, η_2=0,98,
P_{ab}=45,1 kW.
To find: a) Overall efficiency η.
b) Output power P_{zu}.

$\eta = 0{,}92 \cdot 0{,}98 = 0{,}9016$
$P_{zu} = 45{,}1 \text{ kW} : 0{,}9016 = 50{,}0 \text{ kW}$

Energy

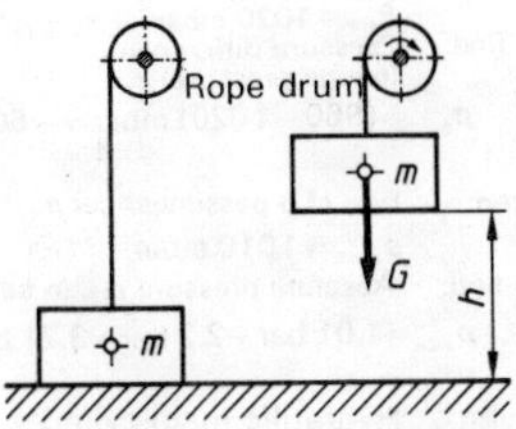

Potential energy (Positional energy) $E_p = m \cdot g \cdot h$

$E_p = G \cdot h$ $h = \frac{E_p}{m \cdot g}$

E_p Potential energy in Nm, J
G Gravitational force in N
g Acceleration due to gravity 9.81 m/s^2
h Ascent, descent in m
m **Mass, Weight in kg**

Given: Drop hammer $G = 3000$ N, $h = 3$ m.
To find: Potential energy E_p in Nm.
$E_p = 3000 \text{ N} \cdot 3{,}0 \text{ m} = 9000 \text{ Nm}$

Given: Passenger car on a lifting platform.
$m = 1631$ kg, $h = 2$ m.
To find: Potential energy E_p in Nm.
$E_p = 1631 \text{ kg} \cdot 9{,}81 \text{ m/s}^2 \cdot 2{,}0 \text{ m} = 32000 \text{ Nm}$

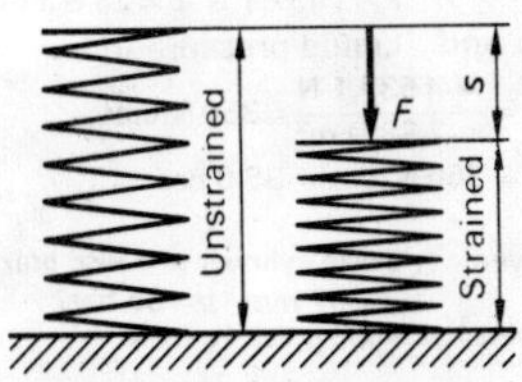

Strain energy $E_{sp} = \frac{c \cdot s^2}{2}$

$c = \frac{F}{s}$ $E_{sp} = \frac{F \cdot s}{2}$

c Spring rate, Spring constant in N/m
E_{sp} Strain energy in Nm, J
F Spring force in N
s Spring deflection in m

Given: Spring loaded cylinder for the locking brake of a truck
$c = 80000$ N/m, $s = 80$ mm.
To find: Strain energy in Nm E_{sp} in Nm.
$E_{sp} = \frac{80000 \text{ N/m} \cdot 0{,}08^2 \text{ m}^2}{2} = 256{,}0 \text{ Nm}$

Given: Helical spring for a truck
$F = 6000$ N. $s = 120$ mm.
To find: Strain energy E_{sp} in Nm
$E_{sp} = \frac{6000 \text{ N} \cdot 0{,}12 \text{ m}}{2} = 360{,}0 \text{ Nm}$

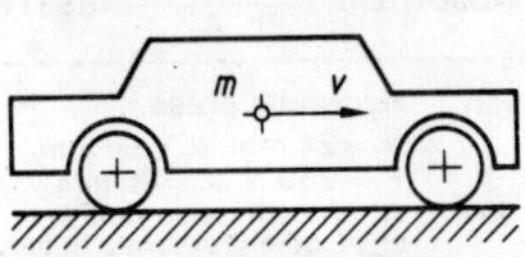

Kinetic energy (Motion energy) $E_k = \frac{m \cdot v^2}{2}$

$m = \frac{2 \cdot E_k}{v^2}$ $v = \sqrt{\frac{2 \cdot E_k}{m}}$

E_k Kinetic energy in Nm, J
m Mass, Weight in kg
v Velocity in m/s

Given: Passenger car
$m = 1200$ kg, $v = 126$ km/h.
To find: Kinetic Energy E_k in Nm.
$v = 126 : 3{,}6 = 35{,}0 \text{ m/s}$
$E_k = \frac{1200 \text{ kg} \cdot 35^2 \text{ m}^2/\text{s}^2}{2} = 735000 \text{ Nm}$

Transformation of energy

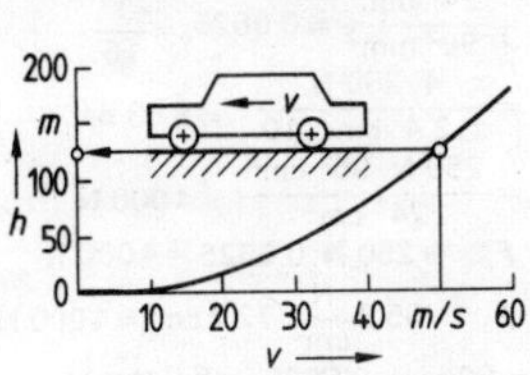

Positional energy = Motion energy

$h = \frac{v^2}{2 \cdot g}$ $s = \frac{m \cdot v^2}{2 \cdot F_B}$

$v = \sqrt{2 \cdot g \cdot h}$ $F_B = \frac{m \cdot v^2}{2 \cdot s}$

g Acceleration due to gravity 9,81 m/s^2
h Height of fall in m
F_B Brake force in N
m Mass, Vehicle weight in kg
s Braking distance in m
v Travel velocity in m/s

Given: Passenger Car
$v = 108$ km/h (30 m/s).
To find: Height of full in m
$h = \frac{30^2 \text{ m}^2/\text{s}^2}{2 \cdot 9{,}81 \text{ m/s}^2} = 45{,}9 \text{ m}$

Given: Truck $m = 10000$ kg, $F_B = 50000$ N,
$v = 72$ km/h (20 m/s).
To find: Braking distance s in m
$s = \frac{10000 \text{ kg} \cdot 20^2 \text{ m}^2/\text{s}^2}{2 \cdot 50000 \text{ N}} = 40{,}0 \text{ m}$

Declined plane

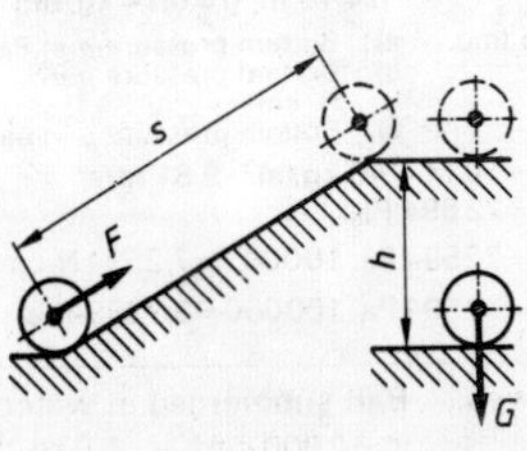

Work expended = work gained

$F \cdot s = G \cdot h$ $F \cdot s = m \cdot g \cdot h$

$F = \frac{G \cdot h}{s}$ $F = \frac{m \cdot g \cdot h}{s}$

$h = \frac{F \cdot s}{G}$ $h = \frac{F \cdot s}{m \cdot g}$

F Thrust, lifting force (without friction)
G Gravitational force of the mass m
g Acceleration due to gravity 9.81 m/s^2
h Height, Lifting height
s Displacement on the inclined plane

Given: Passenger car on a ramp
$s = 10$ m, $G = 14000$ N,
$h = 1{,}50$ m.
To find: Force F in N
$F = \frac{14000 \text{ N} \cdot 1{,}50 \text{ m}}{10 \text{ m}} = 2100 \text{ N}$

Given: Goods wagon on hump
$m = 20388$ kg,
$F = 2500$ N, $s = 120$ m.
To find: Height of the hump in m
$h = \frac{2500 \text{ N} \cdot 120{,}0 \text{ m}}{20388 \text{ kg} \cdot 9{,}81 \text{ m/s}^2} = 1{,}50 \text{ m}$

Screw

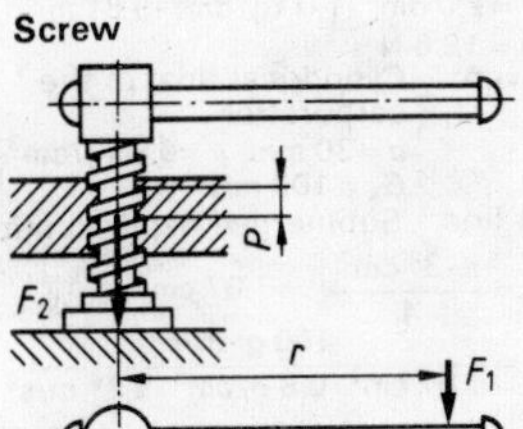

$2 \cdot \pi \cdot r \cdot F_1 = F_2 \cdot P$ (Without friction)

$F_1 = \frac{F_2 \cdot P}{2 \cdot \pi \cdot r}$ $F_2 = \frac{2 \cdot \pi \cdot r \cdot F_1}{P}$

$r = \frac{F_2 \cdot P}{2 \cdot \pi \cdot F_1}$ $P = \frac{2 \cdot \pi \cdot r \cdot F_1}{F_2}$

F_1 Turning force in N
F_2 Pressing force, Lifting force, Gravitational force in N
P Pitch of thread in mm
r Radius, lever arm in mm

Given: Mechanical wagon elevator $F_2 = 7854$ N,
$r = 400$ mm, $P = 8$ mm.
To find: Turning force F_1 in N
$F_1 = \frac{7854 \text{ N} \cdot 8 \text{ mm}}{2 \cdot \pi \cdot 400 \text{ mm}} = 25{,}0 \text{ N}$

Given: Puller $F_1 = 20$ N,
$r = 149{,}2$ mm, $P = 2{,}5$ mm.
To find: Pulling Force F_2 of the screw
$F_2 = \frac{2 \cdot \pi \cdot 149{,}2 \text{ mm} \cdot 20 \text{ N}}{2{,}5 \text{ mm}} = 7500 \text{ N}$

Hydraulic

Pressure

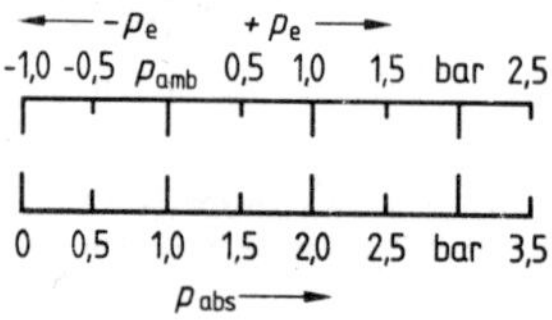

$p_e = p_{abs} - p_{amb}$ $\qquad$ $p_{abs} = p_{amb} \pm p_e$

$-p_e$ for "negative gauge pressure" (vacuum pressure). Instead of the designation vacuum pressure (P_u) used till now for vacuum range, negative values appear now for gauge pressure P_e

- p_{abs} Absolute pressure
- p_{amb} Atomospheric pressure, Ambient pressure
- p_e Gauge pressure, Atmospheric pressure difference

Given: carburettor
P_{abs} = 960 mbar
P_{abb} = 1020 mbar
To find: Pressure difference (gauge pressure) p_e

p_e = (960 − 1020) mbar = − 60 mb

Given: Tyre of a passenger car p_e = 2,2
p_{amb} = 1010 mbar.
To find: Absolute pressure p_{abs} in bar

p_{abs} = 1,01 bar + 2,2 bar = 3,21 bar

Liquid pressure

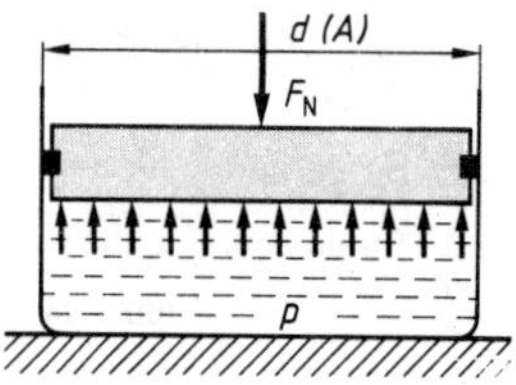

$$p = \frac{F_N}{A} \qquad p = \frac{4 \cdot F_N}{\pi \cdot d^2}$$

$$F_N = p \cdot A \qquad F_N = p \cdot \frac{\pi \cdot d^2}{4}$$

$$A = \frac{F_N}{p} \qquad d = \sqrt{\frac{4 \cdot F_N}{\pi \cdot p}}$$

- A Piston area
- d Piston diameter
- F_N Normal force, Piston force
- p Liquid pressure

Given: Hydraulic main cylinder
F = 1537,1 N. d = 23,8 mm.
To find: Liquid pressure in bar

$$p = \frac{4 \cdot 1537{,}1\ \text{N}}{\pi \cdot 2{,}38^2\ \text{cm}^2} = 350\ \text{N/cm}^2$$

p = 350 N/cm² = 35,0 bar

Given: Brake cylinder of a disc brake
d = 57 mm, p = 35 bar.
Given: Hydraulic press F_N in N.

$$F_N = 350\ \text{N/cm}^2 \cdot \frac{\pi \cdot 5{,}7^2\ \text{cm}^2}{4} = 8931\ \text{N}$$

Hydraulic Press

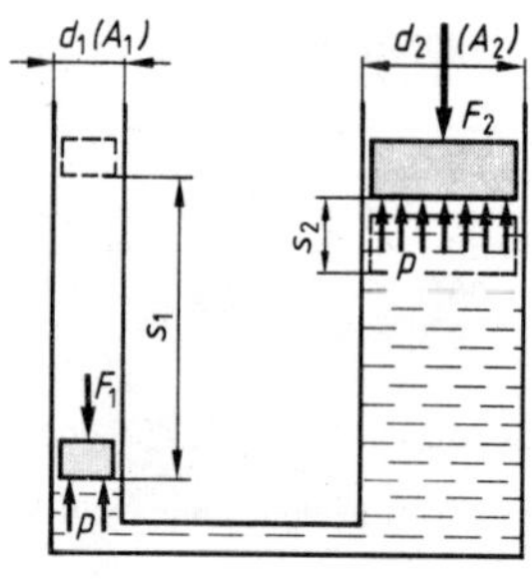

$$p = \frac{F_1}{A_1} = \frac{F_2}{A_2} \qquad i = \frac{F_1}{F_2} = \frac{d_1^2}{d_2^2} = \frac{s_2}{s_1}$$

$$F_1 = \frac{F_2 \cdot d_1^2}{d_2^2} \qquad F_1 = p \cdot A_1 \qquad F_1 = F_2 \cdot i$$

$$F_2 = \frac{F_1 \cdot d_2^2}{d_1^2} \qquad F_2 = p \cdot A_2 \qquad F_2 = \frac{F_1}{i}$$

$$s_2 = s_1 \cdot i \qquad s_1 = \frac{s_2}{i}$$

- A_1, A_2 Piston areas
- d_1, d_2 Piston diameters
- F_1, F_2 Piston forces
- i Force transmission ratio, Distance transmission ratio
- p Liquid pressure
- s_1, s_2 Piston displacements

Given: Hydraulic press
d_1 = 24 mm, d_2 = 96 mm,
F_1 = 250 N, s_1 = 80 mm.
To find:
a) Transmission ratio i
b) Liquid pressure p in bar
c) Piston force F_2 in N
d) Piston displacement s_2 in mm

$$i = \frac{24^2\ \text{mm}^2}{96^2\ \text{mm}^2} = 0{,}0625 = \frac{1}{16}$$

$$p = \frac{4 \cdot 250\ \text{N}}{\pi \cdot 2{,}4^2\ \text{cm}^2 \cdot 10} = 5{,}53\ \text{bar}$$

$$F_2 = \frac{250\ \text{N} \cdot 96^2\ \text{cm}^2}{24^2\ \text{cm}^2} = 4000\ \text{N}$$

or F_2 = 250 N : 0,0625 = 4000 N

or $F_2 = 55{,}3 \frac{\text{N}}{\text{cm}^2} \cdot 72{,}4\ \text{cm}^2 = 4000\ \text{N}$

s_2 = 80 mm · 0,0625 = 5,0 mm

Hydrostatic pressure

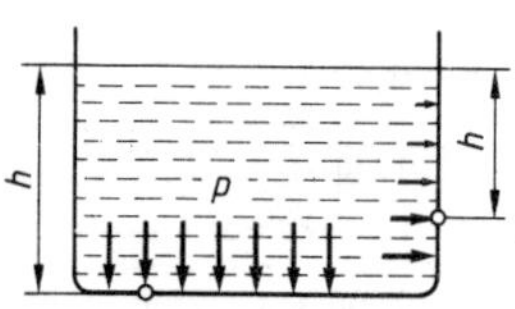

$$p = h \cdot \varrho \cdot g$$

$$h = \frac{p}{\varrho \cdot g} \qquad \varrho = \frac{p}{h \cdot g}$$

- g Acceleration due to gravity 9.81 m/s²
- h Pressure column in m
- p Bottom pressure, side pressure in Pa
- ρ Density of the liquid in kg/m³

Given: Storage tank for petrol
h = 10 m, ϱ = 0,74 kg/dm³
To find:
a) Bottom pressure p in Pa
b) Bottom pressure p in N/cm²
c) Bottom pressure p in bar

p = 10 m · 740 kg/m³ · 9,81 m/s²
= 72594 Pa

p = 72594 Pa : 10000 = 7,2594 N/cm²

p = 72594 Pa : 100000 = 0,72594 bar

Buoyancy

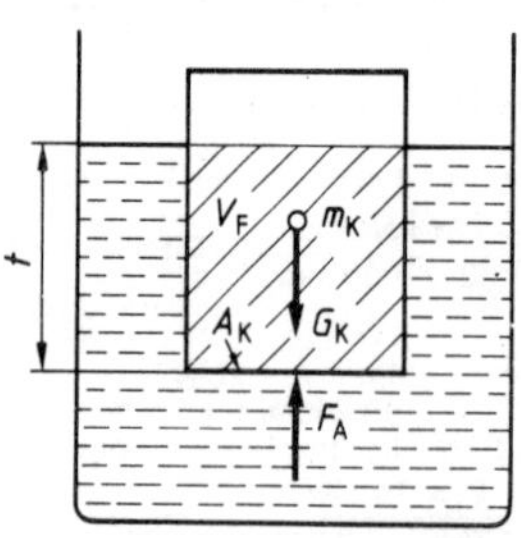

$$F_A = V_F \cdot \varrho_F \cdot g \qquad t = \frac{t = G_k \text{ or } F_A}{A_K \cdot \varrho_F \cdot g}$$

$$V_F = \frac{F_A}{\varrho_F \cdot g} \qquad t = \frac{m_K}{A_K \cdot \varrho_F}$$

$$\varrho_F = \frac{F_A}{V_F \cdot g} \qquad t = \frac{V_F}{A_K}$$

- A_k Base area of the body in dm² (cm²)
- F_A Force of buoyancy in N (mN)
- G_k Gravitational force of the body in N (mN)
- m_k Mass, Weight of the body in kg (g)
- t Submerged depth of the body in dm (cm)
- V_F Displacement volume in dm³ (cm³)
- ρ_F Density of liquid in kg/dm³ (g/cm³)

Given: Ball submerged in water
V_F = 2000 cm³, ϱ = 1,0 kg/dm³
To find: Force of buoyancy F_A in N.
F_A = 2,0 dm³ · 1,0 kg/dm³ · 9,81 m/s²
= 19,6 N

Given: Cylindrical float in the carburettor
d = 30 mm, ϱ = 0,8 kg/dm³,
G_K = 100 mN.
To find: Submerged depth in mm

$$A_K = \frac{\pi \cdot 3^2\ \text{cm}^2}{4} = 7{,}07\ \text{cm}^2$$

$$t = \frac{100\ \text{g} \cdot \text{m/s}^2}{7{,}07\ \text{cm}^2 \cdot 0{,}8\ \text{g/cm}^3 \cdot 9{,}81\ \text{m/s}^2}$$

= 1,80 cm = 18,0 mm

Gas pressure

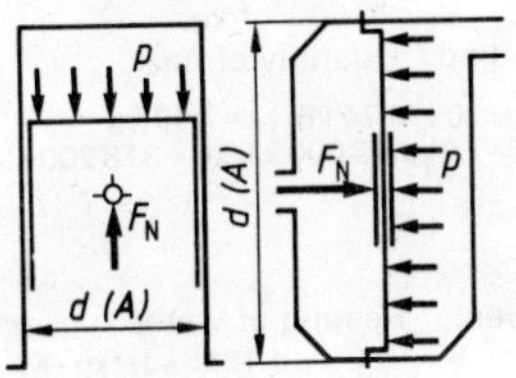

$$p = \frac{F_N}{A} \qquad p = \frac{4 \cdot F_N}{\pi \cdot d^2}$$

$$F_N = p \cdot A \qquad F_N = p \cdot \frac{\pi \cdot d^2}{4}$$

$$A = \frac{F_N}{p} \qquad d = \sqrt{\frac{4 \cdot F_N}{\pi \cdot p}}$$

A Piston area, Membrane area
d Piston diameter, Membrane diameter
F_N Normal force, Piston force
p Gas pressure

Given: Single chamber brake cylinder of a compressed air brake
$d = 120$ mm, $F = 8\,482{,}5$ N.
To find: Gas pressure p in bar

$$p = \frac{4 \cdot 8\,482{,}5 \text{ N}}{\pi \cdot 12^2 \text{ cm}^2} = 75 \text{ N/cm}^2 = 7{,}5 \text{ bar}$$

Given: Piston in an internal combustion engine
$d = 80$ mm, $p = 40$ bar.
To find: Piston force F in N.

$$F = 400 \text{ N/cm}^2 \cdot \frac{\pi \cdot 8^2 \text{ cm}^2}{4} = 20\,106 \text{ N}$$

Gas pressure and volume at constant temperature

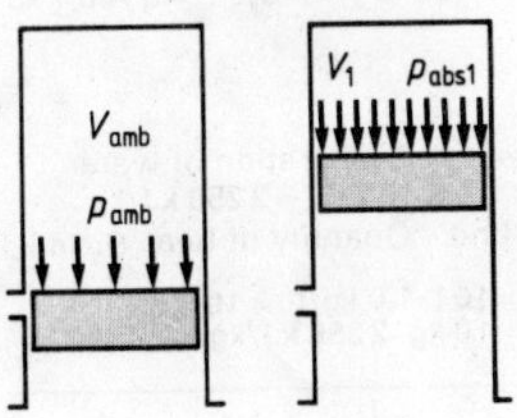

Pressure times volume = constant

$$V_{amb} \cdot p_{amb} = V_1 \cdot p_{abs1} = V_2 \cdot p_{abs2}$$

$$V_{amb} = \frac{V_1 \cdot p_{abs1}}{p_{amb}} \qquad p_{abs1} = \frac{V_{amb} \cdot p_{amb}}{V_1}$$

$$V_2 = \frac{V_1 \cdot p_{abs1}}{p_{abs2}} \qquad p_{abs2} = \frac{V_1 \cdot p_{abs1}}{V_2}$$

p_{abs1} Absolute pressure at volume V_1
p_{abs2} Absolute pressure at volume V_2
p_{amb} Atmospheric pressure
V_{amb} Gas volume at atmospheric pressure
V_1, V_2 Gas volume at P_{abs1} and P_{abs2} respectively

Given: Stock bin of a compressed air brake $V_1 = 40$ l,
$p_{e1} = 8{,}1$ bar, $p_{amb} = 1050$ mbar.
To find: Air volume V_{amb} in l.

$$p_{abs1} = 1{,}05 \text{ bar} + 8{,}1 \text{ bar} = 9{,}15 \text{ bar}$$

$$V_{amb} = \frac{40 \text{ l} \cdot 9{,}15 \text{ bar}}{1{,}05 \text{ bar}} = 348{,}6 \text{ l}$$

Given: Air compressed (temperature constant) from V_{amb} to v_1 $= 560$ cm³ in
$V_1 = 70$ cm³;
$p_{amb} = 1020$ mbar.
To find: Absolute pressure P_{abs1} in bar.

$$p_{abs1} = \frac{560 \text{ cm}^3 \cdot 1{,}02 \text{ bar}}{70 \text{ cm}^3} = 8{,}16 \text{ bar}$$

Gas discharge pressure tanks

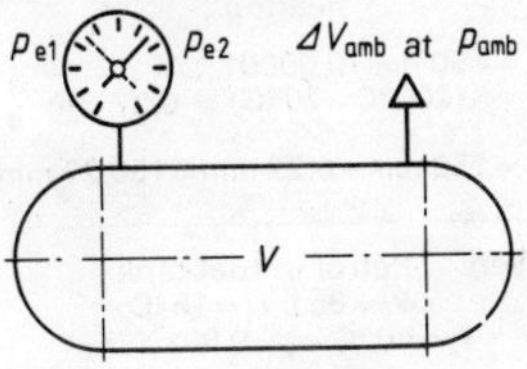

$$\Delta V_{amb} = \frac{V \cdot \Delta p}{p_{amb}}$$

für $p_{amb} = 1$ bar ist $\Delta V_{amb} = V \cdot \Delta p$

$$\Delta p = \frac{\Delta V_{amb} \cdot p_{amb}}{V} \qquad \Delta p = p_{e1} - p_{e2}$$

p_{amb} Atmospheric pressure
p_{e1} Gauge pressure in the tank before discharge
p_{e2} Gauge pressure in the tank after discharge
Δp Pressure difference
V Volume of the gas tank
ΔV_{amb} Volume of the gas discharged at atmospheric pressure

Given: Acid bottle $V = 40$ *l*
$p_{e1} = 135$ bar, $p_{e2} = 118$ bar;
$p_{amb} = 980$ mbar.
To find: a) Pressure difference Δp in bar
b) Volume of the gas discharged ΔV_{amb}

$$\Delta p = 135 \text{ bar} - 118 \text{ bar} = 17 \text{ bar}$$

$$\Delta V_{amb} = \frac{40 \text{ l} \cdot 17 \text{ bar}}{0{,}98 \text{ bar}} = 693{,}9 \text{ l}$$

or $\Delta V_{amb} \approx 40 \cdot 17 = 680{,}0$ l

Flow

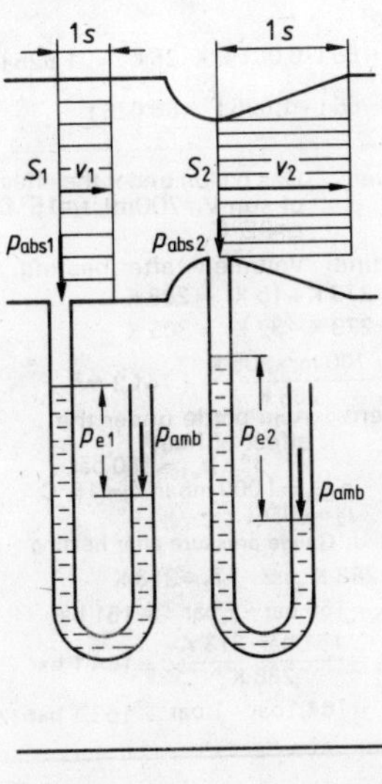

$$\Delta p = \frac{\varrho}{200} \cdot (v_2^2 - v_1^2) \qquad \Delta p = p_{abs1} - p_{abs2} \qquad \Delta p = p_{e2} - p_{e1}$$

$$p_{e1} \approx -\frac{v_1^2}{160} \qquad p_{e2} \approx -\frac{v_2^2}{160}$$

p_{abs1} Absolute pressure, section 1 in mbar
p_{abs2} Absolute pressure, section 2 in mbar
p_{e1} Negative gauge pressure section 1 in mbar
p_{e2} Negative gauge pressure section 2 in mbar
Δp Pressure difference in mbar
v_1, v_2 Gas velocity in the sections 1 and 2 respectively in m/s
ϱ Density of the gas in kg/m³

$S_1 \cdot v_1 = S_2 \cdot v_2$ Cross-sectional area times velocity = constant
d^2 can also be substituted for cross- sectional area S

$$v_1 = \frac{S_2 \cdot v_2}{S_1} \qquad v_2 = \frac{S_1 \cdot v_1}{S_2}$$

$$S_1 = \frac{S_2 \cdot v_2}{v_1} \qquad S_2 = \frac{S_1 \cdot v_1}{v_2}$$

d_1, d_2 Diameters of the sections
S_1, S_2 Cross-sectional area of flow
v_1, v_2 Velocities of flow

Given: Gas flow in the draft tube
$v_1 = 65{,}3$ m/s, $v_2 = 108$ m/s,
$\varrho = 1{,}30$ kg/m³, $p_{amb} = 1{,}0$ bar.
To find: a) Pressure difference
b) Negative gauge pressure
c) Absolute pressure

$$\Delta p = \frac{1{,}3}{200} \cdot (108^2 - 65{,}3^2) = 48{,}1 \text{ mbar}$$

$$p_{e2} \approx -\frac{108^2}{160} = -72{,}9 \text{ mbar}$$

$$p_{abs2} \approx 1000 \text{ mbar} + (-72{,}9) \text{ mbar} = 927{,}1 \text{ mbar}$$

Given: Gas flow in the draft tube
$d_1 = 36$ mm, $d_2 = 28$ mm,
$v_1 = 65{,}3$ m/s.
To find: a) Cross-sectional areas S_1, S_2 in mm²
b) Gas velocity v_2 in m/s and in km/h

$S_1 = 1018$ mm² $\qquad S_2 = 615{,}8$ mm²

$$v_2 = \frac{1018 \text{ mm}^2 \cdot 65{,}3 \text{ m/s}}{615{,}8 \text{ mm}^2} = 108{,}0 \text{ m/s}$$

or

$$v_2 = \frac{36^2 \text{ mm}^2 \cdot 65{,}3 \text{ m/s}}{28^2 \text{ mm}^2} = 108{,}0 \text{ m/s}$$

$$v_2 = 108{,}0 \cdot 3{,}6 = 388{,}8 \text{ km/h}$$

Heat Engineering

Quantity of heat

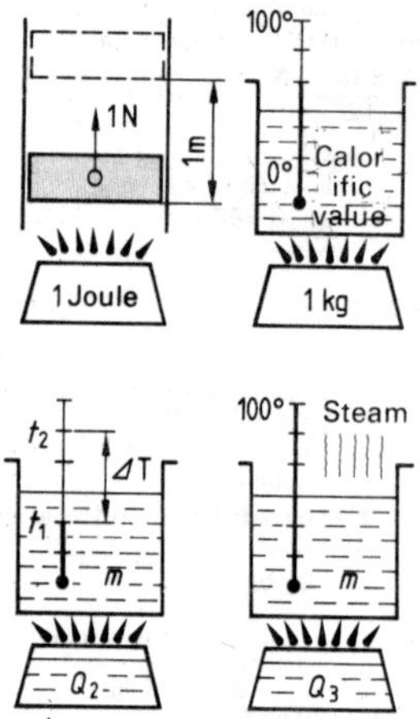

From calorific value $Q_1 = m \cdot H_u$

$$m = \frac{Q_1}{H_u} \qquad H_u = \frac{Q_1}{m}$$

From heat capacity $Q_2 = m \cdot c \cdot \Delta T$

$$m = \frac{Q_2}{c \cdot \Delta T} \qquad \Delta T = \frac{Q_2}{m \cdot c}$$

From evaporation heat $Q_3 = m \cdot Q_v$

$$m = \frac{Q_3}{Q_v} \qquad Q_v = \frac{Q_3}{m}$$

- c Specific heat capacity
- H_u Specific calorific value
- m Mass, weight
- Q Quantity of heat
- Q_v Specific evaporation heat
- ΔT Temperature difference in °C or K

Given: Burning of petrol
$V = 10$ l, $\varrho = 0{,}74$ kg/l,
$H_u = 43\,000$ kJ/kg.
To find: Quantity of heat

$m = 10\ \text{l} \cdot 0{,}74\ \text{kg/l} = 7{,}40$ kg
$Q_1 = 7{,}4\ \text{kg} \cdot 43\,000\ \text{kJ/kg} = 318\,200$ kJ

Given: Heating of water
$V = 10$ l, $c = 4{,}1868$ kJ/(kg · K)
$t_1 = 15$ °C, $t_2 = 90$ °C.
To find: Quantity of heat Q_1 in kJ

$m = 10\ \text{l} \cdot 1{,}0\ \text{kg/l} = 10{,}0$ kg
$Q_2 = 10\ \text{kg} \cdot 4{,}1868\ \text{kJ/(kg·K)} \cdot (90\,°\text{C} - 15\,°\text{C}) = 3140{,}1$ kJ

Given: Evaporation ot water
$V = 10$ l, $Q_v = 2\,256$ kJ/kg.
To find: Quantity of heat Q_2 in kJ.

$m = 10\ \text{l} \cdot 1{,}0\ \text{kg/l} = 10{,}0$ kg
$Q_3 = 10\ \text{kg} \cdot 2\,256\ \text{kJ/kg} = 22\,560$ kJ

Linear expansion

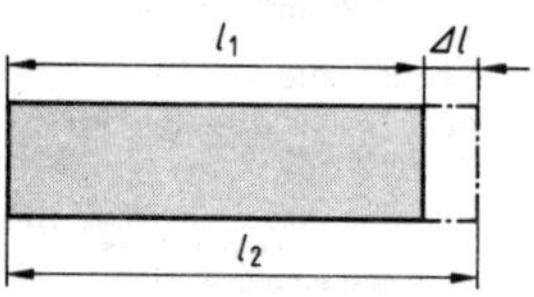

$$\Delta l = l_1 \cdot \alpha \cdot \Delta T \qquad l_2 = l_1 + \Delta l$$

$$l_1 = \frac{\Delta l}{\alpha \cdot \Delta T} \qquad \Delta T = \frac{\Delta l}{\alpha \cdot l_1}$$

- l_1 Length before heating
- l_2 Length after heating
- ΔV Linear expansion
- ΔT Temperature difference in °C or K
- α Coefficient of linear expansion in 1/K (1/°C)

Given: Exhaust valve for truck engine
$l_1 = 150$ mm, $\alpha = 0{,}00001$ 1/K,
$t_1 = 20$ °C, $t_2 = 200$ °C.
To find: a) Linear expansion
b) Length l_2 after heating

$l = 150\ \text{mm} \cdot 0{,}00001\ 1/\text{K} \cdot (200\,°\text{C} - 20\,°\text{C}) = 0{,}27$ mm

$l_2 = 150\ \text{mm} + 0{,}27\ \text{mm} = 150{,}27$ mm

Volumetric expansion

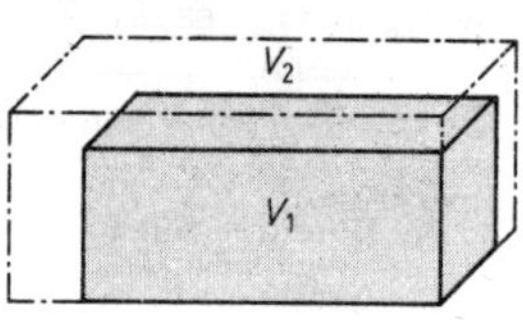

$$\Delta V = V_1 \cdot \alpha_V \cdot \Delta T \qquad V_2 = V_1 + \Delta V$$

$$V_1 = \frac{\Delta V}{\alpha_V \cdot \Delta T} \qquad \Delta T = \frac{\Delta V}{\alpha_V \cdot V_1}$$

- V_1 Volume before heating
- V_2 Volume after heating
- ΔV Volumetric expansion
- ΔT Temperature difference in °C of K
- α_V Coefficient of volumetric expansion in 1/K (1/°C)

Given: Petrol in fuel tank
$V_1 = 65$ l, $t_1 = 15$ °C,
$t_2 = 40$ °C, $\alpha_V = 0{,}001$ 1/K.
To find: a) Change in volume
b) Volume V_2 after heating

$\Delta V = 65\ \text{l} \cdot 0{,}001\ 1/\text{K} \cdot 25\ \text{K} = 1{,}625$ l

$V_2 = 65\ \text{l} + 1{,}625\ \text{l} = 66{,}625$ l

Effect of heat in gases

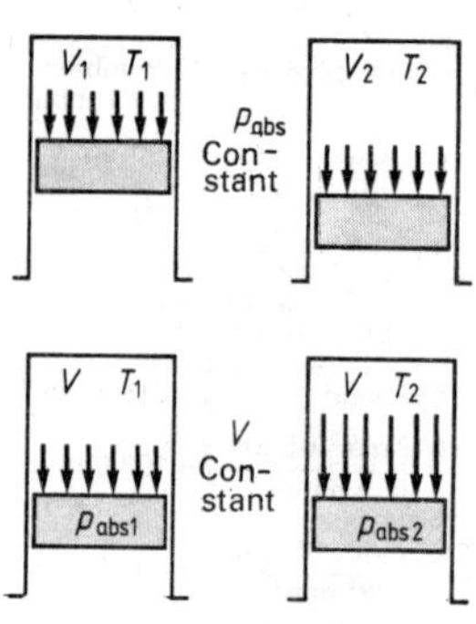

At constant pressure $\frac{V_1}{V_2} = \frac{T_1}{T_2}$

$$V_2 = \frac{V_1 \cdot T_2}{T_1} \qquad T_2 = \frac{V_2 \cdot T_1}{V_1}$$

At constant volume $\frac{p_{abs1}}{p_{abs2}} = \frac{T_1}{T_2}$

$$p_{abs2} = \frac{p_{abs1} \cdot T_2}{T_1} \qquad T_2 = \frac{p_{abs2} \cdot T_1}{p_{abs1}}$$

- p_{abs1} Gas pressure before heating in bar
- p_{abs2} Gas pressure after heating in bar
- T_1 Temperature before heating in K
- T_2 Temperature after heating in K
- V_1 Volume before heating in dm³, m³
- V_2 Volume after heating in dm³, m³

Given Gas ballon under the effect of sun $V_1 = 700\ \text{m}^3$, $t_1 = 15$ °C, $t_2 = 32$ °C.
To find: Volume V_2 after heating
$T_1 = 273\ \text{K} + 15\ \text{K} = 288$ K
$T_2 = 273\ \text{K} + 32\ \text{K} = 305$ K
$V_2 = \frac{700\ \text{m}^3 \cdot 305\ \text{K}}{288\ \text{K}} = 741{,}3\ \text{m}^3$

Given: Acid bottle under the effect of heat
$p_{e1} = 150$ bar,
$p_{amb} = 1\,000$ mbar, $t_1 = 15$ °C,
$t_2 = 40$ °C.
To find: Gauge pressure after heating
$T_1 = 288$ K $\quad T_2 = 313$ K
$p_{abs1} = 150\ \text{bar} + 1\ \text{bar} = 151$ bar
$p_{abs2} = \frac{151\ \text{bar} \cdot 313\ \text{K}}{288\ \text{K}} = 164{,}1$ bar
$p_{e2} = 164{,}1\ \text{bar} - 1\ \text{bar} = 163{,}1$ bar

Piston displacement

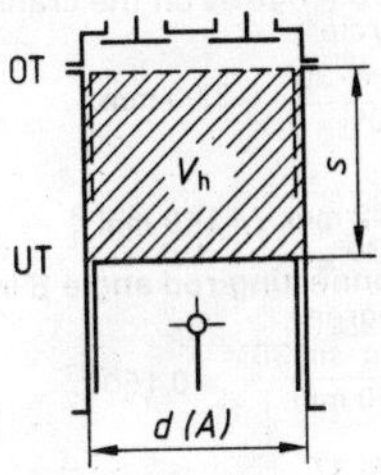

$$V_h = \frac{\pi \cdot d^2}{4} \cdot s \qquad V_h = A \cdot s \qquad V_H = V_h \cdot z$$

$$s = \frac{4 \cdot V_h}{\pi \cdot d^2} \qquad A = \frac{V_h}{s} \qquad V_h = \frac{V_H}{z}$$

$$d = \sqrt{\frac{4 \cdot V_h}{\pi \cdot s}} \qquad s = \frac{V_h}{A} \qquad z = \frac{V_H}{V_h}$$

- A Piston area, Cylinder cross sectional area
- d Cylinder bore, Diameter
- s Piston stroke
- V_h Cylinder displacement
- V_H Engine displacement
- z Number of cylinders

Given: $d = 80$ mm $s = 84$ mm, $z = 4$.
To find: Engine displacement V_H in cm^3.

$$V_h = \frac{\pi \cdot 8^2\ cm^2}{4} \cdot 8{,}4\ cm = 422{,}2\ cm^3$$

$$V_H = 422{,}2\ cm^3 \cdot 4 = 1\,688{,}8\ cm^3$$

Given: $V_H = 1\,688{,}8\ cm^3$, $s = 84$ mm, $z = 4$.
To find: Cylinder bore d in mm

$$V_h = 1\,688{,}8\ cm^3 : 4 = 422{,}2\ cm^3$$

$$d = \sqrt{\frac{4 \cdot 422{,}2\ cm^3}{\pi \cdot 8{,}4\ cm}} = 8\ cm = 80\ mm$$

Compression ratio

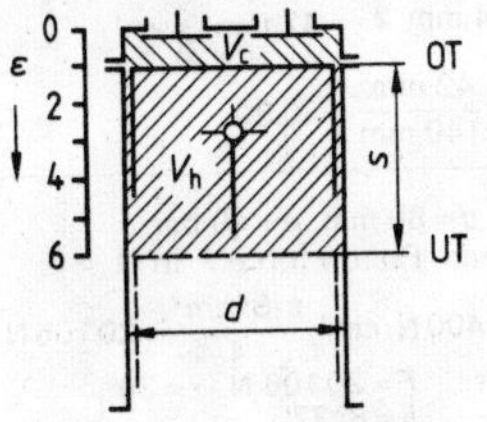

$$\varepsilon = \frac{V_h + V_c}{V_c} \qquad \varepsilon = \frac{V_h}{V_c} + 1$$

$$V_h = V_c \cdot (\varepsilon - 1) \qquad V_c = \frac{V_h}{\varepsilon - 1}$$

- V_c Compression volume, smallest combustion chamber
- V_h Cylinder displacement
- $V_h + V_c$ Largest combustion chamber
- ε Compression ratio

Geg.: $V_h = 422{,}2\ cm^3$, $V_c = 60{,}3\ cm^3$
To find: Compression ratio

$$\varepsilon = \frac{422{,}2\ cm^3 + 60{,}3\ cm^3}{60{,}3\ cm^3} = 8{,}0$$

Given: $V_H = 1\,688{,}8\ cm^3$, $\varepsilon = 8{,}0$, $z = 4$.
To find: Compression volume V_c in cm^3

$$V_h = 1\,688{,}8\ cm^3 : 4 = 422{,}2\ cm^3$$

$$V_c = \frac{422{,}2\ cm^3}{8{,}0 - 1} = 60{,}3\ cm^3$$

Compression change

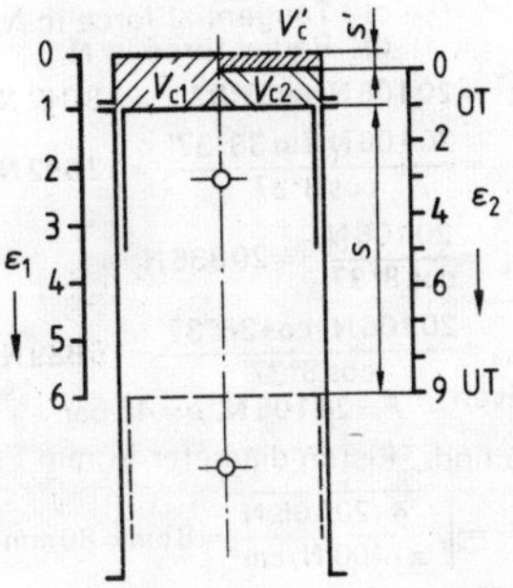

$$s' = \frac{s}{\varepsilon_1 - 1} - \frac{s}{\varepsilon_2 - 1} \qquad \varepsilon_1 < \varepsilon_2 \longrightarrow s \text{ positive} \qquad \varepsilon_1 > \varepsilon_2 \longrightarrow s' \text{ negative}$$

$$s' = \frac{V_{c1} - V_{c2}}{A} \qquad V_c' = A \cdot s'$$

$$\varepsilon_2 = \frac{V_h + V_{c2}}{V_{c2}} \qquad V_{c2} = V_{c1} \pm V_c'$$

$$\varepsilon_2 = \frac{s \cdot (\varepsilon_1 - 1)}{s \pm s' \cdot (\varepsilon_1 - 1)} + 1$$

+ for height increase
- for height decrease

- s Piston stroke
- s' Height change-compression chamber
- V_c' Volume change-compression chamber
- V_{c1} Compression chamber before change
- V_{c2} Compression chamber after change
- ε_1 Compression ratio before change
- ε_2 Compression ratio after change

Given: $s = 84$ mm, $\varepsilon_1 = 8{,}0$, $\varepsilon_2 = 9{,}0$.
To find: Height decrease

$$s' = \frac{84\ mm}{8{,}0 - 1} - \frac{84\ mm}{9{,}0 - 1} = +1{,}5\ mm$$

Given: $s = 84$ mm, $\varepsilon_1 = 8{,}0$, $\varepsilon_2 = 7{,}0$.
To find: Height change s' in mm.

$$s' = \frac{84\ mm}{8{,}0 - 1} - \frac{84\ mm}{7{,}0 - 1} = -2{,}0\ mm$$

Given: $s = 84$ mm, $\varepsilon_1 = 8{,}0$, Height decrease $= -1{,}5$ mm.
To find: Height decrease

$$\varepsilon_2 = \frac{84\ mm \cdot (8 - 1)}{84\ mm - 1{,}5\ mm \cdot (8 - 1)} + 1 = 9{,}0$$

Stroke-bore ratio
Delivery rate

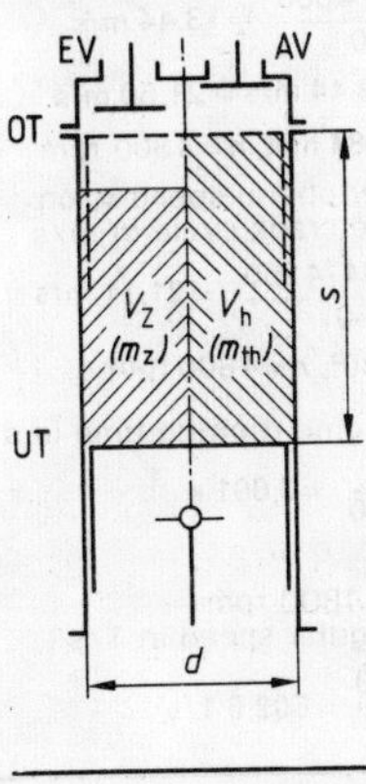

Stroke-bore ratio $\alpha = \frac{s}{d}$

$$s = \alpha \cdot d \qquad d = \frac{s}{\alpha}$$

- d Cylinder bore
- s Piston stroke
- α Stroke-bore ratio

Delivery rate $\lambda_l = \frac{m_z}{m_{th}}$ $\qquad m_z = m_{th} \cdot \lambda_l$

$$m_{th} = V_h \cdot \varrho_{th} \qquad m_z = V_z \cdot \varrho_z$$

- m_{th} Theoretical fresh gas weight in g
- m_z Weight of fresh gas supplied in g
- V_h Cylinder displacement in dm^3
- V_z Fresh gas volume supplied in dm^3
- λ_l Delivery rate, Filling rate
- ϱ_{th} Density of the theoretical fresh gas (mean value for 1,225 g/dm^3
- ϱ_z Density of supplied gas in g/dm^3

Given: $s = 84$ mm, $d = 80$ mm.
To find: Stroke-bore ratio α

$$\alpha = \frac{84\ mm}{80\ mm} = 1{,}05$$

Given: $m_z = 0{,}412$ g, $m_{th} = 0{,}515$ g.
To find: Delivery rate λ_L

$$\lambda_l = \frac{0{,}412\ g}{0{,}515\ g} = 0{,}80$$

Given: $V_z = 329{,}6\ cm^3$, $V_h = 422{,}2\ cm^3$, $\varrho_z = 1{,}25\ g/dm^3$, $\varrho_{th} = 1{,}22\ g/dm^3$
To find: Delivery rate

$$m_z = 0{,}3296\ dm^3 \cdot 1{,}25\ g/dm^3 = 0{,}41\ g$$

$$m_{th} = 0{,}4222\ dm^3 \cdot 1{,}22\ g/dm^3 = 0{,}51\ g$$

$$\lambda_l = \frac{0{,}412\ g}{0{,}515\ g} = 0{,}80$$

Crank gear

Sizes

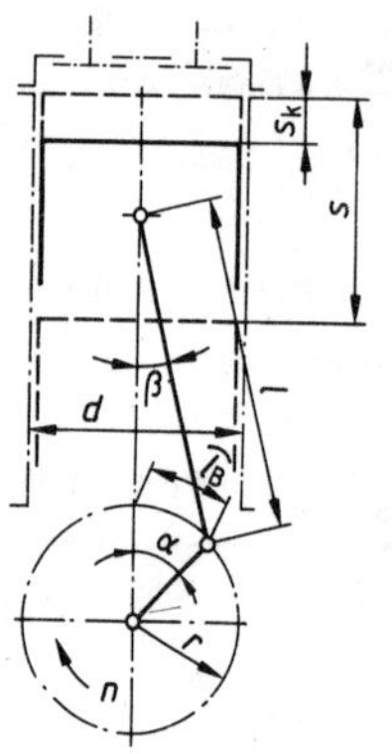

$$l_B = \frac{\pi \cdot s \cdot \alpha}{360°} \qquad \sin\beta = \frac{r \cdot \sin\alpha}{l}$$

$$s = \frac{360° \cdot l_B}{\pi \cdot \alpha} \qquad \sin\alpha = \frac{l \cdot \sin\beta}{r}$$

$$\alpha = \frac{360° \cdot l_B}{\pi \cdot s} \qquad U = \pi \cdot s$$

$$\lambda_{PL} = \frac{r}{l} \qquad l = \frac{r}{\lambda_{PL}} \qquad r = l \cdot \lambda_{PL}$$

$$s_K = r \cdot (1 - \cos\alpha) + l \cdot (1 - \cos\beta)$$

l Connecting rod length
l_B Arc length on the crank circle
r Crank radius, half piston stroke
s Piston stroke, double crank radius
s_K Piston path
U Circumference of crank circle
α Crank angle in degree
β Connecting rod angle in degree
λ_{PL} Connecting rod ratio

Given: $s = 84$ mm, $\alpha = 30°$.
To find: Arc length l_B on the crank circle

$$l_B = \frac{\pi \cdot 84\ \text{mm} \cdot 30°}{360°} = 22{,}0\ \text{mm}$$

Given: $r = 42$ mm, $l = 140$ mm, $\alpha = 30°$.
To find: Connecting rod **angle β** in degree

$$\sin\beta = \frac{42\ \text{mm} \cdot \sin 30°}{140\ \text{mm}} = 0{,}150$$

$$\beta = 8{,}63° = 8°37'$$

Given: $s = 84$ mm, $l = 140$ mm.
To find: Connecting rod ratio λ_{PL}

$$r = 84\ \text{mm} : 2 = 42\ \text{mm}$$

$$\lambda_{PL} = \frac{42\ \text{mm}}{140\ \text{mm}} = 0{,}30$$

Forces

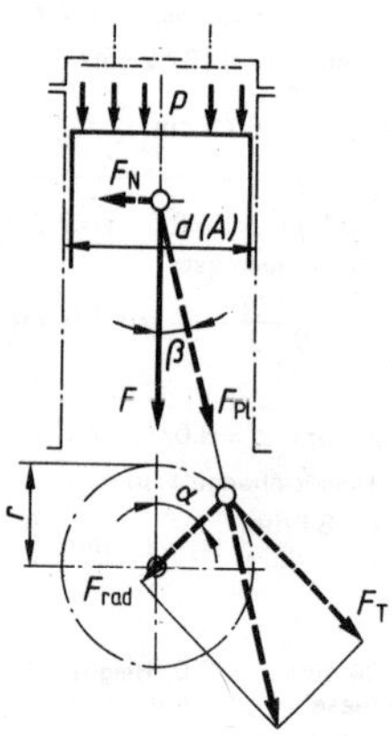

$$F = p \cdot \frac{\pi \cdot d^2}{4} \qquad F_N = F \cdot \tan\beta$$

$$p = \frac{4 \cdot F}{\pi \cdot d^2} \qquad F_{PL} = \frac{F}{\cos\beta}$$

$$d = \sqrt{\frac{4 \cdot F}{\pi \cdot p}} \qquad F_T = \frac{F \cdot \sin(\alpha + \beta)}{\cos\beta}$$

$$F = p \cdot A \qquad F_{rad} = \frac{F \cdot \cos(\alpha + \beta)}{\cos\beta}$$

$$M = F_T \cdot r \qquad F_T = \frac{M}{r}$$

A Piston area
d Piston diameter
F Piston force
F_N Normal force, side force
F_{PL} Connecting rod force, Bearing force
F_{rad} Radial force
F_T Tangential force
M Torque on the crank circle
p Gas pressure, combusition pressure
r Crank radius, half piston stroke
α Crank angle in degree
β Connecting rod angle in degree

Given: $d = 80$ mm, $p = 40$ bar.
To find: Piston **force F in N**

$$F = 400\ \text{N/cm}^2 \cdot \frac{\pi \cdot 8^2\ \text{cm}^2}{4} = 20\,106\ \text{N}$$

Given: $F = 20\,106$ N, $\alpha = 30°$, $\beta = 8°37'$.
To find:
a) Side force in N
b) Connecting rod force in N
c) Tangential force in N
d) Radial force in N

$$F_N = 20\,106\ \text{N} \cdot \tan 8°37' = 3\,047\ \text{N}$$

$$F_T = \frac{20\,106\ \text{N} \cdot \sin 38°37'}{\cos 8°37'} = 12\,692\ \text{N}$$

$$F_{PL} = \frac{20\,106\ \text{N}}{\cos 8°37'} = 20\,336\ \text{N}$$

$$F_{rad} = \frac{20\,106\ \text{N} \cdot \cos 38°37'}{\cos 8°37'} = 15\,889\ \text{N}$$

Given: $F = 20\,106$ N, $p = 40$ bar.
To find: Piston diameter in mm

$$d = \sqrt{\frac{4 \cdot 20\,106\ \text{N}}{\pi \cdot 400\ \text{N/cm}^2}} = 8\ \text{cm} = 80\ \text{mm}$$

Movements

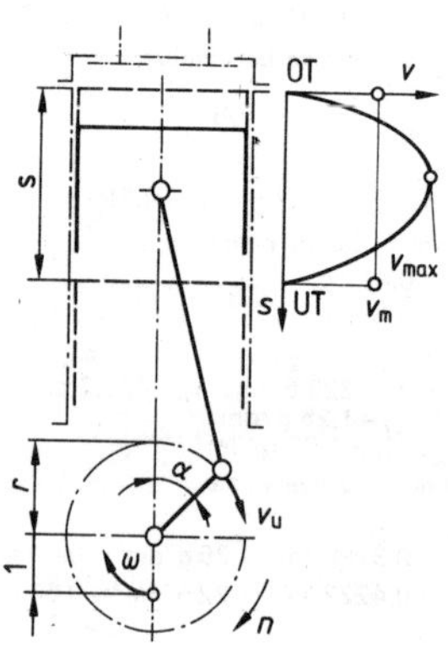

OT=Top Dead Center
UT=Bottom Dead Center

$$v_m = \frac{s \cdot n}{30} \qquad v_u = \frac{\pi \cdot s \cdot n}{60}$$

$$s = \frac{30 \cdot v_m}{n} \qquad s = \frac{60 \cdot v_u}{\pi \cdot n}$$

$$n = \frac{30 \cdot v_m}{s} \qquad n = \frac{60 \cdot v_u}{\pi \cdot s}$$

$$t = \frac{\alpha}{6 \cdot n} \qquad \omega = \frac{\pi \cdot n}{30}$$

$$\alpha = 6 \cdot n \cdot t \qquad n = \frac{30 \cdot \omega}{\pi}$$

$$n = \frac{\alpha}{6 \cdot t} \qquad v_{max} \approx 1{,}6 \cdot v_m$$

n Engine speed in rpm
s Piston stroke, Crank dia. in m
t Engine running time in s
v_m Mean piston speed in m/s
v_{max} Max piston speed in m/s
v_u Peripheral speed in m/s
α Crank angle in degree
ω Angular speed in 1/s

Given: $s = 84$ mm, $n = 4800$ rpm
To find:
a) Mean piston speed v_m
b) Max. piston speed v_{max}

$$v_m = \frac{0{,}084 \cdot 4800}{30} = 13{,}44\ \text{m/s}$$

$$v_{max} \approx 1{,}6 \cdot 13{,}44\ \text{m/s} = 21{,}50\ \text{m/s}$$

Given: $s = 84$ mm, $n = 4800$ rpm
To find: Peripheral speed v_u on the crank circle in m/s

$$v_u = \frac{\pi \cdot 0{,}084 \cdot 4800}{60} = 21{,}11\ \text{m/s}$$

Given: $\alpha = 30°$, $n = 4800$ rpm
To find: Engine running time in s

$$t = \frac{30}{6 \cdot 4800} = 0{,}001\ \text{s}$$

Given: $n = 4800$ rpm
To find: Angular speed in 1/s

$$\omega = \frac{\pi \cdot 4800}{30} = 502{,}6\ 1/\text{s}$$

Control angle

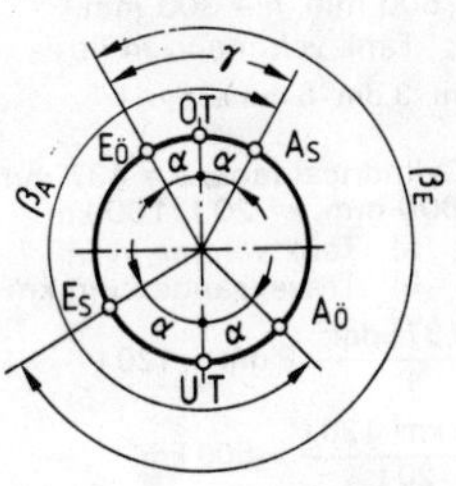

$\alpha_{Eö} = $ Eö to OT $\beta_E = \alpha_{Eö} + 180° + \alpha_{Es}$

$\alpha_{Es} = $ UT to Es $\beta_A = \alpha_{Aö} + 180° + \alpha_{As}$

$\alpha_{Aö} = $ Aö to UT $\beta' = \dfrac{100 \cdot \beta}{720}$

$\alpha_{As} = $ OT to As $\gamma = \alpha_{Eö} + \alpha_{As}$

α Control angle in degree
β Opening angle in degree
β' Opening angle in % per working cycle
γ Angle of overlap in degree

Given: Eö 15° before OT, Es 75° after UT
Aö 60° before UT, As 40° after OT
To find: a) Opening angle of EV
b) Opening angle of AV
c) Angle of overlap

$\beta_E = 15° + 180° + 75° = 270°$

$\beta_A = 60° + 180° + 40° = 280°$

$\gamma = 15° + 40° = 55°$

Given: $\beta_E = 270°$.
To find: Opening angle β' in %

$\beta' = \dfrac{100 \cdot 270°}{720°} = 37{,}5\%$

Arc length on flywheel

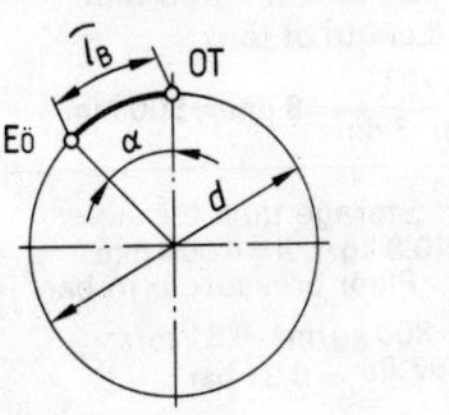

$l_B = \dfrac{\pi \cdot d \cdot \alpha}{360°}$

$d = \dfrac{360° \cdot l_B}{\pi \cdot \alpha}$

$\alpha = \dfrac{360° \cdot l_B}{\pi \cdot d}$

d Diameter of flywheel
l_B Arc length on the flywheel
α Control angle

Given: $d = 260$ mm, Eö 15° before OT.
To find: Arc length on flywheel

$l_B = \dfrac{\pi \cdot 260\ \text{mm} \cdot 15°}{360°} = 34{,}0$ mm

Given: $d = 260$ mm, Eö 34 mm before OT.
To find: Control angle in degrees

$\alpha_{Eö} = \dfrac{360° \cdot 34\ \text{mm}}{\pi \cdot 260\ \text{mm}} = 15°$

Valve opening time

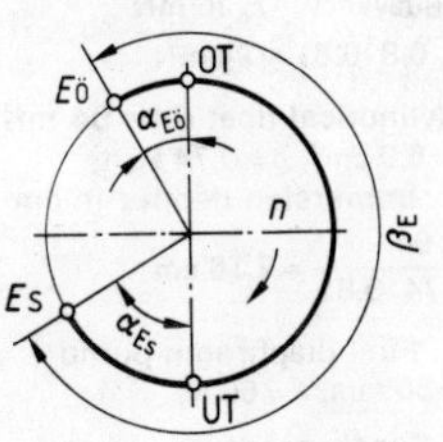

$t_1 = \dfrac{\beta}{6 \cdot n}$ $t_2 = \dfrac{\beta}{12}$

$\beta = 6 \cdot t_1 \cdot n$ $\beta = 12 \cdot t_2$

$n = \dfrac{\beta}{6 \cdot t_1}$ $t_2 = t_1 \cdot \dfrac{n}{2}$

n Engine speed in rpm
t_1 Valve opening time/working cycle in s
t_2 Valve opening time/minute in s
β Opening angle in degrees

Given: $\beta_E = 270°$, $n = 4800$ rpm
To find: a) Opening time/working cycle
b) Opening time per minute

$t_1 = \dfrac{270}{6 \cdot 4800} = 0{,}00937$ s

$t_2 = 270 : 12 = 22{,}5$ s

Given: $\beta_E = 270°$, $t_1 = 0{,}00937$ s.
To find: Engine speed in rpm

$n = \dfrac{270}{6 \cdot 0{,}00937} = 4800$ rpm

Valve opening area

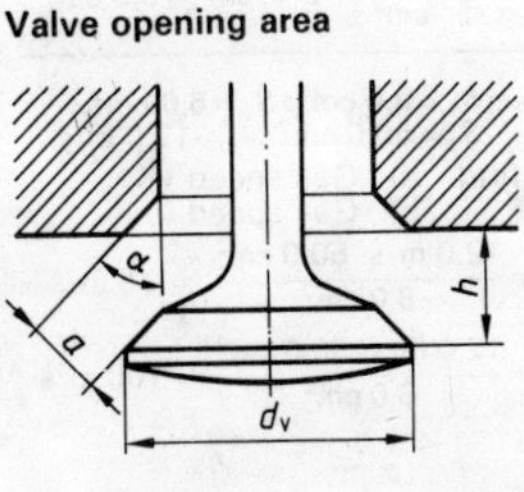

$A_V = \pi \cdot \sin\alpha \cdot h \cdot d_V$ $a = \sin\alpha \cdot h$

$h = \dfrac{A_V}{\pi \cdot \sin\alpha \cdot d_V}$ $\sin 45° = 0{,}7071$

$d_V = \dfrac{A_V}{\pi \cdot \sin\alpha \cdot h}$ $\sin 60° = 0{,}866$

A_V Valve opening area
d_V Diameter of valve disc
h Valve stroke
α Valve seating angle

Given: $d = 42$ mm, $h = 9$ mm, $\alpha = 45°$
To find: Valve opening area in mm²

$A_V = \pi \cdot \sin 45° \cdot 9 \cdot 42 = 839{,}7\ \text{mm}^2$

Given: $A_V = 839{,}7\ \text{mm}^2$, $d = 42$ mm, $\alpha = 45°$
To find: Valve stroke h in mm.

$h = \dfrac{839{,}7\ \text{mm}^2}{\pi \cdot \sin 45° \cdot 42\ \text{mm}} = 9{,}0$ mm

Gas speed

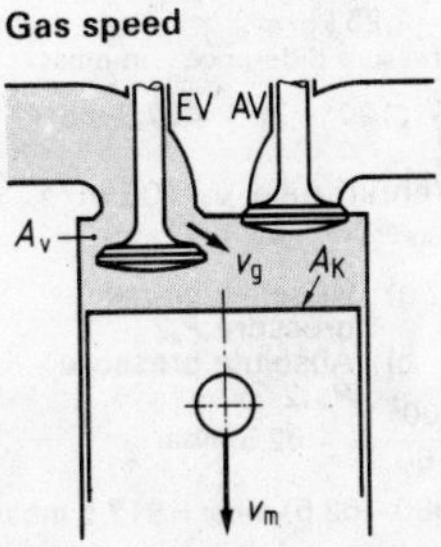

Cross sectional area x speed = constant

$A_V \cdot v_g = A_K \cdot v_m$ Conversion 1 m/s = 3,6 km/h

$v_g = \dfrac{A_K \cdot v_m}{A_V}$ $v_m = \dfrac{A_V \cdot v_g}{A_K}$

$A_V = \dfrac{A_K \cdot v_m}{v_g}$ $A_K = \dfrac{A_V \cdot v_g}{v_m}$

A_K Piston area
A_V Valve opening area
v_g Gas speed at the valve
v_m Mean piston speed

Given: $A_K = 50{,}3\ \text{cm}^2$, $v_m = 13{,}44$ m/s
$A_V = 8{,}397\ \text{cm}^2$.
To find: a) Gas speed in m/s
b) Gas speed in km/h

$v_g = \dfrac{50{,}3 \cdot 13{,}44\ \text{m/s}}{8{,}397} = 80{,}5$ m/s

$v_g = 3{,}6 \cdot 80{,}5 = 289{,}8$ km/h

Given: $v_g = 80{,}5$ m/s, $A_K = 50{,}3\ \text{cm}^2$, $v_m = 13{,}44$ m/s.
To find: Valve opening area in cm²

$A_V = \dfrac{50{,}3\ \text{cm}^2 \cdot 13{,}44}{80{,}5} = 8{,}397\ \text{cm}^2$

EV=Inlet valve; AV=Outlet valve; Eö=Inlet opening; Es=Inlet closing; Aö=Outlet opening; As=Outlet closing.

Fuel equipment, Carburettor

Tank volume Travel range

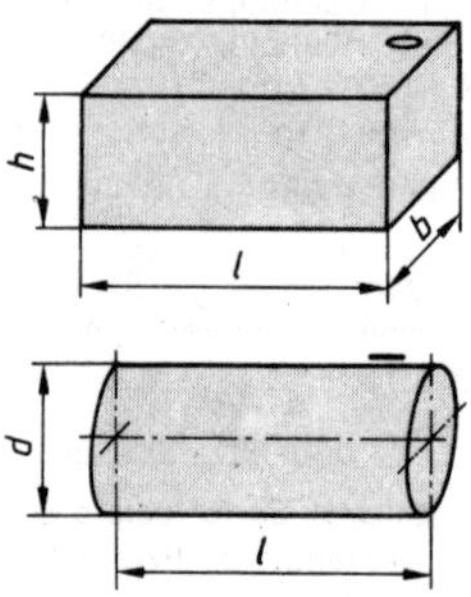

$V_B = b \cdot h \cdot l$ $\qquad V_B = \frac{\pi \cdot d^2}{4} \cdot l$

$l = \frac{V_B}{b \cdot h}$ $\qquad l = \frac{4 \cdot V_B}{\pi \cdot d^2}$

$b = \frac{V_B}{h \cdot l}$ $\qquad d = \sqrt{\frac{4 \cdot V_B}{\pi \cdot l}}$

$h = \frac{V_B}{b \cdot l}$ $\qquad V_B = \frac{k \cdot s_F}{100}$

$s_F = \frac{100 \cdot V_B}{k}$ $\qquad k = \frac{100 \cdot V_B}{s_F}$

- b Width in dm
- d Diameter in dm
- h Height in dm
- k Fuel consumption in 1/100 km
- l Length in dm
- s_F Travel range in km
- V_B Tank volume in l

Given: Prismatic tank, l = 800 mm, b = 500 mm, h = 300 mm.
To find: Tank volume v_B in litres

$V_B = 5\ \text{dm} \cdot 3\ \text{dm} \cdot 8\ \text{dm} = 120\ \text{l}$

Given: Cylindrical tank, d = 437 mm, l = 800 mm, k= 20 l/100 km.
To find: a) Tank volume v_B in l
b) Travel range s_F in km

$V_B = \frac{\pi \cdot 4{,}37^2\ \text{dm}^2}{4} \cdot 8\ \text{dm} = 120\ \text{l}$

$s_F = \frac{100\ \text{km} \cdot 120\ \text{l}}{20\ \text{l}} = 600\ \text{km}$

Given: Cylindrical float, V_B= 120 l, b = 500 mm, h = 300 mm,
To find: Lenqth of tank

$l = \frac{120\ \text{l}}{5\ \text{dm} \cdot 3\ \text{dm}} = 8\ \text{dm} = 800\ \text{mm}$

Hydraulic Pressure, Buoyancy

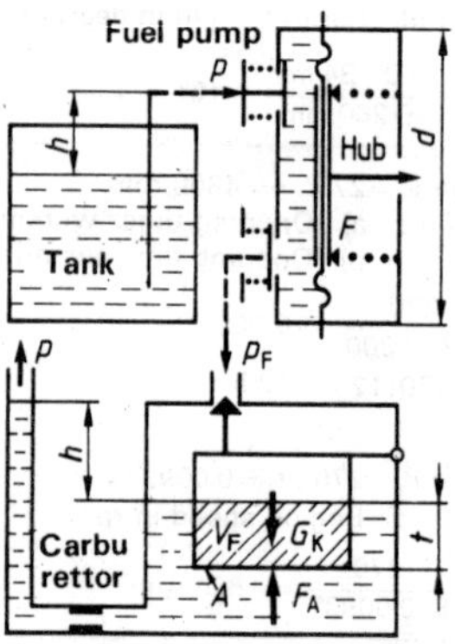

$p = h \cdot \varrho \cdot g$ $\qquad F_A = A \cdot t \cdot \varrho \cdot g$

$h = \frac{p}{\varrho \cdot g}$ $\qquad t = \frac{F_A \text{ or } G_K}{A \cdot \varrho \cdot g}$

$\varrho = \frac{p}{h \cdot g}$ $\qquad F_A = V_F \cdot \varrho \cdot g$

$p_F = \frac{4 \cdot F}{\pi \cdot d^2}$ $\qquad F = p_F \cdot \frac{\pi \cdot d^2}{4}$

- A, S Mean float cross section in cm²
- d Diameter of diaphragm
- F Force of diaphragm spring in N
- F_A Buoyancy, Gravitational force in mN
- g Acceleration due to gravity 9,81 m/s²
- G_k Gravitational force of the float in mN
- h Pressure head, Suction head in m
- p Floor pressure, side pressure in Pa conversion 100000 Pa=1 bar
- p_f Supply pressure of fuel pump in N/cm²
- t Immersion depth of float in cm
- V_f Displaced liquid volume in cm³
- ϱ Density in kg/m³ (pe) in g/cm³ (F_A)

Given: Storage tank for super $\varrho = 0{,}8$ kg/l, $h = 4\,000$ mm.
To find: Floor pressure p in bar

$p = 4\ \text{m} \cdot 800\ \text{kg/m}^3 \cdot 9{,}81\ \text{m/s}^2 = 31\,392\ \text{Pa} = 0{,}31\ \text{bar}$

Given: Cylindrical float, $A = 6{,}0\ \text{cm}^2$, $t = 20$ mm, $\varrho = 0{,}8$ kg/l.
To find: Buoyancy F_A in mN.

$F_A = 6 \cdot 2 \cdot 0{,}8 \cdot 9{,}81 = 94\ \text{mN}$

Given: Cylindrical float G_S= 94 mN, $A = 6{,}0\ \text{cm}^2$, $\varrho = 0{,}74$ kg/l.
To find: Immersion depth t in cm

$t = \frac{94}{6 \cdot 0{,}74 \cdot 9{,}81} = 2{,}16\ \text{cm}$

Given: Fuel diaphragm pump $d = 50$ mm, $F = 60$ N.
To find: Supply pressure in bar

$p_F = \frac{4 \cdot 60\ \text{N}}{\pi \cdot 5^2\ \text{cm}^2} = 3\ \text{N/cm}^2 = 0{,}3\ \text{bar}$

Gas pressure, Gas speed

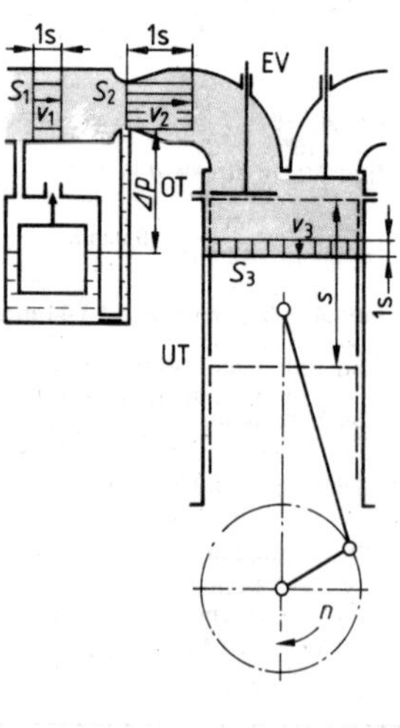

Cross sectional area x speed = constant

$v_2 = \frac{v_3 \cdot S_3}{S_2}$ $\qquad \Delta p = \frac{\varrho}{200} \cdot (v_2^2 - v_1^2)$

$v_1 = \frac{v_3 \cdot S_3}{S_1}$ $\qquad \Delta p = p_{e2} - p_{e1}$

$\Delta p = p_{abs1} - p_{abs2}$

$v_2 = \frac{v_1 \cdot S_1}{S_2}$ $\qquad p_{abs} = p_{amb} - p_e$

$v_3 = \frac{s \cdot n}{30}$ $\qquad p_{e1} \approx -\frac{v_1^2}{160}$ $\qquad p_{e2} \approx -\frac{v_2^2}{160}$

- n Engine speed in 1/min
- p_{abs1} Absolute pressure in suction pipe in mbar
- p_{abs2} Absolute pressure in the ventur tube in mbar
- p_{amb} Atmospheric pressure, Ambient pressure
- p_{e1} Negative gauge pressure in the suction pipe in mbar
- p_{e2} **–ve gauge pressure in Venturi tube, mbar**
- ΔP Pressure difference in mbar
- s Piston stroke in m
- S_1 Cross sectional area of suction pipe
- S_2 Cross sectional area of Venturi tube
- S_3 Cross sectional area of cylinder, piston area
- V_1 Gas speed in the suction pipe
- V_2 Gas speed in venturi tube
- V_3 Mean piston speed
- ϱ Density of fuel+air mixture in kg/m³

Given: $S_1 = 8{,}0\ \text{cm}^2$, $S_2 = 6{,}0\ \text{cm}^2$, $S_3 = 50{,}0\ \text{cm}^2$, $v_3 = 12{,}0$ m/s.
To find: a) Gas speed v_1
b) Gas speed v_1

$v_1 = \frac{12{,}0\ \text{m/s} \cdot 50{,}0\ \text{cm}^2}{8{,}0\ \text{cm}^2} = 75\ \text{m/s}$

$v_2 = \frac{12{,}0\ \text{m/s} \cdot 50{,}0\ \text{cm}^2}{6{,}0\ \text{cm}^2} = 100\ \text{m/s}$

Given: $v_1 = 75{,}0$ m/s, $v_2 = 100{,}0$ m/s, $\varrho = 1{,}25$ kg/m³
To find: Pressure difference in mbar

$\Delta p = \frac{1{,}25}{200} \cdot (100^2 - 75^2) = 27{,}3\ \text{mbar}$

Given: Venturi tube, v_2=100 m/s, p_{amb} = 980 mbar.
To find: a) Negative gauge pressure P_e2
b) Absolute pressure $P_{abs}2$

$p_{e2} \approx -\frac{100^2}{160} = -62{,}5\ \text{mbar}$

$p_{abs2} = (980 - 62{,}5)\ \text{mbar} = \mathbf{917{,}5}\ \text{mbar}$

Air requirement, Air ratio

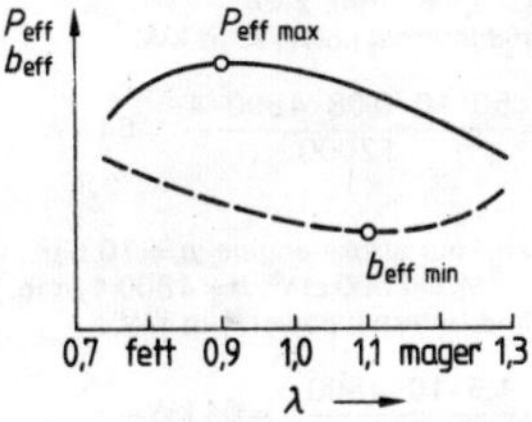

Air requirement for Otto engines

Theor. value	14,2…14,8 kg/kg
Full load	≈13,0 kg/kg
Partial load	≈15,0 kg/kg
Idle running	≈10,0 kg/kg
Cold start	≈ 3,0 kg/kg

$L = \frac{\text{Air quantity in kg}}{\text{Fuel quantity in kg}}$ $\quad L = \frac{m_L}{m_K}$

$m_L = m_K \cdot L$ $\quad m_K = \frac{m_L}{L}$

L Air requirement in kg per kg of fuel
m_K Fuel quantity in kg
m_L Air quantity in kg

$\lambda = \frac{\text{Supplied air quantity}}{\text{Theoretical air quantity required}}$ $\quad \lambda = \frac{L_z}{L_{th}}$

$L_z = L_{th} \cdot \lambda$ $\quad L_{th} = \frac{L_z}{\lambda}$

L_{th} Theoretical quantity of air requred in kg/kg
L_z Supplied quantity of air in kg/kg
λ Air ratio

Given: $m_L = 162{,}8$ kg, $m_K = 10$ kg.
To find: Air required L in kg/kg
$L = 16{,}28$ kg/kg $= 162{,}8$ kg$: 10$ kg

Given: $m_K = 10$ kg, $L = 16{,}28$ kg/kg.
To find: Quantity of air m_L in kg
$m_L = 10$ kg $\cdot 16{,}28$ kg/kg $= 162{,}8$ kg

Given: $L_z = 16{,}3$ kg/kg, $L_{th} = 14{,}8$ kg/kg.
To find: Air ratio λ.
$\lambda = 16{,}3$ kg/kg$: 14{,}8$ kg/kg $= 1{,}10$

Given $L_{th} = 14{,}8$ kg/kg, $\lambda = 0{,}90$.
To find: Supplied air
$L_z = 14{,}8$ kg/kg $\cdot 0{,}90 = 13{,}32$ kg/kg

Fuel consumption

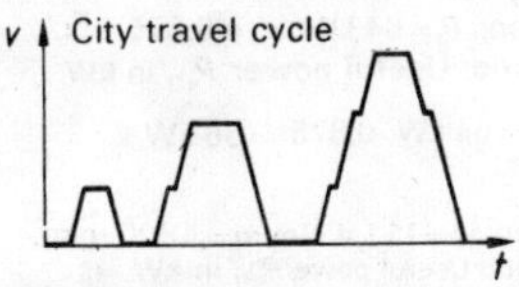

According to DIN 70030-1
1. In the city cycle
2. At v=90 km/h, constant
3. At v=120 km/h, constant

According to DIN 70030-2
Test road plane and dry 10 km long to and fro 3/4 of maximum speed

According to DIN 70030-1 $\quad C = \frac{100 \cdot V_K}{s}$

$C = \frac{100 \cdot m}{\varrho \cdot s}$

According to DIN 70030-2 $\quad k = 1{,}1 \cdot \frac{100 \cdot V_K}{s}$

Mileage Consumption $\quad k_s = \frac{100 \cdot V_K}{s}$

C Fuel consumption in L/100 km
k Fuel consumption in L/100 km
k_s Fuel Mileage consumption in L/100 km
m measured fuel consumption in kg
s Mileage in km
V_K Measured fuel volume in l
ϱ Density of fuel in kg/l

Given: Passenger car, city cycle, m=148 g, $\varrho = 0{,}74$ kg/l, $s = 2000$ m.
To find: Consumption as per DIN 70030-1
$C = \frac{100 \cdot 0{,}148}{0{,}74 \cdot 2} = 10{,}0$ l/100 km

Given: Truck $V_K = 3{,}6$ l, $s = 20{,}0$ km.
To find: Consumption as per DIN 70030-2
$k = 1{,}1 \cdot \frac{100 \cdot 3{,}6}{20} = 19{,}8$ l/100 km

Given: Passenger car, $V_K = 34{,}2$ l, $s = 360$ km.
To find: Fuel Mileage consumption
$k_s = \frac{100 \cdot 34{,}2}{360} = 9{,}5$ l/100 km

Travel range

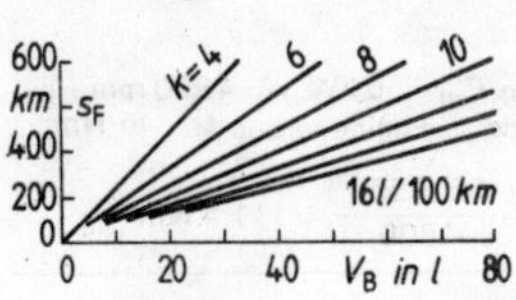

According to DIN 70020 $\quad s_F = \frac{100 \cdot V_B}{k}$

$V_B = \frac{k \cdot s_F}{100}$ $\quad k = \frac{100 \cdot V_B}{s_F}$

Fuel consumption according to DIN 70030-2
s_F Travel range in km
V_B Tank volume in l

Given: $k = 19{,}8$ l/100 km, $V_B = 99$ l.
To find: Travel range According to DIN 70020
$s_F = \frac{100 \text{ km} \cdot 99 \text{ l}}{19{,}8 \text{ l}} = 500$ km

Given: $k = 19{,}8$ l/100 km, $s = 500$ km.
To find: Tank volume in l
$V_B = \frac{19{,}8 \text{ l} \cdot 500 \text{ km}}{100 \text{ km}} = 99$ l

Specific fuel consumption per hour

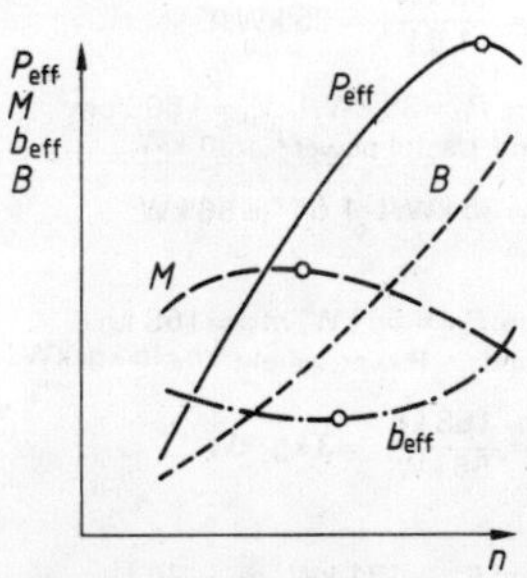

Specific fuel consumption

Otto engine	300…380 g/kWh
Diesel engine	200…270 g/kWh

$b_{eff} = \frac{3600 \times \text{measured volume} \times \text{density}}{\text{flow time} \times \text{usefuel power}}$

$b_{eff} = \frac{3600 \cdot V \cdot \varrho}{t \cdot P_{eff}}$ $\quad V = \frac{b_{eff} \cdot P_{eff} \cdot t}{3600 \cdot \varrho}$

b_{eff} Specific fuel consumption
P_{eff} Useful power of engine in kW
t Flow time in s
V Measured fuel volume in cm³
ϱ Density of fuel in g/cm³

$B = \frac{\text{Specific fuel consumption} \times \text{useful power}}{1000}$

$B = \frac{b_{eff} \cdot P_{eff}}{1000}$ $\quad b_{eff} = \frac{1000 \cdot B}{P_{eff}}$

B Fuel consumption in kg/h
b_{eff} Specific fuel consumption in g/kWh
P_{eff} Useful power of engine in kW

Given: $V = 100$ cm³, $t = 20$ s, $\varrho = 0{,}74$ g/cm³, $P_{eff} = 45$ kW.
To find: Specific fuel consumption
$b_{eff} = \frac{3600 \cdot 100 \cdot 0{,}74}{20 \cdot 45} = 296$ g/kWh

Given: $b_{eff} = 296$ g/kWh, $t = 20$ s, $\varrho = 0{,}74$ g/cm³, $P_{eff} = 45$ kW.
To find: Measured volume V in cm³
$V = \frac{296 \cdot 20 \cdot 45}{3600 \cdot 0{,}74} = 100$ cm³

Given: $P_{eff} = 45$ kW, $b_{eff} = 296$ g/kWh.
To find: Fuel consumption B in kg/h
$B = \frac{296 \text{ g/kWh} \cdot 45 \text{ kW}}{1000} = 13{,}32$ kg/h

Given: $B = 13{,}32$ kg/h, $P_{eff} = 45$ kW.
To find: Specific fuel consumption
$b_{eff} = \frac{1000 \cdot 13{,}32 \text{ kg/h}}{45 \text{ kW}} = 296$ g/kWh

Engine Power

Internal power

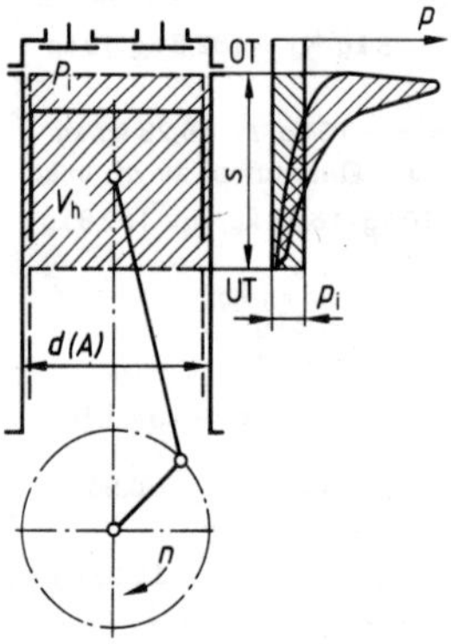

$$P_i = \frac{A \cdot p_i \cdot s \cdot n \cdot z}{12000\ ^{1)}} \qquad P_i = \frac{V_H \cdot p_i \cdot n}{1200\ ^{2)}}$$

$$A = \frac{12000 \cdot P_i}{p_i \cdot s \cdot n \cdot z} \qquad V_H = \frac{1200 \cdot P_i}{p_i \cdot n}$$

$$p_i = \frac{12000 \cdot P_i}{A \cdot s \cdot n \cdot z} \qquad p_i = \frac{1200 \cdot P_i}{V_H \cdot n}$$

$$s = \frac{12000 \cdot P_i}{A \cdot p_i \cdot n \cdot z} \qquad n = \frac{1200 \cdot P_i}{V_H \cdot p_i}$$

$$n = \frac{12000 \cdot P_i}{A \cdot p_i \cdot s \cdot z}$$

A Piston area in cm²
n Engine speed in 1/min.
p_i Mean indicated working pressure in bar
P_i Internal power of engine in kW
s Piston stoke in m
V_H Engine displacement in l
z Number of cylinders
1) 6000 or 2) 600 for two stroke engine

Given: Four-stroke engine, $A = 50$ cm², $p_i = 10$ bar, $n = 4800$ 1/min, $s = 80$ mm, $z = 4$.
To find: Internal power P_i in kW.

$$P_i = \frac{50 \cdot 10 \cdot 0{,}08 \cdot 4800 \cdot 4}{12000} = 64\ \text{kW}$$

Given: Four stroke engine, $p_i = 10$ bar, $V_H = 1600$ cm³, $n = 4800$ 1/min.
To find: Internal power P_i in kW.

$$P_i = \frac{1{,}6 \cdot 10 \cdot 4800}{1200} = 64\ \text{kW}$$

Given: Four-stroke engine, A=50 cm², $s = 80$ mm, $P_i = 64$ kW, $n = 4800$ 1/min, $z = 4$.
To find: Mean indicated working pressure

$$p_i = \frac{64 \cdot 12000}{50 \cdot 0{,}08 \cdot 4800 \cdot 4} = 10\ \text{bar}$$

Useful power

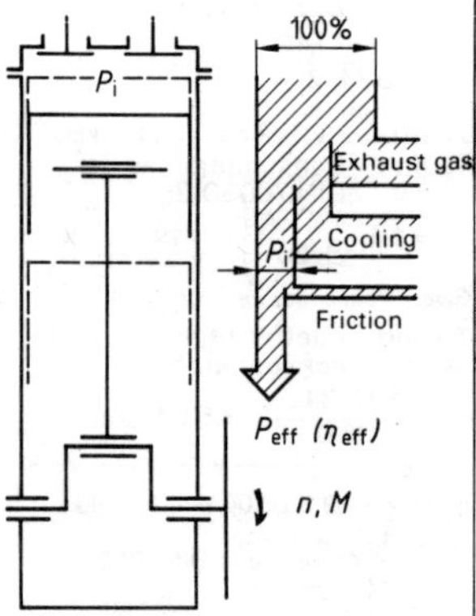

Degree of efficiency

	η_{eff}	η_m
Otto	0,22...0,30	0,80...0,90
Diesel	0,32...0,40	0,75...0,85

From internal power $P_{eff} = P_i \cdot \eta_m$

$$P_i = \frac{P_{eff}}{\eta_m} \qquad \eta_m = \frac{P_{eff}}{P_i}$$

From torque $P_{eff} = \dfrac{M \cdot n}{9550}$

$$M = \frac{9550 \cdot P_{eff}}{n} \qquad n = \frac{9550 \cdot P_{eff}}{M}$$

From heat quantity $P_{eff} = \dfrac{B \cdot H_u \cdot \eta_{eff}}{3600}$

$$B = \frac{3600 \cdot P_{eff}}{H_u \cdot \eta_{eff}} \qquad \eta_{eff} = \frac{3600 \cdot P_{eff}}{B \cdot H_u}$$

B Fuel consumption in kg/h
H_u Specific calorific value in kJ/kg
M Engine torque in Nm
n Engine speed in rpm
P_{eff} Useful power of engine in kW
P_i Internal power of engine in kW
η_{eff} Net efficiency
η_m Mechanical efficiency

Given: $P_i = 64$ kW, $\eta_m = 0{,}875$.
To find: Useful power P_{eff} in kW.

$P_{eff} = 64\ \text{kW} \cdot 0{,}875 = 56\ \text{kW}$

Given: $M = 111{,}4$ Nm, $n = 4800$ rpm
To find: Useful power P_{eff} in kW.

$$P_{eff} = \frac{111{,}4 \cdot 4800}{9550} = 56\ \text{kW}$$

Given: $B = 19{,}2$ kg/h, $\eta_{eff} = 0{,}25$, $H_u = 42000$ kJ/kg.
To find: Useful power P_{eff} in kW.

$$P_{eff} = \frac{19{,}2 \cdot 42000 \cdot 0{,}25}{3600} = 56\ \text{kW}$$

Given: $P_{eff} = 56$ kW, $n = 4800$ rpm
To find: Engine torque M in Nm.

$$M = \frac{56 \cdot 9550}{4800} = 111{,}4\ \text{Nm}$$

Characteristic quantities

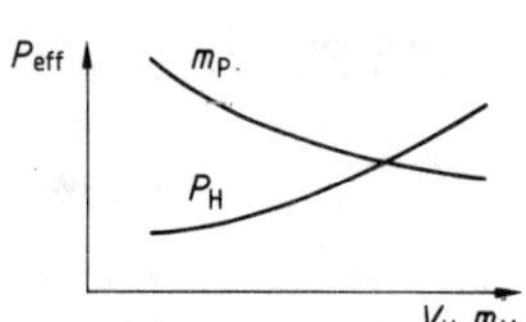

Piston displacement power

Otto engine	
Motor cycle	35...60 kW/l
Passenger car	25...55 kW/l
Diesel engine	
Passenger car	15...22 kW/L
Truck	15...25 kW/L
Power weight	
Otto engine	
Motor cycle	2.0...6.0 kg/kW
Passenger car	2.5...7.0 kg/kW
Diesel engine	
Passenger car	4.0...8.0 kg/kW
Truck	5.0...10 kg/kW
Weight power, minimum	
Truck, Bus	4.4 kW/t
Trains, Trailors and Trams	2.2 kW/t

Piston displacement power $P_H = \dfrac{P_{eff}}{V_H}$

$$P_{eff} = P_H \cdot V_H \qquad V_H = \frac{P_{eff}}{P_H}$$

Power weight $m_P = \dfrac{m_M}{P_{eff}}$

$$P_{eff} = \frac{m_M}{m_P} \qquad m_M = m_P \cdot P_{eff}$$

Weight power $P_m = \dfrac{P_{eff}}{m_G}$

$$P_{eff} = P_m \cdot m_G \qquad m_G = \frac{P_{eff}}{P_m}$$

m_G Permissible total weight in t
m_M Engine weight in kg
m_P Power weight in kg/kW
P_{eff} Useful power of engine in kW
P_H Piston displacement power
P_m Weight power in kW/t
V_H Engine displacement in l

Given: $P_{eff} = 56$ kW, $V_H = 1600$ cm³.
To find: Piston displacement power P_H in kW/L.

$$P_H = \frac{56\ \text{kW}}{1{,}6\ \text{l}} = 35\ \text{kW/l}$$

Given: $P_H = 35$ kW/l, $V_H = 1600$ cm³.
To find: Useful power P_{eff} in kW.

$P_{eff} = 35\ \text{kW/l} \cdot 1{,}6\ \text{l} = 56\ \text{kW}$

Given: $P_{eff} = 56$ kW, $m_M = 168$ kg.
To find: Power weight m_P in kg/kW.

$$m_P = \frac{168\ \text{kg}}{56\ \text{kW}} = 3\ \text{kg/kW}$$

Given: $P_{eff} = 171$ kW, $m_G = 36$ t.
To find: Weight power in kW/t

$$P_m = \frac{171\ \text{kW}}{36\ \text{t}} = 4{,}75\ \text{kW/t}$$

Cooling

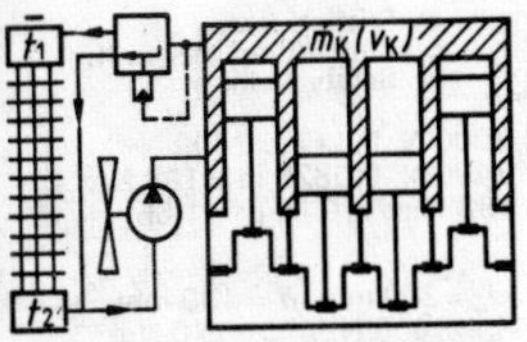

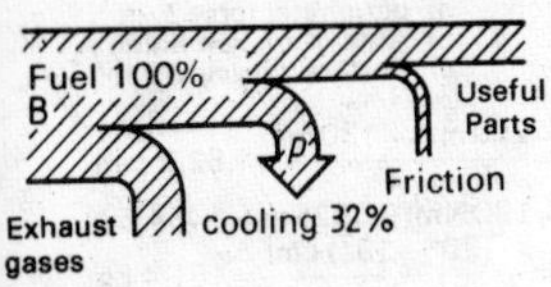

anti-frost freezing point	water	anti-frost
−10°	4 parts	1 part
−20°	2 parts	1 part
−30°	5 parts	4 parts
−40°	8 parts	9 parts

$\Phi_K = B \cdot H_u \cdot p$ $\qquad$ $\Phi_K = \dot m \cdot c \cdot \Delta T$

$p = \dfrac{\Phi_K}{H_u \cdot B}$ $\qquad$ $\dot m = \dfrac{\Phi_K}{c \cdot \Delta T}$

$\dot z = \dfrac{\dot m}{m_K}$ $\qquad$ $V_F = \dfrac{V_K \cdot p_F}{p_G}$

$\dot m = \dot z \cdot m_K$ $\qquad$ $p_F = \dfrac{V_F \cdot p_G}{V_K}$

B Fuel consumption in kg/h
c Specific heat capacity in kJ/kg·K
H_u Specific colorific value in kJ/kg
m Coolant water throughput in kg/h
m_k Filling quantity of coolant system in kg/l
p Heat flow share of coolant water
p_F Anti-frost portion
p_G Total share of anti-frost + water
V_F Volume of anti-frost in l
V_K Filling volume of coolant system in l
Z Circulations of coolant water per hour
ΔT Temperature difference in °C or K
Φ_K Coolant water-heat flow in kJ/h

Given: $B = 9$ kg/h, $H_u = 43000$ kJ/kg, $p = 0{,}33$.
To find: Coolant water-heat flow Φ_K.

$\Phi_K = 9\ \text{kg/h} \cdot 43000\ \text{kJ/kg} \cdot 0{,}33 = 127710\ \text{kJ/h}$

Given: $\Phi_K = 127710$ kJ/h, $\Delta T = 12$ K, $c = 4{,}18$ kJ/kg·K.
To find: Coolant water throughput in kg/h

$\dot m = \dfrac{127710\ \text{kJ/h}}{12\ \text{K} \cdot 4{,}18\ \text{kJ/kg·K}} = 2546\ \text{kg/h}$

Given $\dot m = 2546$ kg/h, $m_K = 6$ kg (l).
To find: Coolant water circulation z per h

$\dot z = \dfrac{2546\ \text{kg/h}}{6\ \text{kg}} = 424/\text{h}$

Given: $V_K = 6$ l, $p_F = 4$, $p_G = 10$.
To find: a) Anti-frost volume in l
b) Water volume in l

$V_F = \dfrac{6\ \text{l} \cdot 4}{10} = 2{,}4\ \text{l}$

$V_W = 6{,}0 - 2{,}4 = 3{,}6\ \text{l}$

Lubrication

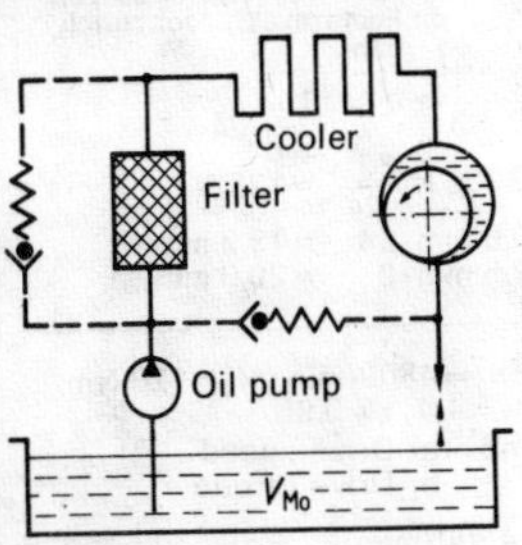

Supply quantity Oil pump	12,5...45 kg/min
Oil consumption per 100 km	
Circulation system lubrication	0,1...0,15 l
Fresh oil lubrication	0,15...0,4 l
Mixing ratio Two stroke	1:20 bis 1:40

$\dot m_{Öl} = \dfrac{60 \cdot V_{Mo} \cdot \varrho}{t}$ $\qquad$ $V_{Öl} = \dfrac{V_{Mi}}{z+1} = \dfrac{V_K}{z}$

$V_{Mo} = \dfrac{m_{Öl} \cdot t}{60 \cdot \varrho}$ $\qquad$ $V_{Mi} = V_{Öl} \cdot (z+1)$

$t = \dfrac{60 \cdot V_{Mo} \cdot \varrho}{\dot m_{Öl}}$ $\qquad$ $V_K = V_{Öl} \cdot z$

$v = \dfrac{100 \cdot m_{Öl}}{\varrho \cdot s}$ $\qquad$ $v = \dfrac{100 \cdot V_{Öl}}{s}$

$s = \dfrac{100 \cdot m_{Öl}}{\varrho \cdot v}$ $\qquad$ $s = \dfrac{100 \cdot V_{Öl}}{v}$

$m_{Öl}$ Oil consumption in kg
$\dot m_{Öl}$ Oil pump supply capacity in kg/min
s Travelled stretch in km
t Time for one circulation of oil in s
v Oil consumtion in
V_K Fuel volume in
V_{Mi} Volume of two-stroke mixture in l
V_{Mo} Oil filling of engine in l
$V_{Öl}$ Oil consumption in l
$1{:}z$ Mixing ratio of oil: fuel
ϱ Density of oil in

Given: $V_{Mo} = 4$ l, $t = 10{,}8$ s, $\varrho = 0{,}9$ kg/dm³.
To find: Supply quantity $m_{Öl}$ in kg/min.

$\dot m_{Öl} = \dfrac{60 \cdot 4 \cdot 0{,}9}{10{,}8} = 20\ \text{kg/min}$

Given: $m_{Öl} = 276$ g, $\varrho = 0{,}92$ kg/l, $s = 300$ km.
To find: Oil consumption v in l/100 km.

$v = \dfrac{100 \cdot 0{,}276}{0{,}92 \cdot 300} = 0{,}1\ \text{l/100 km}$

Given: $V_{Mi} = 20{,}8$ l, $1{:}z = 1{:}25$, $s = 200$ km.
To find: a) Oil consumption $V_{Öl}$ in l
b) Oil consumption v in l/100 km

$V_{Öl} = \dfrac{20{,}8\ \text{l}}{25+1} = 0{,}8\ \text{l}$

$v = \dfrac{100 \cdot 0{,}8}{200} = 0{,}40\ \text{l/100 km}$

Ignition

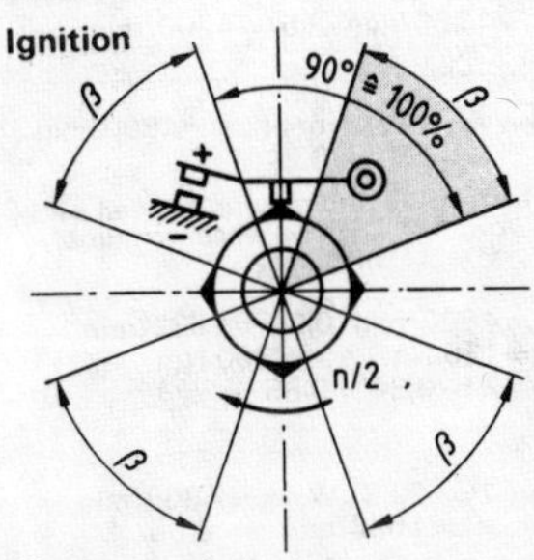

Closing angle in degree

4-Cylinder engine	50° ± 3°
6-Cylinder engine	40° ± 3°

Closing angle in per cent

4-Cylinder engine	52%...59%
6-Cylinder engine	62%...71%

Interrupter closing time $t_U = \dfrac{\beta}{3 \cdot n}$

$p = \dfrac{\beta \cdot z}{3{,}6}$ $\qquad$ $n = \dfrac{\beta}{3 \cdot t_U}$

$\beta = \dfrac{3{,}6 \cdot p}{z}$ $\qquad$ $\beta = 3 \cdot n \cdot t_U$

Number of ignition impulses $f = \dfrac{n \cdot z}{120\ ^{1)}}$

$n = \dfrac{120 \cdot f}{z}$ $\qquad$ $z = \dfrac{120 \cdot f}{n}$

f Number of ignition impulses of the engine in 1/s
n Engine speed in rpm
p Closing angle in per cent
t_u Closing angle of the interrupter in s
z Number of cylinders
β Closing angle in degrees
1) 60 in two-stroke engines

Given: $n = 4500$ rpm, $\beta = 54°$.
To find: Interrupter closing time t_U

$t_U = \dfrac{54}{3 \cdot 4500} = 0{,}004\ \text{s}$

Given: $n = 4500$ rpm, $t_U = 0{,}004$ s.
To find: Closing angle β in degrees

$\beta = 3 \cdot 4500 \cdot 0{,}004 = 54°$

Given: Four-cylinder engine, $p = 60\%$
To find: Closing angle β in degrees

$\beta = \dfrac{3{,}6 \cdot 60}{4} = 54°$

Given: Four-stroke engine, $n = 4500$/min.
To find: Number of ignition impulses f in 1/s.

$f = \dfrac{4500 \cdot 4}{120} = 150/\text{s}$

Clutch, Gear box, Converter

Clutch

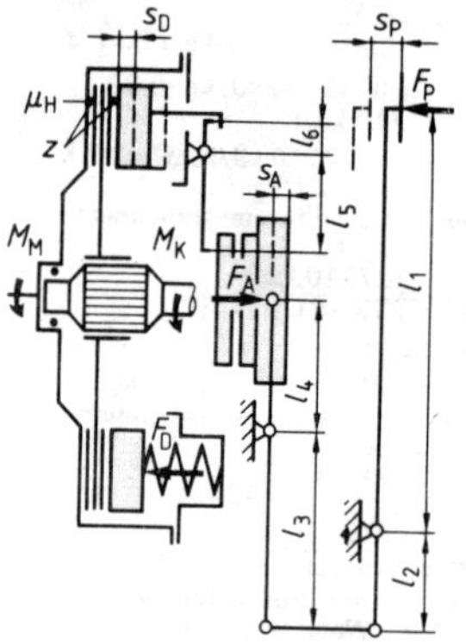

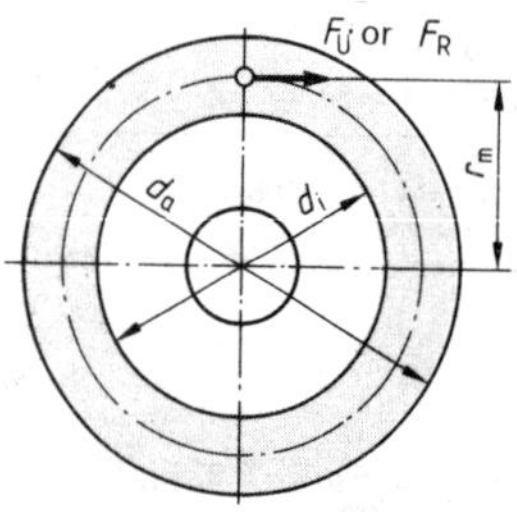

$$F_R = F_D \cdot \mu_H \cdot z \qquad M_K = F_R \cdot r_m$$

$$F_D = \frac{F_R}{\mu_H \cdot z} \qquad F_R = \frac{M_K}{r_m}$$

$$S = \frac{M_K}{M_M} \qquad F_U = \frac{M_M}{r_m} \qquad p = \frac{F_D}{A}$$

$$M_K = M_M \cdot S \qquad M_M = F_U \cdot r_m \qquad F_D = p \cdot A$$

$$i = i_a \cdot i_i \qquad i_a = \frac{l_2 \cdot l_4}{l_1 \cdot l_3} \qquad i_i = \frac{l_6}{l_5}$$

$$F_P = F_D \cdot i \qquad F_P = F_A \cdot i_a \qquad F_A' = F_D \cdot i$$

$$s_P = \frac{s_D}{i} \qquad s_P = \frac{s_A}{i_a} \qquad s_A = \frac{s_D}{i_i}$$

A Area of friction lining
F_A Clutch lever force on the clutch lever
F_D Contact force of pressure plate
F_P Foot force on the Clutch pedal
F_R Friction force of clutch linings
F_U Max. peripheral force on friction lining
i Total transmission ratio
i_{ai} i_a Outer, i_i Inner transmission ratio
l Lever arm of clutch engagement
M_K Transferable clutch torque
M_M Max. engine torque
p Surface pressure on lining
r_m Mean friction radius $(d_a+d_i):4$
S Safety factor
S_A Path at the clutch lever (clutch gap)
S_D Path at the pressure plate (air gap)
S_P Foot travel on the clutch pedal
Z Number of friction surfaces
μ_H Coefficient of static friction of the friction lining

Given: $F_D = 3000$ N, $\mu_H = 0{,}4$, $z = 2$, $r_m = 82{,}5$ mm, $M_M = 120$ Nm.
To find: a) Friction force F_R N
b) Clutch torque M_K in Nm
c) Safety factor

$F_R = 3000\text{ N} \cdot 0{,}4 \cdot 2 = 2400$ N
$M_K = 2400\text{ N} \cdot 0{,}0825\text{ m} = 198$ Nm
$S = 198\text{ Nm} : 120\text{ Nm} = 1{,}65$

Given: $d_a = 200$ mm, $d_i = 130$ mm, $F_D = 3000$ N, $M_M = 120$ Nm.
To find: a) peripheral force F_U in N
b) Area of friction lining A
c) Surface pressure p in N/cm²

$r_m = \frac{200\text{ mm} + 130\text{ mm}}{4} = 82{,}5$ mm
$F_U = 120\text{ Nm} : 0{,}0825\text{ m} = 1454{,}5$ N
$A = \frac{\pi \cdot (20^2 - 13^2)\text{ cm}^2}{4} = 181{,}4\text{ cm}^2$
$p = 3000\text{ N} : 181{,}4\text{ cm}^2 = 16{,}5\text{ N/cm}^2$

Given: $F_D = 3000$ N, $s_D = 0{,}8$ mm, $i_i = 1:3$, $i_a = 1:8$,
To find: Clutch gap $s_A = 2{,}5$ mm
a) Total Transmission ratio i
b) Foot force F_P on the pedal gap
c) Foot Travel S_P for air gap
d) Foot travel S_P for clutch gap

$i = \frac{1}{8} \cdot \frac{1}{3} = 0{,}0417 = \frac{1}{24}$
$F_P = 3000\text{ N} \cdot \frac{1}{24} = 125$ N
$s_P = 0{,}8\text{ mm} \cdot 24 = 19{,}2$ mm
$s_P = 2{,}5\text{ mm} \cdot 8 = 20{,}0$ mm

Gear box

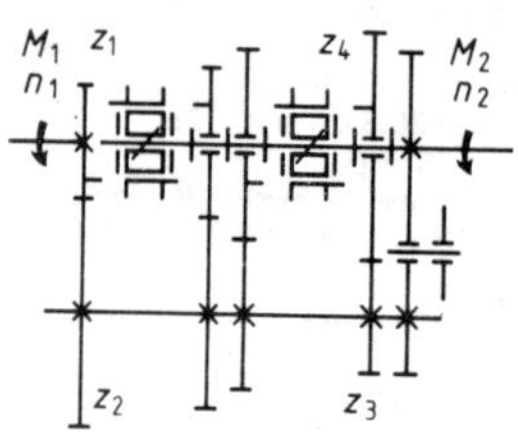

$$n_2 = \frac{n_1}{i} \qquad M_2 = M_1 \cdot i \cdot \eta$$

$$n_1 = n_2 \cdot i \qquad M_1 = \frac{M_2}{i \cdot \eta}$$

$$i = \frac{n_1}{n_2} \qquad i = \frac{z_2 \cdot z_4 \ldots}{z_1 \cdot z_3 \ldots}$$

i Total transmission ratio
M_1 Drive torque, Engine torque
M_2 Drive torque of main shaft
n_1 Drive speed, of Engine speed
n_2 Drive speed of main shaft
z_1, z_3 Number of teeth of driving gears
z_2, z_4 Number of teeth of driven gears
η Efficiency

Given: $n_1 = 4800$/min, $M_1 = 120$ Nm, $i = 4:1$, $\eta = 0{,}95$.
To find: a) Drive speed
b) Drive torque

$n_2 = \frac{4800\text{/min}}{4} = 1200$/min
$M_2 = 120\text{ Nm} \cdot 4 \cdot 0{,}95 = 456{,}0$ Nm

Given: $z_1 = 25$, $z_2 = 50$, $z_3 = 28$, $z_4 = 49$, $n_2 = 1200$/min.
To find: a) Total transmission ratio i
b) Engine speed

$i = \frac{50 \cdot 49}{25 \cdot 28} = 3{,}5$
$n_1 = 1200\text{/min} \cdot 3{,}5 = 4200$/min

Converter

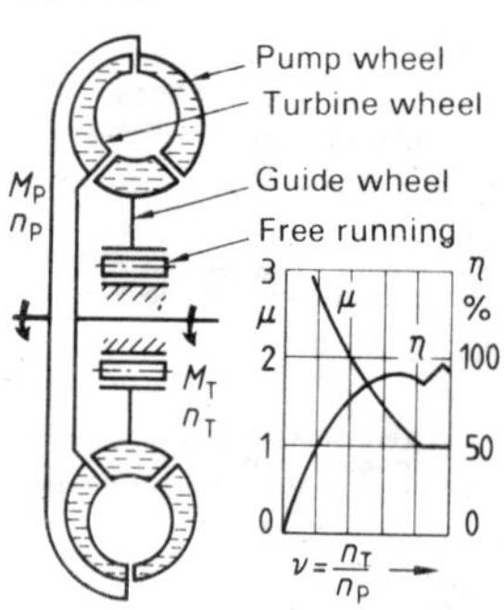

$$\nu = \frac{n_T}{n_P} \qquad \mu = \frac{M_T}{M_P}$$

$$n_T = n_P \cdot \nu \qquad M_T = M_P \cdot \mu$$

$$\eta = \frac{P_T}{P_P} \qquad \eta = \mu \cdot \nu$$

$$P_T = P_P \cdot \eta \qquad \mu = \frac{\eta}{\nu}$$

M_P Torque of pump wheel (engine torque)
M_T Torque of turbine wheel
n_P Speed of pump wheel (engine speed)
n_T Speed of turbine wheel
P_P Power of pump wheel (engine power)
P_T Power of turbine wheel
η Efficiency
μ Conversion torque
ν Speed ratio turbine-pump

Given: $M_P = 120$ Nm at $n_P = 4800$/min, $\mu = 2{,}5$, $\nu = 0{,}34$.
To find: a) Turbine wheel speed n_T
b) Turbine wheel torque M_T
c) Efficiency η

$n_T = 4800\text{/min} \cdot 0{,}34 = 1632$/min
$M_T = 120\text{ Nm} \cdot 2{,}5 = 300$ Nm
$\eta = 2{,}5 \cdot 0{,}34 = 0{,}85$

Given: $P_P = 60{,}3$ kW, $n_P = 4800$/min, $n_T = 1632$/min, $\eta = 0{,}85$.
To find: a) Turbine wheel power P_T
b) Speed ratio ν
c) Torque conversion μ

$P_T = 60{,}3\text{ kW} \cdot 0{,}85 = 51{,}3$ kW
$\nu = 1632\text{/min} : 4800\text{/min} = 0{,}34$
$\mu = 0{,}85 : 0{,}34 = 2{,}50$

Drive Shaft

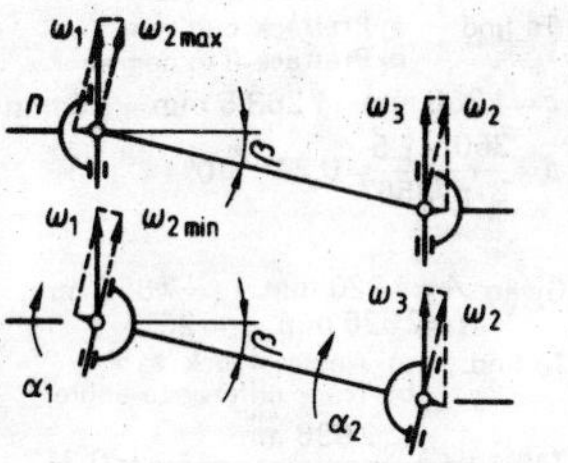

$$\Delta\alpha = \alpha_1 - \alpha_2 \qquad U = \frac{\omega_{2max} - \omega_{2min}}{\omega_1}$$

$$\omega_{2max} = \frac{\omega_1}{\cos\beta}$$

$$\omega_1 = \frac{\pi \cdot n}{30} \qquad \omega_{2min} = \omega_1 \cdot \cos\beta$$

- n Speed of drive shaft
- U Degree of non-uniformity
- α_1 Swivel angle of drive shaft 1
- α_2 Swivel angle of drive shaft 2
- β Inclination angle
- $\Delta\alpha$ Differential angle, Cardan error
- ω_1 Angular speed of driver
- ω_2 Angular speed of driven

Given: $\alpha_1 = 45°$, $\alpha_2 = 40° 53'$.
To find: Cardan error $\Delta\alpha$

$\Delta\alpha = 45° - 40° 53' = 4° 07'$

Given: $\beta = 30°$, $\omega_1 = 400$ 1/s.
To find: a) Minimum angular speed
b) Maximum angular speed
c) Degree of non-uniformity

$\omega_{2min} = 400/s \cdot \cos 30° = 346/s$

$\omega_{2max} = 400/s : \cos 30° = 462/s$

$$U = \frac{462/s - 346/s}{400/s} = 0{,}29$$

Axle drive

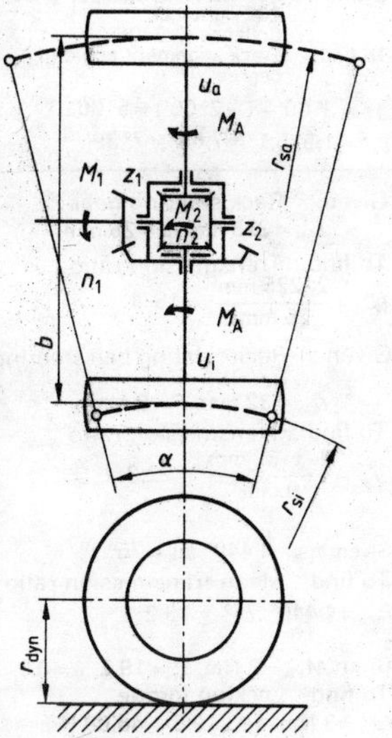

$$n_2 = \frac{n_1}{i} \qquad M_2 = M_1 \cdot i \cdot \eta$$

$$n_1 = n_2 \cdot i \qquad M_1 = \frac{M_2}{i \cdot \eta}$$

$$i = \frac{n_1}{n_2} \qquad i = \frac{z_2}{z_1}$$

$$\Delta u = \frac{b \cdot \alpha}{360° \cdot r_{dyn}} \qquad u = \frac{r_S \cdot \alpha}{360° \cdot r_{dyn}}$$

$$\alpha = \frac{360° \cdot \Delta u \cdot r_{dyn}}{b} \qquad \Delta u = u_a - u_i$$

Without block — With balancing block

$$M_A = \frac{M_2}{2} \qquad M_A = \frac{M_2 \cdot (100\% \pm S)}{200\%}$$

$$M_2 = 2 \cdot M_A \qquad M_{A1} = \frac{M_{A2} \cdot (100\% + S)}{100\% - S}$$

- b Wheel base of drive axle
- i Transmission ratio
- M_1 Drive torque at bevel wheel
- M_2 Driven torque at axle drive plate wheel
- M_A Torque at an axle shaft
- n_1 Drive speed of bevel wheel
- n_2 Drive speed of axle drive plate wheel
- r_{dyn} Dynamic tyre radius
- r_s Track circle radius
- S Block value in per cent
- $U_{a,i}$ Rotations, outer wheel, inner wheel
- Δu Difference in rotation, balance
- α Curve angle in degrees
- η Efficiency

Given: $M_1 = 480$ Nm, $n_1 = 1\,200$/min, $i = 4:1$, $\eta = 0{,}95$.
To find: a) Speed of axle drive plate wheel n_2
b) Torque of axle drive plate wheel M_2

$n_2 = 1\,200/\text{min} : 4 = 300/\text{min}$

$M_2 = 480\ \text{Nm} \cdot 4 \cdot 0{,}95 = 1\,824\ \text{Nm}$

Given: $a = 180°$, $b = 1{,}44$ m, $r_{dyn} = 300$ mm, r_S for outer wheel $= 12$ m
To find: a) Difference in rotation Δu
b) Outer wheel rotations u_a
c) Inner wheel rotations u_i

$$\Delta u = \frac{1{,}44\ \text{m} \cdot 180°}{360° \cdot 0{,}30\ \text{m}} = 2{,}4\ \text{rpm}$$

$$u_a = \frac{12\ \text{m} \cdot 180°}{360° \cdot 0{,}3\ \text{m}} = 20\ \text{rpm}$$

$u_i = (20 - 2{,}4)$ rpm = 17,6 rpm

Given: $M_2 = 1\,824$ Nm, $S = 60\%$.
To find: a) Torque of axle shafts without balancing block
b) Torque of axle shafts with balancing block

$M_A = 1\,824\ \text{Nm} : 2 = 912{,}0\ \text{Nm}$

$$M_{A1} = \frac{1\,824 \cdot (100 + 60)}{200} = 1\,459{,}2\ \text{Nm}$$

$$M_{A2} = \frac{1\,824 \cdot (100 - 60)}{200} = 364{,}8\ \text{Nm}$$

Axle forces

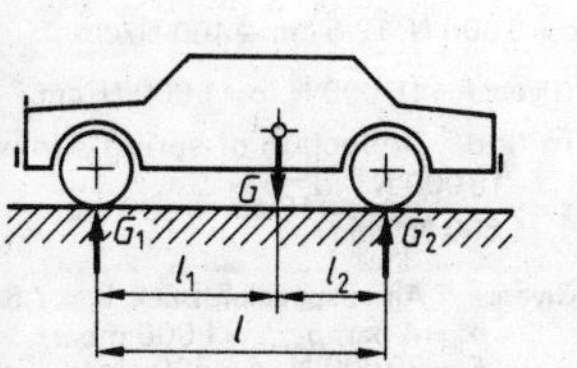

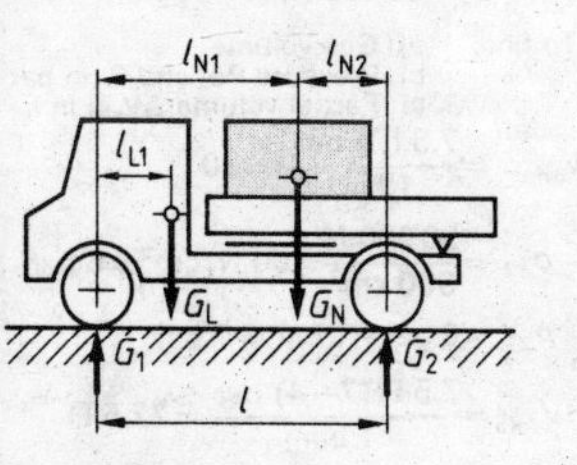

Sum of right-turning torques
=Sum of left-turning torques

$$G_1 = \frac{G \cdot l_2}{l} \qquad G_2 = G - G_1 \qquad G_2 = \frac{G \cdot l_1}{l}$$

$$l_2 = \frac{G_1 \cdot l}{G} \qquad G_1 = G - G_2 \qquad l_1 = \frac{G_2 \cdot l}{G}$$

$$G_1 = \frac{G_L \cdot l_{L2} \pm G_N \cdot l_{N2}}{l}$$

$+F_N$ before rear axle
$-G_N$ after rear axle
e.g. supporting load of trailer coupling

$$G_2 = \frac{G_L \cdot l_{L1} + G_N \cdot l_{N1}}{l}$$

$$G_1 = G_L + G_N - G_2 \qquad G_2 = G_L + G_N - G_1$$

- G Total weight of vehicle
- G_1 Axle force on front axle
- G_2 Axle force on rear axle
- G_L Gravitational force of dead weight
- G_N Gravitational force of pay load
- l Wheel base of vehicle
- l_1 Centre of gravity distance from front axle
- l_2 Centre of gravity distance from rear axle

Given: $G = 12\,000$ N, $l = 2\,500$ mm, $l_2 = 1\,350$ mm.
To find: Axle forces G_1 and G_2

$$G_1 = \frac{12\,000\ \text{N} \cdot 1\,350\ \text{mm}}{2\,500\ \text{mm}} = 6\,480\ \text{N}$$

$G_2 = 12\,000\ \text{N} - 6\,480\ \text{N} = 5\,520\ \text{N}$

Given: $G = 12\,000$ N, $G_2 = 5\,520$ N, $l = 2\,500$ mm.
To find: Centre of gravity distance l_1

$$l_1 = \frac{5\,520\ \text{N} \cdot 2\,500\ \text{mm}}{12\,000\ \text{N}} = 1\,150\ \text{mm}$$

Given: $G_L = 56$ kN, $G_N = 68$ kN, $l = 3{,}2$ m, $l_{L2} = 1{,}8$ m, $l_{N2} = 0{,}5$ m.
To find: Axle forces G_1 and G_2

$$G_1 = \frac{56000\ \text{N} \cdot 1{,}8\ \text{m} + 68000\ \text{N} \cdot 0{,}5\ \text{m}}{3{,}2\ \text{m}}$$

$= 42\,125$ N

$G_2 = 56\,000\ \text{N} + 68\,000\ \text{N} - 42\,125\ \text{N}$

$= 81\,875$ N

Wheel setting

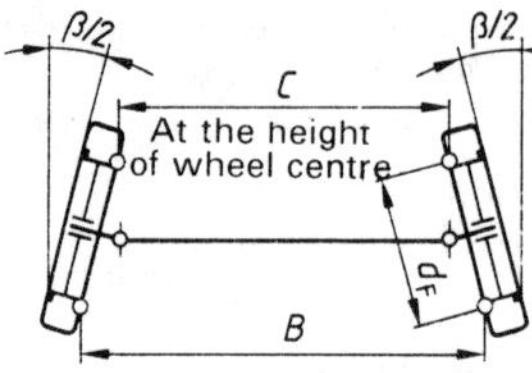

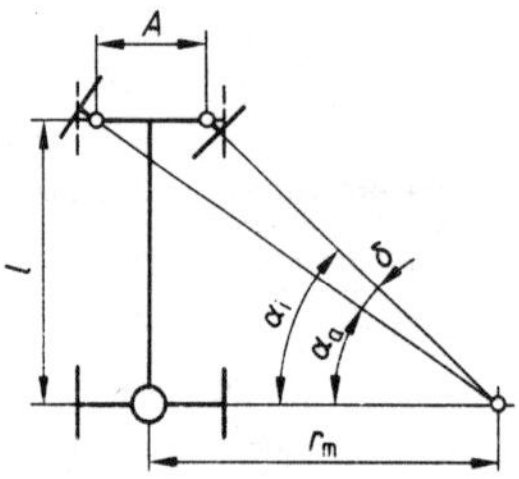

$$c = \frac{2 \cdot \pi \cdot d_F \cdot \beta}{360°} \qquad \delta = \alpha_a - \alpha_i$$

$$\beta = \frac{360° \cdot c}{2 \cdot \pi \cdot d_F}$$

Note the factory Specification

$$d_F = \frac{360° \cdot c}{2 \cdot \pi \cdot \beta}$$

± Pretrack as the case may be

$$\tan \alpha = \frac{l}{r_m \pm A/2}$$

$$c = B - C$$

− Outer wheel
− Inner wheel

$$\varepsilon = \frac{\Delta\gamma \text{ in min}}{40} \qquad \varepsilon = 1{,}5 \cdot (\Delta\gamma \text{ in degrees})$$

- A Steering pivot distance
- B, C Rim flange distance at the height of wheel centre
- c Pre-track (+), Post-track (—) in mm
- d_F Rim diameter in mm
- l Wheel setting (position)
- r_m Mean curve radius
- α_a, α_i Angle of lock outer-inner wheel
- β Pre-track (+), post-track (—) in degrees
- $\Delta\gamma$ Collapse difference between 20° left lock and 20° right lock
- δ Track difference angle
- Track alignment angle in degrees

Given: $B = 1\,265$ mm, $C = 1\,263{,}5$ mm, $d_F = 356$ mm.
To find: a) Pretrack c in mm
b) Pretrack β in degree

$c = 1\,265 \text{ mm} - 1\,263{,}5 \text{ mm} = 1{,}5$ mm

$$\beta = \frac{360° \cdot 1{,}5}{2 \cdot \pi \cdot 356} = 0{,}24° = 0°\,14'$$

Given: $A = 1\,220$ mm, $r_m = 7\,852$ mm, $l = 2\,636$ mm, $\alpha_i = 20°$.
To find: a) Angle of lock α_a
b) Track difference angle δ

$$\tan \alpha = \frac{2\,636 \text{ mm}}{7\,852 \text{ mm} + 610 \text{ mm}} = 0{,}312$$

$\alpha_a = 17{,}30° = 17°\,18'$

$\delta = 17°\,18' - 20°\,00' = -2°\,42'$

Given: 20° left lock, Collapse 3°00′
20° right lock
Collapse — 2°06′
To find: Track alignment angle in degree

$\Delta\gamma = 3°\,00' - (-2°\,06') = 5°\,06'$

$\varepsilon = 1{,}5 \cdot 5{,}1° = 7{,}65° = 7°\,39'$

Steering

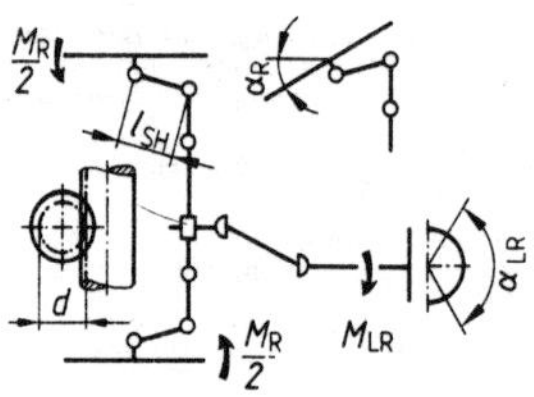

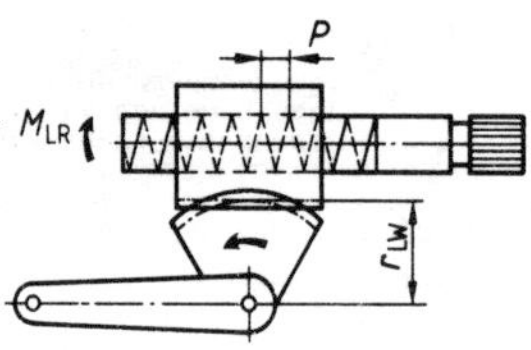

$$i_z \approx \frac{2 \cdot l_{SH}}{d} \qquad i_s = \frac{2 \cdot \pi \cdot r_{LW}}{P}$$

$$d \approx \frac{2 \cdot l_{SH}}{i_z} \qquad P = \frac{2 \cdot \pi \cdot r_{LW}}{i_s}$$

$$M_R = M_{LR} \cdot i_m \cdot \eta \qquad i_m = \frac{\alpha_{LR}}{\alpha_R}$$

$$M_{LR} = \frac{M_R}{i_m \cdot \eta} \qquad \alpha_R = \frac{\alpha_{LR}}{i_m}$$

- d Pitch circle diameter of pinion
- i_M Mean total transmission ratio
- i_S Transmission ratio of screw-type steering gear
- i_Z Transmission ratio of rack-type steering gear
- l_{SH} Length of track rod lever
- M_{LB} Locking torque on steering wheel
- M_R Locking torque of both wheels
- P Pitch of steering spindle, steering screw
- r_{LW} Effective lever arm of steering shaft, Pitch circle radius of tooth segment
- α_{LR} Angle of lock on the steering wheel
- α_R Angle of lock of wheels
- η Efficiency of steering

Given: Rack steering gear
$l_{SH} = 225$ mm, $d = 25$ mm.
To find: Transmission ratio

$$i_z \approx \frac{2 \cdot 225 \text{ mm}}{25 \text{ mm}} = 18:1$$

Given: Recirculating ball steering gear
$r_{LW} = 32$ mm, $P = 9$ mm.
To find: Transmission ratio

$$i_s = \frac{2 \cdot \pi \cdot 32 \text{ mm}}{9 \text{ mm}} = 22{,}3:1$$

Given: $\alpha_{LR} = 1\,440°$, $\alpha_R = 75°$.
To find: Mean transmission ratio

$i_m = 1\,440° : 75° = 19{,}2:1$

Given: $M_{LR} = 3$ Nm, $i_m = 19{,}2$, $\eta = 0{,}8$.
To find: Locking torque

$M_R = 3 \text{ Nm} \cdot 19{,}2 \cdot 0{,}8 = 46{,}08$ Nm

Suspension

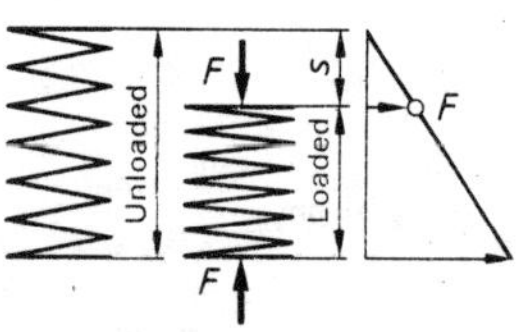

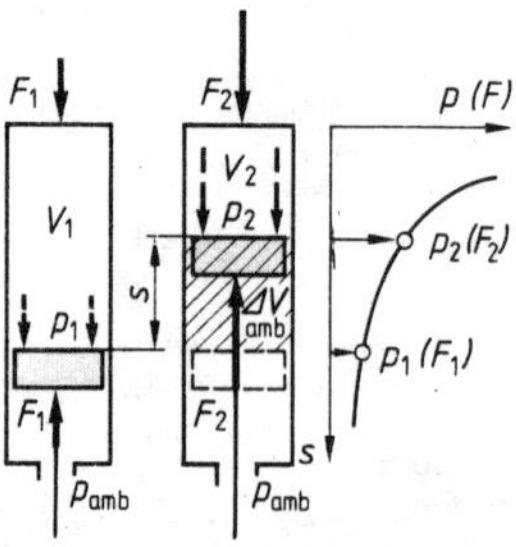

Helical spring and Leaf spring $\quad c = \dfrac{F}{s}$

$$F = c \cdot s \qquad s = \frac{F}{c}$$

Air suspension $\quad p_e = \dfrac{F}{A}$

$$F = p_e \cdot A \qquad A = \frac{F}{p_e} \qquad p_1 = p_{e1} + p_{amb} \qquad p_2 = p_{e2} + p_{amb}$$

$$V_{amb} = \frac{V_1 \cdot p_1}{p_{amb}} \qquad \Delta V_{amb} = \frac{V_1 \cdot (p_2 - p_1)}{p_{amb}}$$

- A Piston area rolled out impression in cm²
- c Spring rate in N/cm²
- F Spring force, load carrying capacity in N
- p_1 Absolute pressure before springing
- p_2 Absolute pressure after springing
- p_{amb} Atmospheric pressure, Ambient pressure
- p_e Gauge pressure in air chamber
- s Deflection of spring during springing in cm
- V_1 Air chamber volume before springing
- V_2 Air chamber volume after springing
- V_{amb} Gas volume at atmospheric pressure
- ΔV_{amb} Partial volume for level regulation

Given: Helical spring, $F = 5\,000$ N, $s = 125$ mm.
To find: Spring rate c in N/cm

$c = 5\,000 \text{ N} : 12{,}5 \text{ cm} = 400$ N/cm

Given: $F = 18\,000$ N, $c = 1\,800$ N/cm.
To find: Deflection of spring s in mm

$$s = \frac{18\,000 \text{ N}}{1\,800 \text{ N/cm}} = 10 \text{ cm} = 100 \text{ mm}$$

Given: Air suspension truck, $V_1 = 7{,}5$ l
$p_1 = 4$ bar, $p_{amb} = 1\,000$ mbar,
$F_2 = 30\,000$ N, $A = 500$ cm²
To find: a) Gas volume
b) Pressure Pe_2 and P_2 in bar
c) Partial volume ΔV_{abm} in l

$$V_{amb} = \frac{7{,}5 \text{ l} \cdot 4 \text{ bar}}{1 \text{ bar}} = 30 \text{ l}$$

$$p_{e2} = \frac{30\,000 \text{ N}}{500 \text{ cm}^2} = 60 \text{ N/cm}^2 = 6 \text{ bar}$$

$p_2 = 6 \text{ bar} + 1 \text{ bar} = 7$ bar

$$\Delta V_{amb} = \frac{7{,}5 \text{ l} \cdot (7-4) \text{ bar}}{1 \text{ bar}} = 22{,}5 \text{ l}$$

Brake accessories

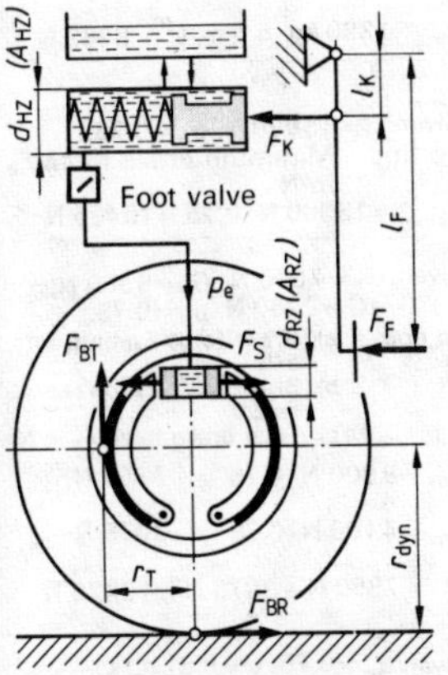

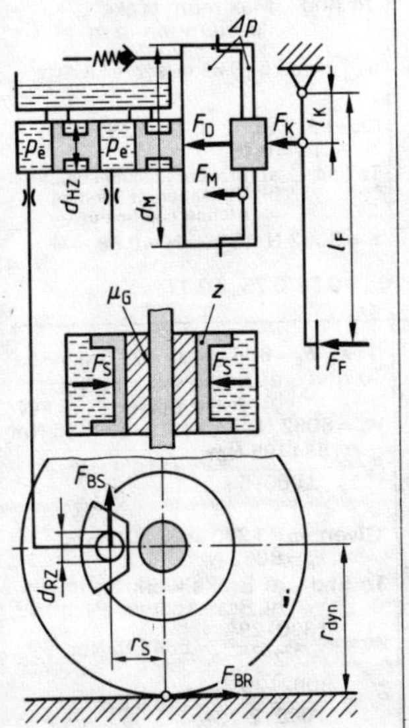

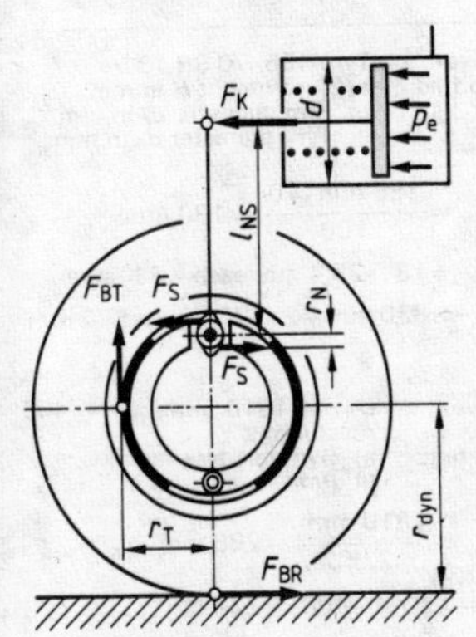

$$F_K = \frac{F_F \cdot l_F}{l_K} \qquad F_K = \frac{F_F}{i_{mec}}$$

$$F_F = \frac{F_K \cdot l_K}{l_F} \qquad F_F = F_K \cdot i_{mec}$$

$$p_e = \frac{F_K \text{ or } F_D}{A_{HZ}} \qquad p_e = \frac{F_S}{A_{RZ}}$$

$$F_K = p_e \cdot A_{HZ} \qquad F_S = p_e \cdot A_{RZ}$$

$$F_D = F_K + F_M \qquad F_D = F_K \cdot v$$

$$F_M = \Delta p \cdot \frac{\pi \cdot d_M^2}{4} \qquad F_K = \frac{F_D}{v}$$

$$F_S = \frac{F_K \cdot d_{RZ}^2}{d_{HZ}^2} \qquad F_S = \frac{F_K \cdot A_{RZ}}{A_{HZ}}$$

$$F_K = \frac{F_S \cdot d_{HZ}^2}{d_{RZ}^2} \qquad F_K = \frac{F_S \cdot A_{HZ}}{A_{RZ}}$$

$$F_{BT} = F_S \cdot C \qquad F_{BS} = F_S \cdot \mu_G \cdot z$$

$$F_S = \frac{F_{BT}}{C} \qquad F_S = \frac{F_{BS}}{\mu_G \cdot z}$$

$$F_{BR} = \frac{F_{BT} \cdot r_T}{r_{dyn}} \qquad F_{BR} = \frac{F_{BS} \cdot r_S}{r_{dyn}}$$

$$F_{BT} = \frac{F_{BR} \cdot r_{dyn}}{r_T} \qquad F_{BS} = \frac{F_{BR} \cdot r_{dyn}}{r_S}$$

$$i_{mec} = \frac{l_K}{l_F} \qquad i_{hy} = \frac{d_{HZ}^2}{d_{RZ}^2} \qquad i = i_{mec} \cdot i_{hy}$$

A_{HZ} Piston area of the main cylinder
A_{RZ} Piston area of the wheel cylinder
C Characteristic value, internal transmission
d_{HZ} Diameter of the main cylinder
d_M Diameter of diaphragm
d_{RZ} Diameter of the wheel cylinder
F_{BR} Brake force at the periphery of tyre
F_{BS} Brake force at the brake disc
F_{BT} Brake force at the brake drum
F_D Plunger rod force
F_F Foot force at the brake pedel
F_K Piston rod force
F_M Diaphragm force
F_S Clamping force at the wheel cylinder
i Overall transmission ratio
i_{mec} Mechanical transmission ratio (i_{hy} hydraulic)
l_F Lever arm of the feet force
l_K Lever arm of the piston rod force
p_e Brake pressure, Δ p. pressure difference
r_{dyn} Dynamic tyre radius
r_S Effective brake disc radius
r_T Radius of brake drum
z Number of cylinders of the fixed saddle
μ_G Coefficient of sliding friction of the lining
v Amplification factor

$$F_K = p_e \cdot \frac{\pi \cdot d^2}{4} \qquad F_S = \frac{F_K \cdot l_{NS}}{2 \cdot r_N}$$

$$p_e = \frac{4 \cdot F_K}{\pi \cdot d^2} \qquad F_K = \frac{2 \cdot F_S \cdot r_N}{l_{NS}}$$

$$d = \sqrt{\frac{4 \cdot F_K}{\pi \cdot p_e}} \qquad r_N = \frac{F_K \cdot l_{NS}}{2 \cdot F_S}$$

d Diameter of the brake cylinder
F_K Output force of the piston rod
F_S Clamping force at the brake cam
l_{NS} Lever arm of the cam setter
p_e Pressure in the brake cylinder
r_N Effective radius of the brake cam

Given: Drum brake $F_F = 200$ N, $l_F = 250$ mm, $l_K = 50$ mm, $d_{HZ} = 19$ mm, $d_{RZ} = 24$ mm.
To find: a) Piston rod force F_K in N
b) Brake pressure p_e in mbar
c) Clamping force F_s in N
d) Transmission ratios i_{mec}, i_{hy}
e) Overall transmission ratio i

$$F_K = \frac{200\text{ N} \cdot 250\text{ mm}}{50\text{ mm}} = 1\,000\text{ N}$$

$$p_e = \frac{4 \cdot 1\,000\text{ N}}{\pi \cdot 1{,}9^2\text{ cm}^2} = 353\ \frac{\text{N}}{\text{cm}^2} = 35\text{ bar}$$

$$F_S = \frac{1\,000\text{ N} \cdot 24^2\text{ mm}^2}{19^2\text{ mm}^2} = 1\,596\text{ N}$$

$$i_{mec} = 50\text{ mm} : 250\text{ mm} = 0{,}2 = 1:5$$

$$i_{hy} = 19^2\text{ mm}^2 : 24^2\text{ mm}^2 = 1:1{,}6$$

$$i = \frac{1}{5} \cdot \frac{1}{1{,}6} = 1:8$$

Given: Disc brake $F_S = 9\,600$ N, $\mu_G = 0{,}35$, $z = 2$, $r_{dyn} = 300$ mm, $r_S = 120$ mm.
To find: a) Brake force at the disc F_{BS}
b) Brake force at the tyre F_{BR}

$$F_{BS} = 9\,600\text{ N} \cdot 0{,}35 \cdot 2 = 6\,720\text{ N}$$

$$F_{BR} = \frac{6\,720\text{ N} \cdot 120\text{ mm}}{300\text{ mm}} = 2\,688\text{ N}$$

Given: Brake force amplification $F_K = 1\,000$ N, $d_M = 230$ mm, $\Delta p = 450$ mbar.
To find: a) Diaphragm force F_M in N
b) Plunger rod force F_D in N
c) Amplification factor v

$$F_M = \frac{\pi \cdot 23^2\text{ cm}^2}{4} \cdot 4{,}5\ \frac{\text{N}}{\text{cm}^2} = 1\,870\text{ N}$$

$$F_D = 1\,000\text{ N} + 1\,870\text{ N} = 2\,870\text{ N}$$

$$v = 2\,870\text{ N} : 1\,000\text{ N} = 2{,}87$$

Given: Drum brake $F_{BR} = 1\,146$ N, $r_{dyn} = 305$ mm, $r_T = 115$ mm, $C = 1{,}9$, $l_K = 50$ mm, $l_F = 250$ mm, $d_{RZ} = 24$ mm, $d_{HZ} = 19$ mm.
To find: a) Brake force at the drum F_{BT}
b) Clamping force at the wheel cylinder F_S
c) Piston rod force F_K
d) Foot force at the pedal F_F

$$F_{BT} = \frac{1\,146\text{ N} \cdot 305\text{ mm}}{115\text{ mm}} = 3\,040\text{ N}$$

$$F_S = 3\,040\text{ N} : 1{,}9 = 1\,600\text{ N}$$

$$F_K = \frac{1\,600\text{ N} \cdot 19^2\text{ mm}^2}{24^2\text{ mm}^2} = 1\,000\text{ N}$$

$$F_F = \frac{1\,000\text{ N} \cdot 50\text{ mm}}{250\text{ mm}} = 200\text{ N}$$

Given Brake cylinder with brake $p_e = 6$ bar, $d = 120$ mm, $C = 2$, $l_{NS} = 135$ mm, $r_N = 15$ mm, $r_{dyn} = 507$ mm, $r_T = 195$ mm.
To find a) Piston force at the brake cylinder F_K
b) Clamping force at the cam F_S
c) Braking force at the drum F_{BT}
d) Braking force at the tyre F_{BR}

$$F_K = \frac{\pi \cdot 12^2\text{ cm}^2}{4} \cdot 60\ \frac{\text{N}}{\text{cm}^2} = 6\,786\text{ N}$$

$$F_S = \frac{6\,786\text{ N} \cdot 135\text{ mm}}{2 \cdot 15\text{ mm}} = 30\,537\text{ N}$$

$$F_{BT} = 30\,537\text{ N} \cdot 2 = 61\,074\text{ N}$$

$$F_{BR} = \frac{61\,074\text{ N} \cdot 195\text{ mm}}{507\text{ mm}} = 23\,490\text{ N}$$

Brake force

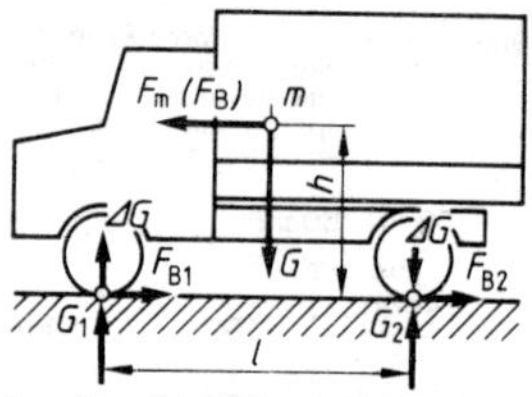

Bonding friction coefficients

Concrete, Asphalt	
a) dry	0,76...0,85
b) wet	0,48...0,52
c) oily	0,35...0,40
Tar macadam	
a) dry	0,58...0,62
b) wet	0,38...0,42
c) oily	0,25...0,30
Small pavement	
a) dry	0,48...0,52
b) wet	0,25...0,30
c) oily	0,15...0,20
Snow	0,10...0,50
Ice surface	
a) dry	0,15...0,20
b) wet	0,05...0,10

Utilisation of bonding friction coefficients

Hydraulic equipment	0,70...0,80
Compressed air brake	0,50...0,75
Mechanical braking equipment	0,60...0,80

$F_B = m \cdot a$ $\qquad$ $F_B = G \cdot z$

$a = \frac{F_B}{m}$ $\qquad$ $z = \frac{F_B}{G}$

$F_{B\,max} = G \cdot \mu_H$ $\qquad$ $\Delta G = G \cdot \mu_H \cdot \frac{h}{l}$

$F_{B1} = (G_1 + \Delta G) \cdot \mu_H$ $\qquad$ $F_{B2} = (G_2 - \Delta G) \cdot \mu_H$

$F_{B1} = \frac{F_B \cdot (G_1 + \Delta G)}{G}$ $\qquad$ $F_{B1} = \frac{F_B \cdot p_1}{100\%}$

$F_{B2} = \frac{F_B \cdot (G_2 - \Delta G)}{G}$ $\qquad$ $F_{B2} = \frac{F_B \cdot p_2}{100\%}$

$a_{max} = \mu_H \cdot g$ $\qquad$ $\varepsilon = \frac{z}{\mu_H} = \frac{a_{max}}{g \cdot \mu_H}$

$a_{max} = z \cdot g$ $\qquad$ $z = \varepsilon \cdot \mu_H$

- a Brake deceleration
- F_B Total braking force of the vehicle
- F_{B1} Brake force on the front axle
- F_{B2} Brake force on the rear circle
- G Weight of the vehicles
- G_1 Static axle force on the front axle
- G_2 Static axle force on the rear axle
- ΔG Axle force displacement while braking
- g Acceleration due to gravity 9.81 m/s²
- h Height of centre of gravity above the road
- l Wheel base of the vehicle
- m Vehicle weight, vehicle mass
- P_1, P_2 Brake force components of the axles in %
- z Braking in decimal value
- ϵ Utilisation of bonding friction coefficient, quality grade
- μ_H Coefficient of bonding friction tyre/road

Given: $m = 1390$ kg, $a = 5{,}8$ m/s².
To find: Total brake force F_B in N

$F_B = 1390\text{ kg} \cdot 5{,}8\text{ m/s}^2 = 8062\text{ N}$

Given: $G = 13900$ N, $\mu_H = 0{,}75$.
To find: Minimum brake force F_B in N

$F_{B\,max} = 13900\text{ N} \cdot 0{,}75 = 10425\text{ N}$

Given: $G_1 = 7540$ N, $G_2 = 6360$ N, $\Delta G = 2260$ N, $\mu_H = 0{,}75$.
To find: a) Brake force F_{B1} on front axle
b) Brake force F_{B2} on rear axle
c) Total brake force F_B in N

$F_{B1} = 9800\text{ N} \cdot 0{,}75 = 7350\text{ N}$

$F_{B2} = 4100\text{ N} \cdot 0{,}75 = 3075\text{ N}$

$F_B = 7350\text{ N} + 3075\text{ N} = 10425\text{ N}$

Given: $\mu_H = 0{,}75$, $g = 9{,}81$ m/s²
To find: Maximum brake deceleration in m/s²

$a_{max} = 0{,}75 \cdot 9{,}81\text{ m/s}^2 = 7{,}4\text{ m/s}^2$

Given: $F_B = 8062$ N, $G = 13900$ N, $\mu_H = 0{,}75$.
To find: a) Braking (decimal value)
b) Utilisation of bonding friction coefficient

$z = 8062\text{ N} : 13900\text{ N} = 0{,}58$

$\varepsilon = 0{,}58 : 0{,}75 = 0{,}77$

Brake work, Brake power

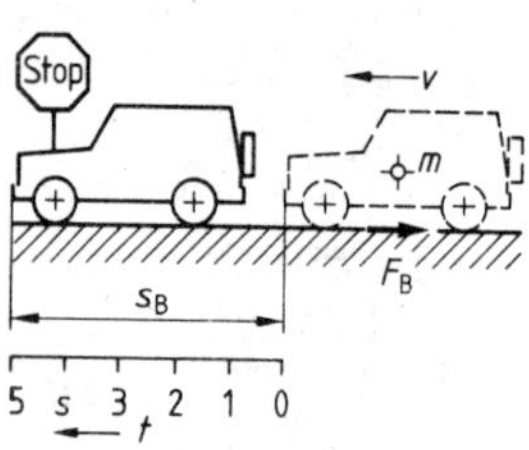

General formulae

$W_B = F_B \cdot s_B$

$P_B = \frac{W_B}{1000 \cdot t}$

Braking up to standstill

$W_B = \frac{m \cdot v^2}{2}$

$P_B = \frac{F_B \cdot v}{1000 \cdot 2}$

- F_B Total brake force of the vehicle in N
- m *Vehicle weight, vehicle mass in kg.*
- P_B Brake power in kW
- S_B Brake distance in m
- t Brake time in s
- v Speed in m/s
- W_B Brake work in Nm (J)

Given: $F_B = 8062$ N, $s_B = 72{,}5$ m, $t = 5$ s
To find: a) Brake work in Nm
b) Brake power F_B in kW

$W_B = 8062\text{ N} \cdot 72{,}5\text{ m} = 584495\text{ Nm}$

$P_B = \frac{584495\text{ Nm}}{1000 \cdot 5\text{ s}} = 116{,}9\text{ kW}$

Given: $m = 1390$ kg, $v = 29$ m/s, $F_B = 8062$ N.
To find: a) Brake work W_B in Nm
b) Brake power P_B in kW

$W_B = \frac{1390 \cdot 29^2}{2} = 584495\text{ Nm}$

$P_B = \frac{8062 \cdot 29}{1000 \cdot 2} = 116{,}9\text{ kW}$

Tyre

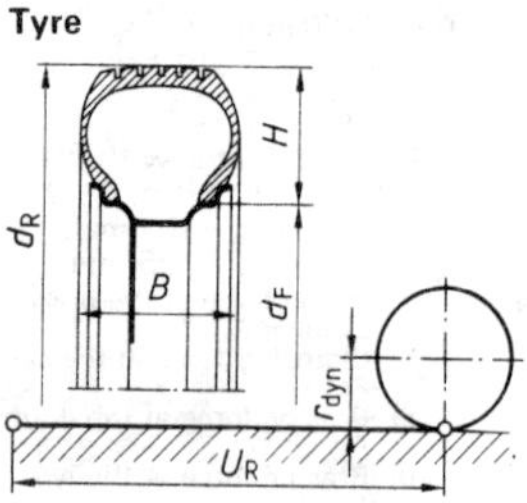

Height-breadth ratio

Super-Ballon tyre	0,95
Low cross-section tyre	0,88
Super low cross-section tyre	0,80
Series 70, 60	0,70, 0,60

$p_{HB} = \frac{100\% \cdot H}{B}$ $\qquad$ $d_R \approx d_F + 2 \cdot H$

$H = \frac{B \cdot p_{HB}}{100\%}$ $\qquad$ $U_R \approx \pi \cdot d_R$

$s = U_R \cdot u$ $\qquad$ $U_R = 2 \cdot \pi \cdot r_{dyn}$

$u = \frac{s}{U_R}$ $\qquad$ $r_{dyn} = \frac{U_R}{2 \cdot \pi}$

- B Nominal breadth, Breadth of cross-section
- d_F Rim diameter (1″=25.4 mm)
- d_R Tyre diameter
- H Tyre height, cross-sectional height
- p_{HB} Cross-sectional ratio in per cent
- s Path length
- r_{dyn} Dynamic tyre radius
- u Revolutions of the tyre
- U_R Developed circumference of the tyre

Given: Tyre 185/70 SR 13
To find: a) Tyre height H in mm
b) Rim diameter d_F in mm
c) Tyre diameter d_R in mm

$H = \frac{185\text{ mm} \cdot 70}{100} = 130\text{ mm}$

$d_F = 13'' \cdot 25{,}4$ mm each = 330 mm

$d_R \approx 330\text{ mm} + 2 \cdot 130\text{ mm} = 590\text{ mm}$

Given: U_R = 1810 mm, u = 400 revolutions
To find: a) Dynamic tyre radius
b) Path length s in m

$r_{dyn} = \frac{1810\text{ mm}}{2 \cdot \pi} = 288\text{ mm}$

$s = 1{,}81\text{ m} \cdot 400 = 724\text{ m}$

Tractive resistance

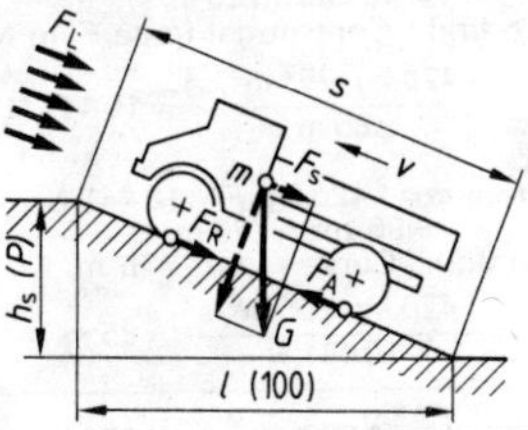

Coefficient of rolling resistance	
Concrete, Asphalt	0,015
Metal rolled	0,020
Earth, strong	0,050
Coefficient of air resistance	
Passenger vehicle	0,2...0,4
Transport Vehicle	0,8...1,5
Omnibus	0,4...0,6
Motor cycle	0,6...0,7
Air density	
in 1000 m height	1,20 kg/m²
in 200 m height	1,11 kg/m²

$F_W = F_R + F_L \pm F_S$ + up-gradient — down-gradient

$F_R = m \cdot g \cdot k = G \cdot k$ on level road

$F_L = 0{,}5 \cdot \varrho \cdot A \cdot v^2 \cdot c_W$ $A \approx 0{,}8 \cdot b \cdot h_F$

$F_S \approx \dfrac{m \cdot g \cdot P}{100\%}$ $F_S = \dfrac{m \cdot g \cdot h_S}{s}$

- A Effective flat area of the vehicle in m²
- b Breadth of vehicle in m
- c_W Coefficient of air resistance in N
- F_L Air resistance in N
- F_R Rolling resistance in N
- F_S Up-gradient resistance in N
- F_W Tractive resistance in N
- g Acceleration due to gravity 9.81 m/s²
- G Weight of the vehicle in N
- h_F Vehicle height in m
- h_S Height difference of slope in m
- k Coefficient of rolling friction
- m Vehicle weight, vehicle mass in kg.
- p Up-gradient, down-gradient in per cent
- s Road length
- v Speed of travel in m/s
- ϱ Density of air in kg/m³

Given: $m = 1420$ kg, $k = 0{,}015$.
To find: Rolling resistance F_R in N
$F_R = 1420\ \text{kg} \cdot 9{,}81\ \text{m/s}^2 \cdot 0{,}015 = 209$ N

Given: $b = 1800$ mm, $h = 1400$ mm, $v = 90$ km/h (25 m/s), $\varrho = 1{,}2$ kg/m³, $c_W = 0{,}4$.
To find: Air resistance F_L in N
$A \approx 0{,}8 \cdot 1{,}8\ \text{m} \cdot 1{,}4\ \text{m} = 2{,}0\ \text{m}^2$
$F_L = 0{,}5 \cdot 1{,}2\ \text{kg/m}^3 \cdot 2{,}0\ \text{m}^2 \cdot 25^2\ \text{m}^2/\text{s}^2 \cdot 0{,}4 = 300$ N

Given: $m = 1420$ kg, $P = 4\%$.
To find: Up-gradient resistance F_S in N
$F_S \approx \dfrac{1420\ \text{kg} \cdot 9{,}81\ \text{m/s}^2 \cdot 4}{100} = 557$ N
or $F_S \approx \dfrac{13930\ \text{N} \cdot 4\ \text{m}}{100\ \text{m}} = 557$ N

Given: $F_R = 209$ N, $F_L = 300$ N, $F_S = 557$ N.
To find: Tractive resistance F_W in N
$F_W = 209\ \text{N} + 300\ \text{N} + 557\ \text{N} = 1066$ N

Tractive power

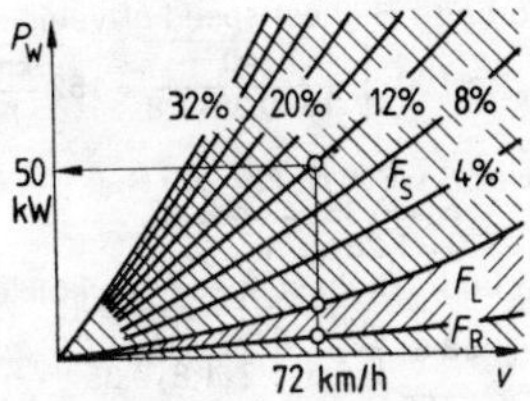

$P_W = \dfrac{F_W \cdot v}{3600}$ $P_{eff} = \dfrac{P_W}{\eta}$

$F_W = \dfrac{3600 \cdot P_W}{v}$ $P_W = P_{eff} \cdot \eta$

$v = \dfrac{3600 \cdot P_W}{F_W}$ $\eta = \dfrac{P_W}{P_{eff}}$

- F_W Tractive resistance in N
- P_{eff} Useful power of the engine in kW
- P_W Tractive power at the wheels in kW
- v Travel speed in km/h
- η Efficiency of force transmission

Given: $F_W = 1066$ N, $v = 90$ km/h, $\eta = 0{,}94$.
To find: a) Tractive power P_W in kW
b) Useful power P_{eff} in kW
$P_W = \dfrac{1066 \cdot 90}{3600} = 26{,}65$ kW
$P_{eff} = 26{,}65\ \text{kW} : 0{,}94 = 28{,}35$ kW

Given: $P_W = 26{,}65$ kW, $v = 90$ km/h.
To find Tractive resistance F_W in N
$F_W = \dfrac{3600 \cdot 26{,}65}{90} = 1066$ N

Drive force

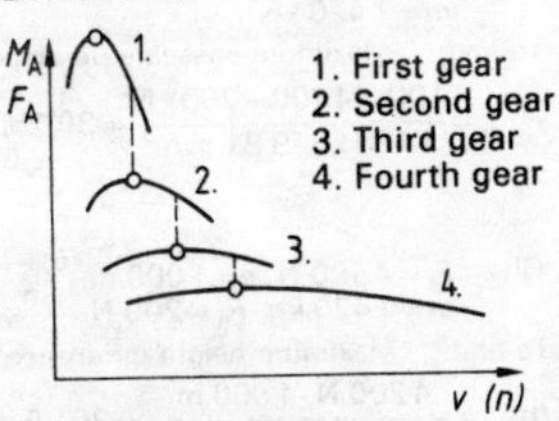

Efficiency	
Coupling, gear box	0,92 ... 0,97
Axle drive, Wheel bearing	0,98 ... 0,99
Overall Transmission	96'0 ... 06'0

$F_A = \dfrac{M_M \cdot i \cdot \eta}{r_{dyn}}$ $M_A = M_M \cdot i \cdot \eta$

$M_M = \dfrac{F_A \cdot r_{dyn}}{i \cdot \eta}$ $M_M = \dfrac{M_A}{i \cdot \eta}$

$i = \dfrac{F_A \cdot r_{dyn}}{M_M \cdot \eta}$ $i = \dfrac{M_A}{M_M \cdot \eta}$

$\eta = \dfrac{F_A \cdot r_{dyn}}{M_M \cdot i}$ $\eta = \dfrac{M_A}{M_M \cdot i}$

- F_A Drive force at the wheels in N
- i Transmission ratio of force transmission
- M_A Drive torque at the wheels in Nm
- M_M Engine torque in Nm
- t_{dyn} Dynamic tyre radius in m
- η Efficiency of force transmission

Given: $M_M = 66{,}5$ Nm, $i = 5{,}2$, $\eta = 0{,}94$, $r_{dyn} = 305$ mm.
To find: a) Drive torque M_A
b) Drive force F_A
$M_A = 66{,}5\ \text{Nm} \cdot 5{,}2 \cdot 0{,}94 = 325$ Nm
$F_A = \dfrac{66{,}5\ \text{Nm} \cdot 5{,}2 \cdot 0{,}94}{0{,}305\ \text{m}} = 1066$ N

Given: $F_A = 1066$ N, $i = 5{,}2$, $\eta = 0{,}94$, $r_{dyn} = 305$ mm.
To find: Engine torque M_M in Nm
$M_M = \dfrac{1066\ \text{N} \cdot 0{,}305\ \text{m}}{5{,}2 \cdot 0{,}94} = 66{,}5$ Nm

Acceleration force

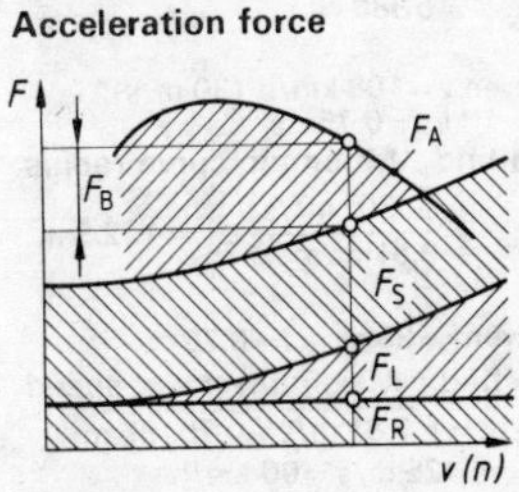

$F_B = m \cdot a$ $F_B = F_A - F_W$

$a = \dfrac{F_B}{m}$ $F_A = F_W + F_B$

$m = \dfrac{F_B}{a}$ $F_W = F_A - F_B$

- a Acceleration in m/s²
- F_A Driving force at the wheels in N
- F_B Accelerating force in N
- F_W Tractive resistance in N
- m Vehicle mass in kg.

Given: $m = 1420$ kg, $a = 1{,}5$ m/s².
To find: Accelerating force F_B
$F_B = 1420\ \text{kg} \cdot 1{,}5\ \text{m/s}^2 = 2130$ N

Given: $F_A = 3196$ N, $F_W = 1066$ N, $m = 1420$ kg.
To find: Acceleration in m/s²
$F_B = 3196\ \text{N} - 1066\ \text{N} = 2130$ N
$a = 2130\ \text{N} : 1420 = 1{,}5\ \text{m/s}^2$

Traction mechanics

Centrifugal force

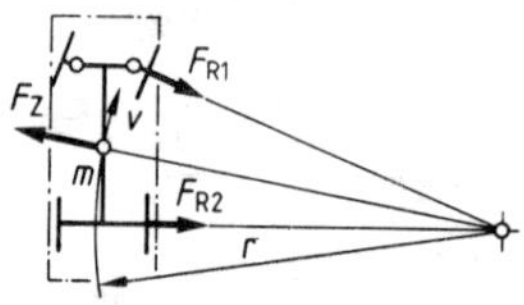

$$F_Z = \frac{m \cdot v^2}{r} \qquad v = \sqrt{\frac{F_Z \cdot r}{m}}$$

$$m = \frac{F_Z \cdot r}{v^2} \qquad r = \frac{m \cdot v^2}{F_Z}$$

- F_R Friction force
- F_Z Centrifugal force in N
- m Vehicle mass in kg
- r Curve radius in m
- v Vehicle speed in m/s

Given: $m = 1\,420$ kg, $r = 85$ m, $v = 90$ km/h (25 m/s).
To find : Centrifugal force F_Z in N

$$F_Z = \frac{1\,420\ \text{kg} \cdot 25^2\ \text{m}^2/\text{s}^2}{85\ \text{m}} = 10\,441\ \text{N}$$

Given: $m = 1\,420$ kg, $F_Z = 10\,441$ N, $v = 90$ km/h (25 m/s).
To find : Curve radius r in m

$$r = \frac{1\,420\ \text{kg} \cdot 25^2\ \text{m}^2/\text{s}^2}{10\,441\ \text{N}} = 85\ \text{m}$$

Vehicle speed

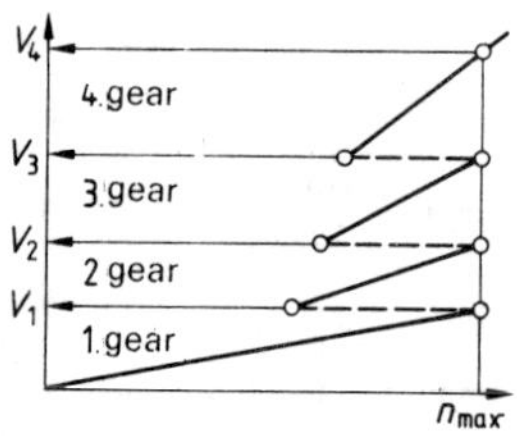

$$v = \frac{60 \cdot U_R \cdot n}{1\,000 \cdot i} \qquad v = \frac{120 \cdot \pi \cdot r_{dyn} \cdot n}{1\,000 \cdot i}$$

$$n = \frac{1\,000 \cdot v \cdot i}{60 \cdot U_R} \qquad n = \frac{1\,000 \cdot v \cdot i}{120 \cdot \pi \cdot r_{dyn}}$$

$$i = \frac{60 \cdot U_R \cdot n}{1\,000 \cdot v} \qquad i = \frac{120 \cdot \pi \cdot r_{dyn} \cdot n}{1\,000 \cdot v}$$

- i Force transmission ratio
- n Engine speed in rpm
- r_{dyn} Dynamic tyre radius in m
- U_R Developed circumference of the tyre in m
- v Vehicle speed in km/h

Given: $U_R = 1\,920$ mm, $n = 4\,070$/min, $i = 5{,}2$.
To find : Vehicle speed v in km/h

$$v = \frac{60 \cdot 1{,}92 \cdot 4\,070}{1\,000 \cdot 5{,}2} = 90\ \text{km/h}$$

Given: $r_{dyn} = 305$ mm, $n = 4\,070$/min, $i = 5{,}2$.
To find : Vehicle speed v in km/h

$$v = \frac{120 \cdot \pi \cdot 0{,}305 \cdot 4\,070}{1\,000 \cdot 5{,}2} = 90\ \text{km/h}$$

Highest speed of vehicle on level road

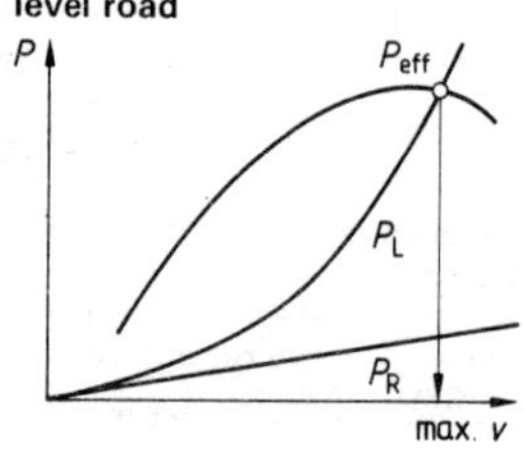

$$v_{max} \approx 100 \cdot \sqrt[3]{\frac{P_{eff}}{18 \cdot c_W \cdot A}}$$

$$v_{max} \approx 3{,}6 \cdot \sqrt{\frac{F_A - F_R}{0{,}5 \cdot \varrho \cdot A \cdot c_W}}$$

- A Effective frontal area of vehicle in m²
- c_W Coefficient of wind resistance
- F_A Drive force at the wheels in N
- F_R Force of rolling resistance
- P_{eff} Useful power of engine in kW
- V_{max} Highest speed of vehicle in km/h
- ϱ Air density

Given: $P_{eff} = 40$ kW, $c_W = 0{,}35$, $A = 1{,}8$ m².
To find : Highest speed of vehicle

$$v_{max} \approx 100 \cdot \sqrt[3]{\frac{40}{18 \cdot 0{,}35 \cdot 1{,}8}} = 152\ \frac{\text{km}}{\text{h}}$$

Given: $F_A = 900$ N, $F_R = 200$ N, $c_W = 0{,}35$, $A = 1{,}8$ m², $\varrho = 1{,}20$ kg/m³.
To find : Highest speed of vehicle

$$v_{max} \approx 3{,}6 \cdot \sqrt{\frac{900 - 200}{0{,}5 \cdot 1{,}2 \cdot 1{,}8 \cdot 0{,}35}} = 155\ \text{km/h}$$

Maximum climbing capacity

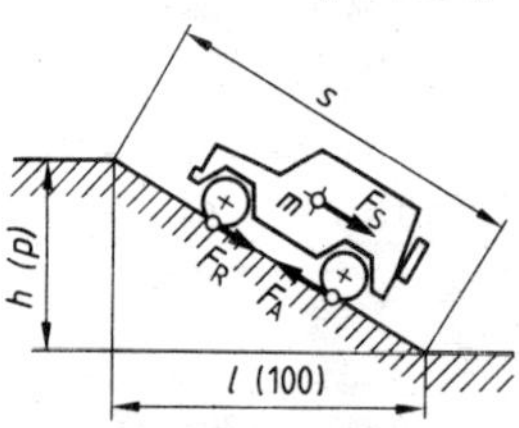

$$P_{max} = \frac{100\% \cdot (F_A - F_R)}{m \cdot g} \qquad h_{max} = \frac{(F_A - F_R) \cdot s}{m \cdot g}$$

$$P_{max} = \frac{100\% \cdot (F_A - F_R)}{G} \qquad h_{max} = \frac{(F_A - F_R) \cdot s}{G}$$

- F_R Force of rolling resistance
- F_A Drive force at the wheels in N
- g Acceleration due to gravity 9.81m/s²
- G Weight of the vehicle in N
- h_{max} Maximum height difference in m
- m Vehicle mass in kg
- P_{max} Maximum passable up-slope in %
- s Road length of the up-slope in m

Given: $F_A = 4\,400$ N, $F_R = 200$ N, $m = 1\,420$ kg.
To find: Maximum passable up-slope

$$P_{max} = \frac{100 \cdot (4\,400 - 200)\ \text{N}}{1\,420\ \text{kg} \cdot 9{,}81\ \text{m/s}^2} = 30{,}2\%$$

Given: $F_A = 4\,400$ N, $s = 1\,000$ m, $m = 1\,420$ kg, $F_R = 200$ N.
To find: Maximum height difference

$$h_{max} = \frac{4\,200\ \text{N} \cdot 1\,000\ \text{m}}{1\,420\ \text{kg} \cdot 9{,}81\ \text{m/s}^2} = 301{,}5\ \text{m}$$

Maximum drive force

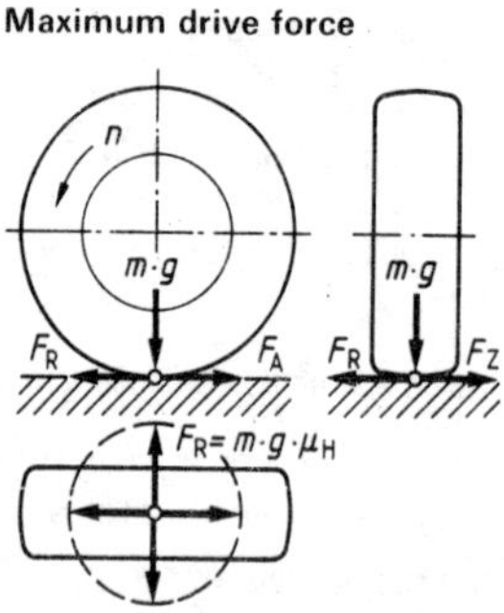

$$F_{A\,max} \leqq F_R \qquad F_{Z\,max} \leqq F_R$$

$$F_{A\,max} = m_A \cdot g \cdot \mu_H \qquad r_{min} = \frac{v^2}{g \cdot \mu_H}$$

$$m_A = \frac{F_{A\,max}}{g \cdot \mu_H} \qquad v_{max} = \sqrt{g \cdot \mu_H \cdot r}$$

- F_{Amax} Maximum transmittable drive force in N
- F_R Friction force, Force closure in N
- F_{Zmax} Maximum permissible centrifugal force in N
- g Acceleration due to gravity 9.81 m/s²
- m_A Axle load of the drive axle in kg.
- r Curve radius in m
- v Travel speed in m/s
- μ_H Coefficient of bonding friction

Given: $m_A = 800$ kg, $\mu_H = 0{,}75$.
To find : Maximum drive force in N

$$F_{A\,max} = 800\ \text{kg} \cdot 9{,}81\ \text{m/s}^2 \cdot 0{,}75 = 5\,886\ \text{N}$$

Given: $v = 108$ km/h (30 m/s), $\mu_H = 0{,}75$.
To find : Minimum curve radius

$$r_{min} = \frac{30^2\ \text{m}^2/\text{s}^2}{9{,}81\ \text{m/s}^2 \cdot 0{,}75} = 122{,}3\ \text{m}$$

Given: $r = 85$ m, $\mu_H = 0{,}75$.
To find : Maximum travel speed

$$v_{max} = \sqrt{9{,}81\ \text{m/s}^2 \cdot 0{,}75 \cdot 85\ \text{m}} = 25\ \text{m/s} = 90\ \text{km/h}$$

Power test

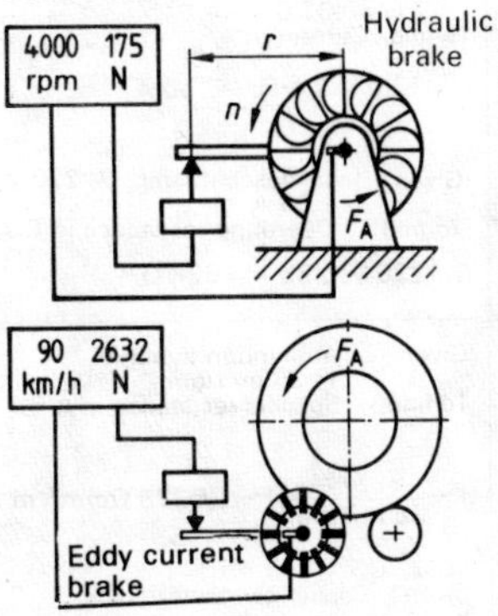

Efficiency	
Gear box, Axle drive	0,93...0,95
Tyre, Wheels	0,81...0,84
Overall efficiency	0,75...0,80

On the engine $P_{eff} = \dfrac{F_A \cdot r \cdot n}{9550}$

$$P_{eff} = \frac{M \cdot n}{9550} \qquad F_A = \frac{9550 \cdot P_{eff}}{r \cdot n}$$

at r = 0,995 m it is

$$M = \frac{9550 \cdot P_{eff}}{n} \qquad P_{eff} = \frac{F_A \cdot n}{10000}$$

On the vehicle $P_{eff} = \dfrac{F_A \cdot v}{3600 \cdot \eta}$

$$P_A = \frac{F_A \cdot v}{3600} \qquad F_A = \frac{3600 \cdot P_{eff} \cdot \eta}{v}$$

$$F_A = \frac{3600 \cdot P_A}{v} \qquad v = \frac{3600 \cdot P_{eff} \cdot \eta}{F_A}$$

- F_A Force, drive force in N
- M Engine torque in Nm
- n Engine speed in 1/min
- P_A Drive power at the roller in kW
- P_{eff} Useful power of the engine in kW
- r Lever arm of the test stand in m
- v Vehicle speed in km/h
- η Overall efficiency at the test stand

Given: $F_A = 175$ N, $r = 955$ mm, $n = 4000$/min.
To find : Useful power P_{eff} in kW

$$P_{eff} = \frac{175 \cdot 0{,}955 \cdot 4000}{9550} = 70{,}0 \text{ kW}$$

or $$P_{eff} = \frac{175 \cdot 4000}{10000} = 70{,}0 \text{ kW}$$

Given: $F_A = 2240$ N, $v = 90$ km/h, $\eta = 0{,}80$.
To find : a) Drive power P_A in kW
b) Useful power P_{eff} in kW

$$P_A = \frac{2240 \cdot 90}{3600} = 56{,}0 \text{ kW}$$

$$P_{eff} = \frac{2240 \cdot 90}{3600 \cdot 0{,}80} = 70{,}0 \text{ kW}$$

Given: $P_{eff} = 70{,}0$ kW, $F_A = 2240$ N, $\eta = 0{,}80$.
To find : Vehicle speed in km/h

$$v = \frac{3600 \cdot 70{,}0 \cdot 0{,}80}{2240} = 90 \text{ km/h}$$

Brake test

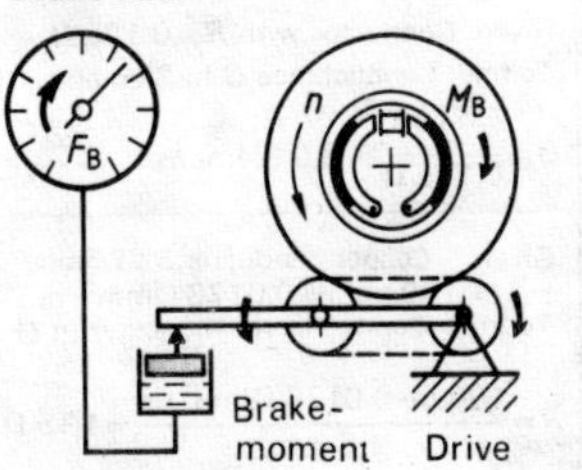

Braking	
Service brake, minimum	
Motor cycle	30%
Other vehicles above 20 km/h	40%
Parking brake	20%
Deviation of the brake forces	
Roller test stand	30%
Platten test stand	20%

$$z = \frac{100\% \cdot F_B}{G} \qquad z = \frac{100\% \cdot F_B}{m \cdot g}$$

$$F_B = \frac{G \cdot z}{100\%} \qquad F_B = \frac{m \cdot g \cdot z}{100\%}$$

$$a_{max} = \frac{z \cdot g}{100\%} \qquad a_{max} \approx \frac{z}{10}$$

$$z = \frac{100\% \cdot a_{max}}{g} \qquad z \approx 10 \cdot a_{max}$$

$$U = \frac{100\% \cdot (F_{Bg} - F_{Bk})}{F_{Bg}}$$

- a Brake deceleration in m/s²
- F_B Total brake force of the wheels in N
- F_{Bg} Greater brake force at the axle in N
- F_{Bk} Smaller brake force at the axle in N
- g Acceleration due to gravity 9.81 m/s²
- G Weight of the vehicle in N
- m Mass of the vehicle in kg
- U Deviation of the brake forces in N
- z Braking in %

Given: $F_B = 5600$ N, $G = 9320$ N.
To find : Braking z in %.

$$z = \frac{100 \cdot 5600 \text{ N}}{9320 \text{ N}} = 60\%$$

Given: $F_{B1} = 1750/1450$ N, $F_{B2} = 1150/1250$ N, $m = 950$ kg.
To find:
a) Total brake force F_B in N
b) Braking Z in %
c) Deviation U in front in %
d) Brake deceleration a

$F_B = 1750 \text{ N} + 1450 \text{ N} + 1150 \text{ N} + 1250 \text{ N} = 5600 \text{ N}$

$$z = \frac{100 \cdot 5600 \text{ N}}{950 \text{ kg} \cdot 9{,}81 \text{ m/s}^2} = 60{,}0\%$$

$$U = \frac{100(1750 - 1450) \text{ N}}{1750 \text{ N}} = 17{,}1\%$$

$$a_{max} = \frac{60 \cdot 9{,}81 \text{ m/s}^2}{100} = 5{,}89 \text{ m/s}^2$$

Axle load test

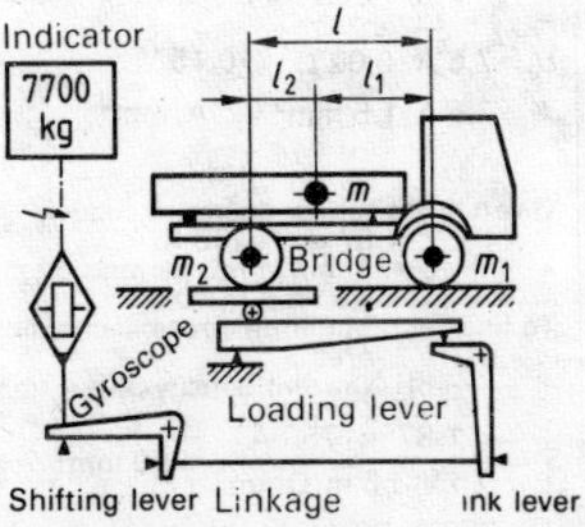

Permissible axle loads	
Single axle	10,0 t
Double axle	16,0 t
Permissible total weight	
Single vehicle	
a) up to 2 axles	16,0 t
b) More than 2 axles	22,0 t
Trailer truck	38,0 t
Saddle vehicle	38,0 t

$$m_1 = \frac{m \cdot p_1}{100\%} \qquad m_2 = m - m_1 \qquad m_2 = \frac{m \cdot p_2}{100\%}$$

$$p_1 = \frac{100\% \cdot m_1}{m} \qquad p_2 = \frac{100\% \cdot m_2}{m}$$

$$m_1 = \frac{m \cdot l_2}{l} \qquad m_1 = m - m_2 \qquad m_2 = \frac{m \cdot l_1}{l}$$

$$l_2 = \frac{m_1 \cdot l}{m} \qquad l_1 = \frac{m_2 \cdot l}{m}$$

$$p_ü = \frac{100\% (m_p - m_z)}{m_z} \qquad m_ü = \frac{m_z \cdot p_ü}{100\%}$$

- l Wheel base of the vehicle
- l_1 Distance from centre gravity to the front axle
- l_2 Distance from centre gravity to the rear axle
- m Total weight, idle weight or useful load
- m_1 Axle load at the front axle
- m_2 Axle load at the rear axle
- m_P Test weight
- $m_ü$ Weight of the overload, Overweight
- m_z Permissible total weight
- p_1 Component of the load on the front axle in %
- p_2 Component of the load on the rear axle in %
- p_u Overload, Overweight in %

Given: $m = 1440$ kg, $p_1 = 40\%$.
To find : a) Axle load m_1 in kg
b) Axle load m_2 in kg

$$m_1 = \frac{1440 \text{ kg} \cdot 40}{100} = 576 \text{ kg}$$

$m_2 = 1440 \text{ kg} - 576 \text{ kg} = 864 \text{ kg}$

Given: $m = 1440$ kg, $l_1 = 1365$ mm, $l = 2275$ mm.
To find : a) Axle load m_2 in kg
b) Axle load m_1 in kg

$$m_2 = \frac{1440 \text{ kg} \cdot 1365 \text{ mm}}{2275 \text{ mm}} = 864 \text{ kg}$$

$m_1 = 1440 \text{ kg} - 864 \text{ kg} = 576 \text{ kg}$

Given: $m_P = 1512$ kg, $m_z = 1440$ kg.
To find a) Overload m_u in kg
b) Overload p_u in %

$m_ü = 1512 \text{ kg} - 1440 \text{ kg} = 72 \text{ kg}$

$$p_ü = \frac{100 \cdot 72 \text{ kg}}{1440 \text{ kg}} = 5\%$$

Ohm's law

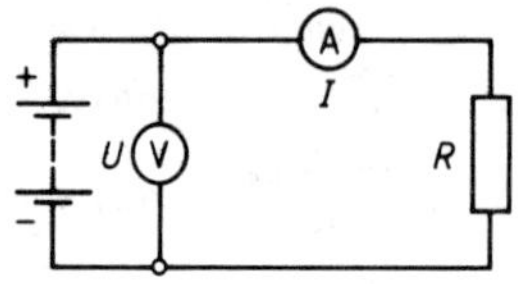

$$\text{Current} = \frac{\text{Voltage}}{\text{Resistance}} \qquad I = \frac{U}{R}$$

$$U = I \cdot R \qquad R = \frac{U}{I}$$

I Electrical current in ampere, A
R Electrical resistance in ohm,
U Electrical voltage in volt, v

Given: Glow plug. $R = 1.6\,\Omega$
$U = 12{,}0$ V.
To find : Current in *A*

$I = 12{,}0\ \text{V} : 1{,}6\ \Omega = 7{,}50\ \text{A}$

Given : Incandescent lamp, $U = 220$ V.
$I = 0{,}35$ A.
To find : Operating resistance in Ω

$R = 220\ \text{V} : 0{,}35\ \text{A} = 628\ \Omega$

Specific resistance, conductivity, conductance

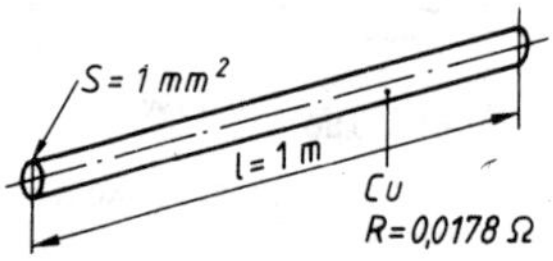

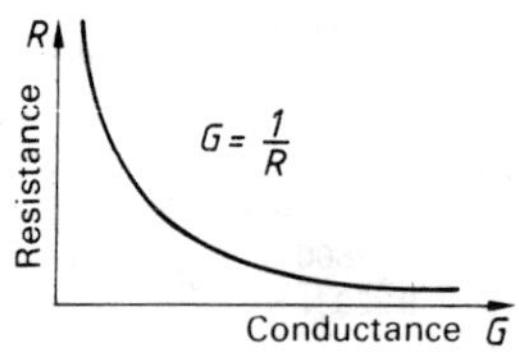

$$\varrho = \frac{1}{\varkappa} \qquad \varkappa = \frac{1}{\varrho}$$

ϱ Specific resistance in $\Omega\,\text{mm}^2/\text{m}$, i.e., electrical resistance of a conductor in Ohm for 1 mm^2 Cross-sectional area and 1 m length
$\varkappa$ Electrical conductivity in $\text{m}/\Omega\,\text{mm}^2$, i.e., conductor length in m for 1 ohm resistance and 1 mm^2 cross-sectional area

Electrical conductance $\quad G = \dfrac{1}{R}$

$$1\,\text{Siemens} = \frac{1}{\Omega} \qquad R = \frac{1}{G}$$

G Electrical conductance in siemens, s
R Electrical resistance in ohm, Ω

Given : Aluminium material,
$\varkappa = 36\ \text{m}/\Omega\text{mm}^2$
To find : Specific resistance in ρ

$$\varrho = \frac{1}{36\ \text{m}/\Omega\text{mm}^2} = 0{,}0278\ \Omega\text{mm}^2/\text{m}$$

Given : Copper conductor grade
$\varrho = 0{,}01786\ \Omega\text{mm}^2/\text{m}$.
To find : Conductivity

$$\varkappa = \frac{1}{0{,}01786\ \Omega\text{mm}^2/\text{m}} = 56\ \text{m}/\Omega\text{mm}^2$$

Given: Conductor with $R = 0{,}125\ \Omega$.
To find: Conductance *G* in Siemens

$$G = \frac{1}{0{,}125\ \Omega} = 8{,}0\ \text{Siemens}$$

Conductor resistance

Conductor material	ϱ $\frac{\Omega\text{mm}^2}{\text{m}}$	$\varkappa$ $\frac{\text{m}}{\Omega\text{mm}^2}$
Silver	0,0167	60
Copper	0,0178	56
Aluminium	0,0278	36
Iron	0,13	7,7
Manganese	0,43	2,3
Constantin	0,49	2,04
Chrome-nickel	1,04	0,96

$$R = \frac{l \cdot \varrho}{S} \qquad R = \frac{l}{S \cdot \varkappa}$$

$$S = \frac{l \cdot \varrho}{R} \qquad S = \frac{l}{R \cdot \varkappa}$$

$$l = \frac{R \cdot S}{\varrho} \qquad l = R \cdot S \cdot \varkappa$$

l Conductor length in m
R Conductor resistance in
S Conductor cross-sectional area in mm^2
χ Conductivity in $\text{m}/\Omega\ \text{mm}^2$
ρ Specific resistance in mm^2/m

Given: Copper conductor, $S = 1{,}5\,\text{mm}^2$
$l = 100$ m, $\varrho = 0{,}0178\ \Omega\text{mm}^2/\text{m}$.
To find: Conductor resistance *R* in Ω.

$$R = \frac{100\ \text{m} \cdot 0{,}0178\ \Omega\text{mm}^2/\text{m}}{1{,}5\ \text{mm}^2} = 1{,}19\ \Omega.$$

Given : Copper conductor, $R = 1{,}19\ \Omega$
$l = 100$ m, $\varkappa = 56\ \text{m}/\Omega\text{mm}^2$.
To find : Cross-sectional areas in mm²

$$S = \frac{100\ \text{m}}{1{,}19\ \Omega \cdot 56\ \text{m}/\Omega\text{mm}^2} = 1{,}5\ \text{mm}^2$$

Conductor calculation

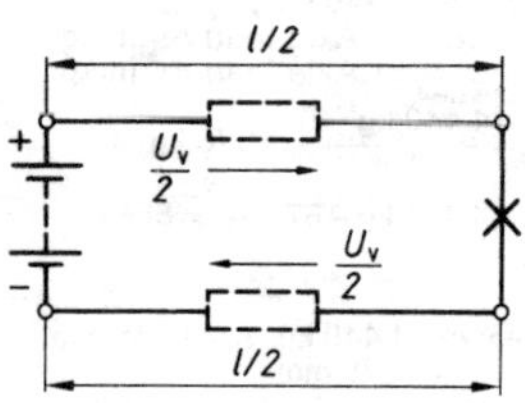

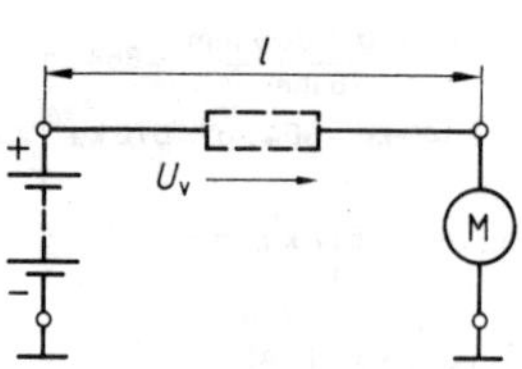

$$U_v = I \cdot R \qquad J = \frac{I}{S}$$

$$R = \frac{U_v}{I} \qquad S = \frac{I}{J}$$

$$I = \frac{U_v}{R} \qquad I = J \cdot S$$

$$U_v = \frac{I \cdot l \cdot \varrho}{S} \qquad U_v = \frac{I \cdot l}{S \cdot \varkappa}$$

$$S = \frac{I \cdot l \cdot \varrho}{U_v} \qquad S = \frac{I \cdot l}{U_v \cdot \varkappa}$$

$$l = \frac{S \cdot U_v}{I \cdot \varrho} \qquad l = \frac{S \cdot U_v \cdot \varkappa}{I}$$

$$I = \frac{S \cdot U_v}{l \cdot \varrho} \qquad I = \frac{S \cdot U_v \cdot \varkappa}{l}$$

I Current in A
J Current density in A/mm^2
L Conductor length in m
R Conductor resistance in Ω
S Cross-sectional area of conductor in mm^2
U Voltage drop in v
χ Conductivity in $\text{m}/\Omega\ \text{mm}^2$
ϱ Specific resistance in $\Omega\ \text{mm}^2/\text{m}$

Given : Lighting conductor $S = 1{,}5\text{mm}^2$
$R = 0{,}02\ \Omega$, $I = 7{,}5$ A.
To find : a) Voltage drop *U* in V
b) Current density *J* in A mm²

$U_v = 7{,}5\ \text{A} \cdot 0{,}02\ \Omega = 0{,}15\ \text{V}$
$J = 7{,}5\ \text{A} : 1{,}5\ \text{mm}^2 = 5\ \text{A}/\text{mm}^2$

Given : Starter conductor of copper
$l = 1{,}87$ m, $I = 750$ A.
$\chi = 56\ \text{m}/\Omega\text{mm}^2$, permissible voltage drop $U_v = 0{,}5$ V.
To find : a) Minimum cross-sectional area *S*
b) Current density *J* in A/mm²

$$S = \frac{1{,}87\ \text{m} \cdot 750\ \text{A}}{0{,}5\ \text{V} \cdot 56\ \text{m}/\Omega\text{mm}^2} = 50\ \text{mm}^2$$

$J = 750\ \text{A} : 50\ \text{mm}^2 = 15\ \text{A}/\text{mm}^2$

Given : Starter conductor $S = 16\ \text{mm}^2$
$l = 1{,}6$ m, $U_v = 0{,}5$ V,
$\varkappa = 56\ \text{m}/\Omega\text{mm}^2$.
To find : Current *I* in A.

$$I = \frac{16\ \text{mm}^2 \cdot 0{,}5\ \text{V} \cdot 56\ \text{m}/\Omega\text{mm}^2}{1{,}60\ \text{m}}$$

$= 280$ A

Series circuit

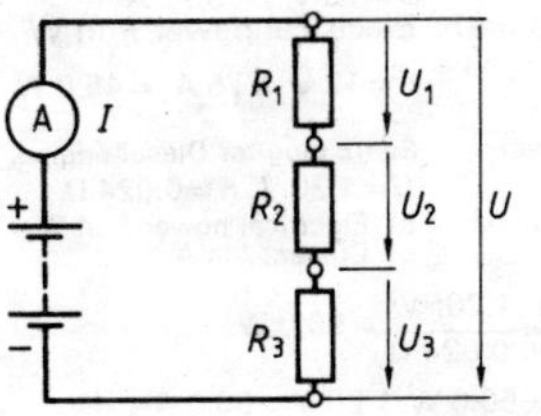

$R = R_1 + R_2 + R_3 + \ldots$ I is same in all resistances

$U = U_1 + U_2 + U_3 + \ldots$ $\qquad I = \frac{U}{R}$

$U_1 = I \cdot R_1 \qquad U_2 = I \cdot R_2 \qquad U_3 = I \cdot R_3$

I Current in A
R Total resistance in Ω
R_1, R_2 Individual resistances in Ω
U Total voltage in V
U_1, U_2 Segmental Voltages in V

Given: Series circuit $R_1 = 4\,\Omega$, $R_2 = 6\,\Omega$, $R_3 = 10\,\Omega$, $U = 12$ V.
To find: a) Total resistance R in Ω
b) Current I in A
c) Segmental voltages U_1, U_2, U_3

$R = 4\,\Omega + 6\,\Omega + 10\,\Omega = 20{,}0\,\Omega$
$I = 12\,\text{V} : 20\,\Omega = 0{,}60\,\text{A}$
$U_1 = 0{,}6\,\text{A} \cdot 4\,\Omega = 2{,}40\,\text{V}$
$U_2 = 0{,}6\,\text{A} \cdot 6\,\Omega = 3{,}60\,\text{V}$
$U_3 = 0{,}6\,\text{A} \cdot 10\,\Omega = 6{,}00\,\text{V}$

Parallel circuit

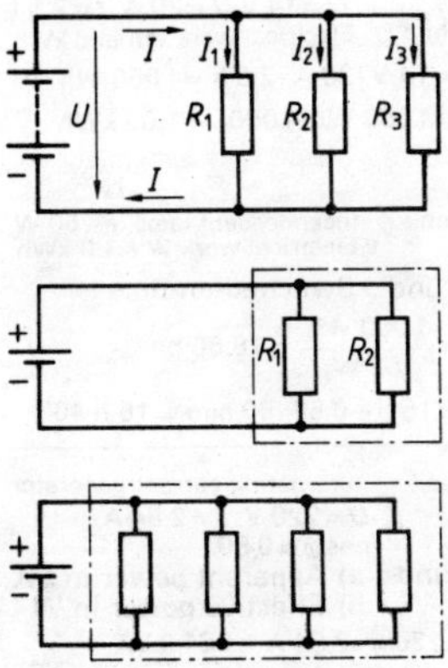

$\frac{1}{R} = \frac{1}{R_1} + \frac{1}{R_2} + \frac{1}{R_3} + \ldots$ U is same across any resistance

$I = I_1 + I_2 + I_3 + \ldots \qquad U = I \cdot R$

$I_1 = \frac{U}{R_1} \qquad I_2 = \frac{U}{R_2} \qquad I_3 = \frac{U}{R_3}$

$R = \frac{R_1 \cdot R_2}{R_1 + R_2}$ Equation for two parallel resistances

$R_1 = \frac{R_2 \cdot R}{R_2 - R} \qquad R_2 = \frac{R_1 \cdot R}{R_1 - R}$

$R = \frac{R_1}{n}$ For n equal parallel resistances

I Total current in A
I_1, I_2 Segmental currents in A
n Number of equal resistances
R Equivalent resistance in Ω
R_1, R_2 Individual resistances in Ω
U Voltage in V

Given: Parallel circuit $R_1 = 10\,\Omega$, $R_2 = 15\,\Omega$, $R_3 = 30\,\Omega$, $U = 12$ V.
To find: a) Equivalent resistance R in
b) Segmental currents I_1, I_2, I_3 in A
c) Total current I in A

$\frac{1}{R} = \frac{1}{10\,\Omega} + \frac{1}{15\,\Omega} + \frac{1}{30\,\Omega} = \frac{6}{30\,\Omega}$

$R = \frac{30\,\Omega}{6} = 5{,}0\,\Omega$

$I_1 = 12\,\text{V} : 10\,\Omega = 1{,}2\,\text{A}$
$I_2 = 12\,\text{V} : 15\,\Omega = 0{,}8\,\text{A}$
$I_3 = 12\,\text{V} : 30\,\Omega = 0{,}4\,\text{A}$
$I = I_1 + I_2 + I_3 = 2{,}4\,\text{A}$
or $I = 12\,\text{V} : 5\,\Omega = 2{,}4\,\text{A}$

Given: *Parallel circuit of two resistances $R_1 = 10\,\Omega$ and $R_2 = 15\,\Omega$*
To find: Equivalent resistance R in Ω

$R = \frac{10\,\Omega \cdot 15\,\Omega}{10\,\Omega + 15\,\Omega} = 6{,}0\,\Omega$

Battery capacity efficiency

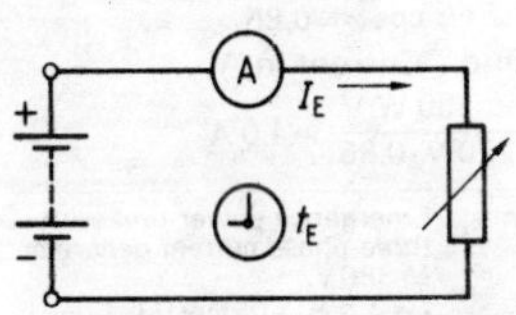

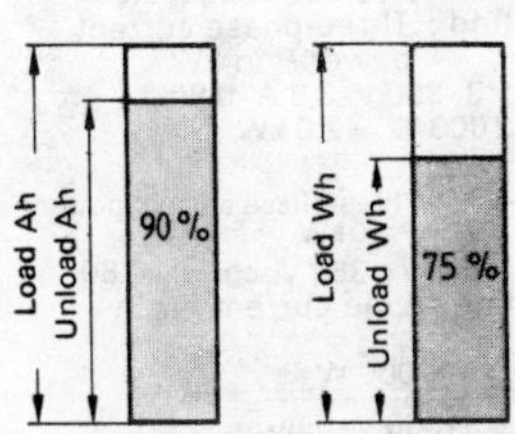

$K_E = I_E \cdot t_E \qquad K_L = I_L \cdot t_L$

$I_E = \frac{K_E}{t_E} \qquad I_L = \frac{K_L}{t_L}$

$t_E = \frac{K_E}{I_E} \qquad t_L = \frac{K_L}{I_L}$

$\eta_{Ah} = \frac{K_E}{K_L} \qquad \eta_{Wh} = \frac{K_E \cdot U_E}{K_L \cdot U_L} = \frac{W_E}{W_L}$

$K_E = K_L \cdot \eta_{Ah} \qquad W_E = W_L \cdot \eta_{Wh}$

$K_L = \frac{K_E}{\eta_{Ah}} \qquad W_L = \frac{W_E}{\eta_{Wh}}$

I_E, I_L Discharge and Charge currents in A
K_E, K_L Discharge and charge capacities in Ah
T_E, T_L Discharge and charge times in h
U_E, U_L Average discharge and charge voltages in V
W_E, W_L Discharge and charge works in Wh
η_{Ah} Charge efficiency
η_{Wh} Energy efficiency

Given: Starter battery $I_E = 2{,}7$ A, $t_E = 20$ h, $I_L = 5{,}0$ A, $t_L = 12$ h.
To find: a) Discharge capacity
b) Charge capacity
c) Charge efficiency

$K_E = 2{,}7\,\text{A} \cdot 20\,\text{h} = 54\,\text{Ah}$
$K_L = 5{,}0\,\text{A} \cdot 12\,\text{h} = 60\,\text{Ah}$
$\eta_{Ah} = 54\,\text{Ah} : 60\,\text{Ah} = 0{,}90$

Given: Starter battery $K_E = 54$ Ah, $U_E = 11{,}5$ V, $K_L = 60$ Ah, $U_L = 13{,}8$ V.
To find: a) Discharge work
b) Charge work
c) Energy efficiency

$W_E = 54\,\text{Ah} \cdot 11{,}5\,\text{V} = 621\,\text{Wh}$
$W_L = 60\,\text{Ah} \cdot 13{,}8\,\text{V} = 828\,\text{Wh}$
$\eta_{Wh} = 621\,\text{Wh} : 828\,\text{Wh} = 0{,}75$

Internal resistance of battery

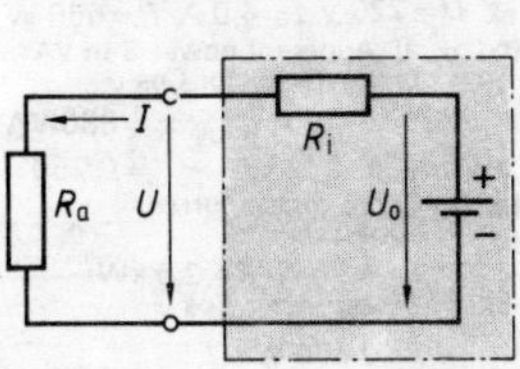

$R_i = \frac{U_0 - U}{I} \qquad R_i \approx \frac{0{,}1 \cdot z}{K_{20}}$

$U = U_0 - I \cdot R_i \qquad I = \frac{U_0}{R_a + R_i} \qquad I_K = \frac{U_0}{R_i}$

I Load current in A
I_K Short circuit current in A
K_{20} Nominal capacity in Ah
R_a, R_i External and internal resistances in Ω
U Terminal voltage, loaded, in V
U_0 Terminal voltage, unloaded, in V
z Number of cells

Given: Starter battery, $I = 200$ A, $U_0 = 12{,}2$ V, $U = 9{,}8$ V.
To find: Internal resistance R_i in Ω

$R_i = \frac{12{,}2\,\text{V} - 9{,}8\,\text{V}}{200\,\text{A}} = 0{,}012\,\Omega$

Given: Starter battery, $I = 200$ A, $U_0 = 12{,}2$ V, $R = 0{,}012\,\Omega$.
To find: Voltage U, loaded, in V

$U = 12{,}2\,\text{V} - 200\,\text{A} \cdot 0{,}012\,\Omega = 9{,}8\,\text{V}$

Electrotechnology

Electrical power

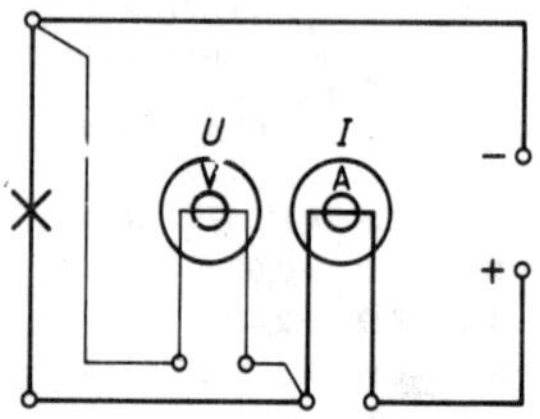

1 W = 1 Nm/s = 1 J/s

$P = U \cdot I$ $\quad P = I^2 \cdot R$ $\quad P = \frac{U^2}{R}$

$I = \frac{P}{U}$ $\quad I = \sqrt{\frac{P}{R}}$ $\quad U = \sqrt{P \cdot R}$

$U = \frac{P}{I}$ $\quad R = \frac{P}{I^2}$ $\quad R = \frac{U^2}{P}$

I Current in A
P Electrical power in W
R Resistance in
U Voltage in V

Given: Incandescent lamp for head light
$U = 12$ V, $I = 3{,}75$ A.
To find: Electrical power P in W

$P = 12 \text{ V} \cdot 3{,}75 \text{ A} = 45{,}0 \text{ W}$

Given: Spark plug for Diesel engine
$U = 1{,}20$ V, $R = 0{,}024\ \Omega$.
To find: a) Electrical power P in W
b) Current I in A

$P = \frac{1{,}20^2 \text{ V}^2}{0{,}024\ \Omega} = 60{,}0 \text{ W}$

$I = 60{,}0 \text{ W} : 1{,}20 \text{ V} = 50{,}0 \text{ A}$

Electrical work

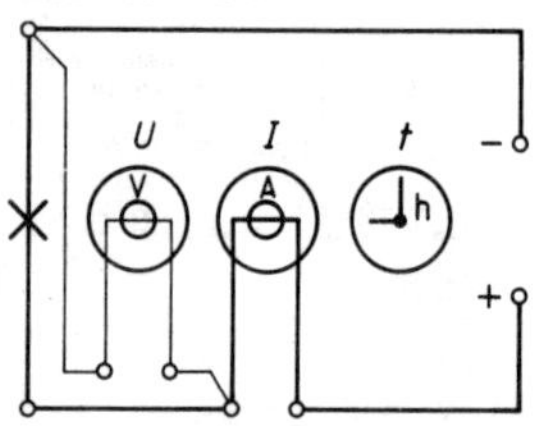

1 Ws = 1 Nm = 1 J
3600 Ws = 1 Wh = 3,6 kJ
1 000 Wh = 1 kWh = 3600 kJ

$W = U \cdot I \cdot t$ $\quad W = P \cdot t$

$I = \frac{W}{U \cdot t}$ $\quad P = \frac{W}{t}$

$t = \frac{W}{U \cdot I}$ $\quad t = \frac{W}{P}$

$U = \frac{W}{I \cdot t}$

I Current in A
P Electrical power in W
t Time, Switched on time in s, h
U Voltage in V
W Electrical work in Ws, Wh

Given: Generator in Vehicle
$U = 14$ V, $I = 30$ A, $t = 2{,}5$ h
To find: Electrical work Wh and kWh

$W = 14 \text{ V} \cdot 30 \text{ A} \cdot 2{,}5 \text{ h} = 1\,050 \text{ Wh}$

$W = 1\,050 \text{ Wh} : 1000 = 1{,}05 \text{ kWh}$

Given: Incandescent lamp, $P = 60$ W, Electrical work $W = 1{,}0$ kWh.
To find: Switched-on time t

$t = \frac{1\,000 \text{ Wh}}{60 \text{ W}} = 16{,}66 \text{ h}$

$t = 16 \text{ h} + 0{,}66 \cdot 60 \text{ min} = 16 \text{ h } 40'$

Electrical power in case of alternating current

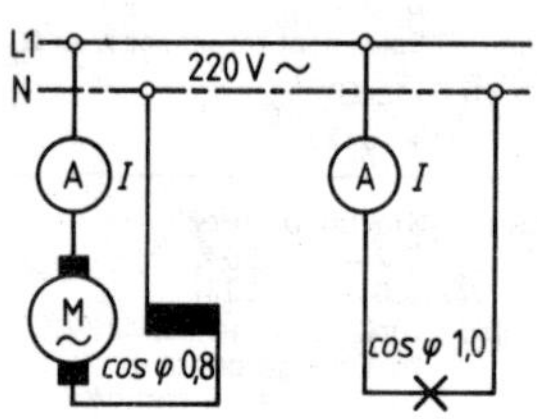

$P = U \cdot I \cdot \cos\varphi$ $\quad S = U \cdot I$

$I = \frac{P}{U \cdot \cos\varphi}$

In case of ohmic load (heating equipment, incandescent lamps), the power factor Cos $\varphi = 1$ so that $P = S$

$U = \frac{P}{I \cdot \cos\varphi}$

I Current in A
P Effective power in W
S Apparent power in VA
U Voltage in V
$\cos\varphi$ Power factor

Given: Alternating current generator
$U = 220$ V, $I = 2{,}84$ A,
$\cos\varphi = 0{,}80$.
To find: a) Apparent power in VA
b) Effective power in W

$S = 220 \text{ V} \cdot 2{,}84 \text{ A} = 624{,}8 \text{ VA}$

$P = 220 \text{ V} \cdot 2{,}84 \text{ A} \cdot 0{,}8 = 500 \text{ W}$

Given: Alternating current power
$P = 750$ W, $U = 220$ V,
$\cos\varphi = 0{,}85$.
To find: Current in A

$I = \frac{750 \text{ W}}{220 \text{ V} \cdot 0{,}85} = 4{,}0 \text{ A}$

Electrical power in case of three-phase current

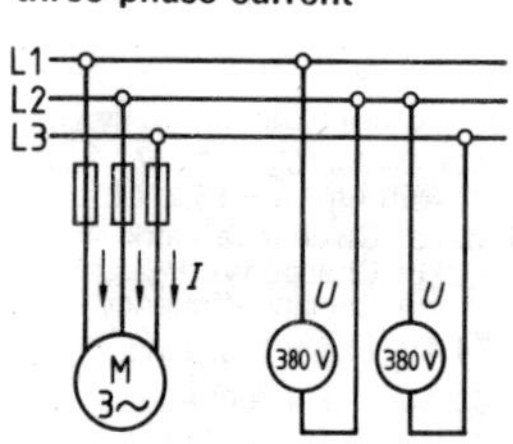

$P = \sqrt{3} \cdot U \cdot I \cdot \cos\varphi$ $\quad S = \sqrt{3} \cdot U \cdot I$

$I = \frac{P}{\sqrt{3} \cdot U \cdot \cos\varphi}$

In case of Ohmic load, Cos $\varphi = 1$ so that $P = S$

$U = \frac{P}{\sqrt{3} \cdot I \cdot \cos\varphi}$

I Line current in A
P Three-phase current power in W
U Line voltage in V
$\cos\varphi$ Power factor
$\sqrt{3}$ Interlink factor

Given: Emergency power unit with three-phase current generator
$U = 380$ V,
$I = 3{,}8$ A, $\cos\varphi = 0{,}80$.
To find: Three-phase current power P in kW

$P = \sqrt{3} \cdot 380 \text{ V} \cdot 3{,}8 \text{ A} \cdot 0{,}80$
$= 2\,000 \text{ W} = 2{,}0 \text{ kW}$

Given: Three-phase current power
$P = 10$ kW,
$U = 380$ V, $\cos\varphi = 0{,}80$.
To find: Line current I in A

$I = \frac{10\,000 \text{ W}}{\sqrt{3} \cdot 380 \text{ V} \cdot 0{,}80} = 19 \text{ A}$

Power factor

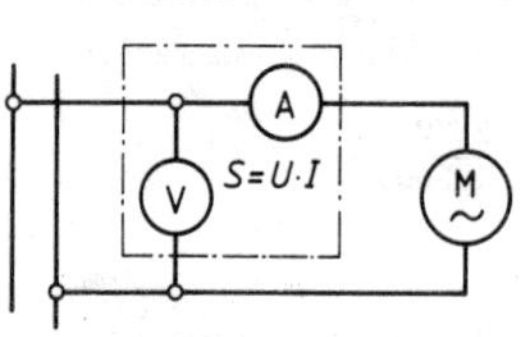

Cos $\varphi = 0{,}7$: the effective power P amounts to 70% of the apparent power S

$\cos\varphi = \frac{P}{S}$

$P = S \cdot \cos\varphi$ $\quad S = U \cdot I$ (for alternating current)

$S = \frac{P}{\cos\varphi}$ $\quad S = \sqrt{3} \cdot U \cdot I$ (for three-phase current)

P Effective power in W, kW
S Apparent power in VA, kVA
$\cos\varphi$ Power factor

Given: Alternating current motor
$U = 220$ V, $I = 4{,}0$ A, $P = 660$ W.
To find: a) Apparent power S in VA
b) Power factor Cos φ

$S = 220 \text{ V} \cdot 4{,}0 \text{ A} = 880 \text{ VA}$

$\cos\varphi = 660 \text{ W} : 880 \text{ VA} = 0{,}750$

Given: Three-phase current generator
$I = 4{,}75$ A, $P = 2{,}5$ kW.
To find: Power factor Cos φ

$\cos\varphi = \frac{2\,500 \text{ W}}{\sqrt{3} \cdot 380 \text{ V} \cdot 4{,}75 \text{ A}} = 0{,}80$

Taper Turning

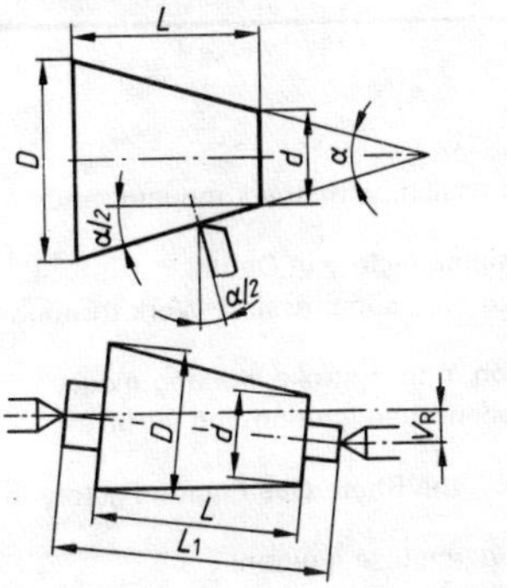

$C = \frac{D-d}{L}$ $\qquad$ $\tan\frac{\alpha}{2} = \frac{D-d}{2 \cdot L} = \frac{C}{2}$

$L = \frac{D-d}{C}$ $\qquad$ $L = \frac{D-d}{2 \cdot \tan\frac{\alpha}{2}}$

$D = d + C \cdot L$ $\qquad$ $D = d + 2 \cdot L \cdot \tan\frac{\alpha}{2}$

$d = D - C \cdot L$ $\qquad$ $d = D - 2 \cdot L \cdot \tan\frac{\alpha}{2}$

$V_R = \frac{D-d}{2}$ $\qquad$ $V_R = \frac{(D-d) \cdot L_1}{2 \cdot L}$

D Large diameter of taper
d Small diameter of taper
L Taper length
L_1 Workpiece length
V_R Tailstock setting
C Taper ratio, taper reduction 1:x
$\alpha/2$ Setting angle

Given: Taper, D=32 mm, d=26 mm, L=60 mm.

To find: a) Taper ratio C
b) Setting angle $\alpha/2$ in degree
c) Tailstock setting V_R

$C = \frac{32\ \text{mm} - 26\ \text{mm}}{60\ \text{mm}} = 0{,}1 = 1{:}10$

$\tan \alpha/2 = \frac{32\ \text{mm} - 26\ \text{mm}}{2 \cdot 60\ \text{mm}} = 0{,}050$

Setting angle $\alpha/2 = 2°\ 51'$

$V_R = \frac{32\ \text{mm} - 26\ \text{mm}}{2} = 3{,}0\ \text{mm}$

Given: Taper 1:20, D=40 mm, L=80 mm.

To find: Diameter d in mm

$d = 40\ \text{mm} - \frac{80\ \text{mm}}{20} = 36{,}0\ \text{mm}$

Gear calculation

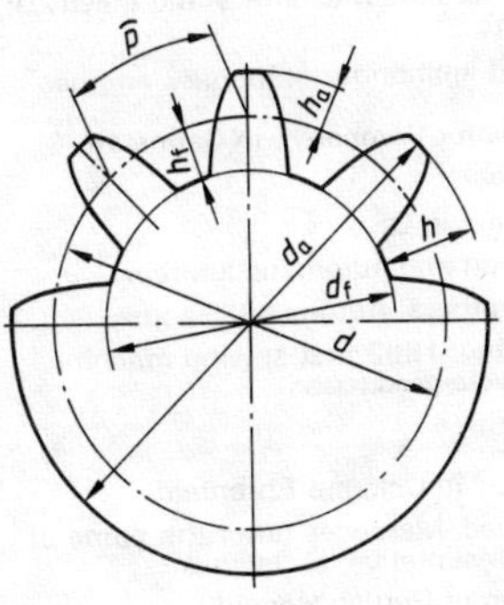

$d = m \cdot z$ $\qquad$ $a = \frac{m \cdot (z_1 + z_2)}{2}$

$m = \frac{d}{z}$ $\qquad$ $m = \frac{2 \cdot a}{z_1 + z_2}$

$z = \frac{d}{m}$ $\qquad$ $z_1 = \frac{2 \cdot a}{m} - z_2$

$d_a = m \cdot (z+2)$ $\qquad$ $d_f = m \cdot \left(z - \frac{7}{3}\right)$

$m = \frac{d_a}{z+2}$ $\qquad$ $m = \frac{d_f}{z - \frac{7}{3}}$

$z = \frac{d_a}{m} - 2$ $\qquad$ $z = \frac{d_f}{m} + \frac{7}{3}$

$m = \frac{p}{\pi}$ $\qquad$ $h = \frac{13}{6} \cdot m$

$p = \pi \cdot m$ $\qquad$ $h_a = m$ $\qquad$ $h_f = \frac{7}{6} \cdot m$

a	Centre distance	h_a	Addendum
d	Pitch circle diameter	h_f	Dedendum
d_a	Tip circle diameter	m	Module
d_f	Root circle diameter	p	Pitch
h	Tooth height	z	Number of teeth

Given: Gear drive, z_1=36, z_2=54, m=3 mm.

To find: a) Pitch circle diameter
b) Tip circle diameter
c) Root circle diameter
d) Centre distance
e) Tooth height

$d_1 = 3\ \text{mm} \cdot 36 = 108\ \text{mm}$

$d_{a1} = 3\ \text{mm} \cdot (36+2) = 114\ \text{mm}$

$d_{f1} = 3\ \text{mm} \cdot (36 - 7/3) = 101\ \text{mm}$

$a = \frac{3\ \text{mm} \cdot (36+54)}{2} = 135\ \text{mm}$

$h = \frac{13}{6} \cdot 3\ \text{mm} = 6{,}5\ \text{mm}$

Given: Gear drive, a=100 mm. m=2,5 mm, i=1,5.

To find: Numbers of teeth Z_1 and Z_2

$z_1 + z_2 = \frac{2 \cdot 100\ \text{mm}}{2{,}5\ \text{mm}} = 80$ teeth

$z_1 = 1$ part $\qquad$ $z_2 = 1{,}5$ parts

$z_1 = \frac{80\ \text{teeth} \cdot 1{,}5\ \text{parts}}{(1{,}5+1)\ \text{parts}} = 48$ teeth

$z_2 = 80$ teeth $- 48$ teeth $= 32$ teeth

Cutting speed

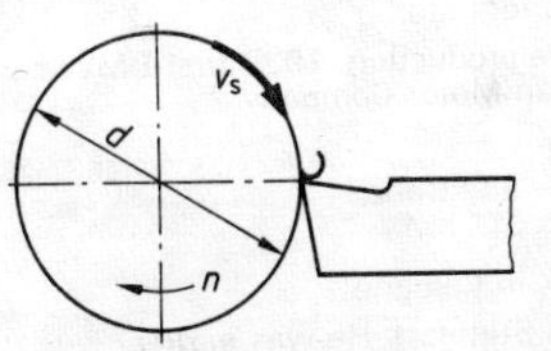

$v_S = \frac{\pi \cdot d \cdot n}{1000}$ $\qquad$ $v_S = 60 \cdot v_u$

$d = \frac{1000 \cdot v_S}{\pi \cdot n}$ $\qquad$ $v_u = \frac{v_S}{60}$

$n = \frac{1000 \cdot v_S}{\pi \cdot d}$

d Diameter in mm
n Speed in rpm
v_S Cutting speed in m/min
v_u Peripheral speed in m/s

Given: Longitudinal turning of shaft, d=80 mm, n=100 rpm

To find: Cutting speed V_s

$v_S = \frac{\pi \cdot 80 \cdot 100}{1000} = 25$ m/min

Given: Boring of hole $v_S = 18$ m/min, $d = 20$ mm.

To find: Speed n in rpm

$n = \frac{1000 \cdot 18}{\pi \cdot 20} = 286$ rpm

Cutting time

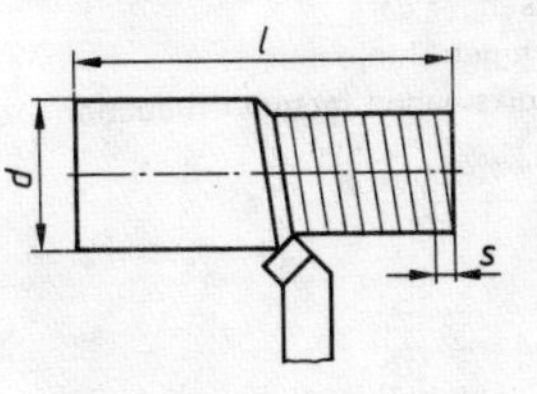

$t_h = \frac{\pi \cdot d \cdot l \cdot i}{1000 \cdot v_S \cdot s}$ $\qquad$ $t_h = \frac{l \cdot i}{n \cdot s}$

d Diameter before turning in mm
Drill diameter in mm
i Number of cuts, of the holes
l Turning length in mm
Turning distance in case of face (d/2) in mm
Boring distance = Hole depth + 0.3.d in mm
n Speed in rpm
s Feed in mm/resolution
t_h Cutting time in min
v_s Cutting speed in m/min.

Given: Longitudinal turning of shaft i=2, d=40 mm, l=250 mm. s = 0.50 mm/revolution v_s=25 m/min, n=199 1/min.

To find: Cutting time t_h in min.

$t_h = \frac{\pi \cdot 40 \cdot 250 \cdot 2}{1000 \cdot 25 \cdot 0{,}5} = 5{,}0$ min

or $t_h = \frac{250 \cdot 2}{199 \cdot 0{,}5} = 5{,}0$ min

History of Automobile

1770	Nicolaus Cugnot	Steam Car
1860	Lenoir	Gas machine without compression, with spark inductor and double-acting piston.
1864	Nicolaus Otto and Eugen Langen	Establishment of the first engine factory in Deutz
1867	Nicolaus Otto (1832 to 1891) with Eugen Langen	Atmospheric gas machine without compression. Work through external air pressure.
1876-78	Nicolaus Otto	Gas engine with compression in four-stroke working mode.
1883	Gottlieb Daimler (1834 to 1900) with Wilhelm Maybach	Petrol engine with incandescent tube ignition and surface carburettor.
	Karl Benz (1844-1929)	Establishment of 'Benz & Co.', the Rhein Gas Engine Factory in Mannhein.
1884	Nicolaus Otto	Low-voltage magneto ignition (impulse ignition)
1885	Gottlieb Daimler	Two-wheeler (Low wheel) with petrol engine.
	Karl Benz	Three-wheeler with petrol engine and electric ignition.
1886	Gottlieb Daimler	4-wheel carriage with petrol engine
	Robert Bosch (1861-1942)	Establishment of 'Workshop for Fine Mechanics and Electrical Engineering of Robert Bosch'.
1887	Robert Bosch	Electromagnetic HV frictional ignition for stationary engines.
1890	Gottlieb Daimler	Establishment of 'Daimler Motor Company', in Cannstatt.
	Dunlop	Air filled tyres for automobiles.
1893	Henry Ford	First vehicle with petrol engine in USA
1893-97	Rudolf Diesel (1858-1913)	Heavy oil engine with injection and automatic ignition
1896	Stoever Brothers	Establishment of 'Stoever Brothers' Automobile Factory.
1898	Adam Opel (1837-1895)	Commencement of car building; 1862 first sewing machine; 1887 commencement of bicycle production.
1899		Establishment of 'Fiat' in Turin.
1900	August Horch	Establishment of 'Horch & Co.' in Cologne-Ehrenfeld.
1900/01		Daimler vehicles are christened 'Mercedes' after the name of the daughter of General Representative, E. Jellinek.
1902	Robert Bosch	High-voltage magneto ignition of Gottlob Honold
	Buessing	First German truck with petrol engine
1903	Henry Ford	Establishment of the 'Ford Motor Company' in Detroit
1905	Borgward	Establishment of 'Hansa Factory' in Bremen
1909	Otto Vollnhals	First tractor with diesel engine
	August Horch	Establishment of 'Audi Factory'
	Wilhelm Maybach	Establishment of 'Maybach Motors Ltd.'
1923	Bavarian Motor Company (BMW)	Commencement of motor cycle production, 1929 first BMW car, 1917 Establishment of Bavarian Motor Company
1923-24	Benz-MAN	Truck with Benz diesel engine
1926		Merger of companies—Daimler and Benz
1928	Rasmussen	DKW two-stroke two-cylinder cars
1930		Establishment of 'Ford Factory' in Cologne
1930-31	Ferdinand Porsche (1875-1951)	Opening of a design bureau in Stuttgart. He was earlier employed at Lohner, Austro-Daimler and Daimler-Benz
1932		Establishment of 'Auto Union' by the merger of Audi, Horch, Zschopauer motor factories and Wanderer factory
1936	Daimler-Benz AG	First mass-produced diesel passenger cars
1938		Internal combustion engine with petrol injectior
		Foundation stone laid for the Volkswagen factory Production commenced in 1945
1939	Brown-Boveri	Stationary gas engines
1950	Rover	Automobile with turbine motor
1957/58	NSU/Wankel	Rotary piston engine.

Concepts in Automobile Engineering

Working cycle	All operations in the cylinder, starting from a definite state till the next repetition of the same state.
Continuous power output or short duration power output	Maximum output the engine can give continuously or for 15 minutes without damage in respect of the purpose of application
Dead weight (mass) of vehicle	Mass of the ready-for-operation vehicle with filled fuel tank inclusive of the mass of all equipment carried in operation. In case of vehicles other than cars and motor cycles, it includes the driver's weight which is taken as 75 kg.
Cylinder filling	Ratio of the actually sucked quantity of fresh gas during a working cycle to the stroke volume of a cylinder.
Total stroke volume	Sum of the stroke volumes of all working cylinders of an engine.
Maximum or continuous speed	Maximum speed that can be maintained by the vehicle on 1 km test stretch or the speed maintained continuously.
Maximum combustion pressure or end pressure of compression	Maximum pressure acting on the piston of an ignited or not ignited operationally warm internal combustion engine.
Stroke volume of a cylinder	Volume swept by the piston during a stroke.
Output power per unit of displacement	Maximum useful power than can be obtained from 1 l of total stroke volume.
Internal power	Power taken from the interior of the cylinder by the crank drive.
Piston stroke	Distance between the two dead centres of a piston.
Fuel consumption as per DIN	Quantity of fuel fed into the engine as determined according to DIN 70030: half useful load, operationally warm engine, 10 km. dry drive stretch, maximum 1.5% gradient and wind speed 3 m/s. To and fro dirve without interruption, 3/4 maximum speed, max. 110 km/h, commercially available fuel, supplement 10%.
Crank angle	Position of the crank pin with reference to the upper or lower dead centre.
Weight per horse power	Necessary mass of the ready-to-drive vehicle or engine for 1 kW.
Air requirement	Air required by 1 kg of fuel for complete combustion
Air ratio	Ratio of air quantity sucked in for the combustion of fuel to the quantity of air required for complete combustion.
Mechanical efficiency	Ratio of useful power to internal power.
Mean working pressure	Mean pressure on the piston during the entire working cycle.
Mean piston speed	Distance covered by the to-and-fro moving piston in one second at constant speed.
Engine speed	Number of revolutions of the crankshaft per minute.
Useful power	Useful power delivered by the engine at the clutch, where the equipment needed for the operation of the engine, e.g. fan, cool air blower, water and fuel injection pumps and generator. (a) are driven by the engine in DIN—PS (b) are not considered in SAE—P
Useful efficiency or internal efficiency	Ratio of useful power or internal power to the maximum power that can be obtained with the available fuel.
Top or bottom controlled engine	Valve disc and gas path lie above, or in or below a level which lies on the piston located on the top dead centre.
Lubricating oil consumption	Lubricating oil quantity consumed in unit time. The drained quantity of oil during oil change is not taken into account.
Specific fuel consumption	Quantity of fuel fed to the engine per unit power and time.
Field consumption	Determination: 100 km circuit, normal streets, driving in traffic, vehicle occupied by 2 persons. Field consumption is approximately 10% above DIN consumption.
Dead centres	Reversal points of the piston at the ends of piston stroke.
Combustion chamber	The chamber covered on all sides by working cylinder, cylinder cover and piston. It changes its size during the working cycle.
Compression chamber	Smallest combustion chamber above the top dead centre.
Compression ratio	Ratio of the maximum combustion cahmber above bottom dead centre to the compression chamber.
Number of cylinders	Number of working cylinders of an internal combustion engine.
Permissible total weight (total mass)	Mass of the vehicle which should not exceed the maximum value fixed by taking into consideration the material stress, permissible axle load and StVZO specifications.

Tyre size	Other designs	Ply rating PR	Rim size	Air hose	Tyre dimensions: Outer diameter mm	Tyre dimensions: Nominal width mm	Tyre dimensions: Operating width mm	Effective radius static mm	Rolled out circumference dynamic radius mm	Load carrying capacity N	Air pressure bar
Belted tyre in super low cross-section design											
135 SR 12			4,00 × 12	1220	522	137	141	236	1585	2900	2,2
145 SR 12			4,00 × 12	1220	542	147	151	245	1645	3550	2,3
155 SR 12			4,00 × 12	1230	550	152	157	249	1665	4000	2,4
135 SR 13			4 J × 13	1320	548	137	141	249	1665	3100	2,2
145 SR 13			4 J × 13	1320	566	147	151	257	1720	3750	2,3
155 SR 13			41/2 J × 13	1330	578	157	162	263	1750	4250	2,4
165 SR 13	HR		41/2 J × 13	1330	596	167	172	271	1800	4700	2,4
175 SR 13			5 J × 13	1340	608	178	183	276	1840	5200	2,5
145 SR 14			4 J × 14	1420	590	147	151	269	1795	3950	2,3
155 SR 14			41/2 J × 14	1430	604	157	162	276	1835	4450	2,4
165 SR 14	HR		41/2 J × 14	1430	622	167	172	284	1885	4950	2,4
175 SR 14	HR		5 J × 14	1440	634	178	183	289	1920	5500	2,5
185 SR 14	HR, VR		51/2 J × 14	1440	650	188	194	296	1965	5900	2,5
195 SR 14	HR, VR		51/2 J × 14	1460	666	198	204	303	2010	6400	2,5
135 SR 15			4 J × 15	1520	600	137	141	275	1830	3500	2,2
145 SR 15			4 J × 15	1520	616	147	151	282	1875	4150	2,3
155 SR 15	HR		41/2 J × 15	1530	630	157	162	289	1915	4750	2,4
165 SR 15	HR, VR		41/2 J × 15	1530	646	167	172	296	1960	5200	2,4
185 SR 15	HR		5 J × 15	1540	674	183	188	308	2040	6150	2,5
Belted tyre of '70' Series											
145/70 SR 12	HR, VR		4,00 × 12	1210	512	144	148	235	1560	2900	2,1
155/70 SR 13	HR, VR		4 J × 13	1320	550	151	155	253	1675	3550	2,2
165/70 SR 13	HR, VR		41/2 J × 13	1330	568	165	170	260	1725	4000	2,3
175/70 SR 13	HR, VR		5 J × 13	1330	580	176	181	265	1760	4450	2,3
185/70 SR 13	HR, VR		5 J × 13	1340	598	186	192	272	1810	4950	2,4
195/70 SR 13	HR, VR		51/2 J × 13	1340	608	197	203	276	1840	5500	2,4
175/70 SR 14	HR, VR		5 J × 14	1430	606	176	181	278	1845	4700	2,3
185/70 SR 14	HR, VR		5 J × 14	1440	624	186	192	285	1890	5250	2,4
195/70 SR 14	HR, VR		51/2 J × 14	1440	636	197	203	289	1925	5800	2,4
185/70 SR 15	HR, VR		5 J × 15	1540	648	186	192	297	1970	5500	2,4
Diagonal tyre in super low cross-section design (According to DIN 7803)											
5,65/135-12		4	4,00 × 12	1220	526	137	145	244	249	2800	2,1
6,15/155-13	S	4	41/2 J × 13	1330	282	157	166	273	278	3850	2,1
6,45/165-13	S, H	4	41/2 J × 13	1330	600	167	177	279	285	4250	2,1
6,95/175-13	S	4	5 J × 13	1340	610	178	189	282	289	4750	2,1
6,45/165-14	S, H	4	41/2 J × 14	1430	626	167	177	290	296	4550	2,1
6,95/175-14	S, H	4	5 J × 14	1440	638	178	189	295	302	5050	2,1
7,35/185-14	H	6	51/2 J × 14	1440	654	188	199	300	308	6150	2,5
7,75/195-14	H	6	51/2 J × 14	1460	670	198	210	305	314	6700	2,5
Diagonal tyre in low cross-section design (According to DIN 7803)											
5,00-12	S	4	3,50 × 12	1220	532	128	136	250	255	2750	2,1
5,50-12		4	4,00 × 12	1220	552	142	151	260	265	3150	2,1
6,00-12		4	4,00 × 12	1230	574	151	160	265	271	3700	2,1
5,50-13		4	4 J × 13	1320	578	142	151	272	277	3350	2,1
6,00-13	S	4	4 J × 13	1330	600	151	160	282	288	3950	2,1
7,00-13	S, H	4	5 JK × 13	1340	644	178	189	299	305	5100	2,1
7,25-13	S, H	4	5 JK × 13	1350	654	184	195	303	312	5450	2,1
7,50-13	S, H	6	51/2JK×13	1350	666	190	201	306	315	6250	2,3
6,00-14	S	4	41/2 J × 14	1430	626	156	165	294	300	4150	2,1
6,50-14		6	41/2 J × 14	1440	650	166	176	304	310	5300	2,5
7,00-14	S, H	6	5 JK × 14	1440	668	178	189	307	313	5850	2,5
7,50-14		6	51/2JK×14	1450	688	190	201	319	325	6550	2,5
6,00-15 L	S	4	41/2 J × 15	1530	650	156	165	304	309	4400	2,1
7,00-15 L	S, H	6	5 J × 15	1540	694	178	189	320	326	6100	2,5
Diagonal tyre in super balloon design (According to DIN 7803)											
5,60-13	S	4	4 J × 13	1330	600	145	154	278	284	3850	2,1
5,90-13	S	4	4 J × 13	1330	616	150	159	285	291	4250	2,1
6,40-13	S	4	41/2 J × 13	1340	642	163	173	298	304	4650	2,1
5,60-15	S	4	4 J × 13	1530	650	145	154	304	309	4250	2,1

Internal combustion engine	Otto four-stroke engine	Diesel four-stroke engine
Dimensions	Smaller	bigger
Fuel		
Type	**Normal petrol** ≈ 92 OZ (Octane number) Super fuel ≈ 98 OZ (Octane number) Liquid gas ≈ 100 OZ (Octane number)	Diesel fuel, in winter 45 CaZ (Cetane number) in summer 55 CaZ (Cetane number) Kerosine, Petrol } for multi-substance engines
Flash point	— 30 . . . — 50°C	Diesel fuel 55 . . . 100°C
Danger class	A 1	A 3
Self ignition temperature	Normal petrol 480 . . . 550°C Super fuel 480 . . . 700°C	≈ 350°C
Fuel consumption	300 . . . 380 g/kWh	230 . . . 370 g/kWh
Suction		
Gases	Fuel-air mixture	Pure air
Filling of cylinder	0.7 . . . 0.9	0.8 . . . 0.9
Mixture formation	Outside the combustion chamber (external mixture formation) a) atomization in the carburettor b) injection into the suction pipe inside the combustion chamber (internal mixture formation) c) injection into the cylinder	inside the combustion chamber a) directly into the cylinder b) in the ball-shaped combustion chamber of cylinder (MAN-M Process) c) in the pre-combustion chamber d) in the whirling chamber e) in the cylinder before air feeder
Air requirement	Theoretical 15 kg/kg Partial load (10 30% excess air) 17 kg/kg Full load (10 0% air shortage) 13 kg/kg Idle running (30 40% airshortage) 10 kg/kg Cold starting 3 kg/kg Hot running 8 kg/kg	Theoretical 16 kg/kg Over the entire load range, excess air because of the smoke limits. Full load up to 15% Partial load 40% Idle running up to 1000%
Air ratio	0.7 . . . 1.3	1.15 . . . 10
Regulation	Modification of the fuel air mixture by turning the throttle flap in the carburettor	Modification of the injected fuel quantity by turning the piston in the injection pump.
Compression		
Gases	Fuel-air mixture	Pure air
Compression temperature	400 . . . 600°C	700 . . . 900°C
Compression ratio	7 . . . 10 : 1 upper limit: anti-knock rating of fuel	14 . . . 22 : 1, lower limit: self ignition temperature of the fuel
Compression pressure	12—18 bar	30—50 bar
Power		
Ignition	Outside ignition through electric ignition spark on the spark plug	Self-ignition through compressed hot air; under certain conditions through spark plug, hot flange or starting pilot.
Ignition or injection point of time	basic setting 10° before to 5° after top dead centre, change through inertia and negative pressure adjuster.	Injection commencement 22—32° before top dead centre, change through injection adjuster, changeable injection end.
Ignition lag	Practically 0 s	1/1000 s
Pressure head, injection pressure	0.3—0.7 bar negative gauge pressure in the air funnel of the carburettor.	90—200 bar positive gauge pressure on the injection jet.
Combustion	Constant volume combustion. Flame speed 10—25 m/s	Constant pressure combustion
Knocking, dieseling	At the end of combustion, the still not burnt fuel—air mixture gets ignited in an explosive manner, flame speed 200—250 m/s	At the commencement of combustion by very great ignition delay (0.002—0.25) and finally explosive ignition of fuel-air mixture.
Maximum combustion pressure	40—60 bar	65—90 bar
Combustion temperature	2000—2500 °C	2000—2500 °C
Exhaust		
Exhaust gas temperature	Full load 700—1000°C Idle running 300—500°C	Full load 500—600°C Idle running 200—300°C
Carbon monoxide content of exhaust gases	Idle running 4—6% Full load 1—4%	Idle running up to 0.05 % Full load up to 0.3 %
Characteristic values		
Mean working pressure	6—10 bar	5—8 bar
Maximum speed	3600—600 rpm	2000—4000 rpm
Torque	30—600 Nm (truck 100—150 Nm)	100—120 Nm (truck 100—700 Nm)
Torque curve	Highly dependent on speed	Uniform, independent of speed
Power per piston. displacement	22—50 kW/l	18—26 kW/l
Weight to power ratio	2 7—6 8 kg/kW	4.8—9.5 kg/kW
Efficiency	Useful work 24 % Loss through cooling 33 % Loss through radiation 7 % Loss through exhaust gases 36 %	Useful work 32 % Loss through cooling 32 % Loss through radiation 7 % Loss through exhaust gases 29 %
Operational cost	**According to local conditions**	**According to local conditions**

Four Stroke and Two Stroke Otto Engines

Mean values

Internal combustion	4-stroke engine	2-stroke engine
General		
Moving parts	Expensive	Simple
Crank housing	Divided, open	Closed
Working cycle	4 strokes over 2 crank shaft rotations	2 strokes over 1 crank shaft rotation; at low or high speeds as well as in the skidding car, "four-stroking" car occur
Control	Valve control with inlet and outlet valve	Split control with inlet, outlet and overflow ducts; no special control organs. Actuation through piston or rotary valve.
Control time	Eö 2—20° before TDC Es 40—60° after BDC Aö 35—60° before BDC As 5—30° after TDC	Symmetric: Eö 55—60° before OT; Es 55—60° after OT; Aö 60—65° before UT; As 60—65° after UT; Üö 55° before UT; Üs 55° after UT Asymmetric: 87° before OT; 60° after OT; 73° before UT; 55° after UT; 48° before UT; 66° after UT
Lubrication	Oil pump or dry-sump lubrication	Mixture lubrication or fresh oil automatic
Lubrication oil consumption	0,10—0,15 1/100 km	0,25—0,40 1/100 km
Fuel consumption	300—380 g/kWh, lowest at full load and medium speeds	410—540 g/kWh, lowest at partial load and medium speeds; maximum at full load and low speeds.
Suction		
Suction	Into the cylinder chamber	Into the crank chamber or charging pump
Cylinder filling	0.7—0.9	0.5—0.7
Compression		
Compression ratio	7—10 : 1	6—8 : 1
Compression pressure	12—18 bar	8—12 bar
Compression temperature	400—600°C	300—450°C
Pre-compression	Without supercharger nil	By crank chamber 0.3—0.8 bar
Combustion		
Maximum combustion pressure	40—60 bar	25—40 bar
Combustion temperature	2000—2500°C	2000—2500°C
Combustion residues	Nil	Incompletely burnt oil vapours can colour the exhaust gases and form residues in the exhaust system
Exhaust		
Dynamic pressure of exhaust gases	80—160 m bar with full load 0—5 m bar in idle running	20—120 m bar
Flushing	Through piston in the exhaust stroke	Through overflowing fresh gas
Flushing losses	Nil	Gross flushing 30—45 % Reverse flushing 25—30 % Direct current flushing 20—25 %
Characteristic data		
Mean working pressure	6—10 bar	4—6 bar
Torque	50—400 Nm	40—110 Nm
Torque curve	Dependent on speed, highest at medium speeds	Uniform, favourable at lower speeds, intermittent idle running
Mean piston speed	9—15 m/s	9—11 m/s
Speed	4200—600 rpm	4000—5400 rpm
Maximum power	18—200 kW	11—40 kW
Piston displacement power	22—50 kW/1	30—40 kW/1
Weight to power ratio	6.1—3.4 kg/kW	5.4—2.7 kg/kW
Working price	According to local conditions	According to local conditions
Efficiency	21—29%	17—19%
Miscellaneous		
Heat load	Lower	Higher
Balancing of masses	Better	Worse
Non-unformity	Worse	Better
Braking action of engine	Better	Worse
Stroke volume of engine	0.25—7.01	0.3—1.01

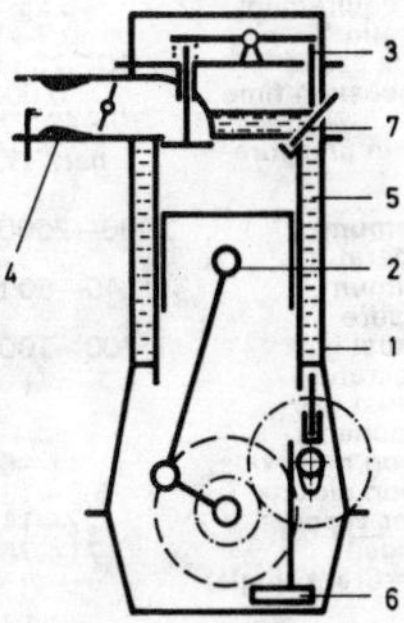

1 Engine housing
2 Crank drive
3 Control
4 Carburettor
5 Cooling
6 Lubrication
7 Ignition

1. According to duty cycle
   a) two-stroke engine
   b) four-stroke engine

2. According to ignition
   a) Otto engine-applied ignition
   b) Diesel engine-compression ignition
   c) Semi-diesel engine-combustion cup

3. According to mixture formation
   a) carburettor engine—outer
   b) injection motor—inner or outer

4. According to fuel
   a) carburettor engine (light oil) petrol, benzene, mixture
   b) Diesel engine (heavy oil)
   c) multi-substance engine
   d) gas engine
   e) coal dust engine
   f) fuel cells

5. According to layout of cylinders
   a) in-line engine
   b) V-engine
   c) opposed cylinder type engine
   d) double piston engine
   e) opposed piston engine
   f) radial engine

6. According to number of cylinders
   a) single-cylinder engine
   b) double or multi cylinder engine

7. According to control
   a) I-head engine
   b) engine controlled from below

8. According to drive
   a) reciprocating piston engine
   b) rotary piston engine (Wankel)
   c) turbine engine (gas turbine)
   d) jet engine (reaction)

9. According to cooling
   a) water-cooled engine
   b) air-cooled engine

10. According to engine filling
    a) suction engine
    b) super charged engine

11. According to direction of rotation
    a) right rotating engine
    b) left rotating engine

12. According to the assembly of engine
    a) front engine
    b) rear engine
    c) under-floor engine

In-line engine (four-stroke) 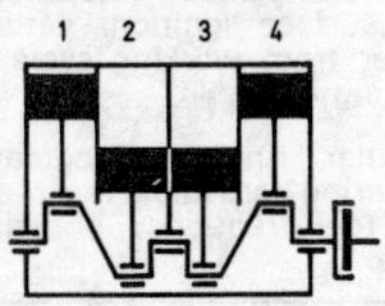 	Cylinders in a line arranged one after the other. Multi cylinder engines lead to unfavourable construction lengths. Designation of the cylinder in the direction of flux of force. Direction of rotation of the crank shaft on the side lying opposite to force output seen in the direction of flux of force (DIN 73021)	4 Cyl. 1 3 4 2 1 2 4 3 6 Cyl. 1 5 3 6 2 4 1 2 4 6 5 3 1 4 2 6 3 5 1 4 5 6 3 2 8 Cyl. 1 6 2 5 8 3 7 4 1 3 6 8 4 2 7 5 1 4 7 3 8 5 2 6 1 3 2 5 8 6 7 4
V-engine  	2 cylinder lines lie parallel one by the side of the other. They form an angle whose vertex passes through the crank shaft axis. Shorter construction lengths. Counting of cylinders begin with the left line of cylinders.	4 Cyl. 1 2 4 3 1 3 4 2 6 Cyl. 1 4 2 5 3 6 8 Cyl. 1 6 3 5 4 7 2 8 1 5 4 8 6 3 7 2 1 8 3 6 4 5 2 7
Opposed cylinder type engine 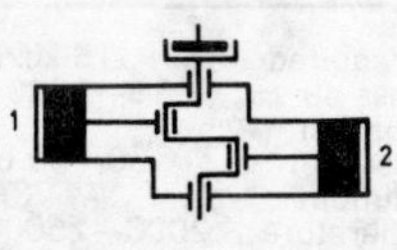 	2 cylinder lines are located on opposite sides of the crank shaft axis and together form a stretched angle.	4 Cyl. 1 4 3 2
Radial engine 	Several cylinders are arranged in a stellar form either in one level or several levels.	4 stroke, odd number of cylinders. 5 cylinders 1 3 5 2 4 7 cylinders 1 3 5 7 2 4 6 9 cylinders 1 3 5 7 9 2 4 6 8 2 stroke, even number of cylinders 6 cylinders 1 2 3 4 5 6

Otto Engine

Normal combustion

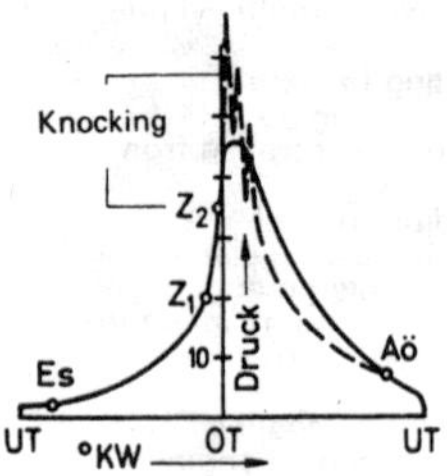

An electric ignition spark ignites the compressed fuel + air mixture. The flame spreads uniformally in the combustion chamber. The mixture burns. Temperature and pressure increase.

The course of combustion is influenced by the fuel, filling, speed, temperature, mixture rate, compression turbulance and the combustion chamber design.

During combustion carbon monoxide (poisonous), carbon dioxide and water vapour are formed. The quantity of not fully burnt carbon monoxide is a measure for the quality of combustion.

Air requirement	15 kg/kg
Air ratio	0.7—1.3
Flame speed	10—25 m/s
Combustion time	0.001 s
or	30—40°/KW
Rise in pressure	1 bar/° KW
Maximum temperature	2000—2500°C
Maximum pressure	40—60 bar
Exhaust gas temperature	700—100°C
Exhaust gas components	
Carbon monoxide	1—6 %
Carbon dioxide	6,5—13 %
Water vapour	7—11 %
Nitrogen	71—76 %
(KW=Crank angle)	

Knocking

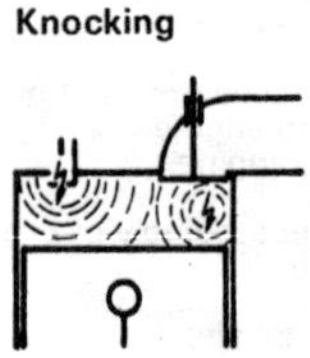

The ignition spark induces the combustion. While the flame spreads, the temperature and the pressure of the still not burning end gases increase. After reaching the ignition temperature, they get ignited and burn with high speed, increasing thereby the temperature in an explosive manner. This irregular combustion causes a knocking noise.

Flame speed 230—300 m/s
Rise in pressure 8 bar/°KW
Pressure vibrations 6000—7000 Hz
Knocking increases bearing load, fuel consumption and thermal engine load. It reduces output and shortens the engine life.
(KW=Crank angle)

Surface ignition

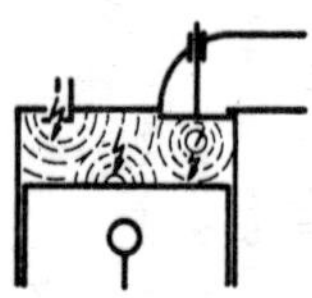

Uncontrolled ignition by glowing spots on over-heated parts of the combustion chamber or due to the foreign matter floating freely in the combustion space. This phenomenon can become faster with each working cycle, can occur irregularly or as multiple ignition; can appear before or after the electric ignition; can be audible or inaudible. It leads to rough engine running and lower output.

Early ignition: before the striking of the ignition spark. Late ignition: after the regular point of time of ignition Accumulative surface ignition: getting earlier from working cycle to working cycle.

Irregular ringing: irregularly appearing hard impacts. Creaking: low frequency booming noise.

After running (Dieseling)

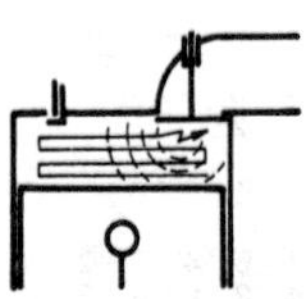

The engine runs further after switching off ignition. It is a self-ignition of compression which aided by the high engine temperature, hot surfaces or residues.

The period of after-running depends on the engine temperature and carburettor setting. Electro-magnetic switch-off valves cut out after-running.

Engine speed 173—375 rpm
Period of after-running variable. Can be reduced by:

higher octane number, aromatic fuels, better cooling, idle running jet with electro magnetic switch-off valve.

Diesel Engine

Normal combustion

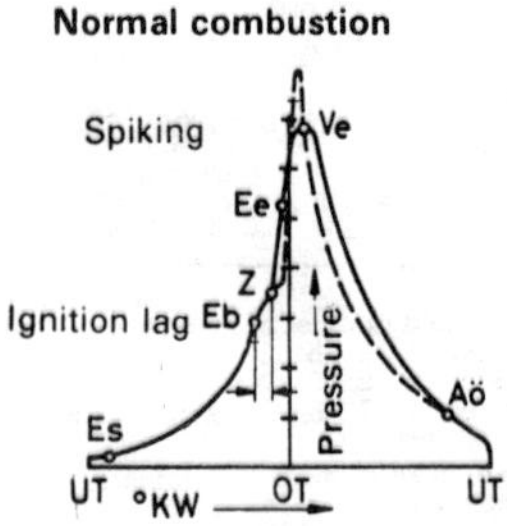

The fuel is injected into pure, comressed hot air. It gets vaporized and mixed with air into an ignitable fuel-air mixture. The time needed for the preparation of this mixture is known as the ignition lag.

After attaining the ignition temperature the mixture gets automatically ignited. The flame, spreads uniformly in the combustion chamber and the mixture burns. Additional fuel is fed for combustion. Thus a softer combustion as well as uniform combustion pressure are obtained.

Air required	16 kg/kg
Excess air	15—1000 %
Maximum pressure	50—80 bar
Maximum temperature	2000—2500°C
Ignition lag	0.001 s
from 0.002 onwards, harder running of the engine	
Exhaust gas temperature	500—600°C
Components of exhaust gas	
Carbon monoxide	0.05—0.3 %
Carbon dioxide	3.5—7 %
Water vapour	3.5—5 %
Nitrogen	77 %
Oxygen O_2	10—16 %

Note: For abbreviations, see page 1.

1st Stroke-Suction 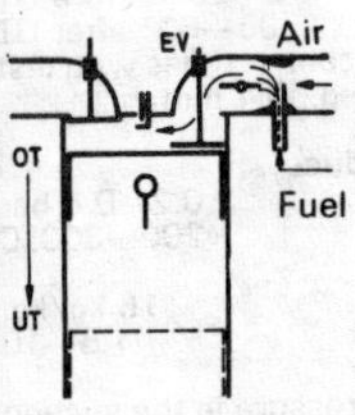 OT=Top dead center UT=Bottom dead center EV=Inlet valve	Piston at the top dead centre, inlet valve opened, outlet valve closed, compression chamber filled with hot exhaust gas residues under low position gauge pressure. The piston moves to the bottom dead centre. The residual gases get expanded. A negative gauge pressure builds up in the cylinder chamber. The fuel air mixture is sucked in from the carburettor. Filling dependent upon air pressure, temperature, speed, throttle control position, compression ratio, engine condition, suction pipe and valve opening time.	Inlet valve opened 2—20° before OT closed 40—60° after UT Measurement of control times with test valve clearance, eg. 0.4 mm. Exhaust gas residue Pressure 0.2—0.6 bar Temperature 400—600°C Air required 15 kg/kg. Air ratio 0.7—1.3 Negative gauge pressure in the suction pipe 0.1—0.3 bar in the air funnel 0.4—0.7 bar Suction temperature 50—100°C Filling efficiency 0.7—0.9
2nd Stroke-Compression 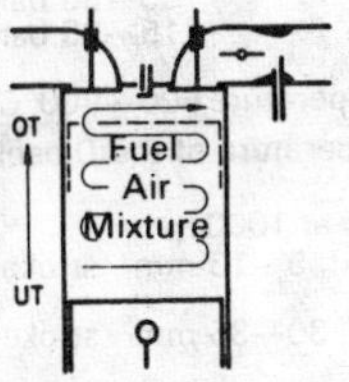	Piston at the bottom dead centre. Inlet and outlet valves closed, cylinder chamber filled with fuel=air mixture. Piston is moved to the top dead centre. Thereby the fuel=air mixture compressed to the compression chamber. The fuel and air particles come closer together. Pressure and temperature increase in the cylinder.	Compression ratio 7—10 : 1 **Burning pressure** 12—18 bar **Working pressure** 6—12 bar Compression temperature 400—600°C Increase in compression from 5 to 8 : 1 increases power by ≈ 20 % reduces consumption by ≈ 18 % Increase in compression from 6 to 7 : 1 increases power by 6 % reduces consumption by 5 %
3rd Stroke-Power 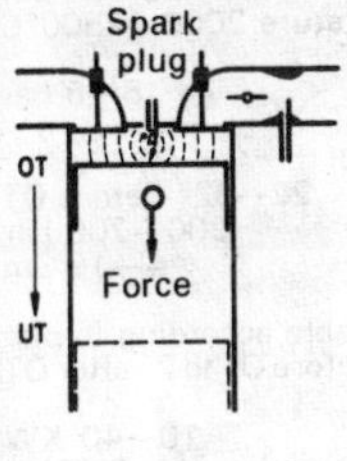	Piston at the top and dead centre, both valves closed, combustion chamber filled with hot compressed fuel=air mixture. An electric ignition spark strikes on the electrodes of the spark plug; it ignites the mixture. The flame spreads out, and the mixture burns fast. The gases become hot. Pressure increases in the cylinder, and the piston is pushed to the bottom dead centre.	Maximum pressure 40—60 bar Maximum temperature 2000—2500°C Mean working pressure 6—10 bar Basic setting of ignition time point 10° before . . . 5° after OT Adjustment by negative gauge pressure upto 20°KW inertia up to 45°KW Flame speed 10 to 25 m/s Combustion time 0 001 s or 30—40° kW Pressure increase 1 at/°KW Useful work 24 %
4th Stroke-Exhaust Exhaust gases 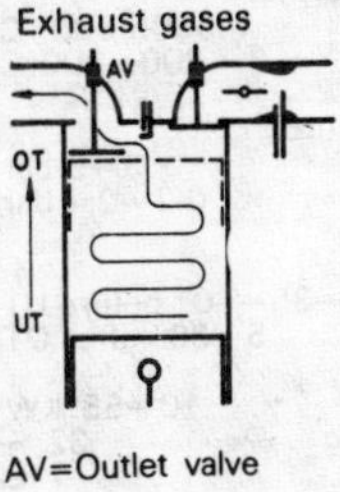 AV=Outlet valve	Piston at the bottom dead centre, outlet valve opened, inlet valve closed, cylinder filled with hot exhaust gas under positive gauge pressure. Piston is moved to the top dead centre. Thereby the exhaust gases are pushed out towards exhaust valve port. During the beginning of exhaust, the gases get puffed out due to the positive gauge pressure prevailing in the piston chamber at high speed. The fluctuating conditions of flow and stress release cause vibrations.	Exhaust gas temperatures full load 700—1000°C idle running 300—500°C Exhaust gas pressure beginning 4.7 bar end 0.2—0.6 bar Outlet valve opens 35—60°C before UT closes 5—30°C after OT Valve overtap 10—50°KW Loss through cooling 33 % radiation 7 % exhaust gases 36 % For composition of exhaust gases, see "Combustion in the engine"

Four Stroke Diesel Engine

1st Stroke-Suction

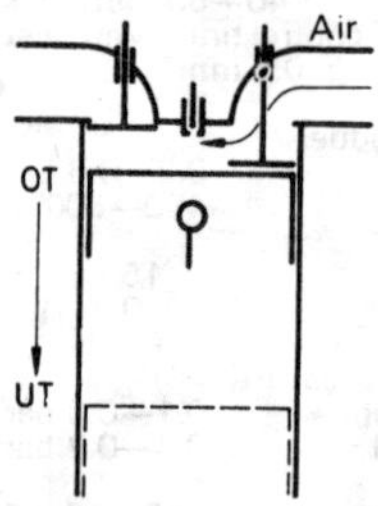

OT=Top dead center
UT=Bottom dead center

Piston at the top dead centre, inlet valve open, outlet valve closed, compression chamber filled with hot exhaust gas residues under a low positive gauge pressure.

Piston moves down to the bottom dead centre. Thereby initially the residual gases get expanded, then a negative gauge pressure occurs in the cylinder chamber, resulting in the suction of pure air.

Filling dependent upon speed, temperature, air pressure, suction line, load, composition and condition of the engine.

Inlet valve
opens 5—25° before OT
closes 35—60° after UT
Measurement of control times with test valve clearance, eg.. 0.4 mm

Exhaust gas residues
Pressure 0.2—0.6 bar
Temperature 200—300°C

Air required 16 kg/kg.
Air ratio 1.5—10

Negative gauge pressure in the suction pipe 0.1—0.2 bar
Filling efficiency 0.8—0.9
Suction temperature 70—100°C

2nd Stroke-Compression

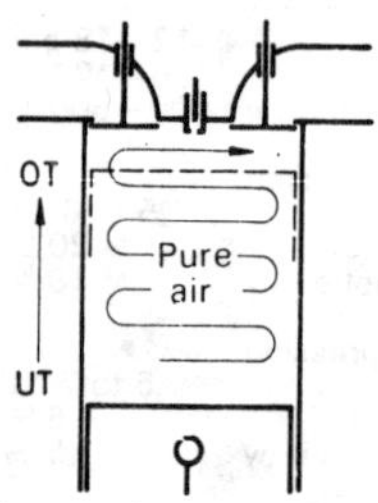

Piston at the bottom dead centre, inlet and outlet valves closed, cylinder chamber filled with pure air.

Piston moves to the top dead centre. Thereby it compresses the pure air to the compression chamber. Pressure and temperature increase in the cylinder.

High compression improves thermal efficiency. Top limit determined by permissible maximum pressure of the bearing and the lower limit by the self-ignition temperature of the fuel.

Compression ratio 14—22 : 1

Burning pressure 30—55 bar
Working pressure 15—25 bar

Compression temperature 600—900°C

Self ignition temperature of the Diesel fuel 400°C

Injection volumes at 1000 rpm
idle running 9—13 mm³/stroke
partial load 30—35 mm³/stroke
full load 38—44 mm³/stroke

3rd Stroke-Power

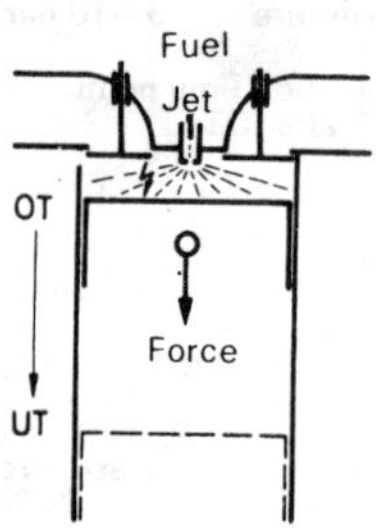

Piston at the top dead centre, both valves closed, compression chamber filled with hot, compressed air.

Diesel fuel, finely atomised, is injected into the compression chamber. It vaporizes and gets ignited by contact with hot air.

The flame spreads out, and the fuel burns rapidly. Thereby the pressure and temperature of the gases rise in the combustion chamber. The gases expand and push the piston to the bottom dead centre.

Maximum pressure 50—80 bar
Maximum temperature 2000—2500°C
Mean working pressure 5—8 bar

Commencement of injection 22—32° before OT
Injection pressure 900—200 bar
Droplet size 4—15 μm

Injection end variable according to piston turning: 20° before OT to 2° after OT

Injection duration 10—40°KW
Ignition lag 0.001 s
Combustion duration 75—85°KW
Useful work 32 %

4th Stroke-Exhaust

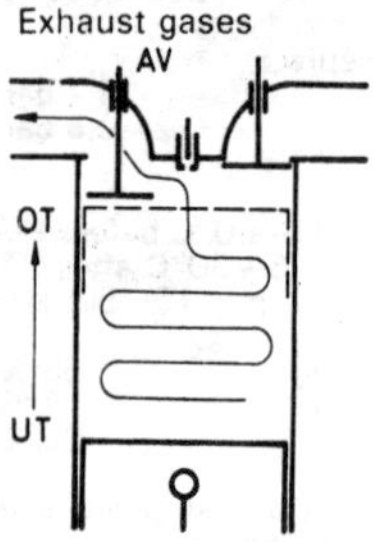

AV=Outlet valve

Piston at the bottom dead centre, inlet valve closed, outlet valve open, cylinder chamber filled with hot exhaust gases under positive gauge pressure.

Piston moves to the top dead centre, pushing the exhaust gases out towards exhaust port.

At the commencement of exhaust the gases are puffed out at high speed due to the excess positive gauge pressure prevailing in the cylinder chamber. The fluctuating conditions of flow cause vibrations.

Exhaust gas temperature
full load 500—600° C
idle running 200—300°C

Exhaust gas pressure
at the beginning 3—5 bar
at the end 0.2—0.4 bar

Outlet valve
opens 35—60° before UT
closes 5—30° after OT

Valve over tap 10—55°KW
Loss due to cooling 32 %
radiation 7 %
exhaust gases 29 %

Composition of exhaust gases, see "Combustion in the engine"

Direct Injection 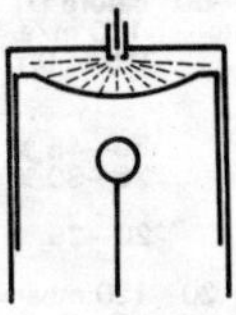	The finely atomized fuel coming through single or multi-hole jets is directly injected into the cylinder chamber. No spark plugs are required to start the engine because the cooling surface is small.	Injection pressure ≈ 200 bar, low heat loss, good starting property, low fuel consumption ≈ 245 g/kWh, greater ignition lag, uniform mixture formation more difficult, harder running of the engine, greater maximum pressures.
Precombustion Chamber Process 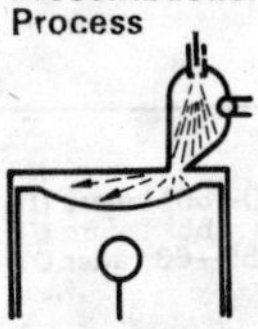	Sub-divided combustion chamber. Fuel is injected through a pintle nozzle into a precombustion chamber. The precombustion chamber is connected to the cylinder chamber through several narrow holes. Spark plugs are necessary for starting the engine because the cooling surface is large.	Injection pressure 90—120 bar, smaller peak pressures, soft and silent running of engine, more uniform mixture formation, higher fuel consumption, smaller ignition lag.
Swirl Chamber Method 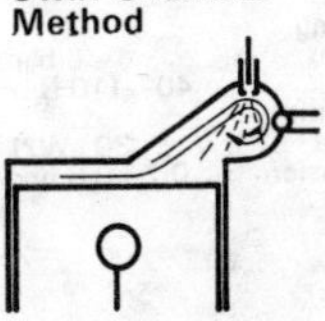	Sub-divided combustion chamber. The fuel is injected into a mostly spherical-shaped swirl chamber through a pintle nozzle in a broad spray. The connecting aperture to the cylinder chamber is wide, and runs mostly tangential. Good turbulence of the fuel with air. Spark plugs are needed for starting.	Injection pressure 100—125 bar, lower peak pressures and combustion shocks, soft and silent running of engine, good mixture formation, small ignition lag, higher fuel consumption.
Air-cell Method 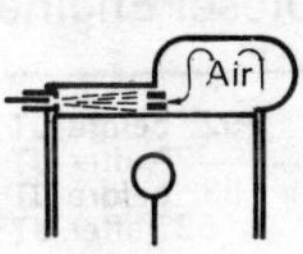	Sub-divided combustion chamber. The fuel is injected into the cylinder against the cell mouth (throttle position). A small portion of that gets into the air cell. The combustion is induced during the power stroke by the outflowing air. Spark plugs are not essential.	Injection pressure ≈ 120 bar, smaller pressure peaks and combustion shocks, soft and silent engine running, smaller ignition lag, more complete combustion, higher fuel consumption.
Centre Sphere Method (Piston chamber) 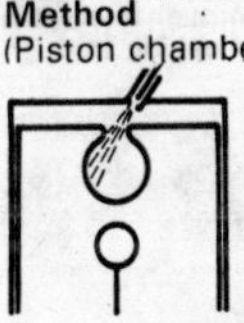	Sub-divided combustion chamber. Spherical shaped combustion chamber in the centre of the piston. A twisted channel in the suction line creates a powerful turbulence. The fuel is injected almost without atomizing through a single or double-hole jet on the surface of the combustion chamber. The fuel film is vaporized by the hot turbulent air. The fuel-air mixture turns layer-wise.	Injection pressure ≈ 175 bar smaller pressure peaks, smaller ignition lag, softer burning, elastic and silent engine running, insensitive to fuel (multi fuel engine), lower fuel consumption 200—230 g/kWh
Hot-bulb engine 	The fuel is injected into the combustion chamber during the compression stroke. There it is ignited by a hot-bulb because the compression temperature is too low for self-ignition. A heating lamp is used for heating the bulb. In bull-dog engines, an electric starting device with spark plug and petrol-heavy oil mixture is used.	Injection pressure ≈ 80 bar Temperature of the glowing shell in the bull dog engines 700—800°C Compression 6—10 : 1 Compression pressure 15 bar Speed 500—600 rpm Compression temperature 400°C Two-stroke method

Otto Engine

I Stroke

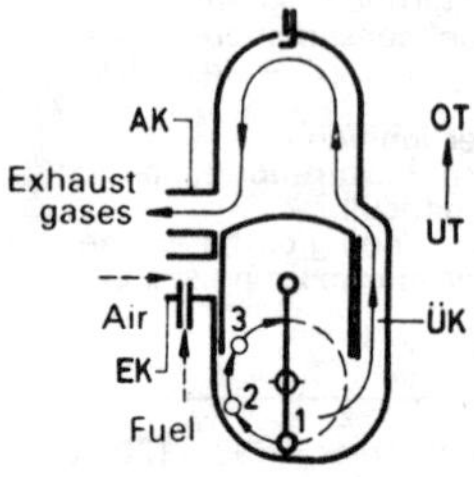

Position 1: Piston at UT, AK and UK open, EK closed. The fuel-air mixture flows out of the crank chamber into the cylinder chamber, flushes the cylinder and pushes out the exhaust gases towards AK.

Position 2: AK, UK and EK closed. The piston compresses the mixture in the cylinder chamber. In the crank case a negative gauge pressure is generated.
Position 3: AK and UK closed. EK open. The mixture is compressed in the cylinder chamber. The crank case sucks in fresh fuel-air mixture.

AK closes	60—65° after UT
UK closes	55° after UT
EK opens	55—60° before OT
Overflow speed upto	100 m/s.
Filling efficiency	0.5 to 0.7
Flushing loss at full load	
Cross flushing	30—45 %
Reverse flushing	25—30 %
Direct current flushing	20—25 %
Dynamic pressure	20—120 mbar
Compression	6—8 : 1
Compression pressure	8—12 bar
Compression temperature	300—450°C

II Stroke

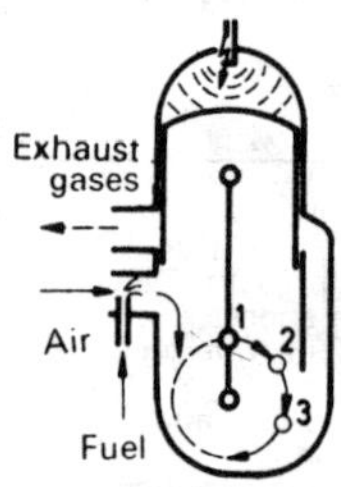

Position 1: Piston at OT, AK and UK closed. EK open. In the cylinder chamber, an electric ignition spark initiates combustion. The fuel air mixture burns.

The gases become hot, get expanded and press the piston towards UT. The mixture flows into the crank case.

Position 2: AK, UK closed. In the cylinder chamber, the gas pressure pushes the piston further towards UT. In the crank case, the fresh gases are pre-compressed.

Position 3: AK and UK open. EK closed. The fresh gases flow from the crank case into the cylinder chamber, flush and eject the exhaust gases.

AK opens	60—65° before UT
ÜK opens	55° before UT
EK closes	55—60° after OT
Ignition point of time	5° before OT
Maximum pressure	25—40 bar
Maximum temperature	2000—2500°C
Mean working pressure	4—6 bar
Torque	40—110 Nm
Power per unit displacement	30 kW/l
Pre-compression	0.2—0.8 bar

Diesel Engine

I Stroke

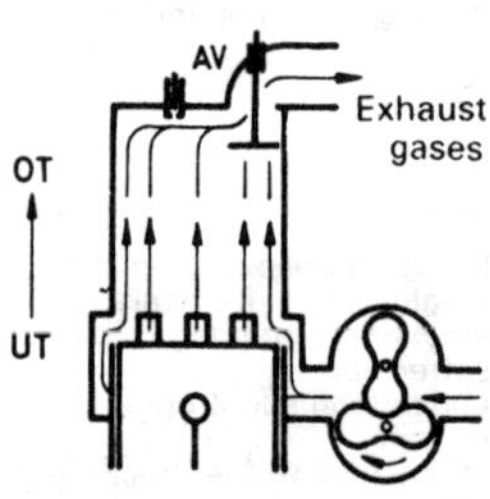

Position 1: Piston at UT. Scavenging port and outlet valve open. A Roots blower sucks in pure air and presses it through the scavenging port into the cylinder. The tangential layout of the scavenging port brings the air into a turbulent motion. The cylinder is completely flushed out in the direct current and filled with fresh air. The exhaust gases flow out towards the outlet valve.

Position 2: Scavenging port and outlet valve closed. The piston compresses the fresh air to the compression chamber. The air temperature increases intensely.

AV opens	92° before UT
closes	72° after UT
ES opens	53° before UT
closes	53° after UT
Compression	15...18:1
Compression pressure at starting speed	25...30 bar

Speed 1900...2500 rpm Direct current flushing through inlet valve and outlet valve flushing through Roots blower.

II Stroke

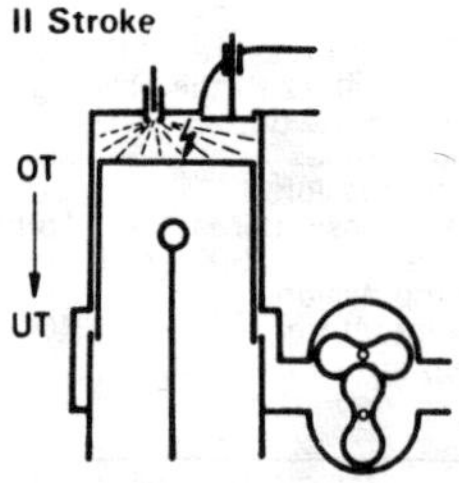

Position 1: Piston at OT. Scavenging port and outlet valve closed. The fuel is directly injected into the cylinder. The fuel gets vaporised into an ignitable fuel-air mixture by the hot air. After attaining the ignition temperature the mixture gets automatically ignited and burns. The heat increases the pressure in the combustion chamber. The gases get expanded and push the piston to the bottom dead centre.

Position 2: Scavenging port and outlet valve open. The gases get released. The cylinder is filled for the next working cycle.

Direct spray injection

Injection pressure	175—230 bar
Beginning of discharge	34 before OT
End of discharge	13° before OT
Torque	500—800 Nm
Consumption	235—255 g/kWh
Power per litre	20—27 kW/l

Cam shaft located above actuates the outlet valve and individual injection pump (pump jet).

Note: For abbreviations, see page 1.

Cross current scavenging (cross-draught) 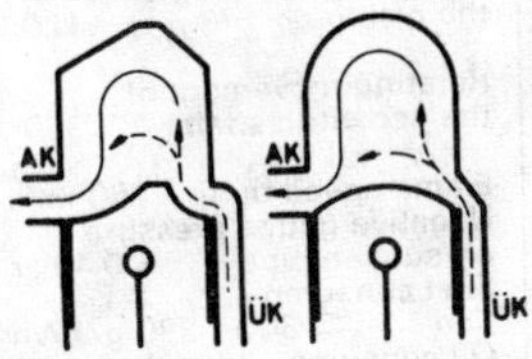 AK=Outlet port ÜK=Transfer port	Scavenging and outlet slots are located opposite. Pistons with deflector (nose) or cylinders with cross-mouthed and flushing channels direct the flushing current against the cylinder head. The nose forms an unfourable combustion chamber, therefore, lower compression necessary. Unequal lengths of current paths cause flushing loss and incomplete flushing of the cylinder head.	Scavenging loss 30—45 % Filling is adequate Higher fuel consumption Symmetric control diagram AK opens 60—65° before UT closes 60—65° after UT UK opens 55° before UT closes 55° after UT EK opens 55—60° before OT closes 55—60° after OT
Reverse scavenging (cross-draught) 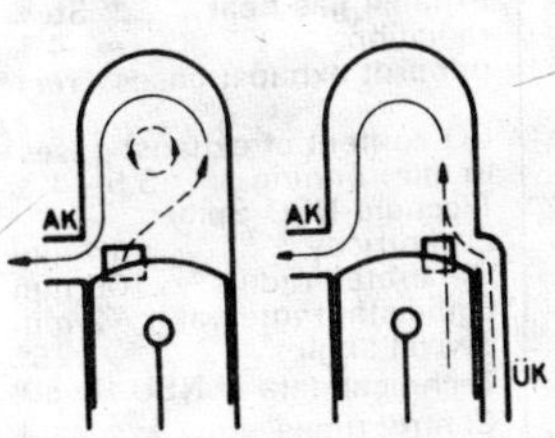 	Scavenging channels are by the side (laces) or below the outlet slot (MAN). They lead the fresh gases to the opposite cylinder wall. The scavenging current gets oriented there and returns into the cylinder head. The fresh gases and exhaust gases flow in opposite directions one by the side of the other. Thereby turbulences can be generated which reduce filling. In the case of updraught scavenging with 3 flushing slits (Zuendapp), the fresh gas current is led upwards through 2 supporting streams appearing by its side in the cross direction.	Scavenging loss 25—30 % Satisfactory filling Lower fuel consumption Symmetric control diagram Control angle for DKW EK 125—146° KW UK 96—105° KW AK 135—148° KW With offset cylinder layout asymmetry of control angle (DKW 250)
Direct current scavenging (longitudinal scavenging) 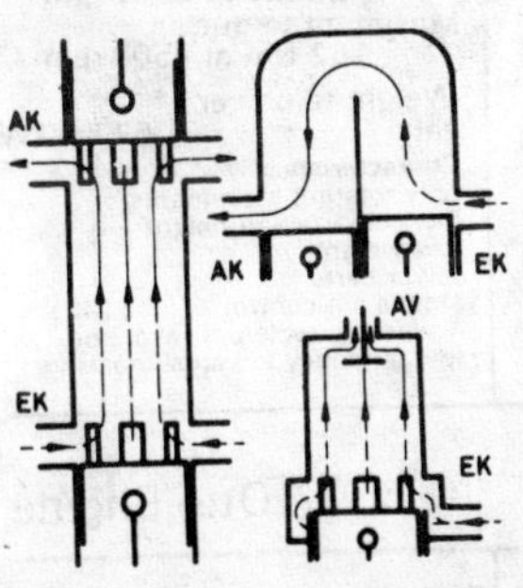 EK=Inlet port	Opposed piston engine: Scavenging slot and outlet slot are locatd at the opposite ends of the cylinder. The lower piston controls the inlet slot and the upper piston the outlet slot. The outlet piston runs forward. The scavenging current flows in the direction of the cylinder axis. Double-piston engine: Both pistons lie side by side, parallel or perpendicular to the crank shaft, separated by a cylinder wall. One piston controls the inlet slot and the other the outlet slot. If the cylinders are designed perpendicular to the crankshaft, then the outlet piston runs forward. Scavenging with outlet valve: The flushing slots are controlled by the top edge of piston and the outlet by the mechanically controlled taper valve.	Scavenging loss 20—25 % Filling good Low fuel consumption In the opposed piston engine lead of the outlet piston by 10—45° KW. Asymmetric control diagram EK opens 87° before OT closes 60° after OT AK opens 73° before UT closes 55° after UT UK opens 48° before UT closes 66° after UT
Scavenging with piston charging pump 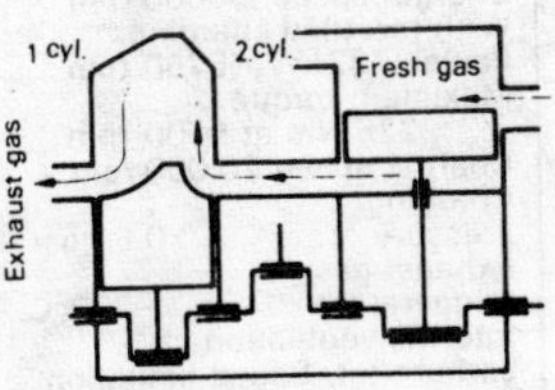 	The piston charging pump takes over suction and pre-compression of the 2 working cylinders. The crank case remains free from pressure and can be employed for economic central lubrication. During the action of the downward going piston the precompression is done and the 1st cylinder is filled, in the top pump chamber a negative gauge pressure is generated whereby fresh gas is sucked from the carburettor.	In the place of piston charging pump, rotary pumps (Roots blowers) are employed now-a days for charging and scavenging of the cylinder.

Rotary Piston Engine

Wankel Engine

Chamber 'A' is connected with the outlet and inlet channels. After the last residual gas is pushed out, the piston edge 1 closes the outlet channel and the suction stroke starts.

Chamber 'B' is filled with fresh fuel air mixture and compressed.

In chamber 'C' the hot gases expand and drive the eccentric shaft through the piston.

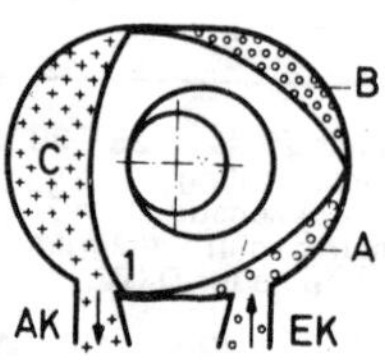

Chamber 'A' stays in contact with the inlet channel and sucks further. Higher compression takes place in chamber 'B'. Thereby the piston turns by 30° in its trochoidal path and the eccentric shaft makes a ¼ ratation. The piston edge 1 releases the outlet control aperture and the burnt gases of chamber 'C' flow out through the outlet channel.

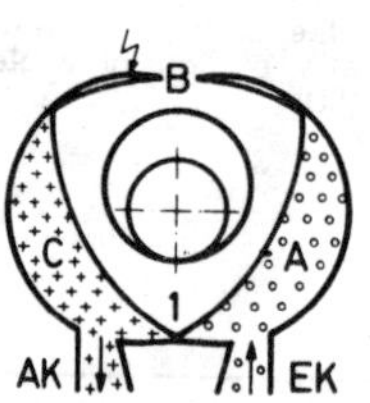

Chamber 'A' continues the suction stroke. The movement of the piston is controlled by its internal gearing which rolls over a stationary pinion.

In chamber 'B' the combustion of the compressed fuel-air mixture is initiated by an ignition spark. Chamber 'C' is connected with the outlet channel, and the piston pushes out the exhaust gases.

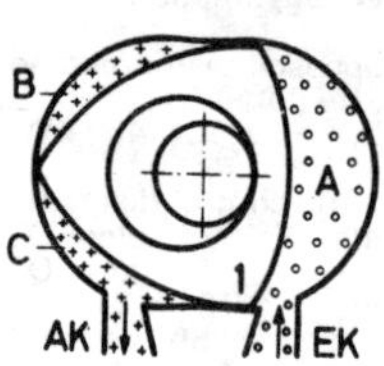

Chamber 'A' is filled with fresh gas. The piston edge 1 closes the inlet opening so that compression begins.

In the chamber 'B' the hot gases expand and drive the eccentric shaft through the piston.

The piston pushes out the exhaust gases of chamber 'C' into the open through the opened outlet channel.

Working cycle 4 Strokes

Angular movement of the piston 120°

Rotating movement of the eccentric shaft 360°

Flame speed up to 60 m/s
Negative gauge pressure on suction pipe ≈ 0.4 bar
Fuel consumption 300—380 g/kWh
Oil consumption 0.5 l/1000 km

Efficiency
Useful work ≈ 28 %
Loss through cooling
water ≈ 13 %
cooling oil ≈ 7 %
exhaust gas heat ≈ 36 %
radiation ≈ 4 %
unburnt exhaust gases rest.

CO-content of exhaust gases in idle running 3.5—4 %
Trochoid NSU-Spider:
Eccentricity 14 mm
Generated radius 100 mm
Equidistant interval 2 mm
Swivel angle ±25°
Technical data of NSU-Ro 80:
Control times:
Aö 63° before UT
Eö108° before OT
As 71° after OT
Es 40° after UT
Chamber volume 2×498 cm
Compression 9 : 1

Engine power
85 kW at 5500 rpm
Maximum torque:
162 Nm at 4500 rpm

Weight to power ratio: 1.52 kg/kW

Characteristics:
only rotating movements
low construction height
low weight
fewer parts
simple slit control
1 working cycle per revoluton
low tendency towards knocking.

Exhaust Turbocharger

Otto Engine

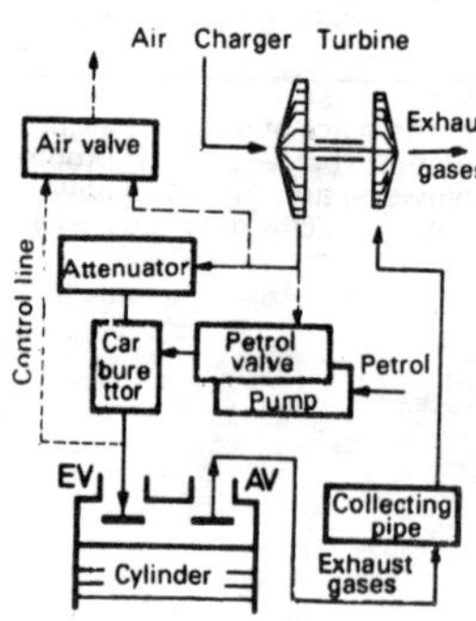

Exhaust turbo chargers serve to enhance power and improve efficiency. The energy of exhaust gases drives a charger through a gas turbine. The charger compresses the combustion air and increases the filling of the engine.

The turbine wheel is impacted by the hot exhaust gases via a spiral housing. The charger located on a common shaft sucks air through a filter and compresses it through the charger spiral into an attenuator and through a modified carburettor into the cylinder. Valves match the air quantity and discharge pressure of the petrol pump to the operating conditions of the Otto engine. The maximum speed of the engine is limited.

Without charging: Power
92 kW at 5500 rpm
Maximum torque:
187 Nm at 3500 rpm
Commencement of charging
Engine speed: 3500 rpm
Charger speed 30000 rpm
With throttled charging:
Power 132 kW at 5750 rpm
Maximum torque
228 Nm at 5100 rpm
Charger speed 70,000 rpm
Charging pressure 0.65 bar
Exhaust gas temperature 700°C
Internal ventilation of carburettor, Speed limitation through ignition interruption to 6200 rpm

Note: For abbreviations, see page 1.

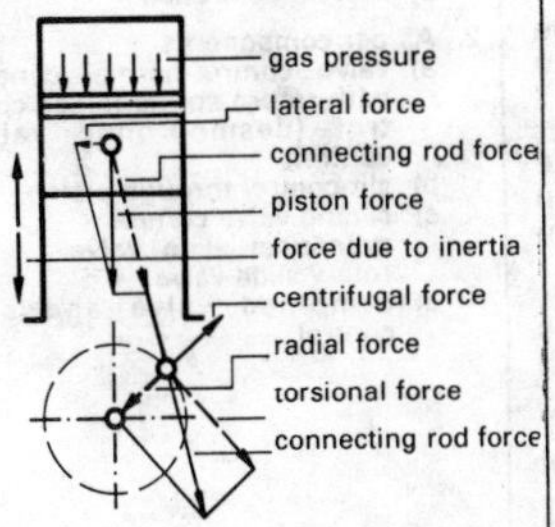

1. Compression chamber to be sealed in a movable manner against the crank case.
2. The pressures generated during combustion to be accepted and the piston force transmitted to the fly wheel.
3. Reciprocatory movement of the piston to be converted into rotary movement.
4. Rotating and vibrating masses of the crank drive to be balanced.
5. Piston to be moved in the idle strokes and gas movement corresponding to respective stroke to be assured.
6. In two-stroke engines, control of gas change to be taken over.

1 Piston
2. Piston ring
   a) sealing ring
   b) oil scraper ring
3. Gudgeon pin
   a) floating
   b) fixed in connecting rod head or in the piston eye.
4. Connecting rod
5. Crank shaft
   a) flywheel
   b) vibration damper
   c) balancing mass
6. Bearing boxes
   a) step and lapped bearings
   b) connecting and bearing

Piston

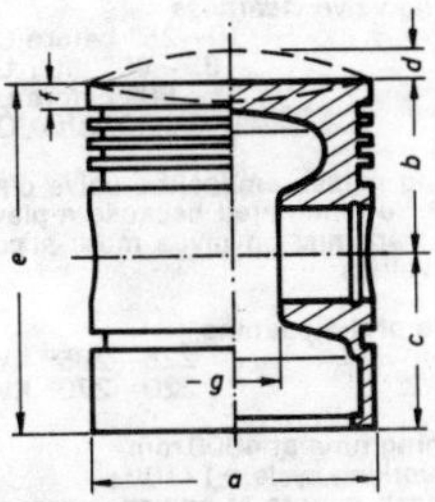

a Piston diameter
b Compression height
c Piston shank
d Hump or Pan
e Total length
f Top land
g Eye distance

1. Single metal piston
   a) solid shank piston
   b) split-skirt piston (U, T slit)
   c) elastic piston (thin-walled shank)
   d) tubular piston
2. Recoil piston
   Shank expansion to be influenced by cast ribbing of low thermal expansion; bimetal effect or shrinking effect
   a) steel ring: circular strip piston Duoflex, Debelack and perimatic piston.
   b) steel band: Nelson-Bohnalite (invar steel), autothermic, autothermatic, punched strip and formed strip piston
3. Special designs
   a) piston with SAP plates
   b) piston with heat arresting ring
   c) piston with circular carrier
   d) Isostatic and Sintal pistons

Requirements:
low weight
good heat conduction
high wear and thermal resistance
good running property
low thermal expansion
good machinability

Wear surface protection:
lead coating (Plumbal method)
tin coating (Stannal method)
graphitizing (Grafal method)
anodizing (Eloxol method)

Piston clearance in ‰ of diameter dependent upon shank form (oval, tapered, bulging), working method and cooling of the engine, at

top land	4—6
top shank end	0.7—2.5
bottom shank end	0.3—0.8

Piston Rings

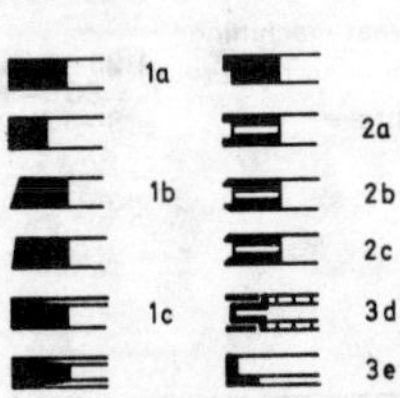

1. Sealing rings
   a) rectangular ring with/without internal chamfer
   b) tapered compression piston ring, normal/weak
   c) trapezoidal ring, single/double sided
2. Oil scraper ring and oil return ring
   a) grooved piston ring, normal/broad
   b) grooved piston ring, top chamfered ring.
   c) grooved piston ring, uniform chamfered ring
3. Special designs
   a) nose ring with expander spring
   b) lamellar rings
   c) grooved piston ring with hose ring
   d) oil ring, 2 plates and circular spring
   e) L-ring.

Dimensions in mm:

Impact gap	0.3—0.7
Axial play	0.04—0.11

Surface treatment:
ferrox, phosphate or tin coating on all sides.
Wear surface hard chrome plated or molybdenum filed.
Clamping of rings:
1. Thermal clamping (heat treatment)
2. Natural clamping (form turning)
3. Hammering the inner side.

Sliding Bearing

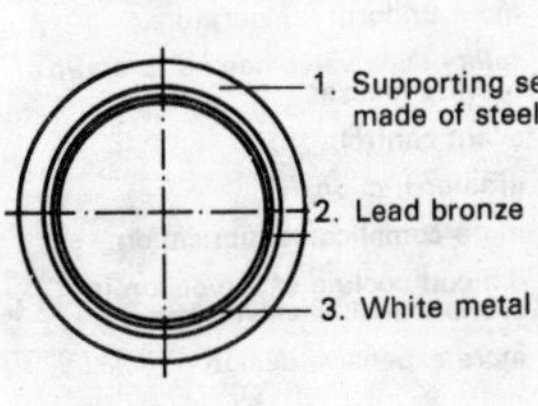

1. Single substance bearing
   a) solid bearing box made of lead bronze, light metal or plastic.
   b) bore covered by white metal
2. Double-substance bearing
   a) supporting seat made of steel, tin bronze or gun metal
   b) filling with sliding material: white metal, lead bronze or aluminium alloy.
3. Three-substance bearing
   a) supporting seat made of steel
   b) lead bronze filling
   c) white metal cladding (electrolysis)

Dimensions of seats:
Ready for assembly, interchangeable, bearing clearance H6/S6
1 normal step, accurate
4 repair steps, each of 0.25 mm

Bearing clearance:
Dependent upon bearing metal and bearing diameter

radial	0.03—0.06 mm
axial	0.10—0.30 mm

Layer thickness in mm:

white metal	0.025—0.5
lead bronze	0.300—1.5

Engine Control

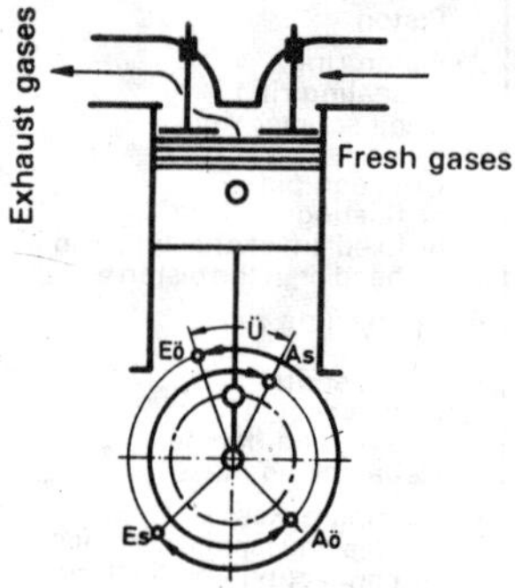

1. To lead the fresh gases during the suction stroke into the combustion chamber.
2. To seal the gas channels gas-tight in the combustion chamber during the compression stroke and power stroke.
3. To conduct the exhaust gases out during the exhaust stroke from the combustion chamber and to scavenge the combustion chamber.
4. Good filling of the cylinder demands gas paths of suitable flow and identical length with smooth surface.
5. Constant cross-sections with big inlet opening.
6. Regularly to-and-fro vibrating gas columns.
7. Good heat conduction of outlet valve.

1. As per the location of gas paths
   a) top controlled
   b) bottom controlled
2. As per components
   a) valve control-normal control with return spring forced control (desmodromic valve control)
   b) slit control through piston
   c) sliding valve control cylindrical slide valve rotary slide valve.
   d) combined valve and slit control.

Valve Control

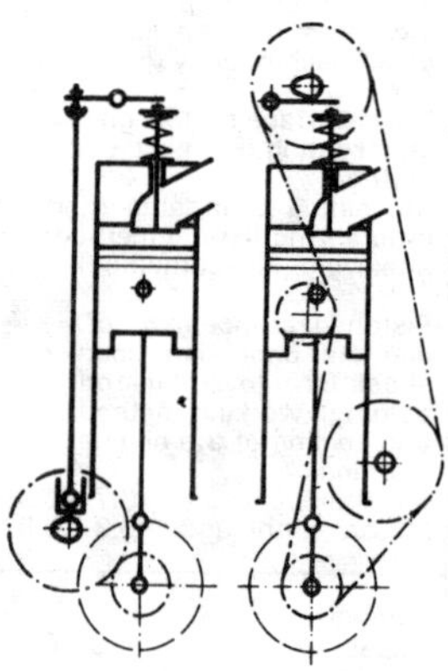

1. Valves (inlet and outlet valves)
   a) valve disc, shank with sealing
   b) special designs: armoured seating face and shank end, shank end with rear-release lever, outlet valve with hollow base and filled with sodium, aluminium coated disc, valve with deflector or oscillating valve.
2. Valve seat
   a) directly in the cylinder block/head.
   b) interchangeable seating ring.
3. Valve spring
   a) cylindrical helical spring, single/double
   b) hairpin spring
   c) torsion-bar spring
4. Spring plate with lock
   a) simple spring plate
   b) plate with turning fixture (rotocap)
   c) lock: valve-spring cotter, takeup washer, cap and check nut
5. Valve guide
   a) bore directly in the cylinder
   b) interchangeable valve guide
6. Intermediate parts
   a) valve tappets: plate type, mushroom type, roller type, pot-type, dancing type and hydraulic type
   b) valve push rods: pipe with ball head
   c) valve rocker with rocker axle and bearing or swinging lever
   d) setting screw
7. Cam shaft with cam shaft bearing Drive through control gears, control chain, king-pin or push-rod.

Control times during operating and testing valve clearness

Eö	10—25° before OT
Es	35—60° after UT
Aö	35—60° before UT
As	5—30° after OT

During measurement the valve play must be eliminated because a play-free transmission gives more accurate values.

Angle of the opening

EV	225—265° KW
AV	220—270° KW

Opening time at 4000 rpm
per working cycle ≈1/100 s
per each minute of engine running ≈20 s

Valve play (cold)

outlet	0.25—0.35 mm
inlet	0.08—0.15 mm

Transmission ratio
crankshaft : cam shaft=2 : 1

Thermal load

AV	650—800°C
EV	300—400°C

Valve seat machining:

width	1.25—3 mm
angle	90°—30′
valve stroke	8—12 mm

Slide valve control

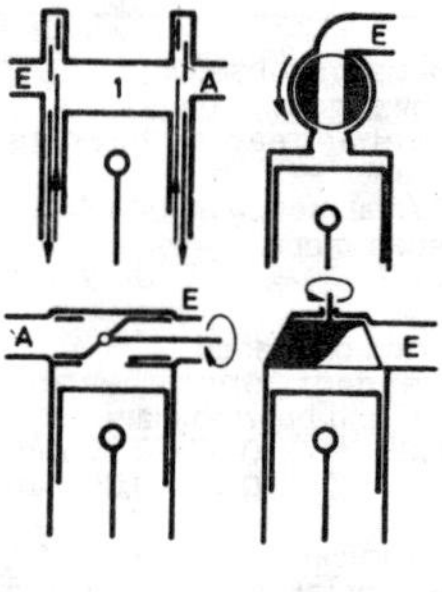

1. Knight cylindrical slide valve control. Two to and fro moving hollow cylinders fitted into each other in the working cylinder.
2. Oscillating slide valve control Burt-Mac Callum
   One to-and-fro moving, simultaneously rotating hollow cylinder in the working cylinder.
3. Rotary slide valve
   a) flat slide valve: rotating face plate with inlet and outlet openings.
   b) rolling slide valve control separate rollers for inlet and outlet, common roller for inlet and outlet (Baer or Cross slide valve control)
   c) taper slide valve control (Aspin)
   1 control channel in the rotary slide valve, 2 openings in the head, ignition in the control channel.
   d) plate-type slide valve control (2 stroke)

Slits open more rapidly and move uniformally,

bigger cross sections,

better filling,

positive control,

more uniform temperature,

rotary slide valve has no to-and-fro moving masses,

silent control,

greater friction,

more complicated lubrication,

difficult cooling of the piston in cylindrical slide valve control.

more expensive design

Note: For abbreviations, see page 1.

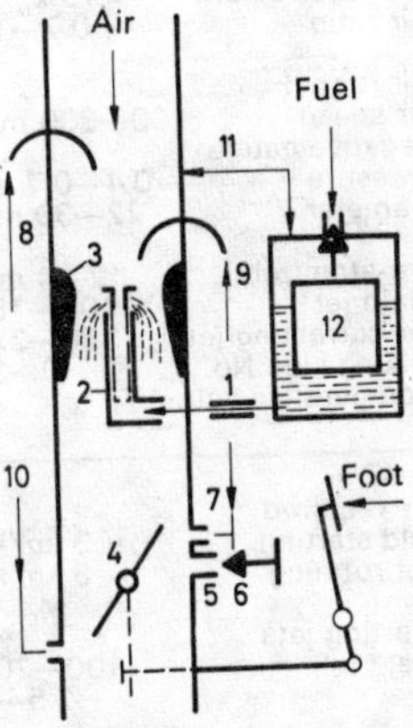

1. Main jet
2. Mixing pipe
3. Air funnel
4. Throttle valve
5. Idling
6. Idling mixture regulating screw
7. Transition
8. Enrichment
9. Acceleration
10. Starter
11. Internal ventilation
12. Float

1. To atomize the fuel
2. To mix fuel with air
3. To establish an ignitable mixture ratio.
4. To regulate the mixture quantity, and to influence power and speed of the engine.

Types

1. As per the layout of the induction port
   a) down-draught carburettor
   b) up-draught carburettor
   c) cross-draught carburettor
2. According to the number of induction ports:
   a) simple carburettor
   b) duplex carburettor
   c) step-type carburettor
3. According to the regulation of fuel level
   a) floating-type carburettor
   b) float-less carburettor nagative gauge pressure membrane with needle valve
   c) spill-type carburettor.
4. According to the preparation of fuel
   a) surface carburettor
   b) atomizing carburettor

1. Float attachment
   Float and pin, chamber with internal, external and changeover ventilation.
2. Main jet unit
   air funnel pre-atomizer main jet air correcting jet and mixing pipe.
3. Mixture and air regulation throttle-valve, rotary slide valve, piston slide valve, regulating carburettor.
4. Starting fixture,
   starter rotary slide valve, piston slide valve, starter valve, carburettor
5. Idling attachment
   mixture or air regulation, dependent or independent
6. Enrichment system
   for full or partial load in the main jet system or intake pipe.
7. Acceleration pump, piston or diaphragm pump, mechanical or pneumatic, with or without enrichment or dilution.
8. Pre-heating attachment
   cooling water or exhaust gases
9. Special attachments
   level corrector, fuel return valve.

Simple Down-draught Carburettor

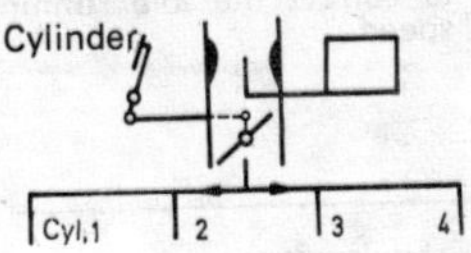

The simple down-draught carburettor is placed over the cylinders. It has only one suction channel for the connected fuel-air mix-cylinder. The current of the fuel-air mixture oriented downwards supports suction and aids filling. Suitable for smaller and medium sized engines of up to 4 cylinders.

Simple construction, favourable accommodation, easy setting and good assembling possibility.

Suction lines for engines with several cylinders are unequal in length and bent to varying degrees, therefore, the filling is uneven. Air throughput limited by the width of the suction channel.

Duplex Carburettor

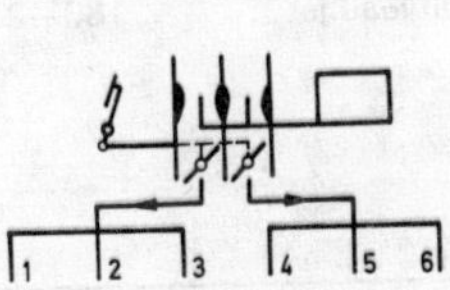

In the duplex carburettor 2 suction channels lie side-by-side in parallel. They have a common float unit. Each suction channel feeds through separate suction pipes, one or several cylinders of the engine. Both the throttle valves open simultaneously. they are together actuated by the driving foot lever.

Depending on the number of cylinders, air throughput is divided into 2 parallel suction channels. Length and curvature of suction lines same in the individual cylinders, more uniform current conditions and better filling of the cylinders, especially in multi-cylinder engine.

Step-type Carburettor (Register)

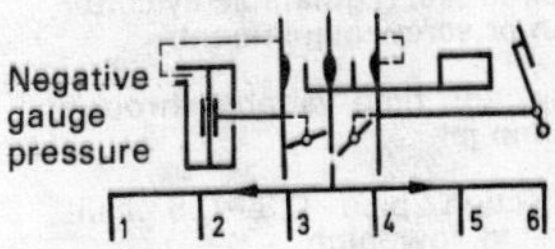

The step-type carburettor has 2 suction channels (steps) which end in a common suction pipe for the connected cylinder. They are successively opened or closed depending on the load. The first step works with the driving float lever and the second step opens automatically due to negative gauge pressure with negative gauge pressure valve or socket.

Air throughput is distributed as per load and speed into 2 steps functioning one after the other, therefore, more uniform current and mixing conditions, better filling of cylinder especially in large-volume engines. Difficult assembly and setting.

Carburettor

Main Jet Unit

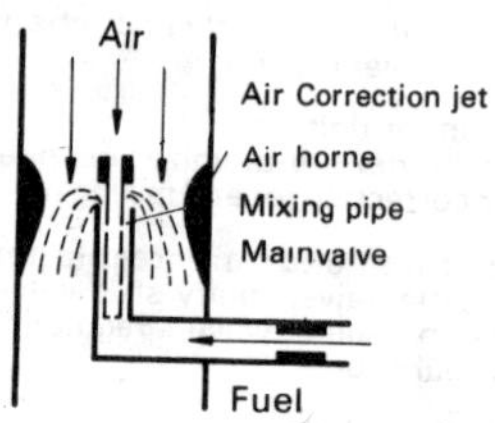

Air funnel and pre-atomizer narrow down the through-flow cross-section, increase the air speed and raise the negative gauge pressure. The negative gauge pressure pulls the fuel from the mixing pipe. With increasing speed the negative gauge pressure gets greater than the air speed so that the mixture becomes too fat. Therefore, uniform state of mixture ratio at higher speed and load requires thinning through the mixing pipe and air correction jet. In setting the main jet to full-load the partial load must be enriched.

Air requirement	≈ 15 kg/kg
Air ratio	0.7—1.3
Air funnel	
air speed	100—300 m/s
negative gauge pressure	0.4—0.7 bar
diameter	22—30 mm
pre-atomizer	2—4 mm
Main jet	105—150
Air correcting jet	120—220
Mixing pipe No. (comparative value)	0—37

Starter Unit

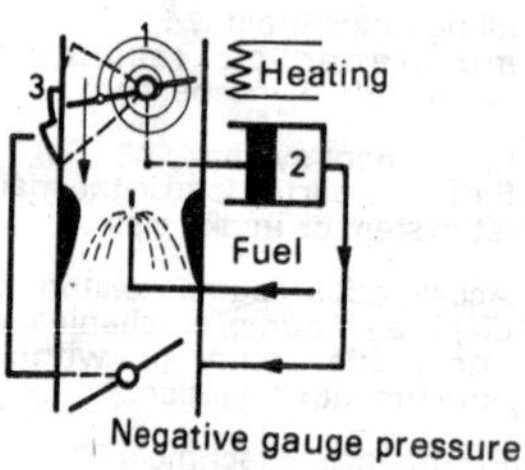

In a cold engine, the fuel condenses on the cylinder wall and suction manifold. The mixture gets thinned and does not ignite. Therefore, a fat starting mixture is necessary.

1. Rotary slide valve with 1 or 2 steps, progressive (stepless) and air valve.
2. Starting piston valve
3. Choke with choke poppet valve or eccentric bearing (semi-automatic), bimetal spring (1) low pressure piston (2) and stepped pulley (3) (automatic starter).
4. Negative gauge pressure of the float with carburettor.

Air required	
cold starting	≈ 3 kg/kg
hot running	≈ 8 kg/kg
Starting jets	
fuel	100—200
air	2.5—5
Cold starting	
speed	≈ 1600 1/min
throttle valve	≈ 8—9°

Opening of the choke poppet valve after 3—4 min.

Idling Unit

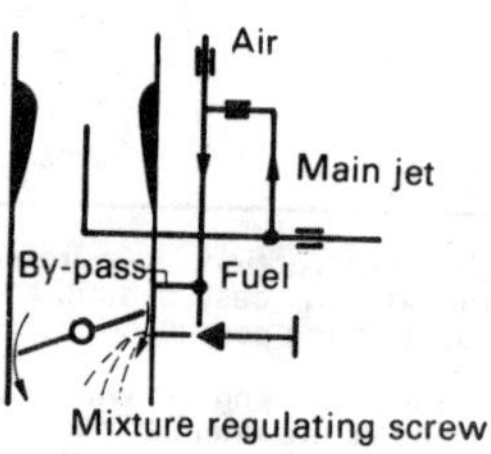

In idling, the negative gauge pressure in air funnel is too small to raise the fuel from the mixing tube. Further, fuel in the suction pipe may fail because of slow flow. Therefore, a fat mixture is prepared in an idling unit and mixed with additional air in the suction pipe.

1. Regulating hole for idle running
2. By-pass hole for transition from idle running to lower partial load range
3. Mixture or air regulation
4. Idle running dependent or not dependent upon main jet.

Air requirement	≈ 10 kg/kg
Speed	600—800 rpm
Mixture regulating screw	1—2 turns
Hot running speed	1000 rpm
Throttle valve	5—6°

Setting
1. throttle valve to be lightly set
2. to set the mixture
3. to correct the idle running speed.

Enrichment System

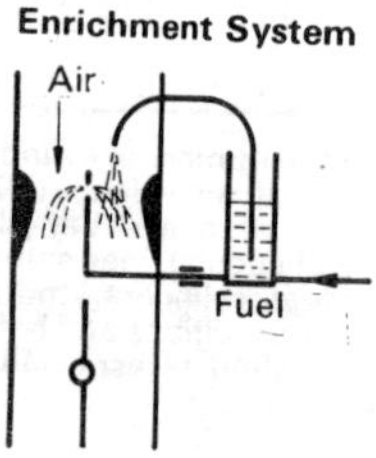

Full load at high speeds, because of higher power output, demands a fatter fuel-air mixture than partial load. The mixture in the top speed range is, therefore, automatically enriched.

1. Full load enrichment or partial load enrichment.
2. Enrichment through the main jet with additional enrichment jet.
3. Simplified enrichment with an enrichment tube in the suction channel to which the fuel flows from the float chamber.

Air requirement	
full load	≈ 13 kg/kg
Partial load	≈ 17 kg/kg
Partial load jet	50—60
Full load jet	180—220

Acceleration Pump

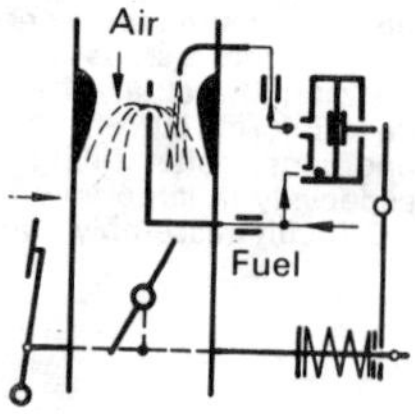

Fast input of gas reduces the negative gauge pressure for a moment, the fuel release from the mixing tube is stagnated and the engine power falls precipitously. Therefore, the fuel is additionally injected through a pump in the mixing chamber.

1. Diaphragm or piston pump
2. Mechanical or pneumatic actuation
3. With or without enrichment, or thinning at full load (fat, poor, neutral)
4. High or low injection pipe

Negative gauge pressure goes below 0.1 bar

Injection volume 0.5—2.5 cm^3 per stroke, regulatable by cotter pin or screw adjustment.

Injection time variable through pump jet

Injection pipe 0.3—0.5 calibrated, low, high

Pump jet	50—80

Stromberg Carburettor

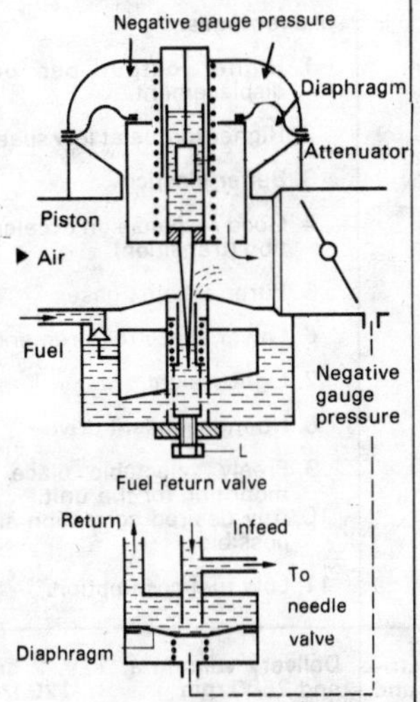

Construction
The Stromberg carburettor has an air funnel cross-section that can be varied by a piston. A taper needle fixed on the piston plunges into a needle jet. Over the piston is a negative gauge pressure chamber. The float chamber with double float and needle valve is below on the carburettor. The external ventilation is switched over to internal ventilation during travel with the help of a throttle valve rod. A negative gauge pressure operated fuel return valve prevents vapour bubbles.

Working Method
On opening the throttle valve, the negative gauge pressure in the suction channel will be effective in the negative gauge pressure chamber via the balancing hole of the piston. The piston, therefore, is raised and the air funnel cross-section is enlarged proportional to the through-flowing air quantity. Air speed and negative gauge pressure remain almost equal under all operating conditions, therefore, no special idle running attachment is needed. The correct fuel dose determination is done by the taper needle which releases a bigger through-flow cross-section with increasing stroke.

Acceleration
A fast opening of the throttle valve makes the attenuator to delay the piston movement; therefore, a greater negative gauge pressure is generated for a short while, whereby the mixture is enriched.

Fuel return
With a closed throttle valve, the fuel return valve is opened due to the negative gauge pressure effect. Hence with small consumption more fuel is circulated and consequently the formation of vapour bubbles is prevented.

Idling speed
Adjustment with the help of idle running set screw by setting the choke valve stop pin.

Idle running mixture
By turning in the idle running fuel regulating screw (L), the needle jet is displaced upward and thereby the annular clearance between the needle and the jet is narrowed, and the idle running mixture is thinned.

Jet centering
The annular clearance in the closed piston is only 0.02 mm; therefore, an accurate centering as per the manufacturer's instructions is necessary.

Attenuator
Oil level in attenuator to be checked once in 5000 km.

Mechanical Petrol Injection

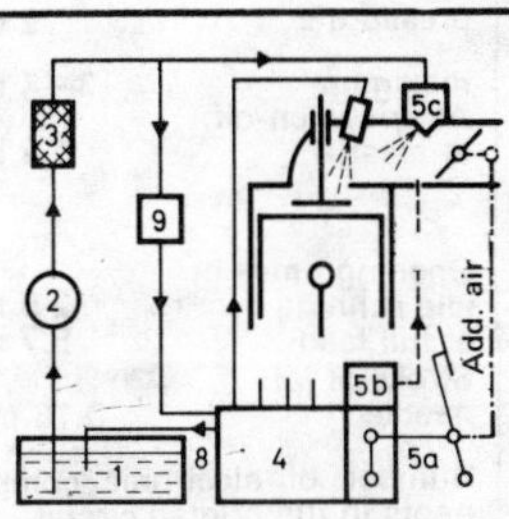

Pump element h Supply stroke

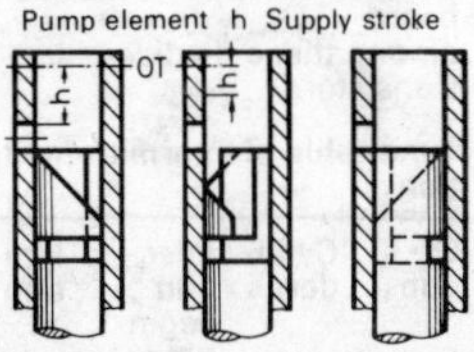

Full supply Partial supply No supply

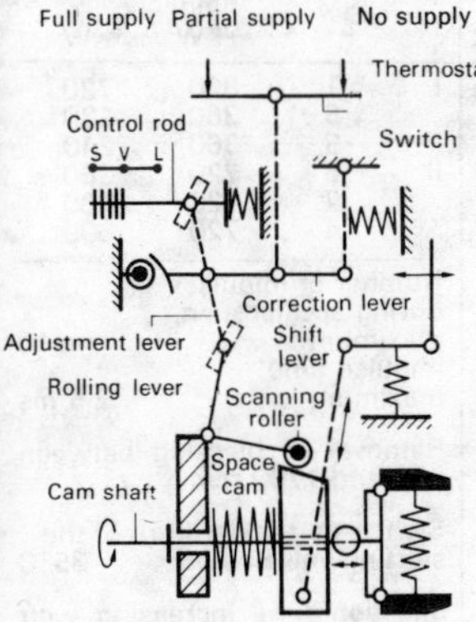

Injection Types:
a) Direct injection into the cylinder
b) Suction channel or head channel injection
c) Intermittent injection (in stroke)
d) Continuous injection (permanent)

Components:
1. Fuel Tank
2. Fuel discharge pump, electric
3. Fine filter for fuel
4. Injection pump: one pump element for one or more cylinders
5. Regulating attachment and correction device.
6. Valve connecting pieces with throttle valve
7. Injection valve, if necessary dosage unit
8. Return manifold with overflow valve
9. Attenuator tank

Injection process:
In the six-punch pump, during the intake stroke. In suction pipe injection with double-punch pump for many cylinders simultaneously, independent of valve position.

Pump elements with single hole cylinder, piston with guiding edge on top; uniform stroke, discharge, quantity varied by turning the pump piston.

Regulation:
The control rod of the injection pump is set by the driving float lever and the throttle valve is opened. Fuel and air are positively arranged on one another.

Control of the injected quantity by cam disc or by space cam which is scanned by a roller and the control rod slides. The special cam is displaced radially by the driving foot lever and axially by the centrifugal force controller.

Starting enrichment through starting valve in the suction pipe and by displacing the control rod with the help of the cold starter magnet. Actuation by thermo timer switch and timer switch.

Hot running enrichment via cooling water thermostat. It releases additional air and pushes the control rod towards fat.

Matching to the level is done by reducing the injected quantity with the help of the level diaphragm.

Fuel is injected in a finely atomized state. The quantity is correctly matched to filling, load and speed.

Properties
better cylinder filling
higher stroke volume power
uniform torque
faster transition
better cold starting properties
low octane number requirement
lower fuel consumption
better exhaust gas conditions
difficult maintenance
higher price

Injection pump, discharge beginning

full load, mechanical	10° before OT
actual	0—30° before OT

discharge end constant

mechanical	20° or 60° after OT
actual	35°—140° after OT

Injection pressure

direct	40—50 bar
suction pipe	12—20 bar

Setting on regulating springs

idle running speed	700—1200 rpm
partial load, bottom	1200—2400 rpm
partial load, top	2400—4200 rpm

Injection volume per stroke

idle running	12—18 mm³
partial load	15—30 mm³
full load	45—60 mm³
cold starting	60—120 mm³

Thermo timer switch.
Below 35°C water temperature closed. Closing time at—20°C max. 12 s; 1 second less for every 5° temperature increase.
Timer switch
One second in action above 35°C water temperature.
Control via starting switch and relay. Cooling water thermostat switch off point 68°C.

Electronic Petrol Injection D-Jetronic

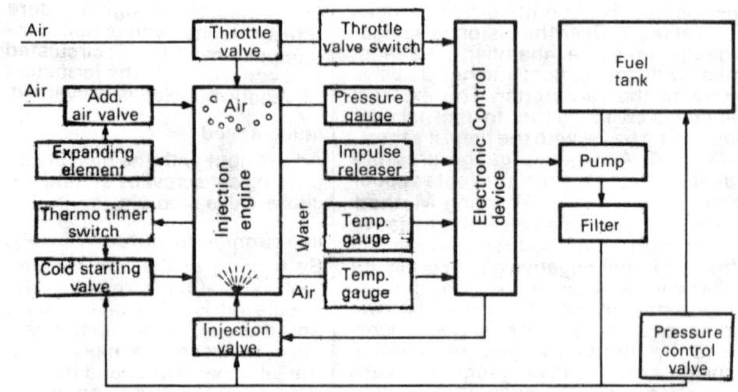

Advantages:

1. Higher output per unit displacement.
2. Higher torque at low speeds
3. Better elasticity
4. Good response on acceleration (transition)
5. Purer exhaust gases
6. Lower space requirement
7. Light weight
8. No independent drive
9. Freely selectablo place of mounting for the unit.
10. Any desired correction size possible
11. Low fuel cosumption.

Fuel pump	The fuel pump is a roller cellular pump which is driven electrically. It pumps the fuel from the tank to injection valves and starter valve and generates at the same time the injection pressure.	Delivery volume at 12V, 2 bar and 2500 rpm 120 l/h Switch-on time without starting 1 s
Pressure control valve	In the pressure control valve there is a springloaded diaphragm which releases an over-flow channel at a constant 2 bar positive gauge pressure, through which the excess fuel can flow back to the tank through the return manifold.	Fuel and injection pressure 2 bar
Injection valve	The injection valves are opened by the impulse signal generator via the control device. Through the excitation of a magnetic coil the jet needle is raised from its seat and the fuel is injected into the suction channel in a finely sprayed form through a calibrated annular port.	maximum 3—4 bar After switch-off, at least 1.2 bar
Electronic control device	The control device determines the injected quantity by altering the injection time by matching to the changing operating conditions of the engine, e.g. speed and suction pipe pressure, and controls the fuel pump. It receives information to carry out this task from various mesuring gauges located inside the engine.	Opening times in idle running 2.5 ms at full load 8.7 ms Stroke of jet needle 0.16 mm
Suction pipe pressure gauge	Two airless siphon diaphragms in a negative gauge pressure chamber push, in the event of pressure change in the suction pipe, the armature of a coil, thus changing its inductance. The inductance is input as a characteristic value into the electronic control device.	Number of electronic components in the printed circuit board 250 among these diodides 40 transistors 30
Impulse releaser	The speed-dependent impulse releaser controls the injection point of time and sets the control device in operation. To achieve this, 2 release contacts at 180° to each other are located in the ignition distributor. These contacts are actuated by a cam on the distributor shaft. One injection group is assigned to each contact.	Time-table of intermittent injection
Temperature gauge	The temperature gauges consist of temperature-dependent resistances. They measure the temperature of cooling water and suction air, and match the injected quntity over the control device to the engine temperature.	
Throttle valve switch	The throttle valve switch blocks the fuel inflow in a sliding mode of operation, and it releases additional injection impulses during acceleration. Below a certain speed the blocking is removed.	
Starting valve	The starting valve is controlled by drive switch and thermotimer switch. On starting a cold engine, additional fuel in finely atomized state is injected into the suction pipe through a turbulance nozzle jet.	Number of impulses during acceleration, maximum ≈10 impulse time maximum 2.5 ms
Additional air valve	For warm running, the engine requires more fuel. The additional air required for this purpose is fed into the suction pipe through an additional air valve. An expanding element alters the cross-section of the valve.	Removal of blocking between 900 and 1700 rpm Switch-on temperature of the starting valve below 35°C Injection time increasing with decreasing temperature at —20°C 12 s Composition of exhaust gases 2—3% CO at 700—750 rpm

Time-table of intermittent injection

Group	Cylinder/s	Injection beginning °KW	Ignition °KW
I	1	360	720
	5	360	120
	3	360	240
II	6	720	360
	2	720	480
	4	720'	600

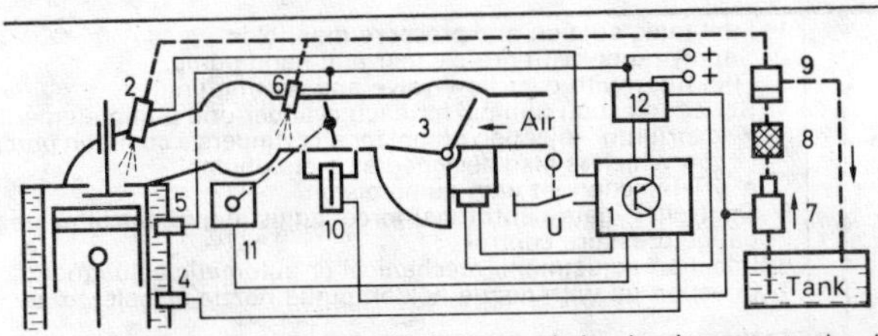

1. Electronic control device with integrated circuits. It converts the data collected on the engine pertaining to air quantity, coolant water temperature, throttle valve position as well as the starting process, engine speed and injection point into electrical impulses and passes it on to the injection valve. Thereby the injection time or the injected fuel quality is determined.
2. The injection valve, electro-magnetically actuated, is located in the suction pipe and always injects fuel before the inlet valve.
3. Air quantity measuring meter gives electric voltage signals (released by the turning of the choke valve) about the sucked air to the control device.
4. Temperature gauges send signals pertaining to coolant water temperature and cylinder head temperature.
5. The thermo-timer switch switches on the starting valve depending on the coolant water temperature.
6. Starting valve injects additional fuel during the start up.
7. Electrical fuel pump (rolling cellular pump) constantly pumps the fuel.
8. Fuel filter and (9) fuel pressure regulating valve for constant pressure.
10. Additional air valve provides for additional air during heat up dependent upon temperature.
11. Throttle valve switch signals to the control device about the idle running and full load.
12. A relay combination switches on the control device, fuel pump and thermo-timer switch.

Working Principle
Injection of the fuel by the electronically controlled injection valve. The electronic control device receives in the form of electrical signals data collected on the engine. The main input data on which the injected fuel quntity depends are the sucked in air quantity and the speed of the engine.

Injection Process
All the injection valves inject simultaneously and half the required quantity two times per cam shaft rotation.
The injection impulses are triggered by the interruptor.

Advantages of air quantity measurement
Engine-based changes (combustion chamber residues, wear) are determined and based on it, uniform exhaust gas values are obtained.
Acceleration enrichment takes place automatically.
Exhaust gas return for combustion chamber cooling is possible, posing no problems.

Petrol Injection K-Jetronic

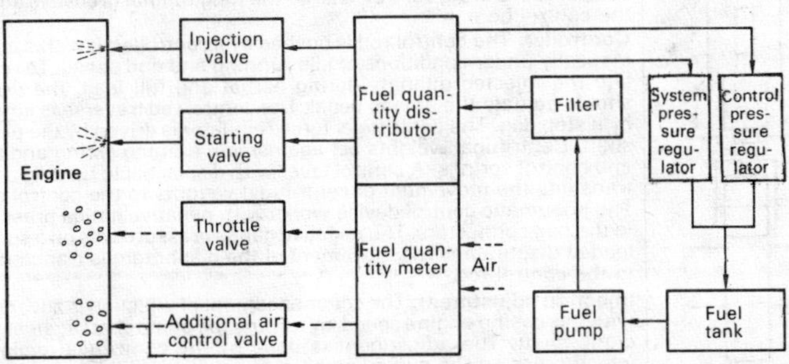

Working principle

Mechanical injection system without its own drive. The finely atomized fuel is continuously injected before the inlet valve. An air quantity measuring meter in front of the throttle valve measures the sucked in air quantity. Corresponding to the measured air quantity, a fuel dosage distributor distributes the correct fuel quantity to the injection valves. The air quantity meter and fuel dosage distributor are combined in the mixture controller.

Air quantity measurement

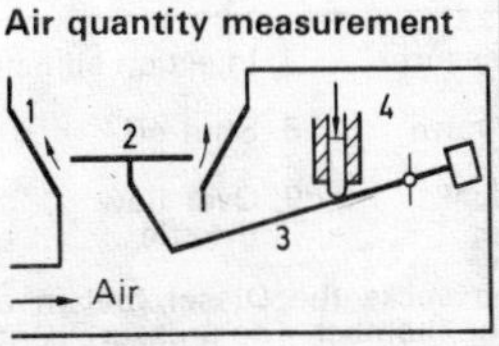

Fuel supply

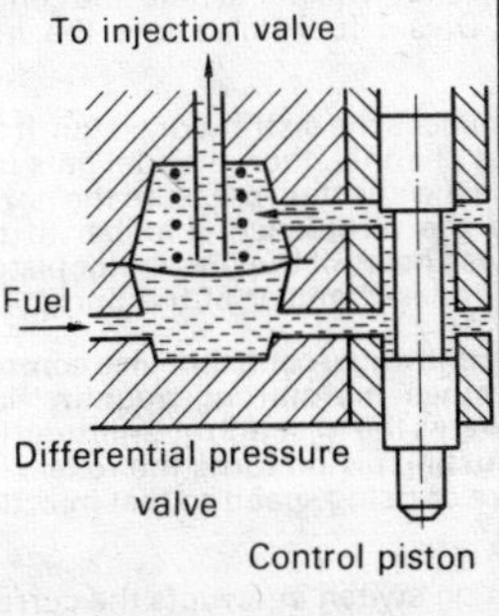

The air quantity meter consists of the air funnel (1) and the baffle plate (2) which is fastened on a lever with counter weight (3). The control piston (4) under control pressure acts on the lever. The intake air raises the baffle as long as the lifting force of air and the pressure force of the control piston are in equilibrium. The baffle plate is raised correspondingly higher in time with the larger quantity of intake air.

The fuel distribution to the injection valves is regulated by the control piston which is fitted in a movable manner inside the slot carrier. A control slot is available for every engine cylinder. The slot is more or less opened depending upon the position of the control position. A differential pressure valve is connected to the individual control throttles. The differential pressure valve keeps the fall in pressure constant. The fuel through-flow quantity is, therefore, proportional to the opening of a control slot. Fuel supply to the system is assured by a roller cellular pump. The fuel supplied in excess is controlled at the pressure regulator and it flows back into the tank.

With cold starting, additional fuel is injected into the intake manifold through an electromagnetic starting valve. The starting valve is controlled by a thermo-timer switch.

Advantages of K-Jetronic

1. Precise dosage of fuel quantity to the sucked-in air quantity.
2. Uniform combustion in all cylinders
3. Lower specific fuel consumption
4. Greater output per unit displacement.
5. Greater torque in low speed range.
6. Greater engine elasticity
7. Less harmful components in the exhaust gas.

Data

Positive gauge pressure in the system pressure circuit constant	4.7 bar
Positive gauge in the control pressure circuit in very cold engine	0.5 bar
at 15°C	1.5 bar
in the warm engine	3.7 bar
Aperture pressure of the injection valve	3.5 bar
Pressure difference in the differential pressure valve	0.1 bar
Supply quantity of fuel pump at 12 V	2 l/min

Injection Unit for Diesel Engines

Construction

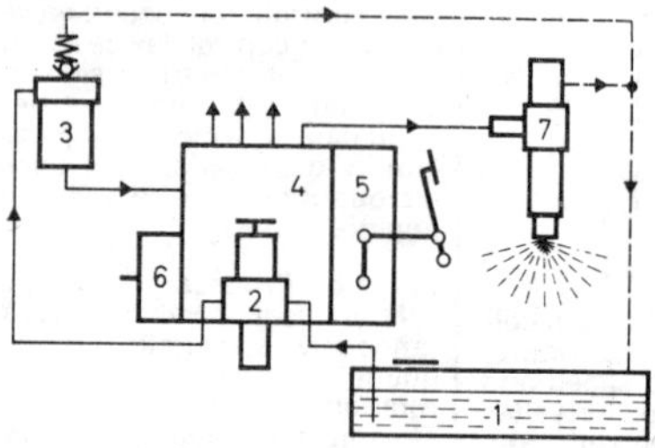

1. Fuel tank, suction and pressure manifolds
2. Supply pump with pre-cleaner and hand pump
3. Fuel filter with over flow-valve and venting line
4. Series injection pumps: for each cylinder one pump element
   - distributor-injection pump: for all cylinders a common pump p ton which is also designed as a distributor
   - PT-Injection jet with pump piston.
5. Controller: gate centrifugal force, adjustment control or negati gauge pressure control.
6. Injection adjustment: mechanical or automatic adjustment
7. Injection jet with nozzle-holder: pintle nozzle or hole nozzle

Tasks

- To inject and distribute the fuel into the combustion chamber
- To match the fuel quantity with load and speed
- To raise the fuel pressure according to the injection method
- To advance the injection point of time with increasing speed

Series injection pump

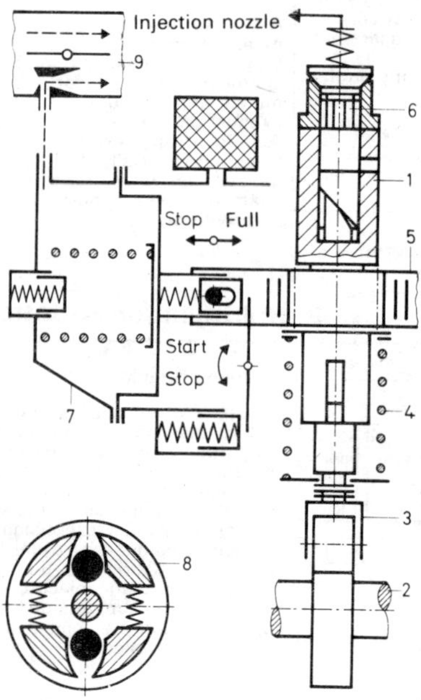

1. Pump element
2. Cam shaft
3. Roller shaft
4. Piston spring
5. Control rod
6. Pressure valve
7. Controller
8. Injection adjuster
9. Flap connectors

Filling: A piston pump discharges the Diesel fuel through the fine filter into the suction chamber of the injection pump. The fuel gets into the pressure chamber through the control bores in the pump cylinder.

Commencement of discharge: The cam shaft moves the pump piston upwards, closing thereby the control bores. The fuel gets into the injection nozzle from the pressure chamber via the pressure valve. If the injection pressure is reached, the fuel is injected into the combustion chamber. The pressure valve prevents the dripping of fuel into the combustion of the engine.

Discharge end: As soon as the piston going upwards releases the control bores with its guiding edge, the fuel is pressed back into the suction chamber through the longitudinal and circular grooves of the piston. The discharge end, and consequently the discharge quantity, is guided by the position of the guiding edge. The discharge quantity is continuously varied by turning the pump piston with the control rod. During zero discharge the longitudinal groove is above the control bore.

Controller: The control rod is pushed by a controller. It works automatically under conditions of idle running and end speed. To regulate the injected quantity during partial and full load, the driver should actuate the driving pedal. The control rod traverse is limited by a stop dog. The centrifugal force regulator is driven by the pump shaft. Centrifugal weights act against idle running spring and two end control springs. A control lever with displaceable turning point transmits the movement of centrifugal weights to the control rod. The pneumatic control device works with negative gauge pressure in the flap connectors. The negative gauge pressure acts on a spring loaded diaphragm. The movement of the diaphragm is transmitted to the control rod.

Injection adjustment: The commencement of injection is advanced with increasing engine speed, by turning the pump shaft against the crank shaft. The adjustment is done by the centrifugal weights which press on the curved paths of the adjusting disc.

Distributor injection pump

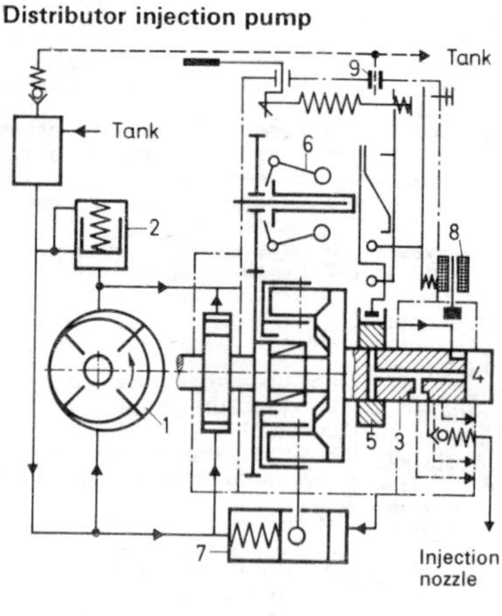

1. Oscillatory cell pump
2. Pressure control valve
3. Distributor piston
4. High pressure chamber
5. Control valve
6. Centrifugal controller
7. Injection adjuster
8. Shut-off
9. Over-flow throttle

Filling: An oscillatory cell pump sucks the Diesel fuel in and discharges it into the pump's inner chamber. The pressure control valve and the overflow throttle limit the discharge pressure. The drive shaft rotates the distributor piston. The filling hole and control slot get covered, and thus the Diesel fuel flows into the high pressure chamber.

Injection: The cam disc axially displaces the distributor piston. If the distributor is covered by the outlet channel, then the fuel gets into the injection nozzle. After attaining the injection pressure the nozzle opens and the fuel springs into the turbulance chamber. If the control slide valve releases the cross holes of the distributor piston, then the fuel flows back into the inner chamber of the pump.

Controller, Injection adjuster: A centrifugal control device controls the control slide valve. It determines the starting quantity, idle running and end speeds, and regulates the injection quantity during partial load and full load. The adjusting piston turns the roller ring with increasing discharge pressure or rising speed so that injection is done earlier.

Shut-off device: If the spark starting switch interrupts the current flow, then a solenoid valve closes the filling channel and blocks the fuel input into the high-pressure chamber. The motor stops.

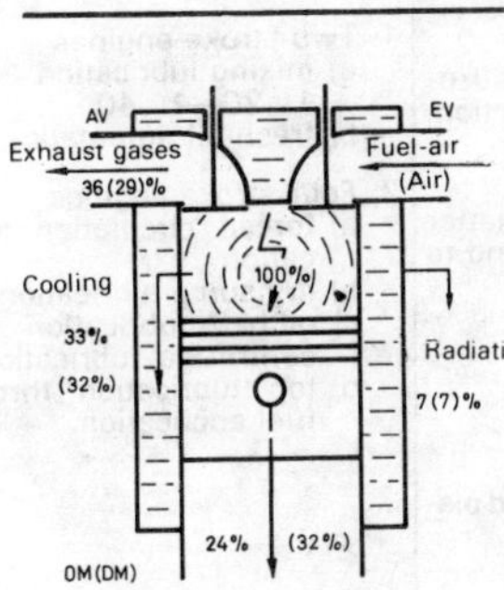

1. Air cooling
   a) travel wind cooling (dynamic pressure)
   b) blower cooling
2. Water cooling
   a) pump circulation cooling positive gauge pressure cooling system, closed cooling system with expansion tank.
   b) heat recirculation cooling (thermo-syphon cooling)
   c) through-flow, evaporation and positive gauge pressure cooling
3. Glycol cooling (hot cooling) operating temperature 120—140°C
4. Oil cooling for piston bottom and bearing spots
5. Internal cooling evaporation heat of fuel

1. To conduct out the excess heat of the engine: 1/3 of the combustion heat.
2. To maintain a certain operating temperature; 80-95°C
3. To protect the lubrication film from destruction and combustion.
4. To avoid surface ignition and fuel knocking.
5. To limit the heat up of the in-flowing fresh gas to maintain the filling loss at a low level.

Pump recirculation cooling

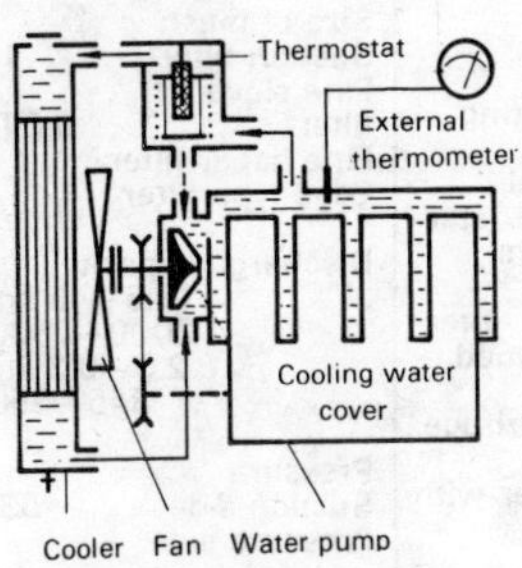

1. Water pump (centrifugal pump)
2. Cooling water cover around the combustion chamber
3. Thermostat (cooling water regulator)
   a) expansion socket with liquid
   b) expanding element regulator with wax
4. Cooler
   a) water-pipe cooler (ribbed pipe cooler) ribs perpendicular to the tubes ribs parallel to the tubes
   b) air pipe cooler and plate cooler
5. Ventilator (fan)
   a) fans that cannot be switched off
   b) electro-magnetic, hydraulic or viscose coupling that can be switched-off
6. Cooling water hose (rubberized cloth)
7. External thermometer (easily boiling liquid)
8. Heat exchanger for heating, expansion tank in the case of closed circulation of liquid, venetian blind of cooler

Positive gauge pressure valve (pressure cooling)
opens at — 0.4—0.9 bar
boiling temperature — 108—116°C

Negative gauge pressure valve opens at — 0.05—0.1 bar

Thermostat (DB)
Main valve
opening begins — 78—79°C
stroke 8 mm at — 91—94°C

Throttle (ball check valve) through-flow volume — 0—10 l/h

Short-circuit valve open stroke — 6—7.5 mm at 78°C

Fan
power required — 3—5% P_e
switch off speed — 25—30%
switching on at the cooling air temperature — 62°C

Blower cooling

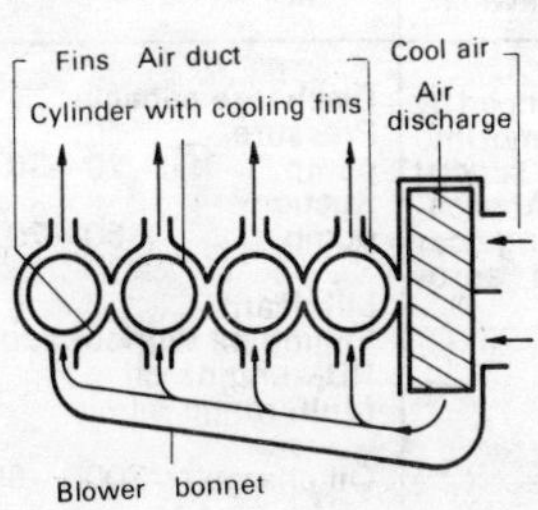

1. Cooling blower with V-belt drive
   a) axial blower
   b) radial blower
2. Cooling ribbing around the combustion chamber.
3. Air conduction
   a) guide bonnet and sheet metal
   b) axial or radial air current to the cylinder.
4. Cold air regulation (thermostat) automatic, dependent upon engine temperature, outer temperature or engine load
   a) throttle attachment for air through-flow two point control: full-zero continuous control-variable
   b) speed control of blower.
5. Heat exchanger for car heating.

Speeds
blower wheel around — 100 m/s
cooling air — 20—30 m/s

Temperature rise of air — 10—80°C

Power consumed by blower — 4—6% P_e
Discharge volume — ≈500 l/min

Ribbing surface per cm³ stroke volume — 10—30 cm

Comparison	Water cooling	Air cooling
Construction	complicated	simple
Place required	large	small
Heat transfer	good	bad
Cooling action	uniform	non-uniform
Temperature control	good	bad
Heat up time	long	short
Engine noise	silent	loud
Operational disturbance	likely	insensitive
Frost protection	necessary	not necessary
Maintenance	required	nor required

Comparison	Pump circulated cooling	Heat circulated cooling
Circulation by	pump	heat (5 g/l)
Circulation speed	fast, depends on speed	slow, depends on fall in temperature
Temperature fall	5—10°C	10—20°C
Water quantity	low	large
Cooler	small	big (25%)
Cooling water control	necessary	not necessary
Cooling action	uniform	not uniform, local overheating
Construction	complicated	simple

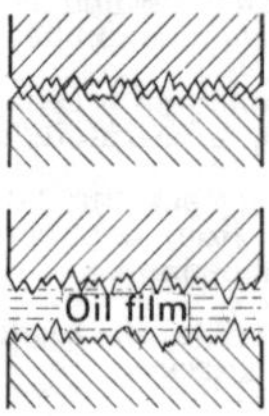

Dry friction

Liquid friction

1. Lubrication
   Formation of an adhesive, pressure-resistant lubricating film to reduce friction.
2. Cleaning
   To wash away metallic dust, foreign matter and combustion chamber residues and to arrest these substances in the filter.
3. Cooling
   To conduct away the friction heat
4. Sealing
   Fine sealing between cylinder wall and piston ring.

1. Two-stroke engines
   a) mixing lubrication
      1 : 20—1 : 40
   b) fresh oil automatic
2. Four-stroke engines
   a) forced circulation lubrication
   b) dry sump lubrication
   c) oil bath lubrication
      centrifugal lubrication
   d) top lubrication through fuel application.

Forced circulation lubrication

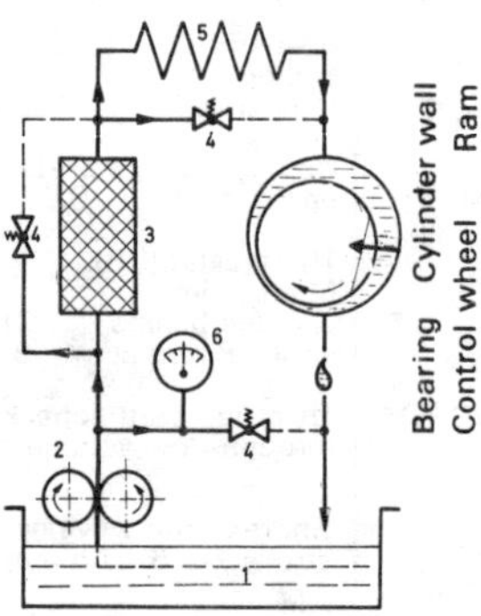

1. Oil pump with gauging rod, ventlation.
2. Oil pump with suction filter
   a) gear pump
   b) piston pump
   c) Eaton pump
   d) eccentric pump
3. Oil filter: built in the
   main current: coarse, complete filtering
   side current: fine, partial filtering
   a) edge type filter: plates, wire spool
   b) sieve type fine filter: disc sieve, star-shaped sieve, cylinder cover filters.
   c) star-shaped felt filter: hose filter.
   d) fine paper filter: cartridge or interchangeable filter, mostly star-shaped.
   e) solenoid filter: oil drain screw
   f) centrifugal filter: free jet centrifuge, catapult filter.
   g) combination filter: fine sieve filter with fine paper filter, depth filter.
   h) depth filter: textile or cellulose
4. Positive gauge pressure valve, short-circuit valve, overflow manifold.
5. Oil coolers with safety valve for air-cooled engines; heat exchanger for water cooled engines.
6. Oil pressure gauge or oil pressure switch.

Oil temperatures

Piston	110—250°C
Conrod bearing	110—115°C
Main bearing	85—100°C
Oil sump	65—80°C

Size of mesh

Suction filter	1—4 mm
Fine sieve filter	0.04—1 mm
Fine paper filter	5 μm
Edge type filter	30 μm

Discharge capacity
12.5—45 kg/min at 2000—2500 rpm
2.3—6.3 kg/min at 345—350 rpm

Pressure

Suction side	≈530 m bar
Pressure side	5 m bar

Port pressure in bar

Positive gauge pressure valve	4—8
Oil pressure switch	0.3—0.8
Oil consumption	≈ 0.1l/100 km

Dry sump lubrication

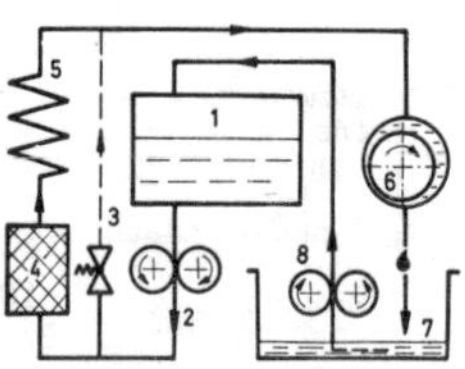

Dripping oil, after its circulation, is pumped by a double-pump from the crank case (sump) into a special reserve tank, and from the special reserve tank to the lubricating spots. Application: cross-country vehicles, racing car engines, under-floor engines and aero-engines.

1. Reserve tank for oil
2. Pressure pump
3. Safety valve
4. Filter
5. Cooler
6. Lubrication spot
7. Dry sump
8. Return discharge pump

Discharge capacity

Pressure pump	29—36 kg/min
Suction pump	50—70 kg/min

Lubricant
Engine oil without additives
HD—engine oil
Multi range oil

Oil charge 3000—5000 km.

Fresh oil automatic

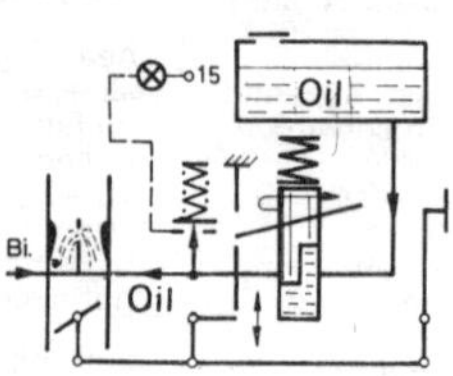

Separate reserve tank for oil. A pump discharges the oil by a cam disc with piston into the carburettor immediately after the main jet. The pump is driven by the crank shaft. Discharge is dependent upon speed. Piston stroke controlled by the throttle reparating rod. Discharge is also dependent upon load. The piston worked with the maximum stroke, if the throttle valve is fully opened.

An electric switch in the oil line lights up a warning light in the event of lack of pressure.

Piston diameter	6 mm
Piston stroke	0.15-1.7 mm
Discharge volume at	5 bar
no load (0°)	5-9 mm³/stroke
partial load (5°)	9-13 mm³/stroke
full load (82°)	43-54 mm³/stroke
Maximum speed	7000 rpm

Lifting of throttle valve

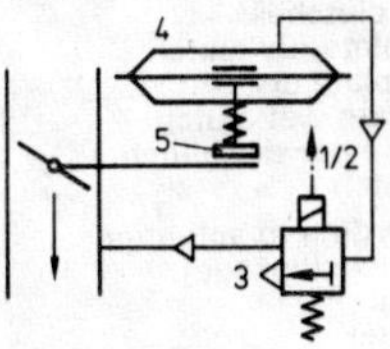

The fuel-air mixture contains too much of burnt residual gases and is excessively fat in the sliding operation with the throttle valve closed. The combustion is incomplete and combustion misfiring occurs. This leads to a high proportion of dangerous components in the exhaust gas, especially the concentration of unburnt hydrocarbons. Its percentage is reduced by lifting the throttle valve. The mild opening is done by the negative gauge pressure control device in the carburettor. At engine speeds below 1800 1/min, the throttle valve goes back to the idle running position.

Lifting of throttle valve in the sliding mode of operation at oil temperatures above 25°C and engine speeds of over approx: 2000 rpm

Construction
1. Temperature switch
2. Speed switch
3. Change-over valve
4. Negative gauge pressure control device
5. Spring with set-screw

Parts 1 and 2 are not shown in the sketch.

Fuel vaporization unit

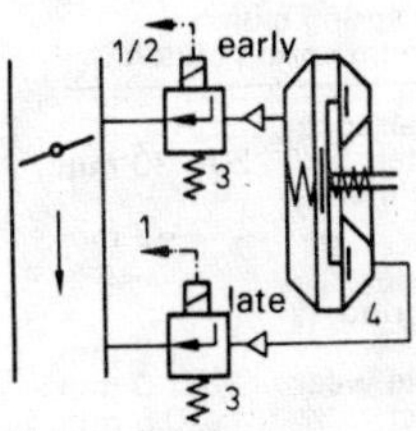

The speed of the engine increases on lifting the throttle valve in the sliding mode. This is not desired. Therefore, with increased lifting of the throttle valve and in idle running state, the ignition point of time is set in the direction of 'late'. Late ignition develops more heat per power stroke in the combustion chamber. The higher thermal load brings about more complete combustion and lower discharge of poisonous exhaust gases. The late setting is done by negative gauge pressure which sets-in in the suction tube between the engine and carburettor, with the throttle valve closed.

In the lower partial-load range, the throttle valve is opened a little, and the filling of the cylinder with fresh gas is bad. The sucked-in fuel-air mixture is more difficult to inflame, and burns slowly. Therefore, the ignition in the partial load region is set in the direction 'early' in addition to the speed dependent centrifugal force setting by negative gauge pressure and quite independent of the position of the throttle valve in the flap connector.

Ignition change-over to the direction of "late", in idle running and sliding mode, with the throttle valve closed and coolant temperatures below 100°C by a ring-diaphragm in the negative gauge pressure socket.

Ignition change-over towards early in the partial load region, at coolant temperatures above 40°C and engine speeds above 2000 1/min by the circular diaphragm in the negative gauge pressure socket.

Construction
1. Temperature switch
2. Speed switch
3. Change-over valve
4. double negative gauge pressure socket for the two settings, viz. 'late' and 'early'

Parts 1 & 2 not drawn in the sketch.

Ignition change-over

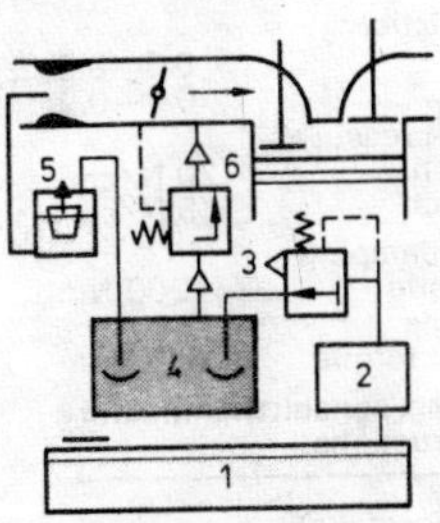

The fuel vaporization unit prevents the exit of vaporized gases from the fuel unit into the atmosphere. The vaporized gases from the fuel tank and float chamber are fed into the active carbon container when the engine is switched off. During driving, they are sucked in to the carburettor with adequate negative gauge pressure in the suction pipe.

Construction
1. Fuel tank
2. Expansion tank
3. Valve system for ventilation and air exhaust, and positive gauge pressure.
4. Active carbon container
5. Float chamber
6. Suction valve

After-burning

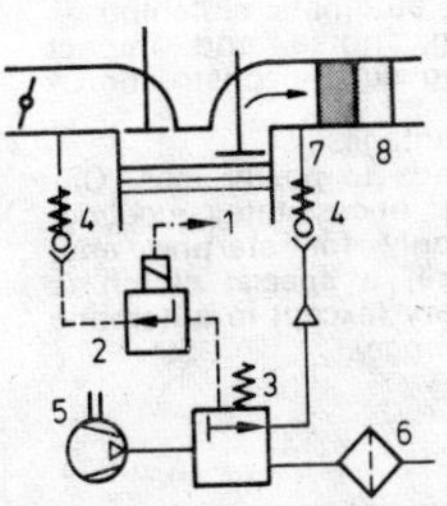

The stringent exhaust gas emission rules require limiting values for the emission of carbon monoxide and unburnt hydro carbons. By an after-burning of the exhaust gases the stipulated values can be maintained. For puposes of after-burning, fresh air is blown into the outlet channels of the cylinder head when the engine is running. Oxygen contained in the air mixes with hot exhaust gases and can react in the catalyst or reactor. The catalyst accelerates the oxidation of CO and hydrocarbons. To maintain the catalyst in a good functional state, the engine should be run only with lead-free fuel. In USA the catalyst is changed after a certain number of miles.

Air blown in at oil temperatures above 17°C; the air is blown out at oil temperatures below 17°C.

Construction
1. Temperature switch
2. Change-over valve
3. Blow-out valve
4. Non-return valve
5. Air pump
6. Silencer
7. Blow-in channel
8. Catalyst or Reactor

Part 1 not drawn in the sketch.

Exhaust gas return

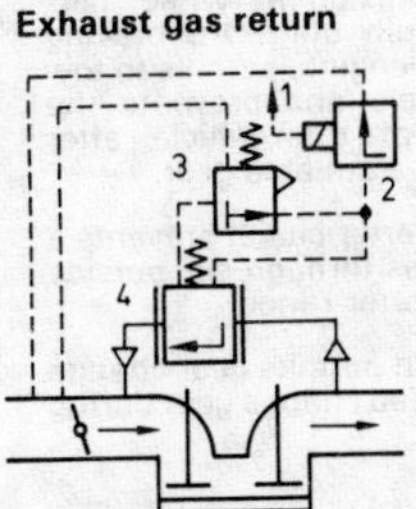

Because of extremely thin fuel-air mixture or by the reduction of the combustion temperature, the concentration of nitrogen in the exhaust gas is reduced. While driving and during acceleration, therefore, a portion of the exhaust gases is returned into the suction pipe from the exhaust manifold via the exhaust gas return valve. The addition of exhaust gases to the fuel-air mixture reduces the combustion temperature. The exhaust gas return works in 2 steps: a small step while driving and a big step while accelerating. The control of the 1st step is done electrically, and the 2nd step pneumatically. On accelerating, the negative gauge pressure switch vents the 2nd step. The exhaust gas return valve can open fully.

Exhaust gas return

a) 1st step in the dirving mode at coolant temperatures above 40°C.

b) 2nd step during acceleration, if coolant temperatures are above 40°C and negative gauge pressure< 270 m bar.

Construction
1. Temperature switch
2. Electric change-over valve
3. Pneumatic change-over valve
4. Exhaust gas return valve

Part 1 not drawn in the sketch.

Clutch System

General

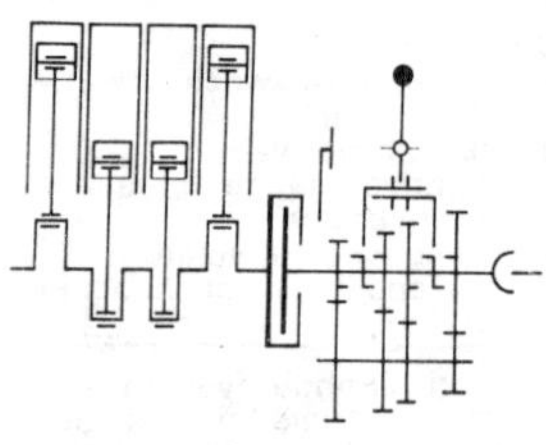

The automobile clutch is in the power flux between engine and change-speed gear. It has the task of

- connecting the crank shaft of the engine with change-speed gear in a detachable and elastic manner.
- transmitting the engine torque to the change-speed gear.
- reducing the engine torque during starting by slip, and in that way connecting the mass of the vehicle uniformally and without jolt, and facilitating acceleration.
- interrupting the power flux during the drive, and guaranteeing easy and silent connection of gear links.
- protecting the engine or gear against overload and reducing gear noise.

Types according to force connection
a) friction clutches:
single-plate dry clutch
double plate dry clutch
multi plate wet clutch
b) magnetic powder clutch
c) fluid clutch

Types according to actuation
a) mechanical (linkage)
b) hydraulic
c) pneumatic
d) electromagnetic
e) centrifugal force

Types according to adjustment
a) centrally adjustable
b) non adjustable

Types according to clutch spring
a) helical spring clutch
b) diaphragm spring clutch

Single-plate dry clutch
Double-plate dry clutch

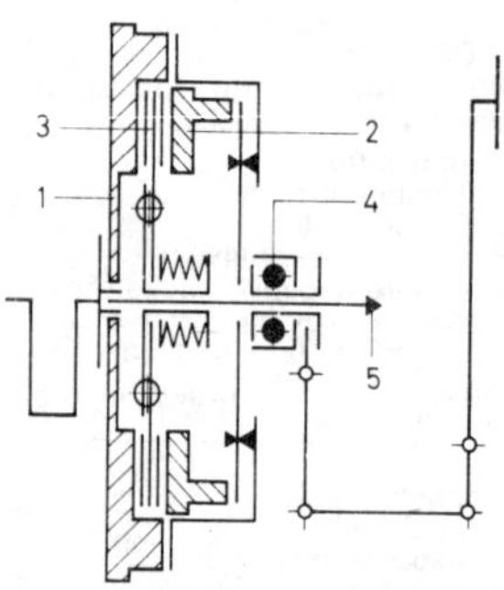

Flywheel 1, Pressure plate 2, Clutch plate 3, Releasing lever 4, Clutch shaft 5.

The engine torque is taken up by flywheel and pressure plate. The clutch springs press the pressure plate on the clutch plate, and the latter on the flywheel. This leads to the generation of linkage through friction. Plate and shaft are driven. To transmit a higher torque double plate clutches are employed. They have 2 clutch plates and 2 pressure plates, thus doubling the friction surface and the transmitted torque.

While disengaging the clutch, the movement of the foot pedal is transmitted mechanically or pneumatically through the clutch release fork, declutch unit, disengaging level and pin to the pressure plate. The latter is drawn back against the spring force. The clutch plate can run freely. On declutching of a double plate clutch double the disengagement path results.

Clutch clearance at the pedal	20—30 mm
at the declutch unit	2—3 mm
Clutch plate coating spring path	1.2 mm
permissible wear	1.0 mm
side impact	0.5 mm
permissible imbalance	0.05 mm
Static friction dry	0.3—0.5
wet	0.1—0.2
Surface pressure dry clutch	20 N/cm²
wet clutch	5 N/cm²
Torsion damping stop torque	200 Nm
stop angle	5°
frictional torque	10 Nm

For setting, consult manufacturer's instructions.

Hydraulic fluid clutch
Fluid clutch

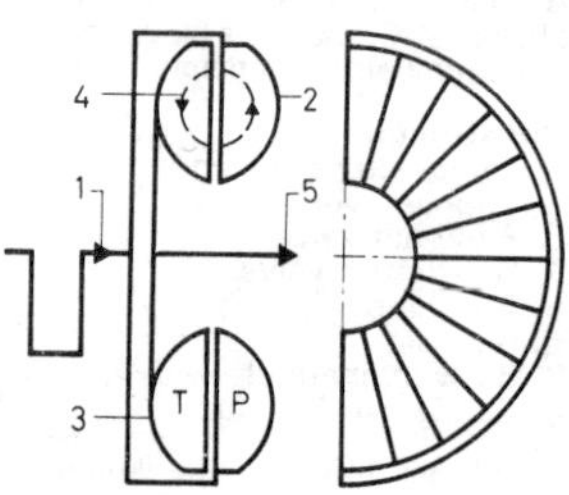

Parts; Drive 1, Pump wheel (driver) 2, Turbine wheel (rotor) 3, ATF-oil 4, Drift 5.

Pump wheel is driven by the crank shaft. The fluid is pushed outward by the centrifugal force, steered on the turbine wheel blades and brought into flow. The driver is moved.

A counter-centrifugal force is developed in the turbine wheel with increasing speed, that slows down the circulation. In synchromesh the current and consequently the flux of force stops. There must, therefore, always be a certain slip, i.e. speed difference between the rotor and driver to transmit the engine torque.

Advantages:
No maintenance and wear-free soft starting, no stalling of the engine, automatic clutching on starting, noise and impact damping, simple construction.

Disadvantages:
Slip leads to power loss. Oil-heating necessitates cooling. Used only for starting and, therefore, a special clutch is necessary (except in automatic gear change).

Torque converter clutch

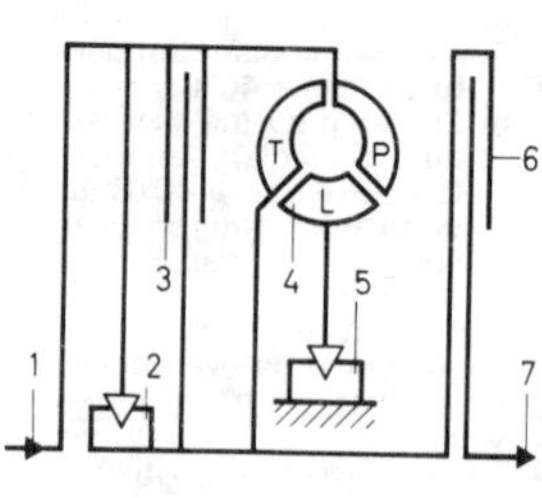

Parts: Drive 1, propulsion free wheel 2, converter clutch 3, converter 4, converter free wheel 5, clutch 6, drift 7.

The propulsion free-wheel bridges the converter in the propulsion. The converter clutch bridges the converter at a certain speed ratio in the range of clutch point.

The connecting clutch interrupts the flux of force on inserting a speed gear. It is actuated mechanically with compressed air support or electro-pneumatically.

The converter with free wheel increases the torque automatically and steplessly, and interrupts the flux of force on stopping.

The propulsion fly wheel permits to fully utilize the braking action of engine, permits to tow the vehicle and prevents the roll-out of the vehicle after inserting a suitable gear.

The converter clutch prevents a power loss through slip outside the converter range.

The clutch permits gear change in the speed change gear during the drive.

Centrifugal Clutch

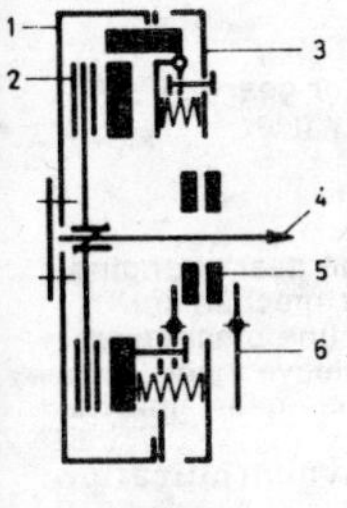

1. Engine flywheel, screwed with crank shaft
2. Drive disc with spring-action lining made of asbestos, synthetic resin and copper or brass wires; torsion damper stop disc and splined hub.
3. Spring housing or clutch cover with contact plate, 3 centrifugal weights, return springs main springs, set screws, disengaging lever and stop lever for parking and towing.
4. Drive shaft with splines.
5. Thrust bearing of clutch.
6. Declutching fork.

On giving gas the engine speed increases. The centrifugal weights go outwards and overcome the return springs. The contact plate is pressed on the clutch disc by the main springs.

In case of standstill of the engine or the engine running below 300 rpm a stopping attachment directly connects the drive shaft with the clutch cover in the turning direction of the engine.

Automatic Clutch

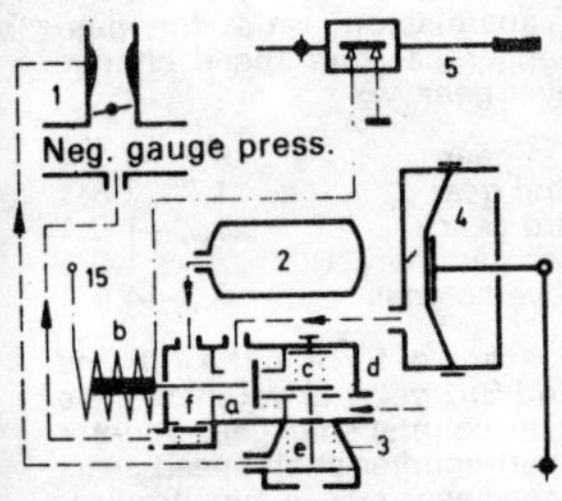

1. Suction pipe with throttle valve and air funnel.
2. Reserve tank for negative gauge pressure.
3. Electromagnetic control valve
   a) Main valve with spring
   b) Electromagnet with iron core
   c) Reduction valve with spring and set screw for 1st stage.
   d) Balancing hole for 2nd stage
   e) Negative gauge pressure chamber with spring-loaded diaphragm and pin. Acceleration of coupling during the 2nd stage.
   f) Non-return valve.
4. Servo motor with diaphragm, linkage and actuating lever for the clutch.
5. Gear engagement lever with contact maker

On getting into gear, the clutch is disengaged with the help of negative gauge pressure in the intake pipe through the electromagnetic control valve by the movement of the diaphragm in the servo motor.
The re-engagement of clutch is done, after releasing the gear lever, by a two-stage reduction of negative gauge pressure over the reduction valve and the jet, by correspondingly matching to the driving conditions.

Electromagnetic clutch

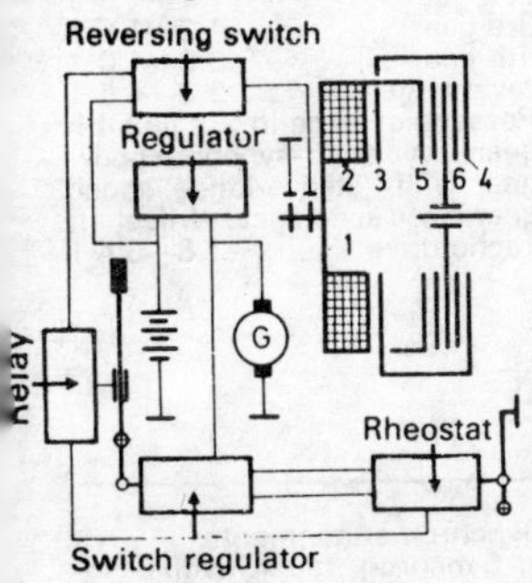

On engaging the clutch the magnet core 2 receives progressively increasing current. The magnetic field attracts the armature plate 3, and the pressure plate 4 presses the clutch plate 6 on the intermediate plate 5.

On disengaging the clutch the current flow is interrupted, the return springs push out the armature plate from the flywheel 1 and the clutch disc runs freely.

Progressive current supply is achieved by the speed-dependent dynamo voltage and a load-dependent rheostat. The disengagement of clutch on connecting the gear is done by the gear change lever, and automatically in idle running. With the dynamo off, during skidding and during shorter stops of the vehicle (parking) the clutch is fed by the battery.

Voltage	6 V
Maximum current intensity	4.8 A
Coil resistance	1.2 Ω
Air gap	0.44 mm
Contact force of the clutch plate	2100—2400 N
Rheostat	
1st and reverse gear	
$R_1+R_2+R_3$	2.5 Ω
R_1+R_2	1.3 Ω
R_1	0.6 Ω
2nd and 3rd gear	
$R_4+R_5+R_6$	1.25 Ω
R_4+R_5	0.8 Ω
R_4	0.4 Ω
Parking resistance	1.5 Ω
Setting resistance	0—1.5 Ω

Magnetic powder clutch

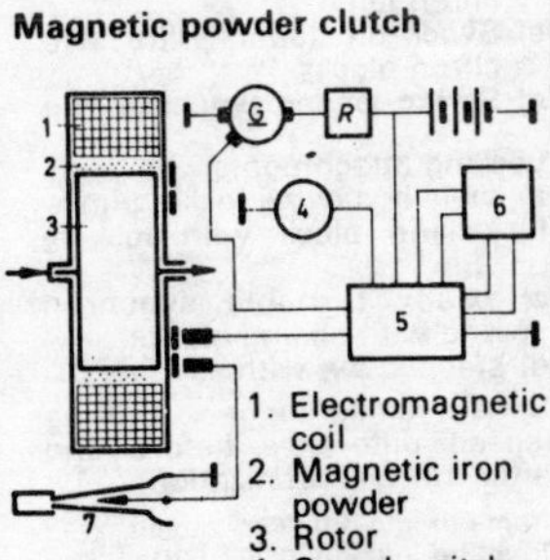

1. Electromagnetic coil
2. Magnetic iron powder
3. Rotor
4. Speed switch
5. Relay
6. Carburettor switch
7. Switch lever

A centrifugal switch feeds, during starting, a slowly increasing dynamo voltage to the coil There a magnetic field is formed which firmly holds the iron powder. That makes the clutch to be positively connected and the rotor is driven. Starting from a certain speed or load, the coil receives the full battery voltage.

The contact with the gear shift lever actuates a cut-off switch which interrupts the current circuit. The magnetic field is cancelled and the clutch is released.

Magnetic powder	110 g
Current consumption	
Idle running	1 A
Partial load up to 2/3	2.5 A
Full load	4.5 A
Resistors	
Indle running	7.5 Ω
Spare resistance	2.5 Ω
Demagnetizing	125 Ω

Connection of idle running and spare resistances through the throttle valve axle.

Change Gear Mechanism

General

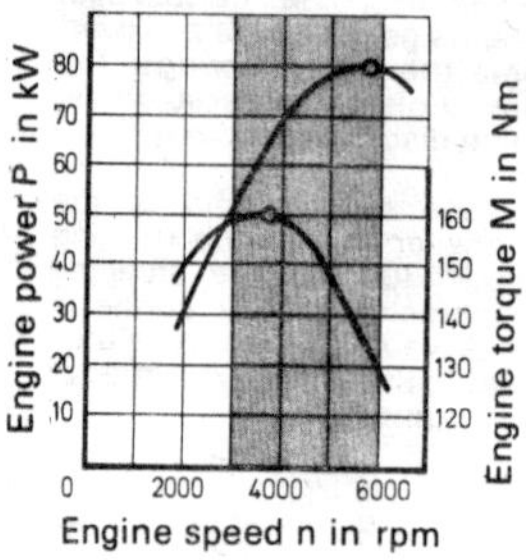

Internal combustion engines provide an adequate, utilizable torque only in a small, limited speed range. To match the intensely changing speed of the driving wheels and the encountered drive resistance to this range, the automobile needs a change gear mechanism. Therefore, it should have the following:

- to convert the engine torque and to match it to the generated drive resistance
- to interrupt the flux of force in no-load
- to facilitate reverse driving of the automobile
- to guarantee a light, fast and silent gear engagement.

Types based on flux of force
a) Gear drive
- change-over gear, stepped gear
- auxiliary gear
- distributor gear
- planetary gear

b) V-belt drive
c) Chain drive
d) Fluid drive

Types based on gear changing
a) sliding gear mechanism
b) pull type spline mechanism
c) gear shift sleeve mechanism
d) synchronous gear mechanism
e) locking synchronisation mechanism
f) semi-automatic gear mechanism
g) automatic gear mechanism

tooth flank curve — involute

Four speed changeover gear

with counter shaft

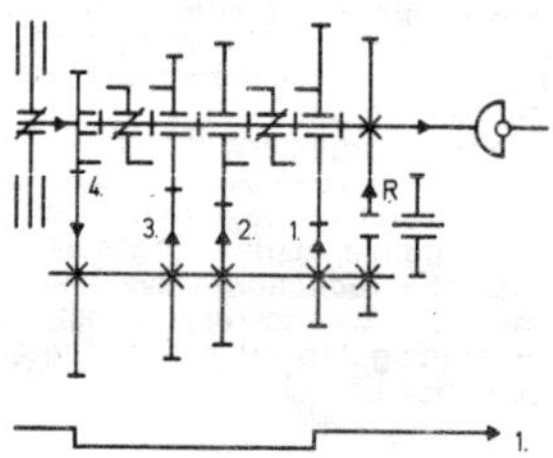

Parts: Drive shaft, counter shaft, main and reverse shafts, driving gear, counter shaft gears, speed gear, reverse gear, synchronization attachment, gear shift sleeve, gear shift fork, gear shift rod and gear shift lever.

Gear shift procedure: By the selector movement of the gear shift lever the respective gear shift fork is located. The shift fork transmits the shifting movement of the lever to the gear sloeeve. This is pushed from the short teeth of speed gear 2 and via the netural central position and the synchronization attachment. The gear 1 is now loose and gear 2 is connected rotationally firmly with the main shaft. The gear is engaged.

Transmission ratio for passenger car four-speed change over gear

1st gear	3.2—4.1:1
2nd gear	1.7—2.5:1
3rd gear	1.2—1.6:1
4th gear (direct)	1.0:1
reverse gear	3.3—4.5:1

Passage of force in the 1st, 2nd and 3rd gear: drive shaft-drive gear-counter shaft gear-counter shaft-counter shaft speed gear-speed gear-sleeve-synchronization body-main shaft.

Four-speed changeover gear

without counter shaft

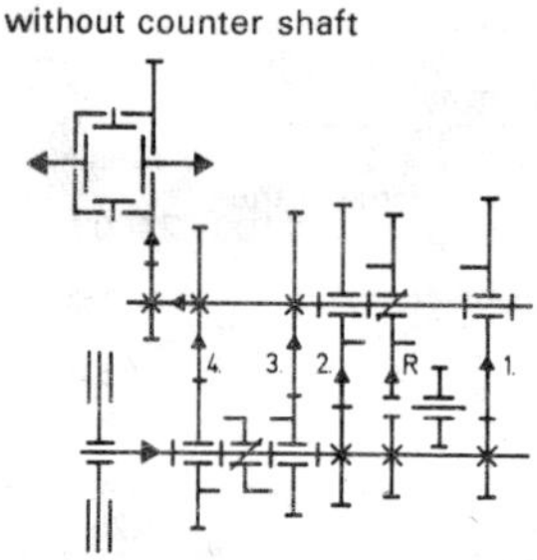

Parts: Drive shaft, drift shaft, reverse shaft, loose and fixed speed gears, reverse gear, synchro attachment, gear shift sleeve, fork, rod and lever.

For gear shift procedure see four-speed changeover gear with counter shaft.

Specialities
- only 'simple' transmission; therefore diameter difference of mutually meshed gears is bigger.
- no direct gear is present; therefore in the 4th gear transmission into fast is possible.
- change over gear without counter shaft and drive gear; therefore the construction is simpler.

Transmission ratio for cars of 4-speed changeover gear

1st gear	3.2—4.1:1
2nd gear	1.7—2.5:1
3rd gear	1.2—1.6:1
4th gear	0.8—1.0:1
reverse gear	3.3—4.5:1

Passage of force in 3rd and 4th gear: drive shaft-synchron body-gear shift sleeve-loose speed gear-fixed speed gear-drift shaft

Tacho drive 1.8—3.4:1

Synchronisation attachment

Synchromesh mechanism with spring loaded synchron ring

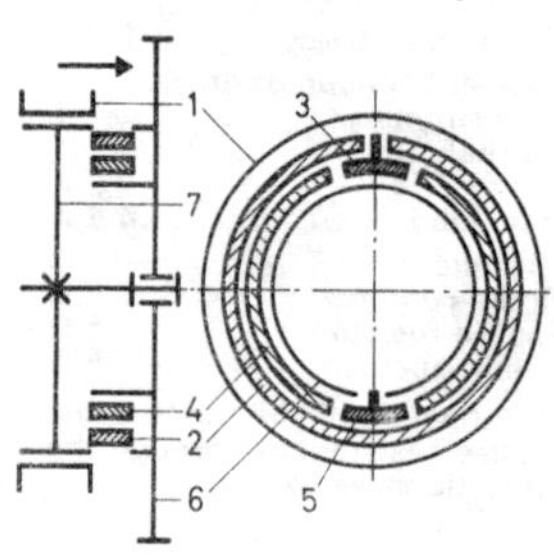

The synchronization should equalize the speed difference between the gear shift sleeve and the gear of the speed to be connected with the help of the friction clutch, and eliminate the possibility of the sleeve touching the short teeth of the gear before attaining synchromesh. The gear shift procedure can thus be performed in a light, low-noise and wear-free manner.

Parts: gear shift sleeve 1, spring loaded synchron ring 2, locking block 3, locking strip 4, stop 5, speed gear 6, and guide sleeve 7.

Synchron position: The gear shift sleeve touches the synchron ring. The synchron ring turns, and its end presses with the locking block the locking strip on the inner sode of the synchron ring, whereupon the stop serves as support. The ring diameter becomes bigger and the gear shift sleeve does not allow itself to slide over. In synchron mesh the locking system relaxes.

Synchron attachments
a) Synchron taper with synchron balls
b) Synchron taper with synchron latch
c) Synchron taper with synchron blocks
d) Spring-loaded synchron ring

Locking attachments
a) radially moving locking links
b) locking block with locking strip
c) axially turnable synchron ring with sliding blocks.
d) stepped pin with circlip

Speed jumps in gear shifting (speed difference before and after the synchronization)

from 4th into 3rd gear — about
from 3rd into 2nd gear — 600 to
from 2nd into 1st gear — 1000 1/min

at engine speed of 3200 rpm

Twelve-speed change-over gear

With series connected change-over gear

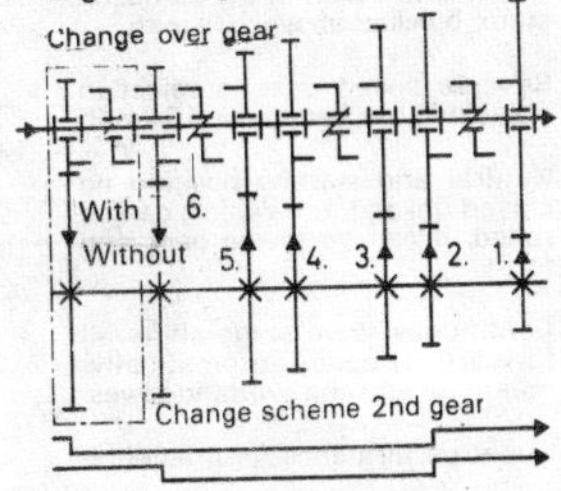

Parts: Six-speed change over gear, series connected auxiliary transmission with drive gear, counter shaft gear and synchronization attachment.

The more number of gear-shift stages a drive mechanism has, correspondingly more accurate will be the maintained utilizable speed range of the engine. Normal gear mechanisms are, therefore, extended by series connected or subsequently connected auxiliary transmission.

The series connected auxiliary transmission acts as a second input stage of the drive shaft on the counter shaft. Both input changes can be alternately clutched in with the help of the lock-synchronized dog clutch engagement. The gear shift of the series connected auxiliary is done electro pneumatically by the tumbler switch. The gear shift process takes place only after the clutch release.

Transmission ratio:

1st input stage	1.00:1
2nd input stage	1.25:1
1st gear	7.65:1
2nd gear	6.07:1
3rd gear	4.40:1
4th gear	3.50:1
5th gear	2.66:1
6th gear	2.12:1
7th gear	1.77:1
8th gear	1.40:1
9th gear	1.26:1
10th gear	1.00:1
11th gear	0.89:1
12th gear	0.71:1
Reverse gear 1	7.18:1
Reverse gear 2	5.71:1

Simple planetary gear

Rotary gear box

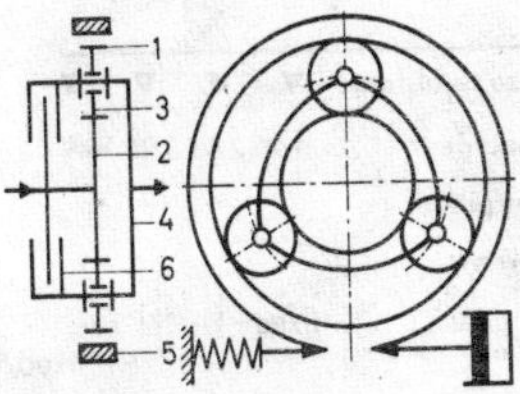

Parts: internal gear 1, sun gear 2, planetary gear 3, carrier 4, band or multi-disc brake 5, multi-plate clutch 6, free wheel (not shown)

To transmit a torque either one part—internal gear, sun gear or carrier—must be firmly held with a band or multi-disc clutch or with a free-wheel and supported against the gear box housing, or two parts viz. carrier and sun gear, must be connected together by a multi plate clutch. Various transmissions are obtained by 'firmly holding' a component; 'connecting' two components with each other provides the direct speed gear. Clamping or releasing of the brake is done hydraulically and mechanically by spring force.

Transmission ratios with internal gear of 76 teeth and sun gear of 50 teeth:

1st speed gear	2.52:1
2nd speed gear	1.66:1
3rd speed gear	1.00:1
Reverse gear	1.52:1

Gear shift	1st speed gear	2nd speed gear
Internal gear	fixed	drive
sun gear	drive	fixed
Carrier	drift	drift

Gear shift	3rd speed gear	R. speed gear
Internal gear	—	drift
Sun gear	clutched	drive
Carrier		fixed

Double planetary set

Ravigneaux planet set

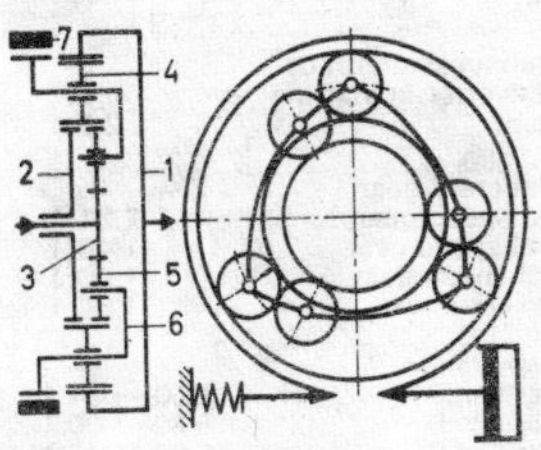

Parts: Internal gear 1, big sun gear 2, small sun gear 3, long planetary gear 4, short planetary gear 5, carrier 6, band or multidisc brake 7, multi plate clutch 8.

Several gear sets connected one after the other provide any number of transmissions and a small building space: two different sun gears have always 3 planetary gears which sit on a common carrier. The planet gears of the big sun gear mesh directly into the internal gear.

Gear shift	1st speed gear	2nd speed gear	3rd speed gear	Reverse speed gear
Small sun	drive	drive	drive	free
Big sun	free	fixed	drive	drive
Carrier	fixed	free	free	fixed
Internal gear	drift	drift	drift	drift

Transmission ratios with

internal gear	60 teeth
big sun gear	30 teeth
small sun gear	20 teeth
long planetary gear	15 teeth
short planetary gear	15 teeth
1st speed gear	3.00:1
2nd speed gear	1.66:1
3rd speed gear	1.00:1
Reverse speed gear	2.00:1

Simpson planet set: both planet sets have an equal number of teeth. The planet set has one sun gear and 2 hollow gears. The torque is conducted through an internal gear.

Hydraulic torque converter

Fluid drive

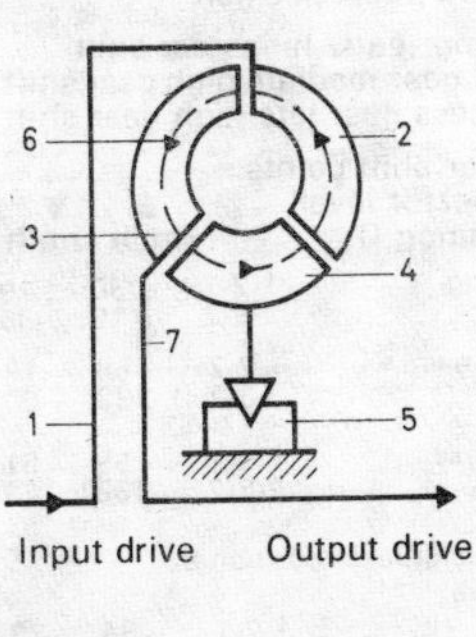

Parts: Fly wheel 1, driver 2, rotor 3, leading wheel 4, free wheel 5, special gear fluid 6, driven shaft 7.

On starting the gear fluid in the driver is accelerated by centrifugal force, i.e. brought into current, and pressed in to the rotor. The force of the current sets the rotor in motion. The fluid current generated from the rotor is decelerated by the leading wheel which supports itself on the housing over a free wheel. Consequently, the speed of the rotor is reduced and the utilizable torque is increased. The slower the rotor rotates, correspondingly higher is the torque generated by it.

If the driver and rotor run almost identically, then the fluid current rotates the leading wheel. Thereby no more conversion of torque takes place. The converter works like the clutch.

Starting conversion	2.0—2.8
Slip	3—20 %
Clutch point at v	≈ 0.60
Fixed brake speed	≈ 1800 rpm
Oil pressure in converter	3—7 bar

- automatic, stepless and load-related conversion
- soft starting
- no stalling of the engine
- low noise, impact damping
- no wear, no maintenance
- slip necessary
- heating of oil
- cooling necessary
- no reverse driving
- automatic clutching in starting and stopping
- small conversion range
- compact construction

Automatic Gear Mechanism

General

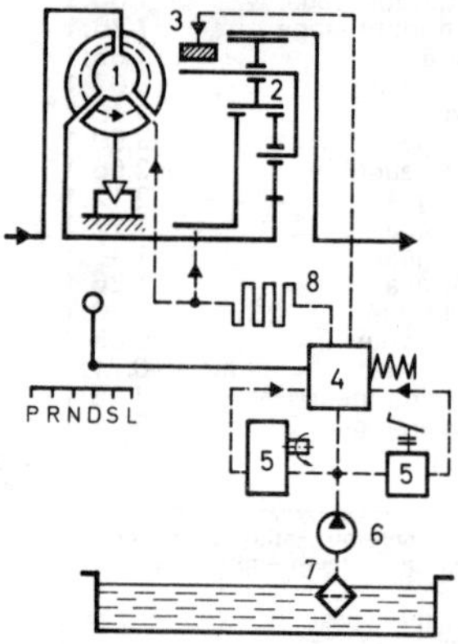

Parts: fluid drive 1, planetary gear 2, gear shift ekements 3, control valve 4, regulating valve 5, oil pump 6, oil filter 7, oil cooler.

Automatic gear mechanism is a fully automatic three speed or four-speed planetary gear with series connected fluid drive. There is no clutch pedal, and a selector lever replaces the gear-shift lever. All forward speed gear shifts are made automatically within the corresponding driving stages without interrupting traction dependent upon speed and engine moment. The automatic gear gear shift can be over-modulated by the kick-down of gas pedal or by manual gear shift with the selector lever. Selectable are three forward driving ranges, idle running position, reverse gear and parking position.

- Automatic gear drives maintain the engine in favourable, economic speed range.
- The gear shift for driving range is done automatically without interruption of traction and without gear shift intervals.
- Clutching is automatic and soft, without the stalling of engine during starting and stopping.
- Protects the engine drive elements, brakes and tyres by the soft elastic flux of force.

Selector gear shift lever positions:

P Parking and starting position. Driven shaft mechanically blocked, to select in the standstill state; blocked above 10 km/h.

R Reverse speed gear to select in standstill, blocked above 10 km/h

N Neutral and starting position no closed linkage, the vehicle can be rolled; it can be towed back and forth.

D Continuous drive stage shifts all forward speeds automatically; meant for city and overland drives.

S Super performance for gradients.

drive for mean gradients; third speed gear is blocked.

L Load; slow (low) for drives on steep passes. 2nd and 3rd gears are blocked.

Flux of force

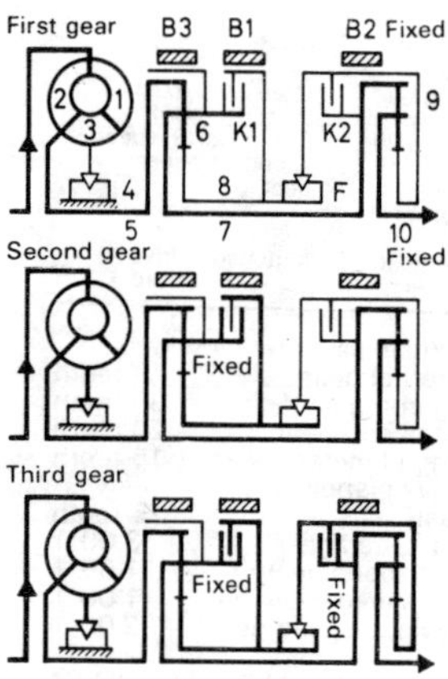

Pump wheel 1, turbine wheel 2, leading wheel 3, converter's free wheel 4, drive shaft 5, front planet set 6, intermediate shaft 7, hollow shaft 8, rear planet set 9, driven shaft 10, brake band B1, B2, B3, clutch K1, K2; free wheel F.

The pump wheel driven by the engine sets the gear fluid in circulation and pumps it into the turbine wheel through centrifugal force. The turbine wheel starts moving and drives the drive shaft. Through the current deflection in the leading wheel the torque on the turbine wheel gets intensified, and the speed is reduced.

The drive shaft, intermediate shaft or the hollow shaft drives the planetary gear set. To transmit a torque on the one hand the internal gear, the carrier or the sun wheel is held firmly, or on the other hand, two wheel parts are coupled together; the wheel set is blocked. Firmly holding a wheel part provides a transmission. By blocking the wheel set, the drive takes place in the direct gear.

If the carrier is held firmly, then the planetary wheels rotate around their own axis. Then the internal gear drives the sun gear in the opposite direction to provide the reverse gear.

Passage of force	▼	▼	▼	▼
Speed gear	1st	2nd	3rd	Rev.
front planet set		+	+	
internal gear	+	‖	‖	+
sun gear	fixed	‖	‖	+
carrier	+			fixed
intermediate shaft	+	+	+	
hollow shaft				+
free wheel				+
rear planet set			+	
internal gear	+	+		fixed
sun gear	fixed	fixed	‖	+
carrier	+	+	‖	+

Transmission ratios

Converter in starting point	2.5:1
Planetary gear	
1st speed gear	2.31:1
2nd speed gear	1.46:1
3rd speed gear	1.00:1
Reverse gear	1.84:1

Hydraulic system

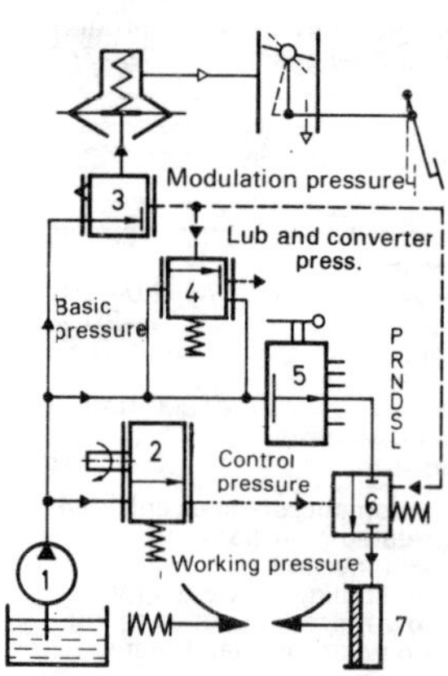

Parts: oil pump 1, regulating valve 2, throttle valve 3, working pressure valve 4, speed selector sliding valve 5, shifting valve 6, working piston 7.

The gear shift elements of the planetary drive are actuated hydraulically. Oil pumps provide the system with basic pressure. The primary pump is driven by the engine and the secondary pump by the driven shaft. The converter oil is routed through a cooler.

A centrifugal controller moved by the driven wheel supplies the control pressure dependent upon the driving speed. A diaphragm control valve connected with the suction pipe of the engine or a throttle valve connected to the throttle flaps supplies the load dependent modulation pressure. The pressure level is determined by the through-flow cross-section.

Shifting valves take over the hydraulic control. The most important shifting valve—the commando slider—is influenced by the control and modulation pressures. Both pressures work against each other and slide the shifting valve corresponding to the pressure fall in the direction of high shifting (control pressure) or reverse shifting (modulation pressure).

Drive range	
Position R	4.60—1.84:1
Position D	5.80—1.00:1
Position S	5.80—1.46:1
Position L	5.80—2.31:1

Actuated gear shift elements:

Speed gear	B1	K1	B2	K2	B3	F
1st	+		+			
2nd		+	+			
3rd		+		+		
Reverse					+	+

Drive pedal position

idling: early high gear shift
full gas: medium high gear shift
excess gas: late high gear shift

Gear shift points:

Selector lever position D		▲ km/h	▼ km/h
idling gas	1-2-1	30	19
	2-3-2	45	35
full gas	1-2-2	68	19
	2-3-2	138	60
excess gas	1-2-1	68	51
	2-3-2	138	123

Selector lever position S			
idling gas	1-2-1	34	23
full gas	1-2-1	77	27
excess gas	1-2-1	77	62

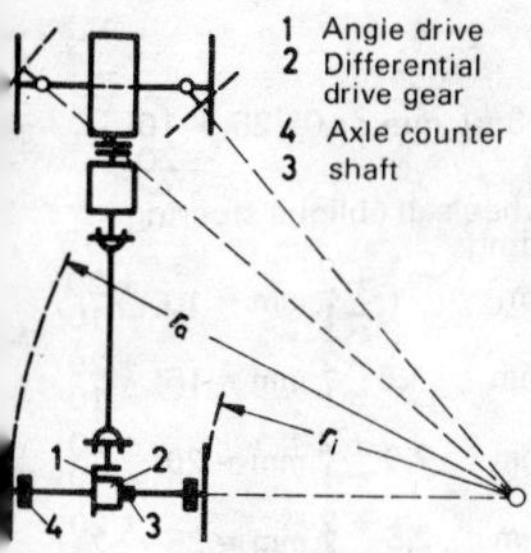

1. Angle drive
   a) bevel gear drive
      drive shaft in the middle position drive shaft in the deep position
   b) worm drive
2. Diffential gear
   a) bevel gear differential
   b) spur gear diffential
   c) worm gear differential
   d) differential gear with fulcrum plates.
3. Differential lock
   a) external actuation with dog clutch
   b) self-lock through friction disc, free wheel, worm gear and fulcrum plate
4. Axle counter shaft
   a) between angle drive and differential
   b) on the wheel hub

1. Speed of the engine reduced by fixed transmission, and thereby the torque is increased.
2. Director of torque passage changed by the angle drive.
3. Torque to be distributed equally on both wheels
4. Different speed of wheels in curves to be equalized.

Bevel gear drive

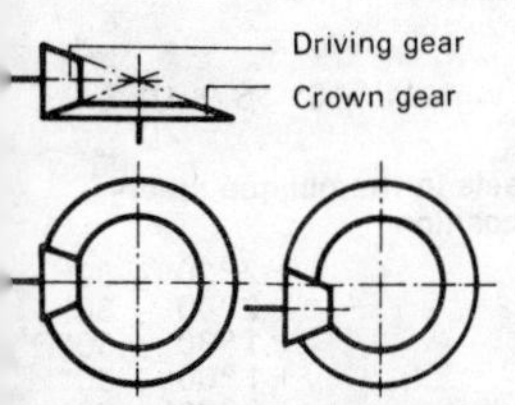

Palloid gearing Hypoid gearing

Longitudinally located engines need transversely lying axle shaft angle gear for purposes of drive. They are mostly built as bevel gear mechanisms with fixed transmission.

The bevel gears have curved teeth according to Gleason or Klingelnberg. In many passenger cars Gleason hypoid gear teeth with low located universal shaft are employed.

Hypoid gear teeth: low universal shaft tunnel, low centre of gravity of the car, good driving characteristic, longer teeth engagement, silent running and longer transmission. Thrust between the gear teeth makes the formation of a cohesive oil film difficult; therefore, more adhesive oil is necessary.

Gleason or Klingelnberg curved teeth.

Gleason palloid gearing
Gleason hypoid gearing

Number of teeth	
Pinion	8—12
Crown wheel	35—49
Transmissions	
Passenger car	2.8—5.12:1
Truck	6—8:1
Back lash of gear	0.1—0.2 mm
Oil filling	1—2.5:1

Universal shaft in hypoid drive is mostly 25.4 mm outside the centre.

Bevel gear differential mechanism

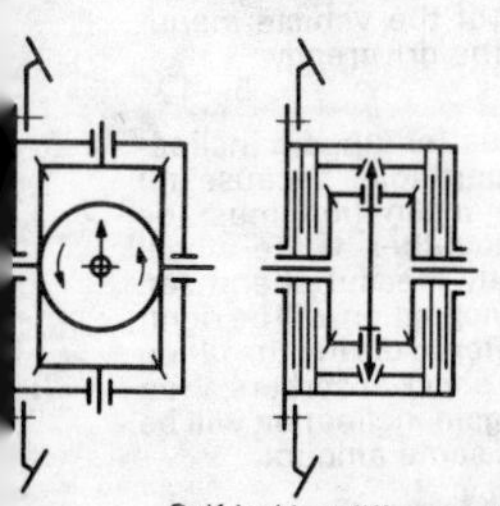

Self locking differential

The differential gear should equalize the path difference of wheels in driving around curves and distribute the torque on the 2 sides.

The differential wheel acts as two-sided, equal-arm lever which exerts equal peripheral forces on both axle shaft wheels, but permits different speeds through rotation. The speed difference of both wheels vis-a vis the housing are equal but opposite.

In the diverse linkage of wheels only one torque can be exerted which corresponds to the least friction. In many vehicles, therefore, a differential lock is necessary.

Straight teeth	
number of teeth	10—16
diameter of differential pin	15 mm

Friction discs for differential lock with sinter coating on one or both sides.

n_a outer wheel speed
n_i inner wheel speed
n_u speed difference
n_m crown wheel speed

$n_a = n_m + n_u$
$n_i = n_m - n_u$

Axle counter shaft

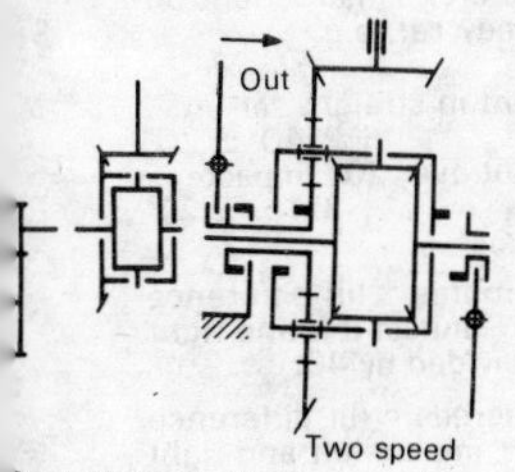

Spur gear hub drive Two speed rear axle

Employed in pay load vehicles:

1. to increase fixed transmission
2. to maintain the angle drive small
3. to obtain an additional gear shift stage, i.e. variable transmission.

It is laid out as spur gear countershaft or planetary gear counter shaft 'before' or 'after' the differential. In the first case, differential and axle shaft are loaded with full torque. In assembling after the differential, for each wheel a counter shaft is required, which is mostly located in the wheel hub. Differential and axle shaft are stressed with smaller torque; non spring-loaded masses are enlarged.

DB two-speed rear axle with differential lock (double gear shift axle).

Number of teeth of bevel gear drive (input axle)	
bevel gear	6
crown wheel	43
Transmissions:	
input axle	7.17:1
planetary gear	1.4:1
two-speed axle	10:1

Gear shift with compressed air

shift pressure ≈ 4 bar

shift lever on steering column
shifting cylinder on the rear axle.

Wheel Position

Track

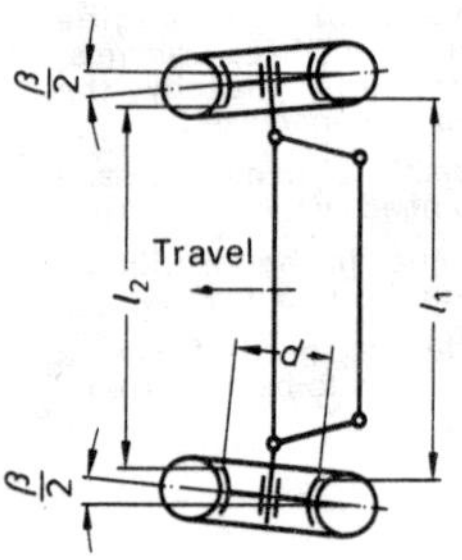

Oblique position of the wheel plane to the driving direction or difference between the wheel flange distance of front and rear at height of the wheel centre.

There is a positive track (toe-in) and a negative track (retrack). Generally, vehicles with rear wheel drive have a positive track and those with front wheel drive have a negative track.

The track should:

- set the wheels parallel during driving
- eliminate play in the joint
- improve track holding
- reduce wobbling of wheels
- equalize lateral force of tilt

Defective setting generates greater oblique running, more intense tyre wear and increased road resistance. Track changes during the assembly of telescopic shock absorbers can lead to self-steering.

Track setting to be done as per the instructions of the vehicle manufacturer on the drive-ready car, e.g.

the front wheels, rolled 3 ± 1 mm or $0° 25^{+10'}_{-20'}$

the rear wheels in oblique steering wheel setting:

0 ... 35 mm	1^{+2}_{-1} mm	$= 10'^{+20'}_{-10'}$
30 ... 50 mm	$1{,}5^{+2}_{-1}$ mm	$= 15'^{+20'}_{-10'}$
50 ... 60 mm	$2{,}0^{+2}_{-1}$ mm	$= 20'^{+20}_{-10'}$
60 ... 70 mm	$2{,}5^{+2}_{-1}$ mm	$= 25'^{+20'}_{-10'}$
70 ... 80 mm	$3{,}0^{+2}_{-1}$ mm	$= 30'^{+20'}_{-10'}$

Tilt

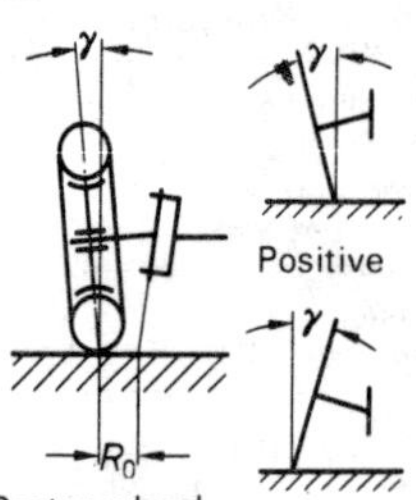

Castor wheel radius

Oblique position of the wheel plane to the vertical, transverse to the vehicle's longitudinal axis.

There is a positive tilt and a negative tilt. Generally, the front wheels have a positive tilt and the rear wheels have a negative tilt.

The tilt should:

- reduce the caster wheel radius
- improve the lateral guidance in driving along curves.
- remove the load on axle nut and eliminate wheel bearing clearance
- reduce wobbling of wheels

The tilt causes a component force through the oblique running wheels. Tilt alteration through steering impact produces gyroscopic couple, wobbling of wheels and higher tyre wear.

Tilt setting to be done as per the instructions of the manufacturer on the drive-ready vehicle.

e.g.

tilt of front wheels $+0° 15^{+10'}_{-20'}$

of rear wheels in the oblique steering wheel position

+ 80 mm	+ 2°30′ ± 30′
+ 70 mm	+ 2°00′ ± 30′
+ 60 mm	+ 1°30′ ± 30′
+ 50 mm	+ 1°00′ ± 30′
+ 40 mm	+ 0°30′ ± 30′
+ 30 mm	0°00′ ± 30′
+ 20 mm	− 0°30′ ± 30′
+ 10 mm	− 1°00′ ± 30′
0 mm	− 1°30′ ± 30′
− 10 mm	− 2°00′ ± 30′

King-pin inclination

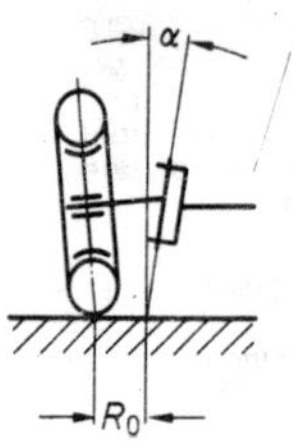

Caster wheel radius

Oblique position of the steering swivel bolt or the axis of rotation of steering to the vertical in the direction of transverse axis of the vehicle.

The king-pin inclination should:

- reduce the radius of caster wheel
- automatically reset the steering after curvacious drive by the weight of the vehicle.
- reduce the wobbling of wheels.

The king-pin inclination lifts the vehicle, during steering impact, uniformally on both sides because the circle described by the wheel up point does not lie in the drive path plane. Tilt and king-pin inclination are dependent upon each other; therefore, they are applied together.

King-pin inclination as per the instructions of the vehicle manufacturer on the driveready car 5—12°

Setting values for kingpin inclination are superfluous because tilt and kingpin inclination must be considered together. If the wheel tilt is correctly measured and set, the kingpin inclination will be right. If the tilt is altered during the filling of telescopic shock absorbers, then even the kingpin inclination will be upset by the same amount.

Caster

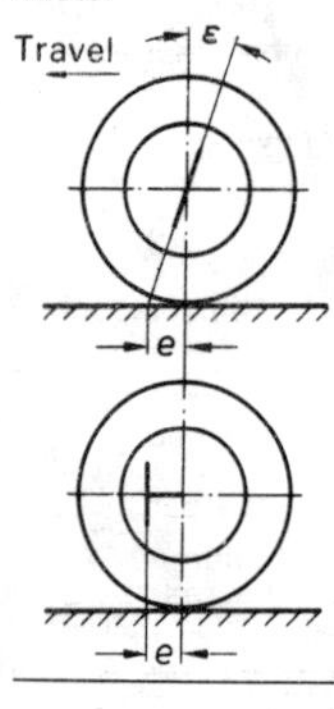

Oblique position of steering swivel bolt to the vertical in the longitudinal axis of the vehicle or the distance from the wheel up face centre to the through-impact point of the elongated axis of rotation of steering on the ground.

There is a positive caster and a negative caster. The negative caster is present in a few vehicles with front wheel drive and city omnibuses.

The caster should:

- stabilize track-holding
- reset the steering after a curvacious drive by the generated lateral forces
- reduce the wobbling of wheels.

Caster setting is to be made as per the manufacturer's instructions on the drive-ready car, e.g.

Measurement in straight set-up
Caster + 2°40′ ± 20′
Measurement over 20° impact
Caster + 2°15′ ± 20′

Caster in minutes = tilt difference between 20° impact left and right in minutes divided by 40.

Caster in degrees = tilt difference between 20° impact left and right in degrees multiplied by 1.5.

Caster wheel radius

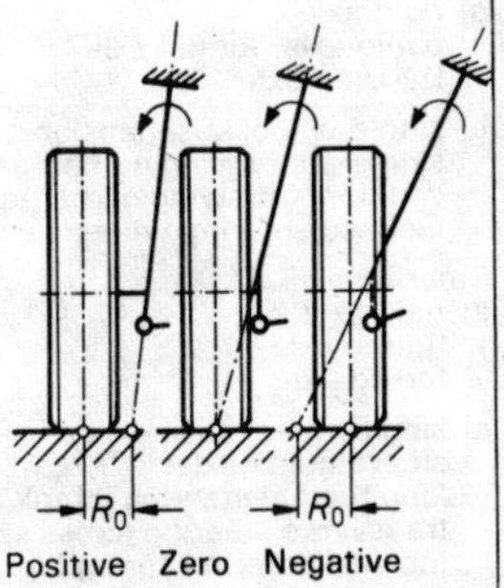

Distance from the centre of wheel-up face to the through impact point of the extended axis of rotation of the steering also called the Kingpin inclination axis—on the ground.

There is a positive caster wheel radius, a negative caster wheel radius and a zero caster wheel radius.

The caster wheel radius should:
- reduce the required steering force
- relieve the steering parts and the steering trapeze from hindrances while overtaking
- improve track-holding in brakes pulling askew.

On braking, a moment of force is generated around the kingpin inclination axis which is dependent upon the magnitude of the caster wheel radius. The smaller the caster wheel radius, the better is the tracking behaviour in one-sided pulling brakes.

$M_b = Fb$ braking force level arm
$M_m = m \cdot g \cdot$ track width/2

R_o positive:	$M = M_m + M_B$
R_g zero:	$M = M_m$
R_o negative:	$M = M_m - M_B$

Caster wheel radius	
rear wheel drive	30 ... 70 mm
front wheel drive	10 ... 35 mm

M_B Moment of the braking force
M_m Moment of the vehicle mass
F_g Braking force difference
m Vehicle mass
g Acceleration due to gravity
M Effective moment
R_O Caster wheel radius

Track difference angle

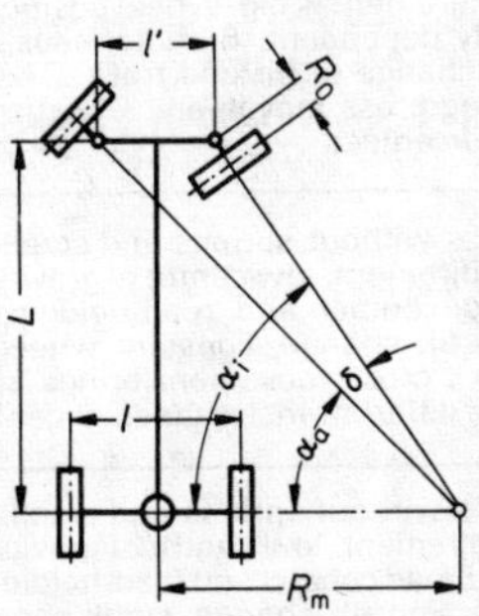

Track at 20° impact, or angular difference between the wheel in the interior of the curve and that on the exterior of the curve at 20° impact of the wheel in the interior of the curve.

The track difference angle is designed in such a manner that
- both the extensions of the wheel hub intersect on the extended rear axle at each steering impact
- the wheels roll impeccably in curves
- the track circles of the front wheels and rear wheels have a common centre during steering impact.

Measurement of the track difference angle at 20° impact of the inner wheel of the curve. The track value contained in the measurement must be taken into consideration.

Track difference angle
Impact angle of the inner wheel of the curve α_i—impact angle of the outer wheel of the curve α_a

$$\cot \alpha_i = \cot \alpha_a - \frac{f}{L}$$

$$\cot \alpha_a = \cot \alpha_i + \frac{f}{L}$$

f = track width — 2 R_o
L = wheel base

Setting as per the instructions of the vehicle manufacturer in the drive-ready vehicle, and 20° impact of the inner wheel of the curve about -0°30′ ± 40′

Steering trapeze

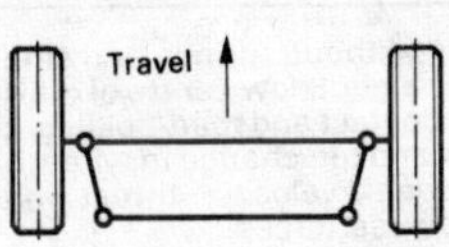

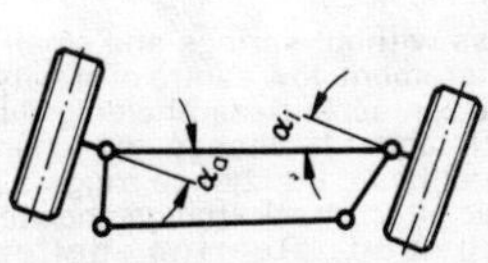

In the kingpin steering assembly, the track rod and both the track rod levers form a trapezoid with the front axle.

The steering trapeze should
- support the wheels during the drive
- set the track, if required
- transmit the steering impact to both wheels
- guarantee the track difference angle, i.e., more strongly impact the inner wheel of the curve than the outer wheel of the curve.

In the calculation of the steering trapeze, the varying track of wheels in the straight position must be considered.

double the rated value = double the actual value

track 20° left impact
+ track 20° right impact
± double the track value track 0

= double the actual value

left impact = right impact

Tolerances:	
double the actual value	± 30′
left impact, right impact	∓ 40′

Errors on the steering trapeze

a)	double the rated value = double the actual value		The track arm is not bent. The differences between the left impact and right impact are caused by an oblique position of track rod central arm or by the steering gear arm.		Right impact smaller: arm is oblique towards right Left impact smaller: arm is oblique towards left
	actual left impact ≠ valve right impact		The track arm is bent outwards. If the track rod lies before the axle, then the arm is bent inward.		Right impact smaller: error on the right side
			The track arm is bent inward. If the track rod lies before the axle, then the track arm is bent outward.		Left impact smaller: error on the left side
b)	double the actual value < double the rated value				Right impact bigger: error on the right side
c)	double the actual value > double the rated value				Left impact bigger: error on the left side

Wheel Suspension

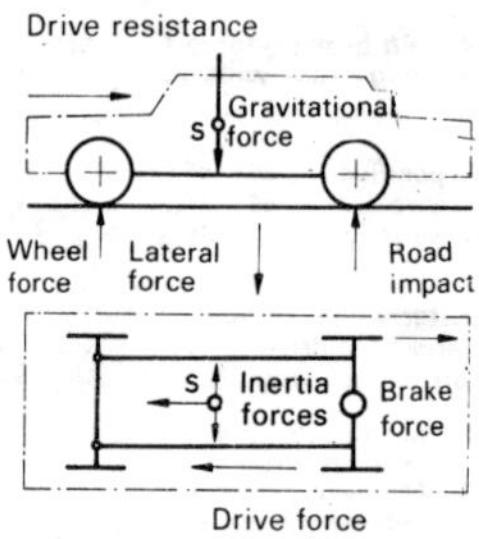	1. Vehicle weight as well as component forces and forces of inertia to be suitably distributed over the wheels. 2. To transmit peripheral forces and component forces of wheels to the chassis. 3. Centre of gravity and instantaneous centre to be located low and provide the vehicle with secure hold on roads and curves. 4. Mass without springs to be maintained small, and to limit the drive path impact on the impacted wheel. 5. Road uneveness to be neutralized without lifting the wheels and without distortion. 6. In making the steering wheels resilient, the wheel position should not be affected.	1. Drive axle a) rigid axle banjo axle, funnel axle De-Dion axle b) single wheel susphension single-joint swinging half-axle double-joint swinging half-axle inclined steering wheel 2. Steering axle a) rigid axle thumb axle forked axle b) single wheel suspension sleeve guide axle longitudinal steering (crank axl transverse steering (cross sprin inclined steering spring leg
Rigid drive axle 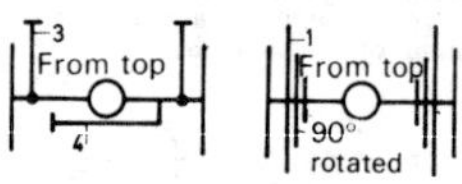	Rigid axle with 1. longitudinal spring 2. 1transverse spring (floating axle) 3. pull rod 4. control arm 5. ball joint or universal joint	Mass without springs and spa requirement is big. Wheel position m ually dependent. Simultaneous lifti No change in tilt and track. One-sid lifting cross movement. Unsatisfacto road contact.
Swinging half-axle for drive axle 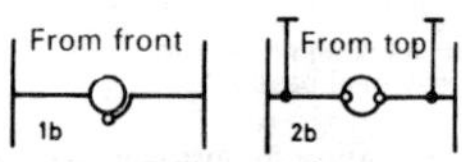	1. Single-joint swinging half-axle a) point of rotation in the centre b) low located point of rotation 2. Double-joint swinging half-axle a) without pull rod b) with pull rod	Mass without springs and small spa requirement, low centre of gravity; go road contact and road-holding; tra and tilt change; constant wheel bas tyre's cross movement brings inclin run; different instantaneous centre.
Inclined steering for drive axle 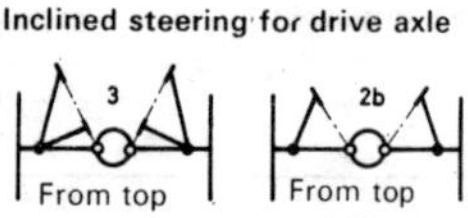	1. Triangular inclined steering 2. Swinging half-axle with a) parallel or b) inclined pull rod 3. Control arm with inclined axis of rotation	Mass without springs and small spa requirement, low centre of gravity; ve good road contact and road-holding; lo track and tilt change; small change wheel base; absorption of longitudin and transverse forces.
Pull rod (crank axle) 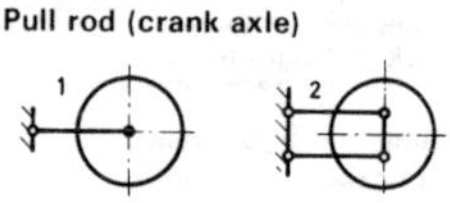	1. Simple pull rod 2. Double pull rod (pure crank axle)	Mass without springs and small spa requirement; low centre of gravity; go road contact and road-holding; consta track and tilt; change in wheel base a rotational velocity; thrust intake; r transverse forces.
Control arm 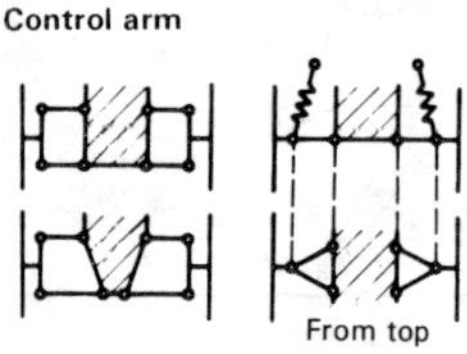	1. Simple triangular control arm a) with spring leg (Mc Pherson) b) with cross spring 2. Double triangular control arm a) steering's axis of rotation parallel or limited (anti-dive effect) b) parallelogram or trapezoidal form.	Mass without springs and small spa requirement; low centre of gravity; go road contact and road-holding; differe track and tilt change; no change wheel base; tilt change triggers gyro copic effect; track change, inclined ru ning and steering deflectio impeccable steering if the divided tra rod has identical arm lengths; anti-di effect causes inclined resilience.
Transverse spring 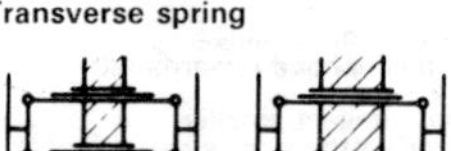	1. Simple transverse spring 2. Double transverse spring 3. Transverse spring with simple control arm	Mass without springs is small; som what bigger space requirement; simp assembly; good road contact and roa holding; low track and tilt change; co stant tilt in double transverse spring, r change in wheel base.
Inclined steering for steering axle 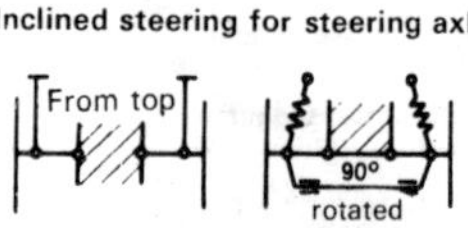	1. Control arm with pull rod 2. Control arm with pull rod and spring leg (Mc Pherson)	Mass without springs and small spa requirement; low centre of gravity; ve good road contact and road-holding; lo change in track and tilt; low wheel ba alteration; good intake of longitudin and transverse forces.

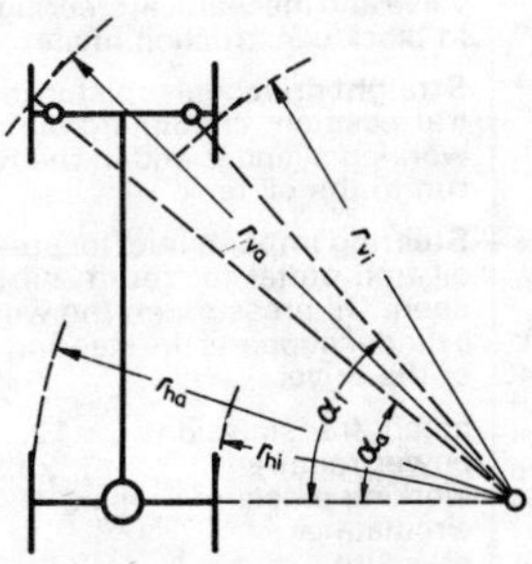	1. To influence direction of travel 2. To arrest and damp drive path impacts (jarring blows) 1. According to the axle design a) fifth-wheel steering (trailer) b) kingpin steering assembly for rigid axle and individual wheel suspension 2. According to the number of steered wheels a) single wheel steering (tractor) b) front-wheel steering (cars, lorries) c) rear-wheel steering (construction machinery) d) all-wheel steering (cross-country vehicles)	1. Steering wheel 2. Steering column 3. Steering gear a) screw and nut steering b) worm and sector steering c) rack and pinion steering d) servo steering 4. Steering gear arm 5. Track rod a) through-going b) divided in two or three 6. Steering track rod 7. Steering arm 8. Intermediate arm 9. Steering damper or back-stroke safety
Steering gear 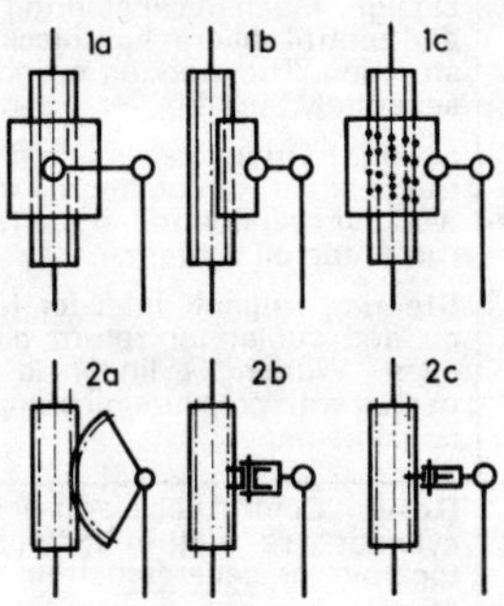	1. Screw and nut steering a) steering nut b) nut segment, screw steering gear, can be executed as double steering c) recirculating ball screw steering 2. Worm and sector steering with uniform or changing pitch a) worm wheel or steering worm sector b) caster wheel (Gemmer steering) with double or three-teeth wheel and globoid worm. c) steering finger (ZF-horse steering, with fixed or roller-supported finger 3. Rack and pinion steering 4. Steering gear for crawler-type vehicles a) differential steering gear b) clutch and brake steering c) recirculating ballscrew gear mechanism	Transmission 12. . . 22:1 Screw and nut steering: insensitive, linear transmission, not back-acting, steering nut cannot be reset, greater friction, therefore recirculating ballscrew steering. Worm and sector steering: permits progressive transmission, in straight-stretch driving the steering is very steady; faster steering impact in curves, cannot be reset, small friction. Rack and pinion steering: simple, back-acting, smaller transmission
Steering linkage 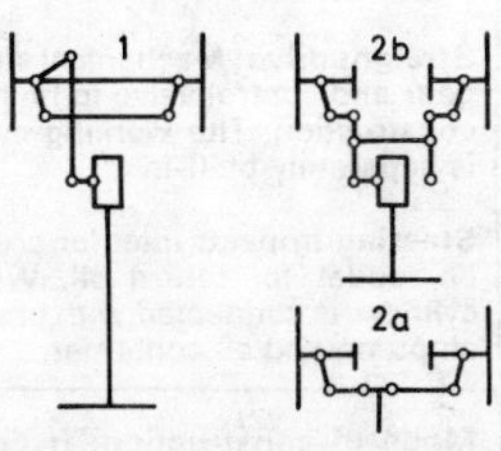	The wheels should roll at each steering impact, and they should not slide. Resilient wheels should not put any torsional force on the steering. The oblique running angle should be matched to the load of the wheels in the curve. The construction type, therefore, depends on the axle design, steering geometry and dynamic behaviour of steering. 1. In rigid axles, elastic track rod 2. In single wheel suspension a) two-piece track rod b) three-piece track rod	Steering to be free of clearance, as far as possible; the average 10°, maximum 30° steering wheel turning Dead travel: clearance of steering or sum of individual clearances of all steering parts. Length of track rods equal to the length of control arm, as far as possible.
Steering damper 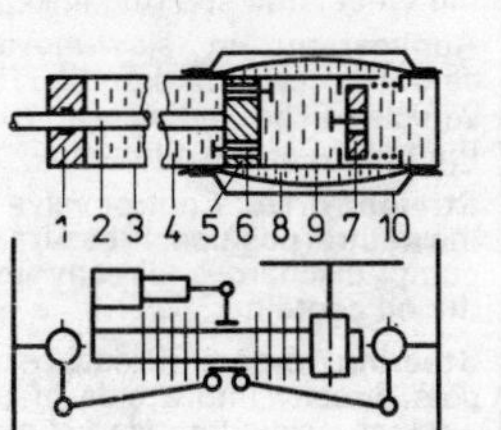	1. guide 6. piston valve 2. piston rod 7. bottom valve 3. cylinder 8. equalizer-chamber 4. work chamber 9. plastic sleeve 5. piston 10. tubular shell In the **traction stage**: the piston rod is taken out of the cylinder. Oil in the work chamber flows through the piston valve and damps the movement. The oil is resucked through the bottom valve from the equaliser chamber. In the **pressure stage** the damper is pressed together. The oil, driven out by the piston rod, flows through the bottom valve into the equalizer chamber and expands the plastic sleeve.	**Tasks** • to eliminate imbalance in steering • to hinder the wobbling of wheels • to protect the wheel suspension from wear **Type of Construction** Single pipe damper with va[illegible]ble equalizer chamber. **Damping** • piston valve for [illegible]ction stage • bottom valve fo[illegible] [illegible]ressure stage. **Damping Force** traction and [illegible]ssure stages are equal.

Ball nut hydrosteering

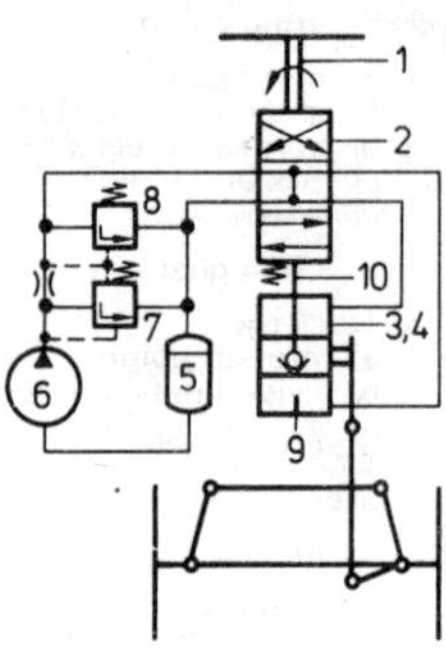

1. steering column, steering wheel
2. control valve, valve piston
3. steering gear, mechanical
4. working cylinder, working piston
5. oil container with oil filter
6. pressure oil pump
7. current limiting valve
8. pressure limiting valve
9. steering limiting valve
10. torsion bar

Steering worm and steering column are connected together by a torsion bar. On actuating the steering wheel a related rotation of both parts takes place, causing the adjustment of the valve pistons which conduct the oil into the working cylinder. The force generated thereby supports the turning movement on the steering wheel. After releasing, the valve pistons are guided back into the neutral position by the torsion bar.

Design: Working cylinder, control valve and mechanical steering gear in block construction mode.

Straight drive: Valve piston in neutral position, circulating oil in the working cylinder and in the return run to the oil tank.

Steering impact: Inlet for pressure oil and outlet for return oil to be open. Oil pressure on the working cylinder supports the steering work of the driver.

mech. transmission	12...15:1
power required	2...3 kW
working pressure	max. 100 bar
circulating pressure	2,5...3 bar
fluid	ATF

(automatic transmission fluid)

Half-block hydro-steering

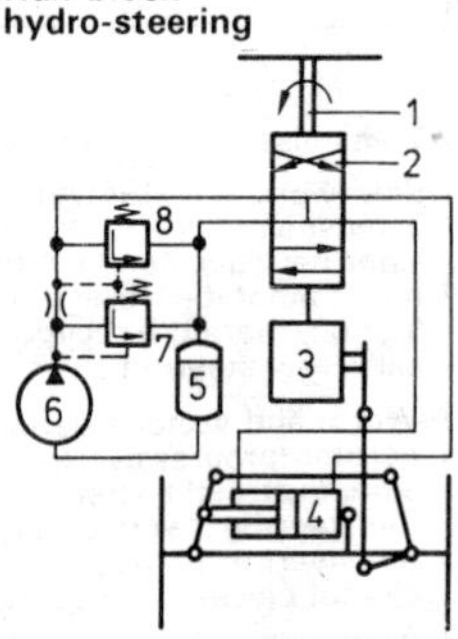

1. steering column, steering wheel
2. control valve, valve piston
3. steering gear, mechanical
4. working cylinder, working piston
5. oil container with oil filter
6. pressure oil pump
7. flow current limiting valve
8. pressure limiting valve.

The control valve sits on the worm spindle and it is moved axially to-and-fro along with the worm spindle on turning the steering wheel, causing the adjustment of the guiding edges. This leads to the pressure oil to enter one side of the working cylinder from the pump; the return oil gets to the oil container from the other side of the working cylinder.

On releasing the steering wheel, the control valve is moved back into the neutral position by spring force.

Application: Payload vehicles of over 8 t steering-axle load with big steering forces and a high work volume.

Design: Mechanical steering gear and control valve in half-block construction. The working cylinder is separately built-in.

Straight Drive: Valve piston neutral position. Circulating oil in the working cylinder and in the return run to the oil container.

Steering Impact: Inlet for forced oil and sublet for return oil are open. Working cylinder is connected with pressure oil pump and oil container.

Hydraulic servo steering unit

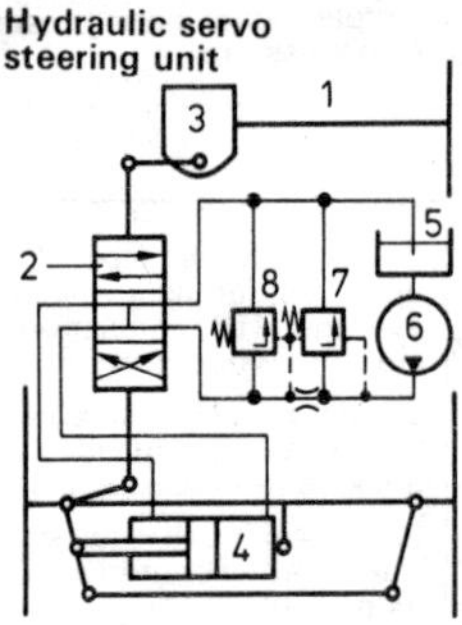

1. Steering column, steering wheel
2. Control valve, valve piston
3. Steering gear, mechanical
4. Working cylinder, working piston
5. Oil container with oil filter
6. Pressure oil pump
7. Flow current limiting valve
8. Pressure limiting valve.

The drag link moves the control valve during the steering impact, or during the road impact, it moves the valve housing axially to-and-fro. Thereby the corresponding control parts are opened or closed so that the pressure oil is conducted to one side of the working cylinder. The force generated in the working piston supports thereby the steering movement or it acts against the drive path impact.

Design: Control valve and working cylinder are built-in-additionally, together or separated from each other.

Assembling the control valve
- in the divided drag link
- between the shortened drag link and wheel steering arm.

Straight drive: Mechanical steering gear and control valve in half-block construction. The working cylinder is separately built-in

Steering impact: Inlet for pressure oil, outlet for return oil. Working cylinder is connected with pressure oil pump and oil container.

Hydrostatic servo steering unit

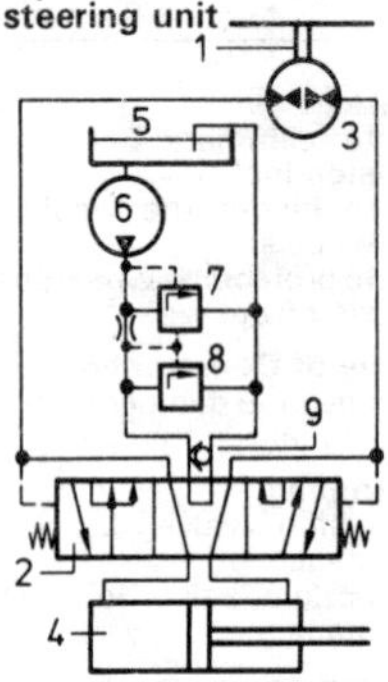

1. Steering column, steering wheel
2. Control valve, valve piston
3. Hand pump, planetary gear, rotors
4. Working cylinder
5. Oil container with oil filter
6. Pressure oil pump
7. Flow current limiting valve
8. Pressure limiting valve.
9. Check valve

The steering wheel moves the piston valve and turns the rotors of the hand pump over a planetary gear. The hand pump discharges pressure oil of the engine-driven pump to one side of the working cylinder. The work volume is therein determined by angle of rotation on the steering wheel. The return of the working cylinder is directly connected with the oil container.

Mode of construction: Hydrostatic transmission of the steering force. No mechanical connection between the steering wheel and steering linkage.

Application: In slow-moving cars (V max<62 km/h). The vehicle can be steered even with the failure of the oil pump.

Straight drive: Control valve in the neutral position. Pressure oil pump discharges directly into the oil cantainer.

Steering impact: Pressure oil gets directly into a side of the working cylinder via control valve and hand pump. Discharge volume of the hand pump is 65—400 cm^3/r.

General

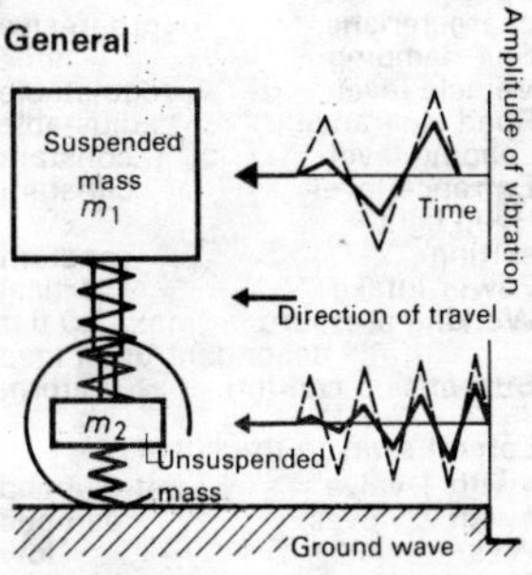

Tasks

Forces of the vehicle due to weight and inertia to be transmitted to the wheels; to absorb peripheral and component forces of the wheels. Dynamic forces due to gravity must be uniformly distributed.

Driving comfort: To arrest road impacts and convert them into soft vibrations; to quickly damp the spring movements; to reduce tilting vibrations and to equalize them and reset; to regulate and constantly maintain the vehicle level.

Driving safety: to suppress the jumping of the wheels from the road by spring prestressing; to guarantee ground-holding during drive, braking and steering in every driving state.

Types of springs
a) steel spring
leaf spring
helical spring
torsion-bar-spring
plate spring
b) rubber spring
c) air spring
d) hydro-pneumatic spring

Shock absorbers

a) double-pipe shock absorber
b) single-pipe shock absorber

Stabilizers

a) torsion-bar stabilizer
b) fluid stabilizer

Leaf spring

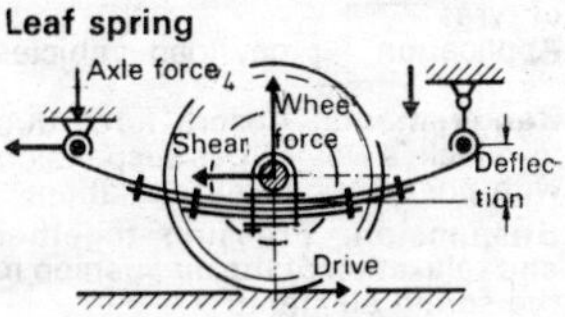

Types: Full, half or quarter elliptic spring, uniform or non-uniform arm, parabolic spring, spring end open or closed; bearing in or on rubber, longitudinal or transverse springs.

Components: Spring leaves, tensioning screw, spring U-bolts, spring fastening plate, spring clip, spring eye, spring seat (bronze or rubber), spring shackle bolt, shackle (sliding plate), spring bracket.

Torsion spring, known self-damping, spring characteristic mostly linear; absorption of transverse and longitudinal forces; low construction, possible to repair, heavy, joints necessary, needs maintenance, small spring deflection, simple manufacture.

Helical spring

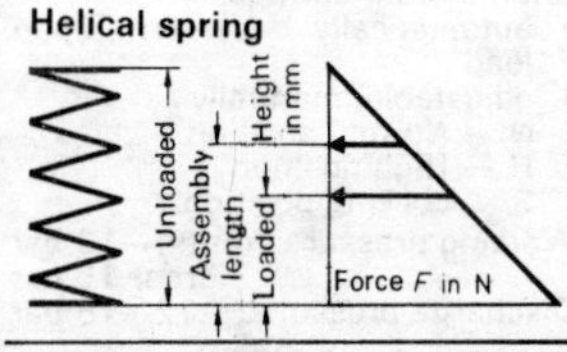

Types: Cylindrical helical springs with circular cross-section, conical spiral spring with circular or rectangular cross-section, springs with uniform or diverse pitch, mutually encased helical springs as spring set.

Components: Load-carrying coils, effective coils, spring plate, rubber bearing.

Torsion spring, spring characteristic mostly linear, partially progressive, stepped in spring set, light, no joints, maintenance-free, large spring deflection. No self-damping, does not transmit shear stress, higher construction, inter-changeable.

Torsion bar spring

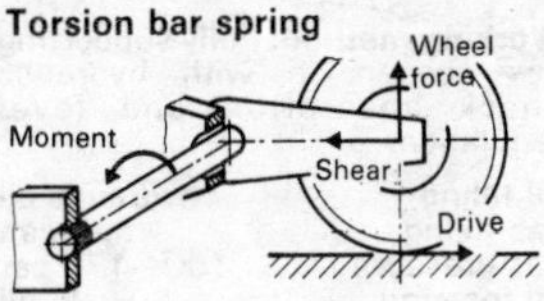

Types: Drilled tube, round or square individual rod, several rods of rectangular cross-section in a spring packet, height regulation possible by turning the clamping spot, arranged longitudinally or transversely. Components: torsion bar, clamping ends with notched or spline teeth, rocker arm or suspension arm, bearing and clamping spot.

Torsion spring, spring characteristic mostly linear, other spring characteristic is possible by arm arrangement or additional springs; light; maintenance-free, small space requirement, transmits wheel forces.

No self-damping, surface sensitive, interchangeable.

Rubber spring

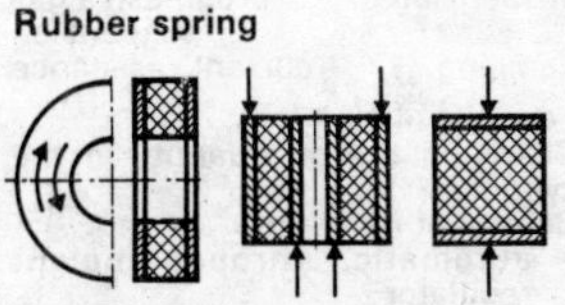

Types: tension, thrust or bolt spring; sleeve type, hollow-type or disc spring; rubber shackle. Suitable for engine, gear and spring suspension; can be employed as additional spring on the front and rear axle.

Components: Rubber element alone or vulcanized on metallic plate (silent block). Metal plate facilitates better screw-joint.

Strong self damping; spring characteristic is progressive; damps noise; absorbs wheel forces; light; maintenance-free

Small spring deflection; high frequency; sensitive to temperature, oil and petrol (natural rubber); possible to repair by exchanging.

Hydro-elastic suspension

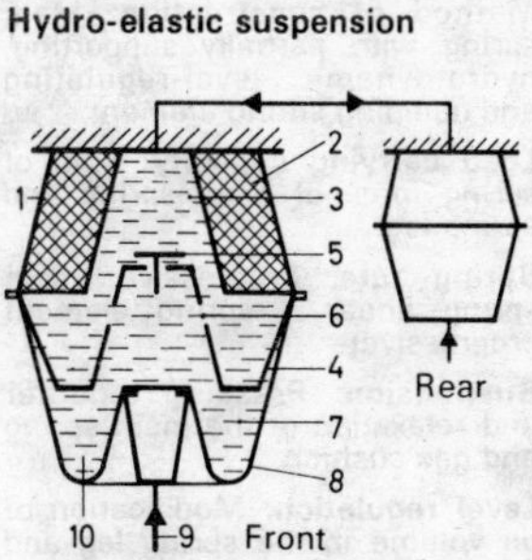

1. Case
2. Rubber thrust spring, progressive characteristic
3. Upper work chamber
4. Separating case
5. Damping valve
6. Damping hole
7. Lower work chamber
8. Rubber diaphragm
9. Taper with ram
10. Filling: 70% water, rest alcohol and anticorrosion agent, filling valve.

The front and rear spring elements are connected together longitudinally. In the suspension of a wheel, the fluid is driven out over a rubber valve into the spring element of the other wheel. That increases the work volume, causing the equalisation of inclinations of the front part or the rear part. The generated pitching vibrations and transverse inclinations are eliminated by the built-in stabilizers.

Requirements: Absorption of the gravitational force; rubber spring support; shock absorption; spring equalization between front and rear axles.

Suspension: Pressing together and releasing of rubber thrust springs.

Damping: By the release of spring stroke the damping valve is closed. The fluid is to flow through a small damping hole.

Spring equalization: During the spring stroke of a front wheel, the fluid flows through the open damping valve into the spring element of the corresponding rear wheel.

Air suspension
with open air circuit

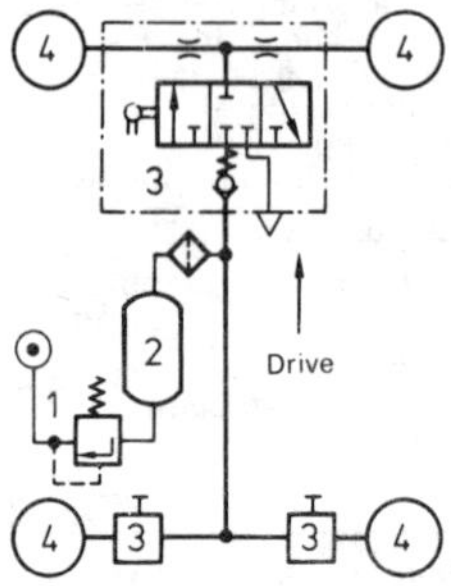

with semi-closed
air circuit

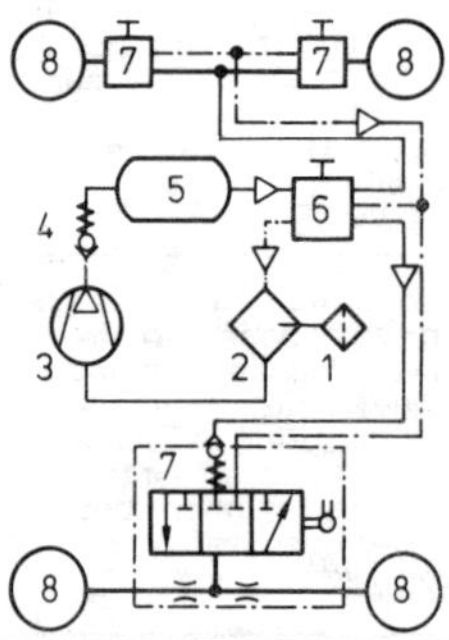

1. Overflow valve without return current
2. Auxiliary compressed air tank
3. Air spring valve (level control valve)
4. Air spring: air chamber, spring bellows, spring piston, possibly pilot valve for container operation.

The compressed air flows through the overflow valve into the air tank and from there to the pneumatic spring valve. In relation to the mechanical steering of the valve, the compressed air gets into the air springs. On reaching the specified operation level, the pneumatic spring valve closes.

On loading the vehicle, the body is lowered; the steered pneumatic valve opens and additional compressed air flows into the air tank. On unloading, the control is reversed, and air escapes from the pneumatic valve. The used compressed air comes out into the open.

1. Intake muffler
2. Antifrost protection
3. Air compressor (buffer operation)
4. Check valve
5. Air tank
6. Valve unit: safety valve, pressure-relief valve, pressure maintaining valve and pilot valve with pull-knob, electrical pressure indicator for warning lamp.
7. Level controlling valves: front 2, rear 1.
8. Pneumatic spring, air chamber, spring bellows and spring bellows and spring piston.

In air suspension with semi-closed air circuit, on unloading the vehicle, the air flowing out of air springs is led over valve-unit and anticorrosion device again into the air compressor. Thereby the discharge capacity of the air-compressor and the incidence of condensate can be held at a low level.

Spring characteristic	progresive
Self-damping	none
Vehicle level	regulatable
Road clearance	adjustable
Loading level	constant
Entrance level	constant
Head light setting	constant
Power intake	vertical
Working pressure	max. 7.0 bar dependent upon load
Suspension comfort	uniform
Lateral sway in the curve	
•with 1 valve	not damped
•with 2 valves	damped
Wear	low
Maintenance	inspection jobs
Closed linkage of tyres	good
Application	pay-load vehicles

Requirements: Absorb force due to vehicle's weight; gas suspension with pneumatic level regulation.

Suspension: Pressing together and relaxation of the air cushion in the spring elements.

Level regulation: Changing the gas volume of air springs.
- automatically dependent upon load
- adjustable, manually
  N — Normal position
  H — High position
  S — Locking position

Working pressure: front 4—10 bar
rear 16 bar
Discharge pressure 12—18 bar

Hydro-pneumatic suspension, fully supporting

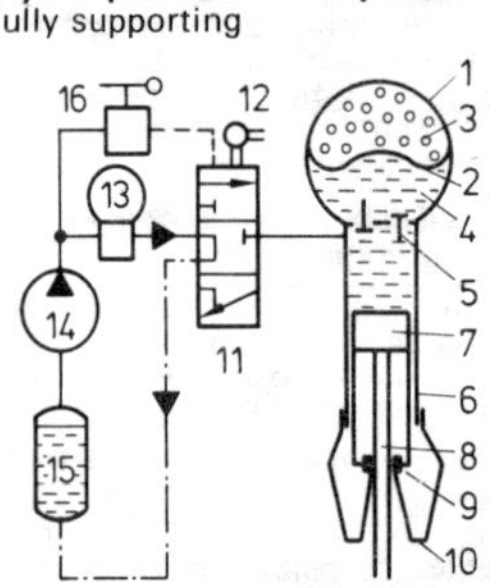

1. Ball casing
2. Rubber diaphragm
3. Gas chamber
4. Oil chamber
5. Damping valves
   - pressure valves
   - pull valve
6. Damping cylinder
7. Piston
8. Spring ram
9. Seal
10. Sealing sleeves
11. Height regulator
12. Linkage
13. Pressure reservoir
14. High pressure pump
15. Reserve tank
16. Adjusting lever

During the spring-stroke of the wheels, oil is pushed by the piston into the lower half of the sphere by the pressure valve. It leads to the compression of the prestressed nitrogen gas in the upper half of the sphere above the diaphragm. On stroke release the gas presses the fluid through the pull-valve set very hard back into the cylinder.

Working method: Fully supporting gas suspension with hydraulic shock absorption and level regulation.

oil filling	hydraulic oil
gas filling	nitrogen
gas pressure	150—175 bar
oil foaming	nil
space required	small
suspension	progressive gas suspension
damping	current resistance

Level regulation
Changing the oil quantity in the cylinder
Control of regulation
- automatic, through height regulator
- manual, through adjusting lever

partially supporting

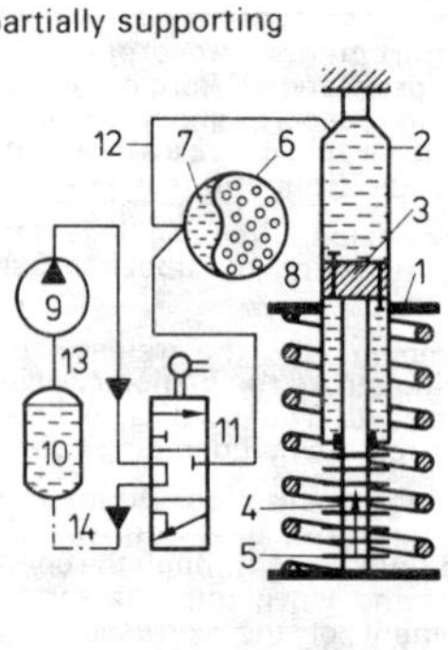

1. Helical spring
2. Spring leg
3. Damping piston
4. Piston rod
5. Rubber sleeve
6. Reservoir
7. Diaphragm
8. Gas chamber
9. Pressure oil pump
10. Reserve oil tank
11. Height regulator valve
12. Pressure line
13. Intake manifold
14. Return flow line

The hydro-pneumatic spring element is applied only in a partially supporting manner. It has a soft spring rate with the base load. The rate rises progressively with increasing load. The cross-section of the piston rod and the gas pressure in the reservoir are decisive for the spring force. The level of the gas pressure is related to the oil volume. In setting the level higher, the helical spring relaxes. The pressure off-take must be additionally absorbed by the spring leg.

Method of construction: Main spring with partially supporting, hydro-dynamic, level-regulating and damping spring element.

Load carrying capacity: sum of spring force of main spring and spring leg.

Spring rate: Progressive (main spring linear + spring element progressive)

Suspension: Pressing together and relaxation of the main spring and gas cushion.

Level regulation: Modification of oil volume in the spring leg and increasing the gas pressure.

Hydro-pneumatic suspension

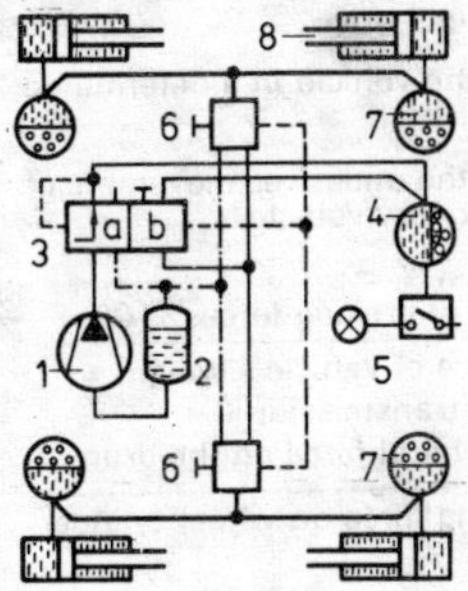

1. Pressure oil pump
2. Reserve oil tank
3. Valve unit
   a) pressure regulator (storage pressure is 200 bar)
   b) cut-out with pull knob
      N Normal position
      H High position
      S Locking position
      M Assembly position
4. Central reservoir
5. Electric push-button for warning lamp
6. Level regulating valve 1 each in front and rear
7. Spring storage, without oil filling 60—75 bar
8. Spring leg, working pressure 120—160 bar

The force of the vehicle's weight is absorbed by the four spring elements consisting of one spring leg and one spring storage each. The load carrying capacity is dependent upon the cross-section of the piston rod and the gas pressure.

Requirements: Absorption of the force of weight; gas suspension with hydraulic shock absorber and level regulator.

Power intake: $F = S \cdot p$

Suspension: Compression and relaxation of gas cushion in the spring storages.

Level regulation: Modification of oil volume in the spring legs
- automatic, load dependent
- manual, through pull knob

Shock absorption: Current resistance in both the piston valves of the spring leg during stroking and non-stroking of springs.

Double-pipe shock absorber

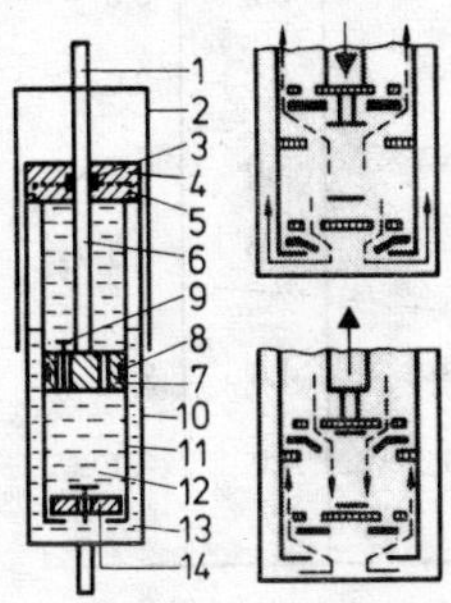

1. Suspension
2. Protection pipe
3. Bar sealing
4. Bar guide
5. Air drainage
6. Piston rod
7. Piston
8. Piston ring
9. Piston valve
10. Outer pipe
11. Inner pipe
12. Work chamber (full)
13. Reserve chamber (1/2 full)
14. Bottom valve

Push-stage: The piston pushes the oil from the bottom work chamber upwards through the open piston valve. The oil driven out by the descending piston rod flows through the bottom valve into the reserve chamber. The current resistance in the bottom valve damps the descending movement.

Full stage: The ascending piston sucks the driven out oil back, without hindrance, into the work chamber. Oil in the upper work chamber flows down via the piston valve. The resistance in the piston valve damps the descending movement.

Task: Vibrations spring supported and unsupported masses to decay quickly (body damping and wheel damping)

Control	automatic
Damping	double-acting
• push stage	in bottom valve
• pull stage	in piston valve
Damping ratio push stage : pull stage	≈ 1:3
Damping force dependent upon	
• the piston stroke	
• the piston speed	
• the current resistance	
Heat dissipation	indirect
Assembling position	piston rod on top
Foaming of oil	possible
Maintenance	inspection jobs

Single-pipe shock absorber, gas pressure shock absorber

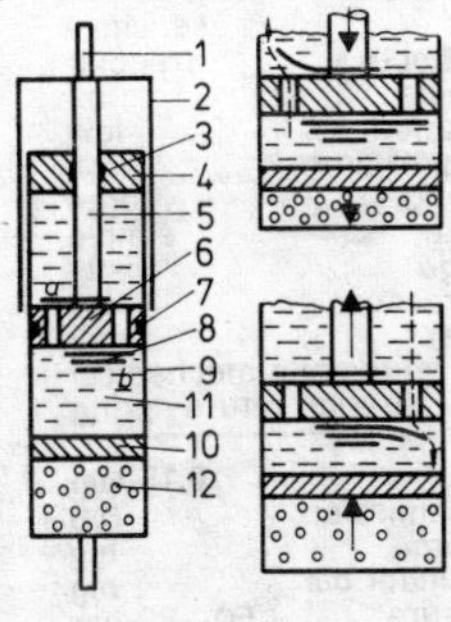

1. Suspension
2. Protective pipe
3. Rod sealing
4. Rod guide
5. Piston rod
6. Piston
7. Piston ring
8. Piston valve
   a. push valve
   b. pull valve
9. Cylinder tube
10. Separating piston
11. Oil chamber
12. Gas chamber

Push stage: The piston moves inwards. The oil from the bottom side to the top side of the piston. The valve on the top side takes over the damping. The volume of the moving-in piston rod reduces the gas chamber.

Pull stage: The piston moves outwards. The oil flows from the top side to the bottom side of the piston. The valve on the bottom side determines the damping. The volume of the moving-out piston rod releases the gas chamber.

Damping due to flow resistance in the piston valve.

Control	automatic
Damping	double acting
Push/pull	in the piston valve
Damping ratio push stage : pull stage	≈ 1:3
Damping force dependent upon	
• the piston stroke	
• the piston speed	
• the flow resistance	
Characteristic	progressive
Heat dissipation	direct
Oil foaming	nil
Gas pressure	20 to 30 bar
Assembly position	any

Force produced by rod = piston rod cross-sectional area × gas pressure

Level lift

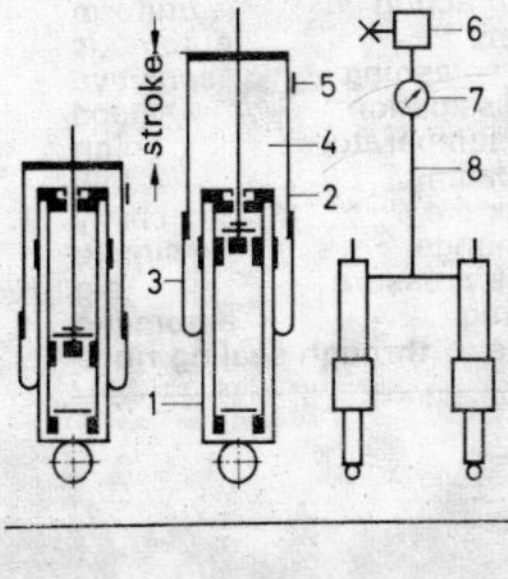

1. Shock absorber
2. Protection pipe
3. Rubber roller bellows
4. Spring chamber
5. Air connection
6. Filling valve
7. Manometer
8. Pressure line

Corresponding to safe loading, air is led in through a filling valve till the design level of the vehicle body is reached. The supporting force of the spring element is decided by the pressure in the interior of the spring chamber. After unloading the vehicle, the original level should be re-established by bleeding the compressed air. Vibrations of vehicle's suspension are braked by the shock absorber. The permissible pay-load of the vehicle should not be exceeded even after the assembly of the unit.

Tasks: To balance the level difference between empty and loaded vehicle and to damp the vibrations of vehicle suspension.

Shock absorber system for rear axles with steel spring support and suitable for frequently changing load. Subsequent assembly is possible.

Excess pressure minimum 1 bar, maximum 8 bar

Carrying capacity = filling pressure × cross-section of air spring chamber

Brake Unit

General

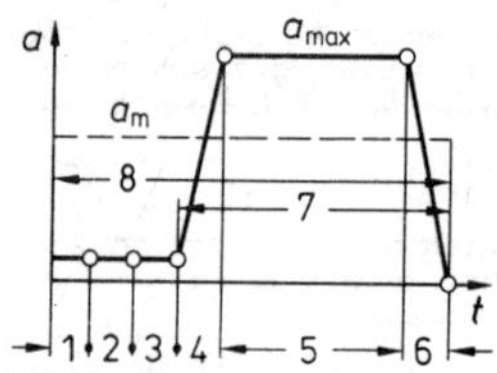

1. Chill time
2. Reaction time
3. Application time
4. Increasing time
5. Full braking time
6. Releasing time
7. Braking time
8. Stopping time

Types

according to function:
service brake
hand brake
auxiliary brake
permanent brake

according to actuation:
muscle force brake
- mech. brake
- hydr. brake

auxiliary force brake
- mech. supporting force
- with braking device

external force brake
- compressed air brake
- suction air brake

according to braking force:
friction brake
engine brake
eddy-current brake
fluid current brake

according to braking circuits:
single-circuit brake
double-circuit brake

according to construction:
- internal jaws
- external jaws

disc brake
- full disc
- partial disc

Tasks:
to decelerate speed
to stop unwanted acceleration in down-hill drive

to stop the vehicle in a determined place

to avoid the undesired movement of the stationary vehicle

$$\text{Braking in \%} = \frac{\text{sum of braking forces} \times 100}{\text{force of vehicle's weight}}$$

$$\text{Internal transmission } C = \frac{\text{peripheral force on the drum}}{\text{clamping force on wheel cylinder}}$$

Wheel brakes

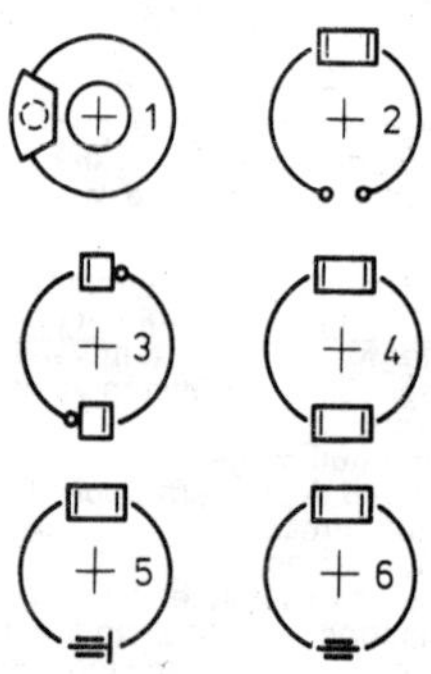

Types	Wheel cylinder		Brake jaws				Moving stay bolts		Internal transmission	
	number	action	forward		backward		forward	backward	forward	backward
1. Disc brake	2	single	–	–	–	–	–	–	0,8	0,8
2. Simplex brake	1	double	1	1	1	1	–	–	2,0	2,0
3. Duplex brake	2	single	2	–	–	2	–	–	3,0	0,9
4. Duo-Duplex brake	2	double	2	–	2	–	–	–	3,0	3,0
5. Servo brake	1	double	2	–	1	1	1	–	4,0	2,0
6. Duo-servo brake	1	double	2	–	2	–	1	1	4,0	4,0

Drum brake

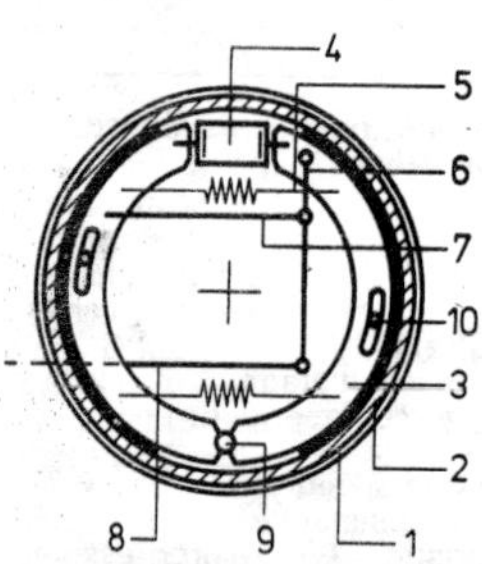

1. Brake mounting plate
2. Brake drum
3. Brake jaw with brake lining
4. Wheel cylinder with air draining valve
5. Return spring
6. Lever for hand brake
7. Plunger for hand brake
8. Brake cable
9. Step bearing, single or double point of rotation
10. Reset disc with lobes, reset screw

Every brake jaw of mounting speed shows self-reinforcement because the friction force strengthens the clamping force. On the brake jaw of decreasing speed self-weakening occurs because the friction force reduces the clamping force. Drum type brakes are sensitive to friction coefficient variations, for the internal transmission C* varies with changing friction co-efficients.

Air gap	0.3—0.5 mm
internal transmission	2—4
clamping force	small
wheel cylinder diam.	small
line pressure	25—50 bar
initial pressure	0.5—0.8 bar
friction coeff. variations	sensitive
braking effect	irregular
self-cleaning	nil
contamination	low
heating up of drum	large
brake fading	large
hand brake	simple
jaw change	costly
surface pressure	low
lining wear	small
resetting	automatic, mechanical
release	through return spring

Disc brakes

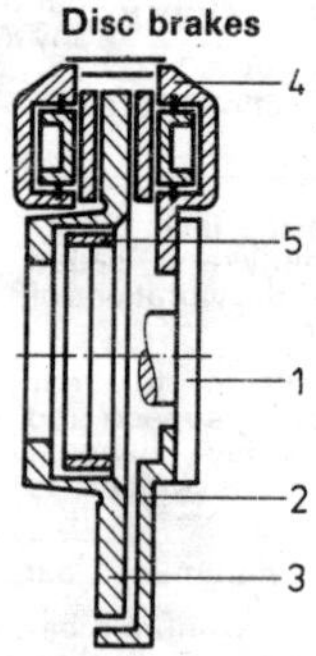

1. Wheel mounting plate, steering knuckle
2. Covering sheet
3. Brake disc
4. Brake saddle, fixed saddle, floating frame saddle: housing, brake cylinder, piston with sealing ring, dust cap, clamping ring, brake lining, cross-spring, air draining valve.
5. Hand brake, duo-servo brake in the disc pot or mechanical vise brake: brake jaws, spacer lever of spacer locking bolt, cable line, spanner, return spring.

Disc brakes have no brake jaws of increasing speed. They have, therefore, no inner transmission, and are insensitive to variations in friction co-efficient. Higher line pressures and bigger wheel cylinder diameters are necessary to obtain adequate braking forces.

Air gap	0,15 mm
Inner transmission	0.8
Contact force	big
Wheel cylinder dia	big
Line pressure	50—80 bar
Initial pressure	0.0 bar
Friction coeff. variation	insensitive
Braking action	uniform
Cleaning	automatic
Water splashing	sensitive
Heat dissipation	good
Lining temperature	high
Brake fading	low
Hand brake	costly
Jaw change	simple
Surface pressure	big
Resetting	automatic
Release	through sealing rings

Single-circuit brake unit
front/rear drum brake

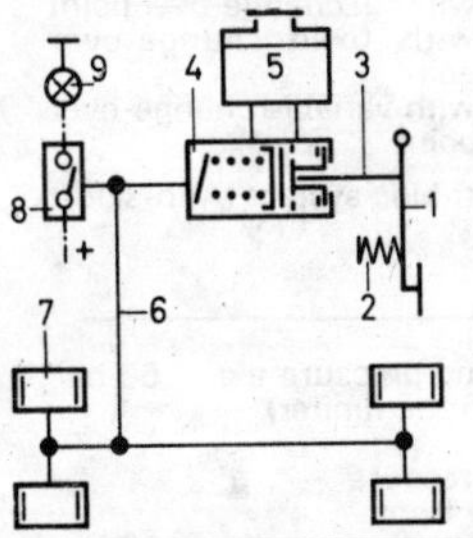

1. Brake pedal
2. Return spring
3. Piston rod
4. Main cylinder
5. Expansion tank
6. Brake lines
   Brake hoses
7. Wheel cylinder
8. Brake light switch
9. Brake light

Simple main cylinder
- housing
- protection cap
- equalizing bore
- after-running bore
- piston, filling hole
- filler plate
- primary sleeve
- secondary sleeve
- bottom valve
- piston spring

Wheel cylinder
- housing
- protection cap
- piston
- sleeve
- pressure spring
- spring plate or filler piece
- air draining valve
- pressure bolt
- pressure mush room head

Brakes of front and rear axle are connected to each other. If one brake fails, then the entire brake will be ineffective.

Number of brake circuits	1
Foot force	max. 800N
Line pressure	25—50 bar
Initial pressure	0.5—0.8 bar
Piston rod clearance	1 mm
Pedal transmission	1:5—8
Hydr. transmission	1:1
Air gap	0.3—0.5 mm
Efficiency in main cylinder	0.92
Actuation path on the wheel cylinder piston	2.4—4.5 mm

Single circuit brake unit
with stepped main cylinder, front/rear drum brake

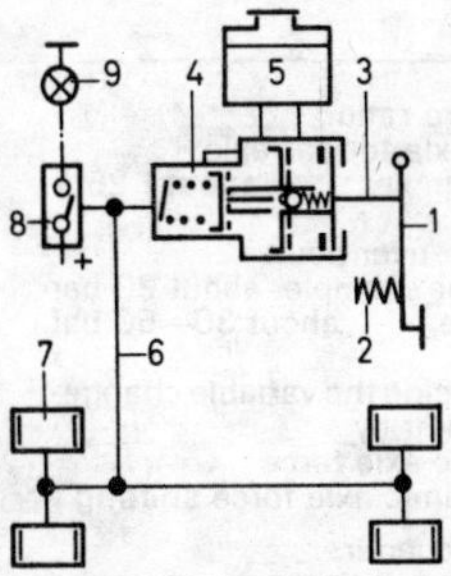

1. Brake pedal
2. Return spring
3. Piston rod
4. Stepped main cylinder
5. Expansion tank
6. Brake lines
   Brake hose
7. Wheel cylinder
8. Brake light switch
9. Brake light

Stepped main cylinder
- housing
  filler stage: big diameter
  push stage: small diameter
- double piston
  filler piston: applying the brake jaws
  pushing piston: pressing the brake jaws
- operating piston: ball valve open
- ball valve: to allow the fluid of the filling stage to flow back after applying the jaws.
- for other parts, see 'simple main cylinder'

Applying (filling)
Filler hole in the pressure piston open, ball-valve closed, fluid of filling stage flows over the pressure piston into the wheel brake cylinder:
- big filling-piston surface
- small line pressure
- big filling volume
- small pedal travel
- short response time

Pressing (braking)
Filler hole in the pressure piston closed, ball valve open, fluid of the filler stage flows into the expansion tank:
- small piston surface
- big line pressure

Double circuit brake unit
disc/drum brake

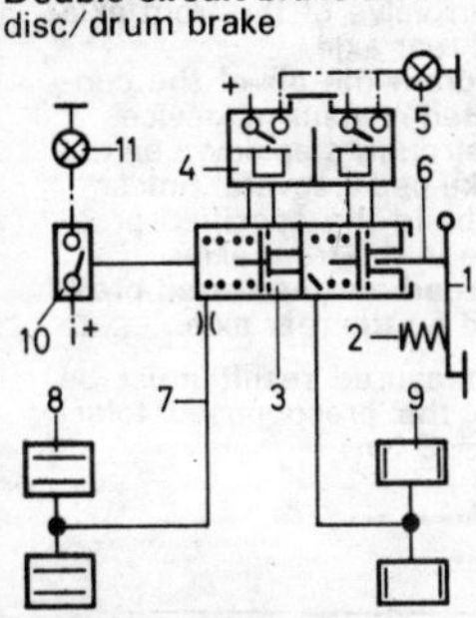

1. Brake pedal
2. Return valve
3. Tandem main cylinder
4. Expansion tank
5. Warning fixture
6. Piston rod
7. Brake lines
8. Brake hoses
   Brake lamps
9. Wheel cylinder
10. Brake light switch
11. Brake light

Tandem main cylinder
- housing, protection cap
- equalizing bores
- after-running bores
- push rod piston
- intermediate piston
- filler discs
- primary sleeve
- secondary sleeve
- piston spring
- special bottom valve
- bottom valve

Brake clamps
- brake saddle
- shaft
- brake cylinder
- piston
- rubber sealing ring
- brake block
- cross-spring
- holding pins
- air drain valve
- dust cap
- clamping ring

Brakes of front and rear axle are separated from each other. If one brake circuit fails, the intact brake will be effective.

Number of brake circuits	2
Foot force	max. 800 N
Line pressure	25—50 bar
Initial pressure	
• disc brake	0.0 bar
• drum brake	0.5—0.8 bar
Piston rod clearance	1 mm
Pedal transmission	1:5—8
Hydr. transmission	1:1—4
Air gap	0.1—0.3 mm
Efficiency in main cylinder	0.88
Actuation path of the wheel cylinder piston	0.6 mm

Double circuit brake unit
with brake force amplification, disc/drum brake

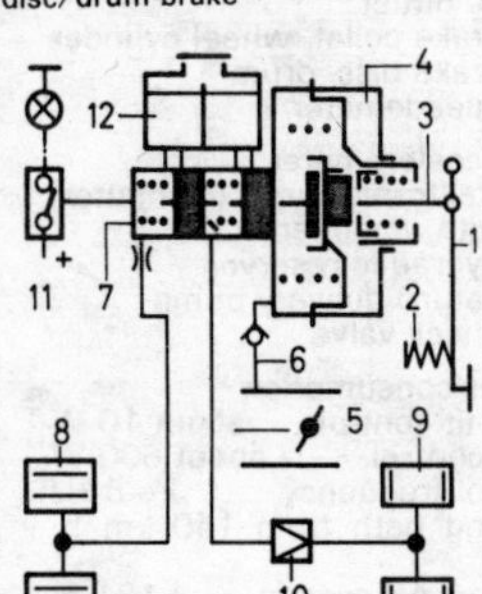

1. Brake pedal
2. Return spring
3. Piston rod
4. Braking device ATE 51
5. Intake pipe
6. Low pressure hose
7. Tandem main cylinder
8. Brake collet
9. Wheel cylinder
10. Pressure limiting valve
11. Brake light switch
12. Expansion tank

Braking device ATE 51, mechanically controlled

• Work piston:	to absorb the pressure difference between chamber 1 and chamber 2
• Push rod:	to pass on the piston force to main cylinder
• Piston spring:	to guide back the work piston
• Piston valve:	to fill chamber with air
• Plate valve:	to suck air from chamber 1
• Reaction disc:	to support reaction force
• Check valve:	to maintain low pressure to protect the diaphragm to lock flame back-fire

Drive position

• Piston valve	closed
• plate valve	open
• chambers 1 & 2	low pressure

All wheels braking position

• piston valve	open
• plate valve	closed
• chamber 2	max. low pressure
• chamber 1	barometric air pressure

Partial braking position

• piston valve	closed
• plate valve	closed
• chamber 2	max. low pressure
• chamber 1	partial pressure

Amplification factor	2.0—4.0
Negative gauge pressure	max. 0.8 bar
Line pressure	50—80 bar

Brake Force Distribution

General

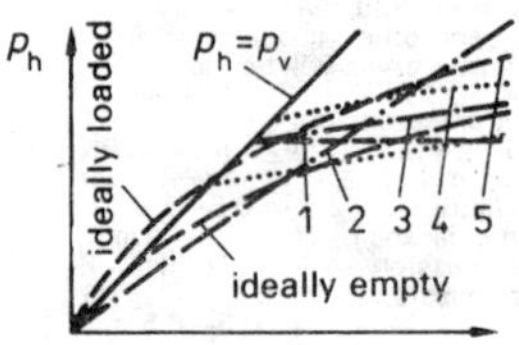

The distribution of the braking force on front and rear wheels can be determined by design measures. The installed braking power distribution is fixed and conceived only for a certain braking.

In the ideal braking force distribution, the braking force should be proportional to the dynamic axle forces and the braking to avoid an excess braking. In such an exercise, depending on the brake pressure, the types shown alongside are different.

Types of construction:
1. Pressure limiter
2. Pressure reducer
   a) without change-over point
   b) with fixed change-over point
   c) with variable change-over point
3. Anti-bloc system (Anti-skid)

Pressure limiter

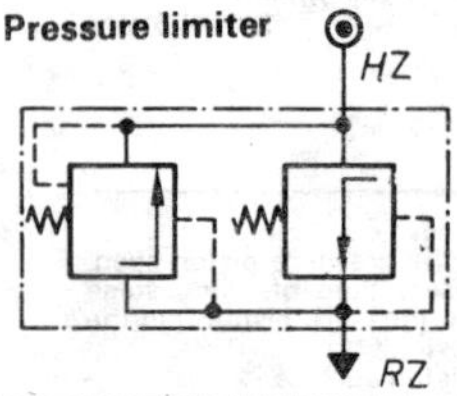

The pressure limiter limits the hydraulic pressure generated by the main cylinder in the rear axle circuit to a determined value. The brakes of the front axle receive full pressure. Change-over pressure is related to the force of valve spring and to the surface difference on the piston valve.

Working pressure e.g. 60 bar
(set on the limiter)

Test pressure
on the front
axle 50—100 bar
on the rear axle
switch-off pressure ± 3 bar

Pressure reducer

Figure 1

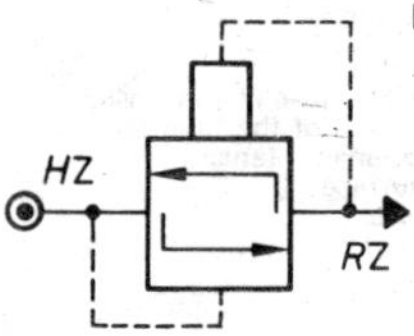

Figure 2

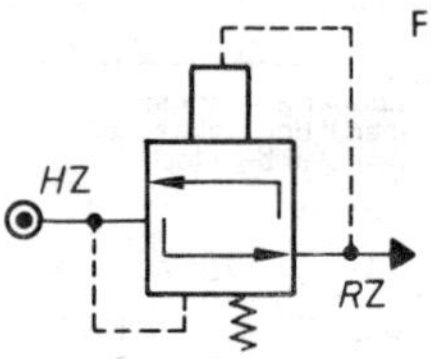

Figure 3

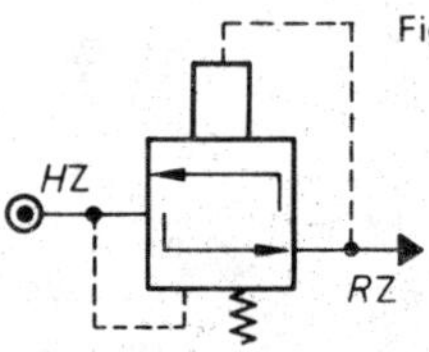

Pressure reducer without change-over point (Fig. 1) reduces the pressure in the rear-axle circuit in a definite ratio to the front-axle circuit. The pressure ratio is related to the surface difference on the valve piston. The transmission ratio is fixed; the pressure distribution curve is linear.

Pressure reducer with fixed change-over point (Fig 2) allows the pressure generated by the main cylinder without hindrance to the rear axle circuit; above the change-over point, it reduces the pressure in a definite ratio to the front axle circuit.

The change-over point is fixed, and it is related to the pre-stressed piston spring. The pressure distribution curve is bent. The transmission ratio is determined by the surface difference on the valve piston.

Pressure reducer with variable change-over point (Fig 3) works like the pressure reducer with fixed change-over point. Above the change-over point it reduces the hydraulic pressure in the rear-axle circuit in a definite ratio to the front axle circuit. The change-over point is variable with load-dependent displacement, e.g. by the modification of spring force over an adjustable linkage. The curve is bent and approaches the ideal pressure distribution in the loaded and empty state.

Pressure ratio
Front axle to rear axle
for example about 1.25:1

Change-over point
fixed for example about 30 bar
variable about 30—50 bar

Influencing the variable change-over point by
1. static axle force
2. dynamic axle force shifting

Test procedure
1. One measuring instrument each connected on the air-drain valve of the front axle and rear axle.
2. To drain the air of the connected measuring device.
3. To strongly step down on the brake pedal several times.
4. To build the specified pressure in the front axle.
5. To read out the tested pressure on the rear axle.

The measured result must be within the presupposed tolerance (± 3 bar)

Antibloc-system (Anti-Skid)

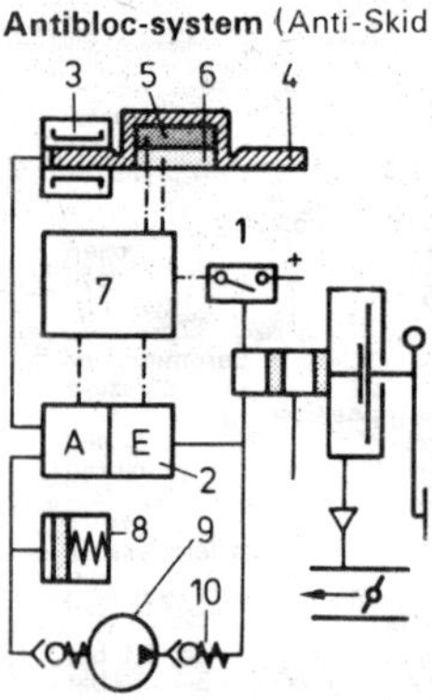

The AB system automatically, controls the brake slip in all driving conditions and sets it to the range of optimal adhesion coefficient. In that way, a blocking of the wheels is avoided and the shortest braking path is achieved.

The brake pressure generated by the main cylinder gets through the inlet valve into the wheel cylinder. If the brake torque and friction torque are not in equilibrium with each other, then the wheel is accelerated or decelerated. The sensors determine the negative or positive accelerations as well as the speed differences, and provide voltage impulses to the electronic unit. There the signals are processed, and the inlet and outlet valves are controlled. The valves match the brake pressure, by pressure reduction or pressure increase, to the changing conditions. The brake fluid drawn out during pressure decrease from the control circuit is guided back by a return delivery pump.

The AB system unit consists of the following parts:
1. Tandem main cylinder
2. Pressure regulator with in/outlet
3. Brake collet, wheel cylinder
4. Brake disc, drum
5. Speedometer
6. Accelerometer
7. Electronic unit (computer) with amplifier
8. Hydraulic reservoir
9. Return delivery pump
10. Check valve

Power consumption
without control about 10 W
with control about 600 W
control frequency 2—8 Hz
Braking path from 140 km/h, wet.
without AB system 181 m
with AB system 112 m

Hand brake 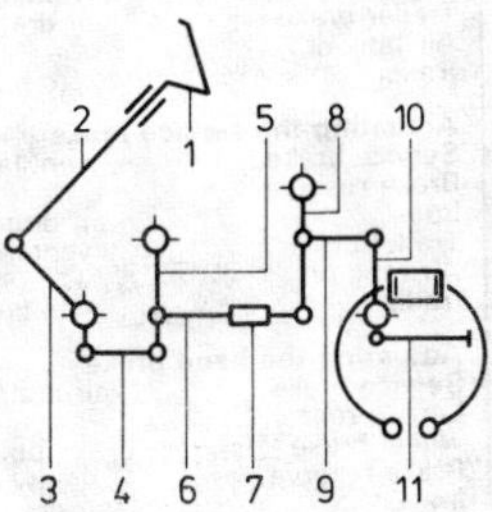	**Lever-operated brake** 1. Grip 2. Halting rod 3. Hand brake lever 4. Pull rod 5. Intermediate lever 6. Pull rod 7. Spanner 8. Intermediate lever 9. Pull rod 10. Brake lever 11. Push rod **Linkageless spring feed hand brake** 1. Air tank 2. Hand brake valve 3. Inspection switch 4. Relay valve 5. Spring feeder 6. Piston rod 7. Brake lever 8. Brake cam In the released hand brake is a reserve pressure in connection with the spring feeder via the hand brake valve and relay valve. The spring is in tension. By actuating the hand lever the air in spring feeder is drained and the spring force will be free.	Working method • mechanical • can be fixed Fixed position with dwells Actuation force max 400 N Lever transmission 1:15—25 Braking at least 20% Mean delay 1.5 m/s Brake power Fully loaded vehicle to be stopped on the biggest gradient covered by it. Braking effect is independent of the service brake. Resetting by spanner. Push rod clearance 1 mm Air gap 0.3—0.5 mm
Engine dynamic pressure brake, engine brake 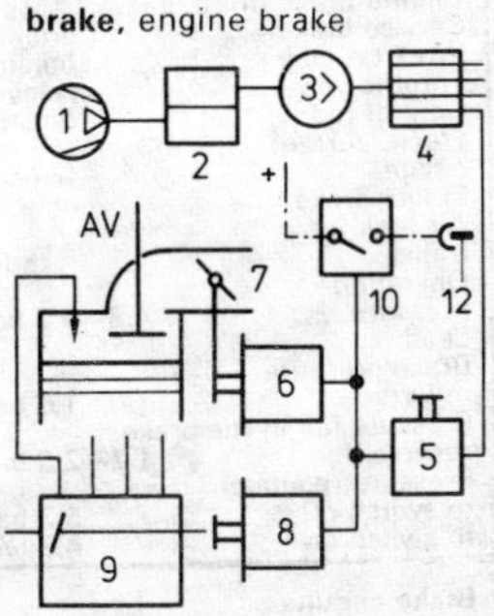	1. Air compressor 2. Pressure regulator 3. Anti frost 4. Protection valve 5. Ventilating valve 6. Working cylinder for throttle valve 7. Exhaust manifold 8. Working cylinder for control rod 9. Injection pump 10. Pressure switch 11. Relay switch 12. Plug socket 13. Solenoid relay valve 14. ANH brake cylinder All buses of over 5.5 t total mass, all lorries, trailers and semi-trailers of more than 9 t total mass must be equipped with a permanent brake. The braking effect is produced mostly by cutting-off fuel supply, damming exhaust gases and brake power of the engine. In the trailers, the already existing wheel brakes are pulled as the 'third brake'.	Working method pneumatic. Engine works as compressor. Brake position • working cylinder ventilated • throttle valve closed • control rod on 'stop' • ANH brake cylinder ventilated • air drained from lorry brake cylinder Brake performance • lorry fully loaded • stretch with 7% fall over 6 km • speed, max 30 km/h • braking, 7% wear-free, low maintenance, soft and impact-free response, service brake without load Parts 11, 13, 14 not shown in the sketch.
Eddy current brake Telma brake 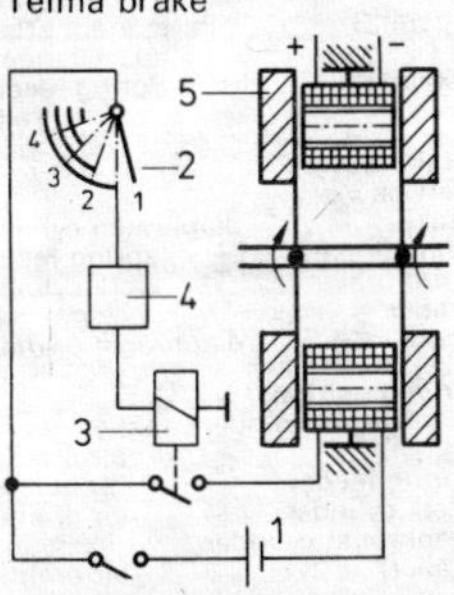	1. Battery 2. Hand switch Pilot lamp 3. Relay valve Cabinet with 4 switching contactors 4. Time-delay relay for very fast through-connection of the 3rd and 4th stages of switching 5. Slowing-down device a) Stator: fixed with 16 regulatable electromagnets. b) Rotor: driven magnetic disc. The battery builds an electric field with magnetic cores (stator) over fixed coils. The electric field brakes the rotors (two) connected with the propeller shaft smoothly and without jerks by electric eddy currents. In that way the service brake is relieved to a great extent.	Working method electro-magnetic Braking 6:5 % Braking torque up to 2500 Nm Weight big Switching stages 4 Electromagnets for each switching stage 4, 8, 12 and 16 Current consumption at 24 V 1st stage 16—26.5 A 2nd stage 32—53.0 A 3rd stage 48—79.5 A 4th stage 64—106 A Air gap 0.9—2.4 mm Response soft, impact-less Time delay 0.7—1.0 s Wear-free, low maintenance, heat dissipation, air cooling
Fluid current brake Retarder, turbo brake, hydro-dynamic brake 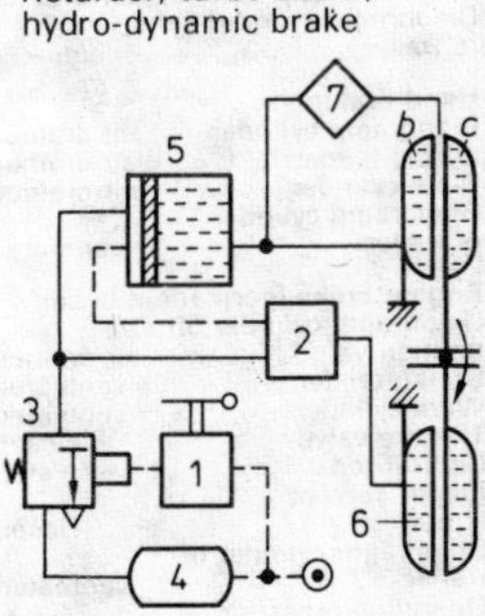	1. Actuating valve 2. Control valve 3. Relay 4. Air tank 5. Charging cylinder (3.–5.) not necessary if the retarder is supplied from the oil circulation of gear mechanism 6. Retarder a) Housing b) Fixed turbine wheel (stator) c) Driven pump rotor (rotor) d) Variable oil filling 7. Heat exchanger, possibly together with the gear The drive energy is converted into flow current energy over a pump rotor and delayed in the bucket chambers of the fixed turbine wheel by fluid friction and converted into heat. The rotor pumps the fluid through heat exchanger to dissipate heat.	Weight low Working method hydro-dynamic Construction simple, small Control through modification of oil filling or oil pressure Brake stage 0 (stepless)—6 Brake fading nil Response fast, smooth Wheel blocking not possible Assembly • on gear drive input • on gear drive output • between converter + gear Service brake relieved Cooling heat exchanger Wear-free, low maintenance Braking 6—8 % Life long

Compressed Air Brake Unit

Single line brake unit

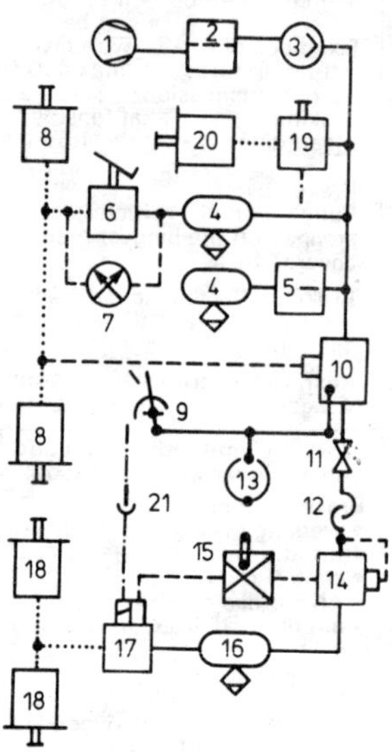

1. Air compressor, intake filter
2. Pressure regulator, tyre filling connection
3. Anti-frost
4. Air tank
5. Overflow valve with return current
6. Control valve (trailer brake)
7. Double air pressure gauge
8. Engine-powered car brake cylinder
9. Hand brake
10. Trailer control valve
11. Shut off cock
12. Clutch head
13. Rear wheel brake
14. Trailer brake valve
15. Release valve, brake force regulator
16. Air tank for trailer
17. Solenoid relay valve
18. Trailer's brake cylinder
19. Ventilating valve, push button switch
20. Working cylinder for throttle valve and control rod
21. Plug-point for electr. conduit from part 17 to part 19

————	supply conduit
·········	working conduit
— — — —	control conduit
—·—·—·—	electr. conduit

Note:
a) to fill the trailer's air tank only in the driving position
b) to brake the trailer by draining the air of brake reserve line

Drive position

Service brake	air drained
Brake reserve line	ventilated
Trailer brake	air drained
Air tank of trailer	to fill up

Actuating the service brake

Service brake	ventilated
Brake reserve line	air drained
Trailer brake	ventilated
Air tank of trailer	to bleed

Actuating the hand brake

Service brake	air drained
Lorry's rear wheel brake	pulled
Brake reserve line	air drained
Trailer brake	ventilated
Air tank of trailer	to bleed

Engine brake (third brake)

Service brake	air drained
Work cylinder	ventilated
Throttle valve	closed
Control rod	on stop
Electr. current circuit	closed
Trailer brake	ventilated
Air tank for trailer	to fill
Operating pressure	4.8—5.3 bar
Lead:	
Brake pressure in lorry	1.0—1.3 bar
Pressure fall in the brake reserve line	1.7—2.5 bar
Pressure regulator	
to switch off	5.3 bar
to switch on	4.8 bar

Double-line brake unit

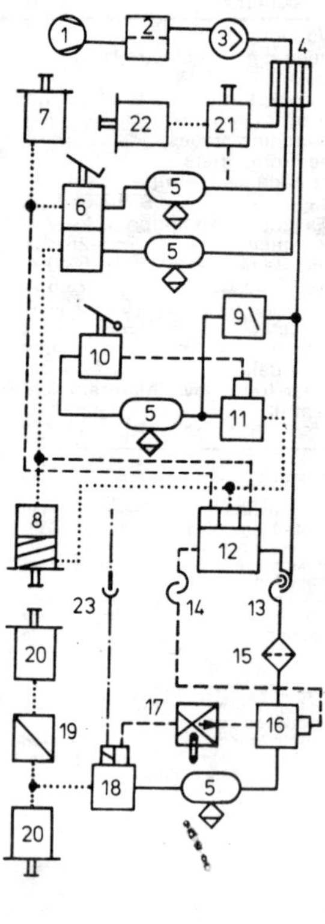

1. Air compressor, intake filter
2. Pressure regulator, tyre filling connection
3. Anti-frost
4. Duplex circuit protection valve
5. Air tank for water drainage
6. Engine-powered car brake valve (double circuit), possibly engine-powered car brake force regulator and load-empty-valve series connected.
7. Diaphragm cylinder
8. Tristop cylinder
9. Overflow valve, limited return current
10. Hand brake valve
11. Relay valve
12. Trailer control valve
13. Clutch head, reserve possibly way-valve-series connected
14. Clutch head brake
15. Line filter
16. Trailer brake valve
17. Trailer brake force controller
18. Solenoid relay valve
19. Control valve
20. Diaphragm brake cylinder
21. Ventilating valve, push button switch
22. Work cylinder for throttle valve and control rod
23. Plug-point for electr. conduit from part 18 to part 21

————	supply conduit
·········	working conduit
— — — —	control conduit
—·—·—·—	electric conduit

Note:
a) To fill the trailer's air tank in driving and braking position
b) To brake the trailer by ventilating the brake line.

Brake circuits

Circuit 1	Service brake front axle
Circuit 2	Service brake rear axle
Circuit 3	Permanent brake, miscellaneous
Circuit 4	Spring feeder, trailer

Brake device

Service brake	diaphragm cylinder
Hand brake	spring feeder
Third brake	work cylinder
Trailer brake	diaphragm cylinder

Drive position

Diaphragm cylinder (traction unit)	air drained
Spring feeder	ventilated
Work cylinder	air drained
Diaphragm cylinder (trailor)	air drained

Actuation of Service brake

Diaphragm cylinder	ventilated
Spring feeder	ventilated
Work cylinder	air drained
Diaphragm cylinder to trailer	ventilated

Hand braking

Diaphragm cylinder	air drained
Spring feeder	air drained
Work cylinder	air drained
Diaphragm cylinder of trailer	ventilated

Engine brake (permanent brake)

Diaphragm cylinder of traction veh.	air drained
Spring feeder	ventilated
Work cylinder	ventilated
Throttle valve	closed
Control rod	on stop
Electr. current circuit	closed
Diaphragm cylinder of trailer	ventilated
Operating pressure	7.3 bar
Pressure regulator	
to switch off	7.3 bar
to switch on	6.2 bar
Protection valve:	safe pressure
is equal to opening pressure	6.0 bar

Arrangement

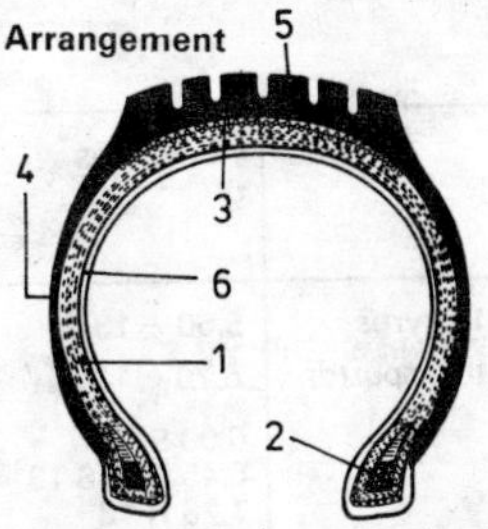

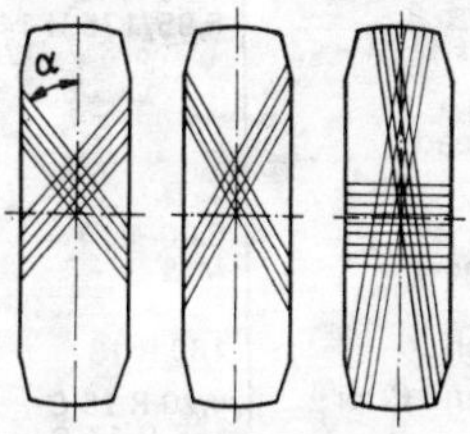

Normal diagonal tyre | S-diagonal tyre | Belt radial tyre

1. **Carcasse**
   Carcasse as the strength carrier; several rubberized cord cloth layers lie cross-wise one above the other.
2. **Bead** with wire core
   Anchoring the cloth layers provides the tyre a firm seat on the rim shoulder. Several rubberized steel wires and a rubber cap, with a covering of rubberized solid cloth.
3. **Sub-structure**
   Protects the carcasse by damping the road impacts; it consists of substructure cloth and rubber cushion.
4. **Side wall** and chafing border
   Side rubber part of tread. It protects the cord tissue and conducts away the heat generated by the rolling tyres through work.
5. **Tread** with shoulder
   Central rubber part of tread (protector) with profiling to increase the holding ability in wet road. It establishes a closed linkage with the road and is subjected to wear.
6. **Center web** and bead covering
   Air-tight, sealed, soft rubber layer below the cord tissue and above the bead in tubeless tyres.

Cord cloth
- without the supporting weft
- cord yarns from 2 twines with 700 capillary yarns each.
- yarns of artificial silk, nylon, polyester or steel.

Yarn angle (Zenith angle α)

Tyres of diagonal construction	
Normal tyres	35—38°
Sports tyres	30—34°
High speed tyres	up to 30°
Belt-type tyres	
sub-structure (2 plys)	58—90°
belt (2 to 4 plys)	0—30°

Tyre air pressure depending on the load carrying capacity of passenger

car	1.3—2.6 bar
lorry	2.2—7.0 bar

Running power

Correct air pressure	100 %
20 % less air pressure	85 %
40 % less air pressure	60 %
60 % less air pressure	25 %

Profile depth

New tyres	about 8 mm
Minimum	1 mm

Increasing load carrying capacity

km/h	60	50	40	30	20	8
in %	10	15	25	35	50	75

Types

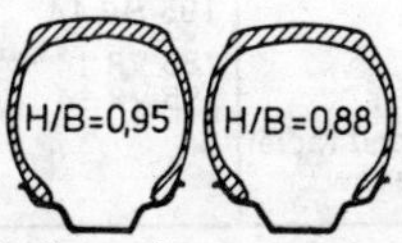

Super balloon | Low cross section

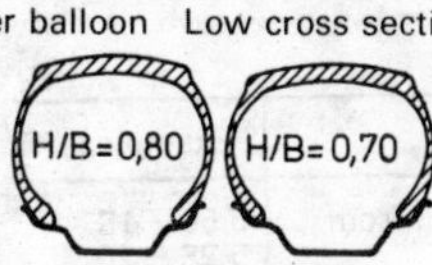

Super low Cross-section tyres | Type 70

Difference	Designation
Tube	Tube type tyres Tubeless tyres
Thread angle	Diagonal tyres Radial belt tyres (R) Semi-radial tyres
Speed	Normal tyres (—) Sports tyres (S) High-speed tyres (H) Very high-speed tyres (V)
Cross-section form	Balloon tyres Super balloon tyres Low cross-section tyres Super low cross-section tyres Type "70" ("60") tyres
Profile	Summer tyres Winter tyres

Tubeless tyres
Simple assembly
Low heating-up
No air escape in the event of foreign bodies sticking; slow softening with outflying foreign objects.

Belt tyres
Long life
Low rolling resistance
Better track holding
Better road grip
No mounting of tyres
Becomes softer above 100 km/h

Flat tyres
Bigger road grip
Better side guidance
Lower rolling resistance
Higher load carrying capacity

Tyre dimensions

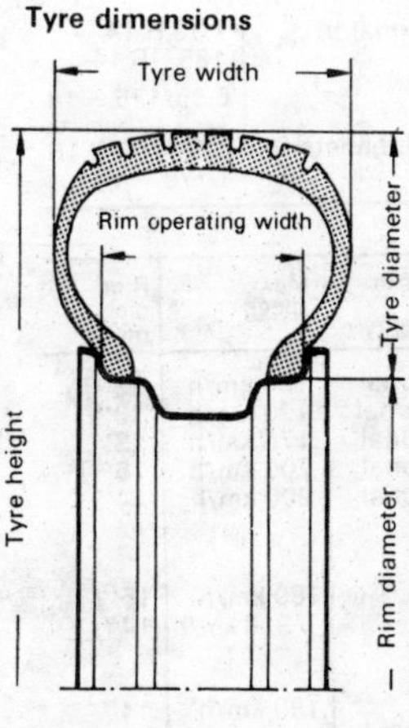

Dynamically effective radius
(r dyn) The stretch of road covered at 60 km/h speed per revolution divided by $2 \times \pi$. In this case, the tyres must be loaded with the maximum permissible, specified load carrying capacity and pumped with the corresponding air pressure. With increasing speed r increases up to about 6%; in belt-type tyres it remains almost constant.

Static radius
Distance between the wheel centre to the stand level of the tyre, loaded to the max. permissible load-carrying capacity and pumped with the respective air pressure.

Tyre diameter
Outer diameter of the pumped up, unloaded tyre in the centre of the running face.

Normal width
Width of the pumped-up, unloaded tyre with flat side wall without chafing border.

Ply rating
International index for the carcasse strength. It does not mention in every case the actual number of cloth plys.

Variation of r dyn % in diagonal-type tyres

	super balloon low cross-section		super low cross-section	
km/h	—	S, H	—	S, H
60	0	+ 1	0,0	+ 0,5
90	+ 1	+ 2	+ 0,5	+ 1
120	+ 2	+ 3	+ 1,5	+ 2
150	+ 3	+ 4	+ 2,5	+ 3
175	—	+ 5	—	+ 4
200	—	+ 6	—	+ 5

Variation of r dyn in % in belt-type tyres

km/h	SR or HR tyres
60	0
90	+ 0,1
120	+ 0,2
150	+ 0,4
180	+ 0,7
210	+ 1,1 (only HR tyres)

Tyres

Maximum permissible speed in km/h

Type of tyre by construction	Tyre profile	Maximum speed Rim diameter 10″		13″—15″	Symbol	Explanation of symbol	Examples
Diagonal tyres	Summer profile	120	135	150	– (×)	Normal or standard 5.60—13 tyres	5,60 – 13
				120	– C	Normal tyres, C stands for transporter tyres	6,70 – 13 C
		150	160	175	S	Sports tyres S stands for speed	6,00 S 14 6,45/165 S 13
		175	185	200	H	High speed tyres H stands for high speed	7,25 H 13 6,95/175 H 14
		more than 175	more than 185	more than 200	V	Very high speed tyres V stands for every high speed	— —
	Winter profile	120	135	150	– M & S	Normal tyres M & S is for sludge and snow	6,00 – 14 M & S
Belt tyre	Summer profile	160	160	160	R	Normal tyres, R is for radial	160 R 13
				120	R C	Normal tyres, R for radial and C for transporter tyres	6,70 R 15 C 165 R 14 C
		180	180	180	SR	sports tyres SR stands for speed radial	165 SR 14 185/70 SR 13
		–	170	170	SR reinforced	Sports tyres in reinforced design High speed tyres	165 SR 14 reinforced
		210	210	210	HR	HR means high-speed radial	165 HR 13 195 HR 14
		more than 210	more than 210	more than 210	VR	Very high-speed tyres VR is for very high-speed radial	185 VR 14 195 VR 14
	Winter profile	160	160	160	SR M & S	Sports tyres; SR stands for speed radial M & S stands for sludge and snow	— —

Height to width ratio (cross-section ratio)

Cross-sectional form	H : W ratio	Marking, width specification	Examples
Balloon tyres	1 : 1	in inch with .00, .25, .50, .75 decimal starting from 15″ rim	5,50 – 18 5,25 – 20
Super balloon tyres	0,95 : 1	in inch with .10, .20, .30, .40, .60, .70, .80, .90, decimal	5,90 S 13 6,40 – 15
Low cross-section tyres	0,88 : 1	in inch with .00, .25, .50, .75 decimal and a succeeding L (low), starting from 15″ rim (differentiation)	6,00 – 13 7,00 – 15 L
Super low-cross-section tyres	0,80 : 1	in inch with .15, .35, .45, .65, .85, .95 decimal in millimeter for belt tyres in inch and millimeter (interchangeability)	7,35 H 14 185 HR 14 6,95/175 – 14
Serie "70" tyres	0,70 : 1	in millimeter with a succeeding "70", in alphabets D to L with a succeeding "70" (USA)	185/70 VR 15 D/70 – 14

Designation of tyres

7.25 S 13 4 PR

- 7.25 – Tyre width and cross-section form
- S – Construction type and max.speed
- 13 – Rim diameter in inch
- 4 PR – International symbol for the carcasse strength of tyre

Designation	Tyre width	Cross-section ratio	Construction type of tyre	Max speed	Rim diameter
5,90 – 13	5,90″	0,95 : 1	diagonal	150 km/h	13″
6,70 – 13 C	6,70″	0,95 : 1	diagonal	120 km/h	13″
7,25 S 13	7,25″	0,88 : 1	diagonal	175 km/h	13″
6,00 H 15 L	6,00″	0,88 : 1	diagonal	200 km/h	15″
6,45/165 H 14	6,45″ 165 mm	0,80 : 1	diagonal	200 km/h	14″
155 SR 15	155 mm	0,80 : 1	belt	180 km/h	15″
185 VR 14	185 mm	0,80 : 1	belt	>210 km/h	14″
195/70 SR 14	195 mm	0,70 : 1	belt	180 km/h	14″
D/70 – 14	200 mm	0,70 : 1	diagonal		14″

Well-base Rim 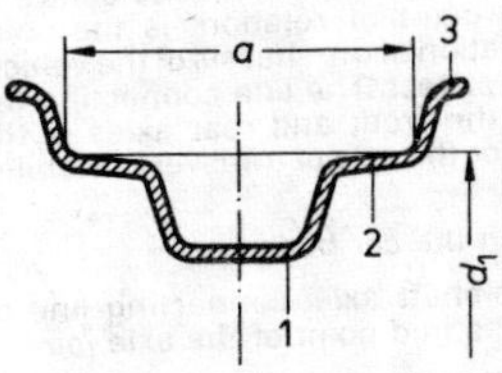	1. Well base 2. Rim shoulder 3. Rim horn Important sizes are : a. width of opening of rim d_1 rim diameter Well-base rims are mostly symmetrically undivided, pressed from steel sheet. The shoulders are inclined inward by 5° for the sake of better seating of the tyre bead. Well-base because it is required for fitting the tyre.	Application in passenger car, station wagon and light lorries, sometimes even for tractors and agricultural machines; frequently wide rim with well base. No bulging rim horns for tubeless tyres; no rust protection application. Sealing faces are rim shoulders. Radial runout of inner side of rim horns is max. 1.5 mm for a diameter of 10—19 inches.
Inclined shoulder rim 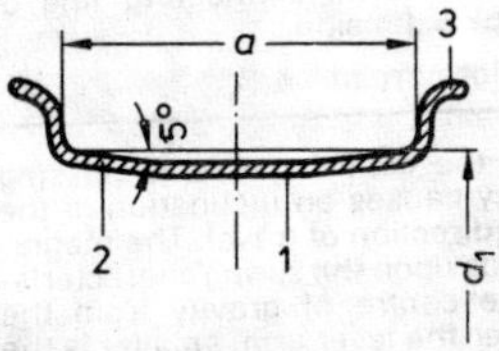	1. Bed 2. Inclined shoulder 3. Rim horn Shoulders inclined by 5° inward. Rim horn can be taken out on one side. Circularly split rim system. There are also peripherally split rims (Trilex rims) with unremoveable rim horns. Rims with cylindrical shoulders (flat-base rims) only occasionally in foreign cars.	Application in lorries, buses, trailers and construction machinery. Occasionally half-flat base rims are employed (only diameters of 16, 20 and 24 inches). Radial runout of shoulders and axial runout of inner side of rim horns (in the finished wheel) is maximum 1.5 mm to 6.5—20, max 2 mm above that.
Safety rim 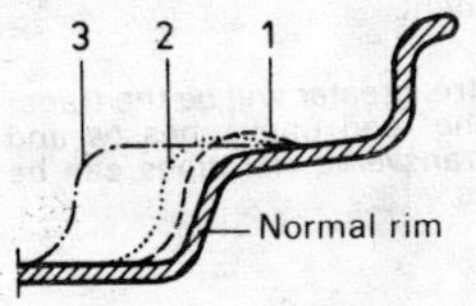	According to designations in USA, the following types are identified as per safety contours: 1. Hump 2. Flat hump 3. Special ledge In Germany the hump design is conventional. The difficult fitting of hump rims is the drawback.	Rims with safety shoulders improve the secure seating of tubeless tyres on rim shoulder to avoid sudden air escape in sharp curves and very low air pressure.
Circularly split rim system 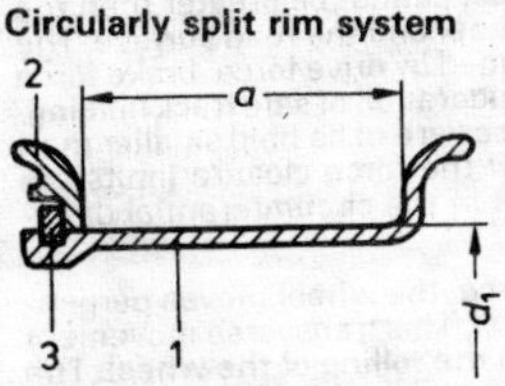	3—piece rim made of 1. Base rim 2. Side ring 3. Locking ring 2—piece rim made of 1. Base rim 2. Combination ring	The side ring (removable horn part) is closed. The locking ring is open. The combination ring (removable horn and shoulder part) is open
Rim designations 41/2J×13	Well-base rim Width of opening a = 4.5 inch Horn shape J Rim diameter d_1 = 13 inch	signifies well-base or wide rim. The horn shape is fixed by capital letters.
7,5-20	Inclined shoulder rim Width of opening = 7.5 inch Rim diameter d_1 = 20 inch	— signifies inclined shoulder, flat-base or half flat base rim. No horn alphabet.
W8×24	Wide rim (with well base) Width of opening a = 8 inch Rim diameter d_1 = 24 inch	W = wide base; for farm tractors
Wheels disc wheel	Rim and wheel pan, welded or rivetted together, build the wheel. Wheel pan consists of pressed steel sheet with cams, holders or humps to fasten the knurled caps and bolt holes with tapered or ball hole.	Application in passenger cars, lorries, buses and farm vehicles. The wheel pan connects the rim with the hub. For passenger car 4—5 fastening holes, for lorries 6, 8 or 10 fastening holes. Wheel nuts with taper or ball collars.
Trilex spoked wheel body	Wheel body or wheel spider together with the wheel hub is a single cast piece. The 3 rim parts are placed on the spoke heads and screwed.	Application in heavy lorries and buses. Very stable, high loadability and easy tyre change.

Instantaneous centre

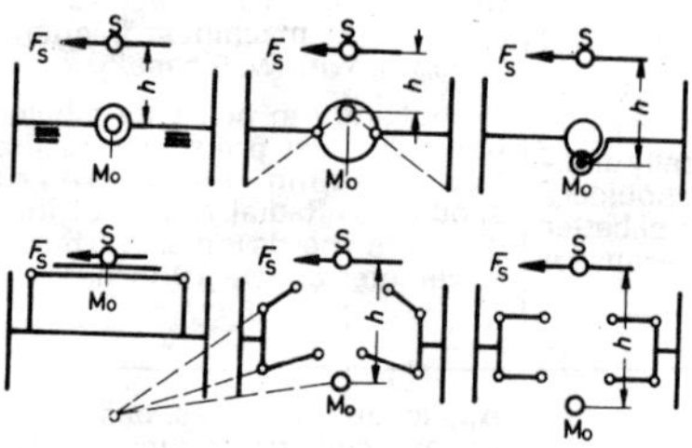

Instantaneous centre, also called instantaneous centre of rotation or instantaneous point of rotation, is the point which at the moment of rotation of the chassis of the vehicle due to centrifugal force is at rest. The line connecting the instantaneous centres of the front and rear axles is the theoretical axis of rotation or the axis of transverse stability (lateral stability).

1. Rigid axle: spring stop on the car body.
2. Shortened jointed cross-shaft axle: connecting line of wheel rising point and the mid-point of the axle joint.
3. Single-jointed cross-shaft axle with low point of rotation: mid point of the axle joint.
4. Suspension axle: spring stop on the car body.
5. Transverse link, trapezoidal form: connecting line of wheel rising point and link extension.
6. Transverse link, parallel form: road.

Distribution of wheel force while negotiating curve

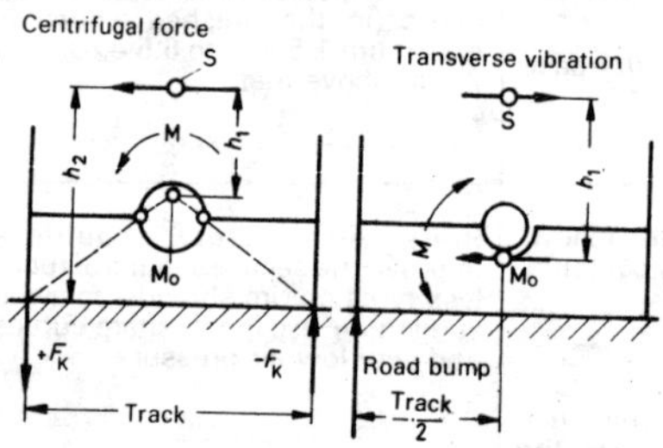

While negotiating a curve, the centrifugal force passing through the centre of gravity causes an inclination of the vehicle perpendicular to the direction of travel. The magnitude of the inclination depends upon the spring characteristic and the distance of the centre of gravity from the instantaneous centre. Smaller the lever arm, smaller is the inclination at the curve and better is the stability at the curve.

$M = F \cdot h_1$ Wheel force variation due to pitching moment. $F_K = \pm \frac{F \cdot h_1}{s}$

Higher the instantaneous centre, greater will be the transverse vibrations caused by the road unevennesses and wheel force variations. The transverse vibrations can be absorbed by transverse springs.

Inclined running (Maximum magnitude 8 to 15°)

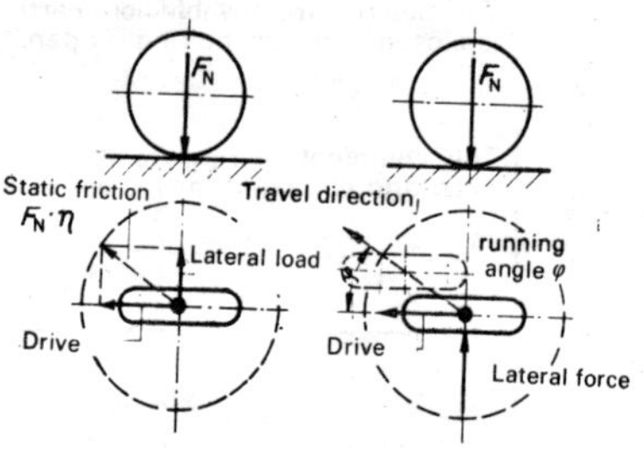

The force exercised by a wheel cannot be greater than the static friction between the wheel and the road surface. The force closure tyre-road is claimed by drive force, brake force or lateral force. From the consideration of safe track holding, the drive forces and brake forces are to be held smaller than the available static friction. If the force closure limits are exceeded the wheel can slide in the circumferential direction as well as laterally.

With the onset of a lateral force, the wheel moves perpendicular to the direction of travel. This transverse movement leads to inclined running with the rolling of the wheel. The angle between the wheel plane and the forward direction of movement is the inclined running angle. It depends upon the wheel force, lateral force, type of tyre, condition of the tyre, inflation of the tyre, wheel camber and the coefficient of static friction.

Overriding or underriding

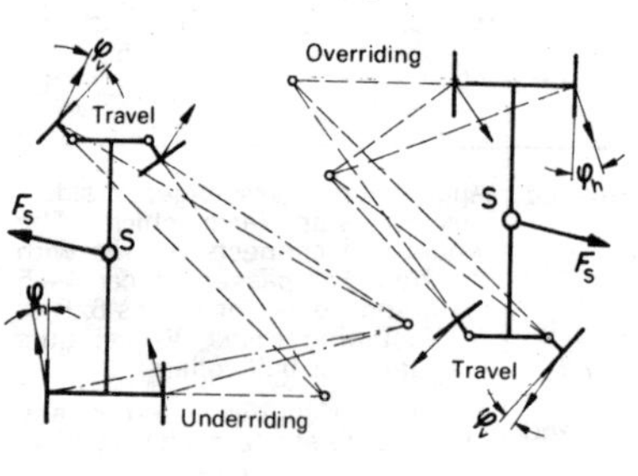

While negotiating a curve, a centrifugal force is caused which passes through the centre of gravity. It is distributed, depending upon the position of the centre of gravity, between the front and rear axles which produces a lateral force at the wheels, which effects a deviation of the direction of travel set by the steering.

The magnitude of the inclined running angle at the front and rear is determined by the lateral force acting at the wheel. If the inclined running angle is greater at the front than at the rear, then the radius of the curve is greater (underriding); if it is greater at the rear than front, the diameter will be smaller (overriding). The influence of the inclined running can be neutralised by suitable steering measures.

Vehicle vibrations 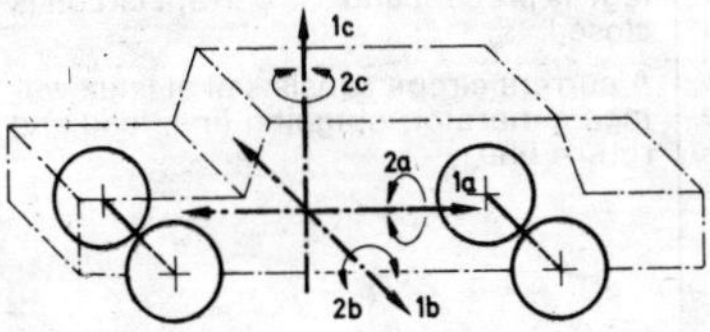	The suspension of the wheels from the vehicle chassis makes undesired vibrations possible. They lesson the travel comfort, produce dynamic auxiliary forces and reduce the travel safety. 1. Linear vibrations parallel to a) Longitudinal axis : jerky movement or sliding b) Transverse axis: shaking or transverse vibrating c) Vertical axis: rising or falling 2. Rotary vibrations about the a) Longitudinal axis: rolling, rotating or tilting b) Transverse axis: pitching or stamping c) Vertical axis: spinning
Wheel vibrations and wheel movements 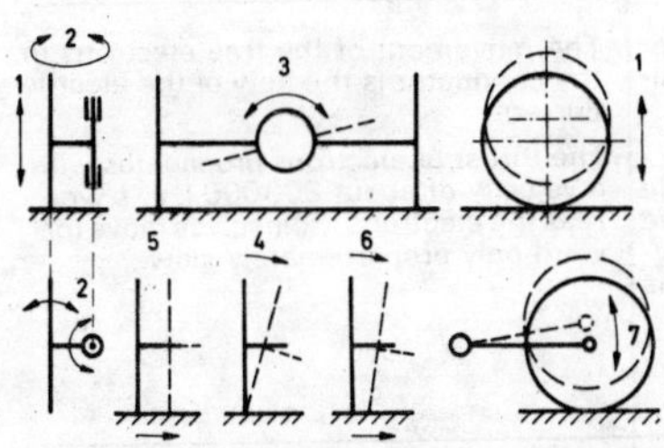	1. Vertical vibratory movement due to vertical component of the road bump or the static unbalance of the wheels. 2. Flutter vibration about a vertical axis (steering rod axis or vertical axis of wheel) due to centrifugal action with change of wheel camber, inclined running and dynamic unbalance of the wheels, horizontal component of road bump, change of the dynamic tyre radius, of the wheel base and of rolling resistance. 3. Beat vibration of a rigid axle. 4. Camber change: centrifugal action, steering deflection 5. Track change: inclined running, tyre wear 6. Track and camber changes small 7. Change of wheel base: rotational speed changing
Distribution of wheel force in standstill condition of the wheel 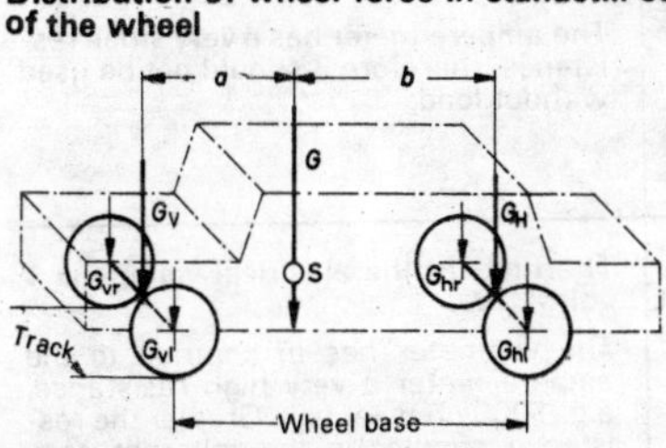	Wheel force is the normal force between wheel and ground. It is decisive for the force closure and hence for drive, brake and track holding. Force at rest — Wheel force in standstill condition of the vehicle. Medium force — Wheel force in a definite condition of travel. Dynamic force — Instantaneous wheel force, sum of the medium force and dynamic additional force. The static wheel force is dependent upon the normal component of the vehicle weight, position of the centre of gravity and the number of axles. It is calculated according to the laws of mechanics.
Distribution of wheel force in braking 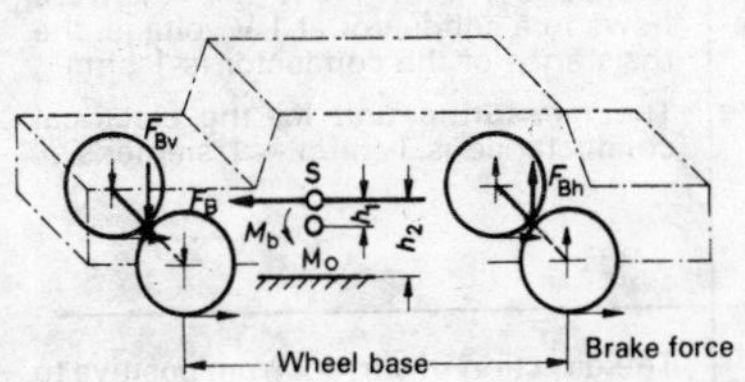	Road unevennesses, braking, accelerations and wind resistance produce longitudinal forces which build auxiliary dynamic forces for the wheels and produce undesired pitching vibrations about the transverse axis of the vehicle. The pitching moment unloads the front axle and loads the rear axle during acceleration; while braking it is reversed. A longitudinal force passing through the centre of gravity of the vehicle does not exercise any pitching moment if the centre of gravity coincides with the instantaneous centre of rotation. The pitching can also be reduced by combination suspension, balancing lever or lowering of the brake. Distribution of the auxiliary force is dependent upon position of centre of gravity, wheel base, wheel suspension and spring characteristic.

Basic concepts in electrotechnology

Current circuit 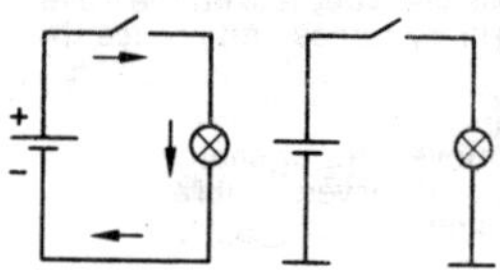	In the external current circuit, the current flows from the voltage generator (battery, generator) through the load and back again to the generator. In automobiles, the vehicle earth (engine, chassis) is used as return line in order to save a line.	Current flows only when electrical voltage is present and the current circuit is closed. A current circuit consists of at least voltage generator, outgoing line, load and return line.
Metallic conductor 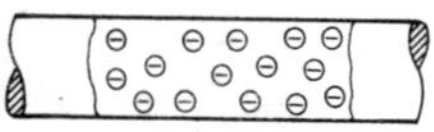	In metals, there are electrons in which the binding forces are raised to the nucleus of the atom. These electrons are freely moving in the conductor and are therefore called free electrons or conducting electrons.	Good conductors like copper and silver contain as many free electrons as atoms. Non conductors like rubber and plastic contain almost no free electrons.
Battery acid 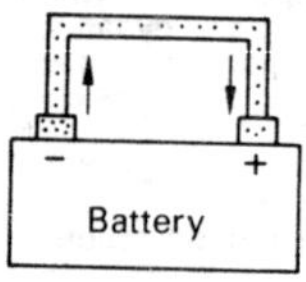	At the negative pole of a voltage generator there is an excess of electrons, at the positive pole a scarcity of electrons. If the two poles are connected by a conductor, the free electrons at the negative pole rush into the conductor and move the free electrons of the conductor. As many electrons are sucked at the positive pole as rush at the negative pole.	The movement of the free electrons in the conductor is the flow of the electric current. The thrust of electrons propagates with a velocity of about 300000 km/s whereas the electrons themselves move forward only proportionately slow.
Measuring of the current 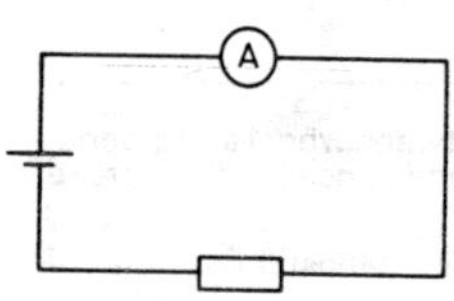	Greater the number of electrons that flow through the conductor cross-section per second, greater is the current or the current intensity. The current is measured with the ampere meter. The ampere meter is connected in the current circuit.	The measuring unit for the electric current is 1 ampere = 1 A. At 1 A, about 6.3 trillion (6.3 $\times$ 10^{18}) electrons flow through the conductor cross-section per second. The ampere meter has a very small resistance, therefore it should not be used without load.
Measuring of the voltage 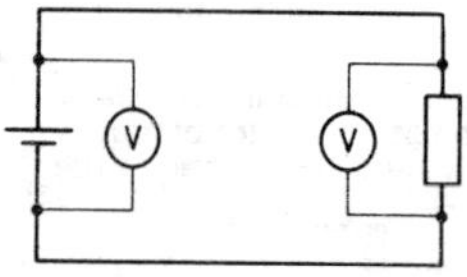	The charge difference between the negative pole and the positive pole brings about a balancing effort of the charges. The balancing effort is the electrical voltage. The voltmeter is connected between positive and negative at the generator or load.	The unit for the electrical voltage is 1 volt = 1V. The voltmeter has in contrast to the ampere meter a very high resistance, e.g. 1000 ohm per volt. Greater the resistance smaller is the inherent consumption of the measuring instrument.
Electrical resistance and conductance 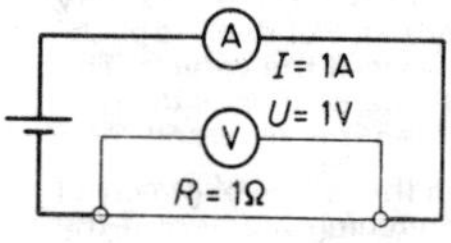	Conductor and load present a resistance opposing the flow of the current since the electrons must move through the intermediate spaces of the atoms. The reciprocal value of the resistance is the conductance.	The measuring unit of the electrical resistance is 1 ohm = 1 Ω. If 1 A current flows in a conductor at 1 V voltage, the resistance of the conductor is 1 ohm. The measuring unit for the electrical conductance is 1/ohm = 1 siemens.
Direction of current 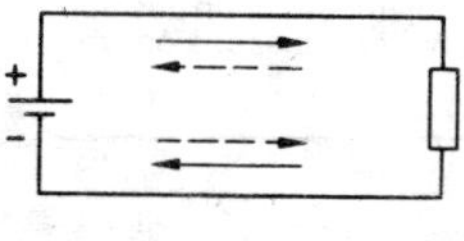	Technical direction of current: The standardized direction of current in the external current circuit is from positive to negative. Direction of electron current: The electrons move from negative pole through the load to the positive pole of the voltage source.	The direction of current from positive to negative was fixed at a time as the electron motion was not yet known. In the voltage generator, the direction of motion is the reverse of that in the external current circuit.

Types of current 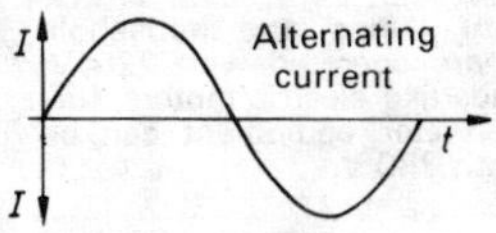	Direct current: In direct current, the conducting electrons of a current circuit always move in the same direction. Alternating current: In alternating current, the conducting electrons oscillate to and fro. The alternating current changes its direction and intensity in regular time intervals.	A complete alternating current oscillation is called a period. The number of periods per second is the frequency. The measuring unit is 1 hertz = 1 Hz. The alternating current of the mains has a frequency of 50 Hz. It changes its direction 100 times per second.
Ohm's law 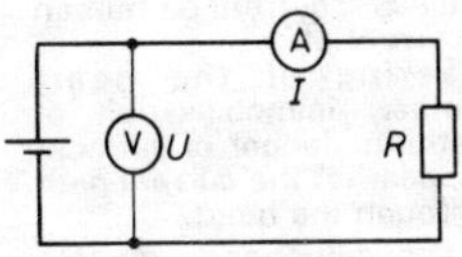	The Ohm's law shows the relation among current, voltage and resistance, for example: The current I in a current circuit is so much larger, as larger the voltage U and as smaller the resistance R are $\text{Current} = \dfrac{\text{Voltage}}{\text{Resistance}}$	Resistance and voltage can be determined from the relations: $\text{Resistance} = \dfrac{\text{Voltage}}{\text{Current}}$ and Voltage = Current×Resistance e.g. 1 V electrical voltage drives a current of 1 A through 1 Ω resistance.
Series circuit 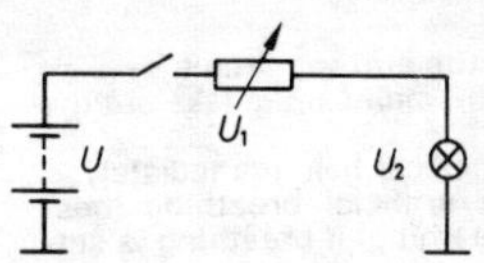	The loads resistors are connected one after the other. The total resistance is given by he sum of the individual resistances. There are partial voltages at the loads which are in the same ratio as the resistances. The current is same at all points of the current circuit.	In a starter battery, the voltage of 12 V is given by 6 batteries each of 2 V connected in series. In the pre-ignition equipment, spark plugs of about 1 V, monitor and preliminary resistors are connected in series. Preliminary resistors in series with the load reduce the voltage at the load (scale lighting).
Parallel circuit 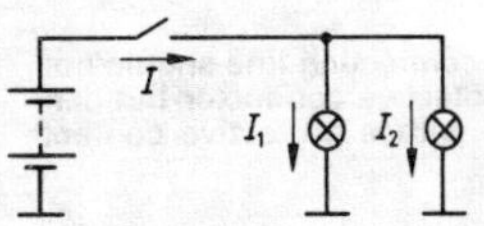	All loads (resistors) are at the same voltage. The individual load currents together give the total current. The individual currents are in inverse ratio as the resistances. The effective resistance is smaller than the smallest individual resistance.	The loads are connected in parallel in the mains so that they are all at the same voltage. They are manufactured for the concerned mains voltage and hence can be fed from the mains independent of one another.
Electromagnet 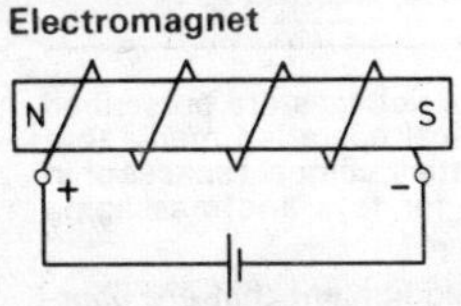	Every conductor through which current flows has a magnetic field around it, therefore a magnetic field is generated in the coil. The magnetic field in the coil is so greater as greater the number of coils is and as greater the current is. The iron core in the coil amplifies the magnetic flux by a multiple.	Rule for the magnetic poles of a coil: If the coil is so held with the right hand that the fingers point in the direction of current (forward +, backward —), the propped thumb points to the north pole of the magnetic field.
Electrical induction 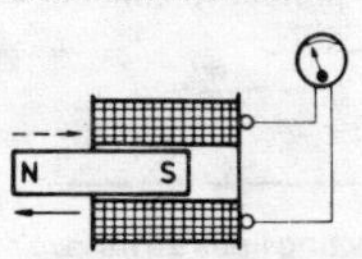	Moving of the magnet or coil induces a voltage in the coil during the movement. The direction of the induced voltage changes when the direction of movement or the direction of magnetic field is changed. With to and fro moving, an alternating voltage is generated.	The induced voltage is so higher as greater the number of coils is, as stronger the magnetic flux is and as faster the movement is. Even without movement, an induction voltage is generated when the magnetic flux inside the coil changes (transformer).
Transformer 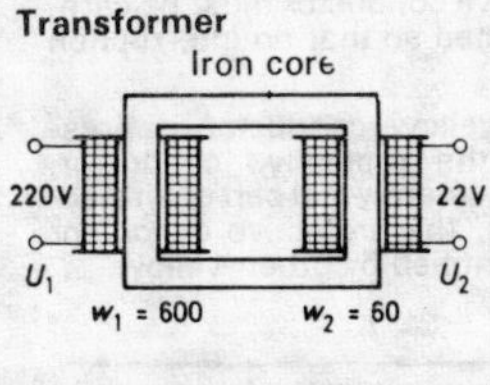	The transformer consists of two coils on a common iron core. The winding I is the input winding (primary winding), the winding II is the output winding (secondary winding). The alternating flux generated by the input winding brings about an alternating voltage of the same frequency as in the input winding in the output winding. The output voltage is guided by the ratio of numbers of coils.	The voltages of the transformer are in the same ratio as the numbers of coils, therefore $U_1 : U_2 = N_1 : N_2$ The currents are in the inverse proportion as the numbers of coils $I_1 : I_2 = N_2 : N_1$ The ratio of the voltages $U_1 : U_2$ is the transformation ratio of the transformer.

Basic concepts in electrotechnology

Four-conductor mains

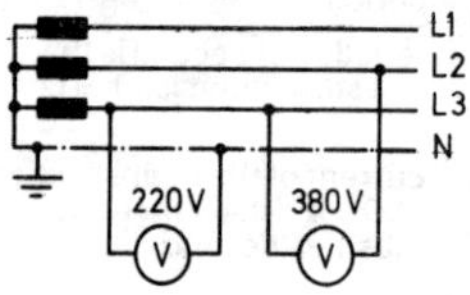

The four-conductor mains is used for the distribution of electrical energy. It consists of the outer conductors L_1, L_2, L_3 and the central conductor (neutral conductor) *N*. There is a voltage of 380 V between any two outer conductors where as the voltage between an outer conductor and the central conductor amounts to 220 V.

By this four-conductor mains there is the possibility to work with two voltages.
Incandescent lamps and household appliances are operated with 220 V. Heavier loads like electric motors, furnaces, hot water equipment can be operated with 380 V.

Dangers through contact with mains

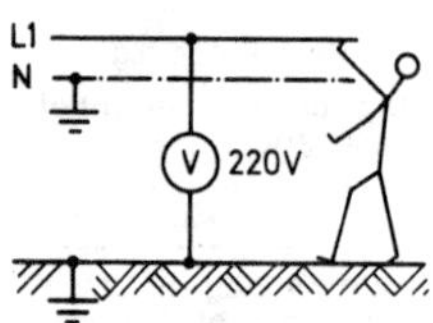

The consequences of an electrical accident are dependent upon

1. the intensity of the current that flows through the body,
2. the duration of action.
3. the path of the current, e.g. the path through the heart is very specially dangerous.
4. the type of current, e.g. the alternating current of the mains is very dangerous.

The intensity of the current is so much greater as smaller the transfer resistances at the places of entry and exit of current are. With moist hands, wet clothing, wet floor the transfer resistance is small and hence the danger is very great.

Effect of the electric current on human beings for a current of

50 mA: flickering of the heart chamber, immobilisation of breathing, danger of suffocation, death if the current path is through the heart.

30 mA: in consciousness, muscle cramps, breathing difficulty.

Contact with mains with voltages above 65 V can give rise to currents over 50 mA which are dangerous to life.

First Aid

1. Interrupt the current circuit: remove the mains plug, take out the fuses.
2. Call for medical help immediately.
3. Introduce artificial breathing (bestowed breathing) if breathing is setting out.

Protective insulation

Symbol for protective insulation

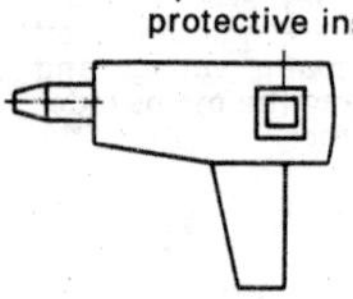

All parts of an equipment that can be contacted, that can conduct voltage in case of a defect are covered permanently with insulation material in addition to the operating insulation.

Driven metallic parts, e.g. drilling spindles, must be separated electrically from the motor by insulation pieces.

There are no connecting points for a protective line on protective-insulated equipment.

The moving connecting line should not have any protective conductor but can be provided with a protective contact plug.

Low protective voltage

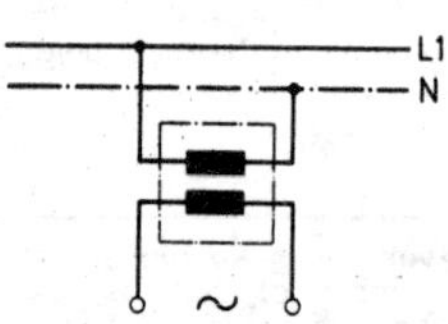

up to 42 V

Low protective voltages are produced with protective transformers, bell transformers and toy transformers.

The input winding of the transformer must be galvanically separated from the output winding.

Low protective voltages are nominal voltages up to 42 V. For toys meant for children only voltages up to 24 V are permissible.

Low protective voltages are prescribed e.g. for electrical operating media that are used in kettles, compact spaces or in damp places, for toys and massaging equipment.

The plugs of equipment that are operated with low protective voltages, should not suit sockets for 220 V. The equipment have no protective conductor terminal.

Protective conductor

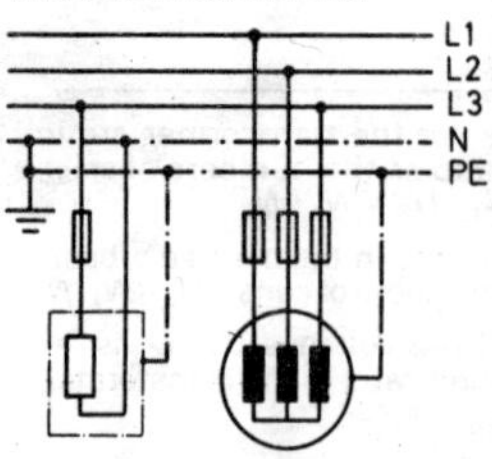

The protective conductor connects the non-conducting metal parts of electrical equipment or machines with central conductor (neutral conductor) N of the mains or with the protective conductor PE in mains with separate laying.

A defective case e.g. connection with body, leads to the flow of a short-circuit current through the protective conductor. The short-circuit current, effects the loading of the fuse and thus the disconnection of the equipment.

In the moving connecting lines an inherent conductor must always be used for the protective conductor.

The protective conductor must be carefully connected so that no interruption can occur.

The green-yellow conductor is prescribed for the protective conductor. Even in permanently laid part of an electrical device, the protective conductor must be identified by green-yellow.

Lead battery

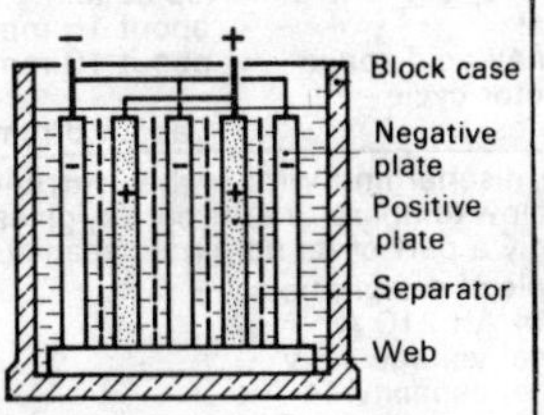

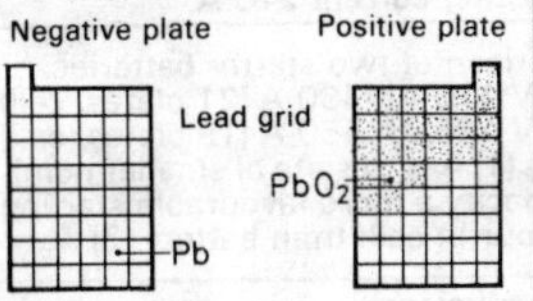

Functions
The battery must store the electrical energy produced by the generator by electrochemical transformation and give it back again on demand, e.g. while starting.

Construction
The basic element of the battery is the cell. It contains the plate block which consists of a set of positive plates and a set of negative plates in which the individual plates are separated from one another by separators placed in between. The cell is filled with mild sulphuric acid.

The block case, e.g. of a 12 V battery, is divided into six cells that are mutually sealed and are tightly closed at the top by the block case cover. The individual cells are connected in series by cell connector. At the first and at the last set of plates, the end poles are welded.

Block case, block cover
From acid-resistant insulation material, partly from hard rubber, mostly from plastic, e.g. polypropylene.

Plates
From lead grid with embedded highly porous effective mass from very small lead particles (negative plate) and lead oxide particles (positive plate). The effective mass is chemically transformed while charging and discharging.

Separators
Microporous insulation plates from acid-resistant plastic. They prevent contact between positive and negative plates, but must allow the electrolyte to pass through.

Electrolyte
Mixture from chemically pure sulphuric acid (H_2SO_4) and desalinated or distilled water. The electrolyte should not be made impure by the mineral salts of water.

Discharging

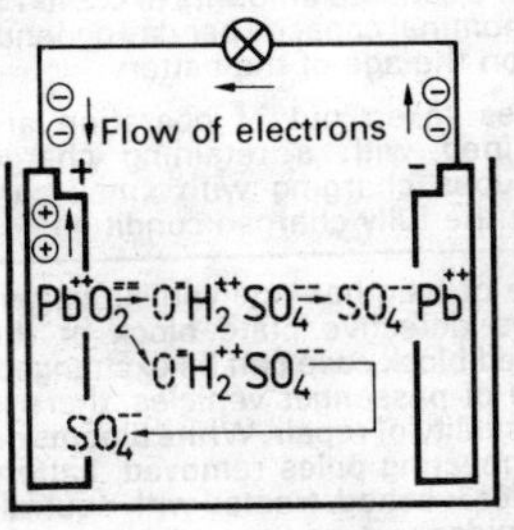

Electrolytic processes
The sulphuric acid is separated into columns of positive H ions and negative SO_4 ions by mixing with water. If the poles of the cell are connected by a load (incandescent lamp), electrons flow from the negative pole to the positive pole. Due to the scarcity of electrons at the negative pole, bivalent positive lead is produced from the neutral lead which combines with the bivalent negative SO_4 group to form lead sulphate $PbSO_4$.

At the positive pole, bivalent positive lead is produced from the quadrivalent positive lead of the lead oxide through the electron supply. The combination with O_2 is therefore ruled out and a combination with SO_4 is introduced, lead sulphate $PbSO_4$ is likewise produced. The oxygen atoms released combine with the hydrogen atoms of the electrolyte to form water. The density of the battery acid decreases.

Discharged condition
positive plates, effective mass lead sulphate $PbSO_4$ negative plates, effective mass lead sulphate $PbSO_4$
Acid density 1.12 kg/dm³
Acid density in degree Baume 16°
Cell voltage, unloaded 1.75 V
High current discharge while discharging with high currents, e.g. while starting, it is to be noted that the water formed in the inside of the plates and on the surface of the plates must mix with the remaining acid. The battery therefore needs relaxation pauses during high current discharges.

Deep discharge
Complete discharge of a battery must be avoided since the resulting lead sulphate has a larger volume and therefore there is the danger of breakage of the effective mass from the plate grid.

Deep discharged batteries should be immediately recharged.

Charging

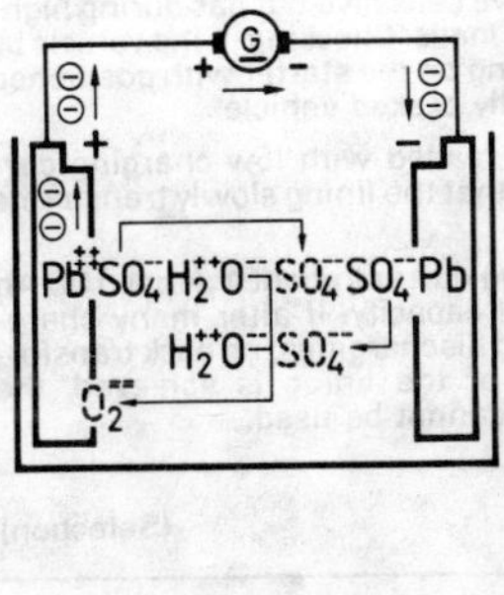

Electrolytic processes
The electrochemical processes take place while charging exactly opposite to those while discharging.
The positive pole of the direct current source is connected to the positive pole of the cell and the negative pole to the negative pole. Electrons are sucked in at the positive pole of the cell and pressed to the negative pole. The lead sulphate compound at both the poles is thereby dissolved. Pure lead is formed at the negative pole. At the positive pole, the quadrivalent lead combines with two bivalent oxygen atoms to form lead-dioxide PbO_2. The SO_4 group goes into the electrolyte by which sulphuric acid is formed. The acid density increases.

Charged condition
Positive plates, effective mass lead dioxide PbO_2
Negative plates, effective mass pure lead Pb
Acid density 1.28 kg/dm³
Acid density in degree Baume 32°
Cell voltage, unloaded 2.1 V
Charging voltage of each cell
at the beginning of gas formation 2.4 V
at the end of charging 2.7 V
Charging current about 0.1 × capacity number
Quick charging about 0.8 × capacity number
Towards the end of the charging water is resolved, hydrogen is formed at the negative plates and oxygen at the positive plates (oxyhydrogen gas). Sparks therefrom can lead to the explosion of a cell.

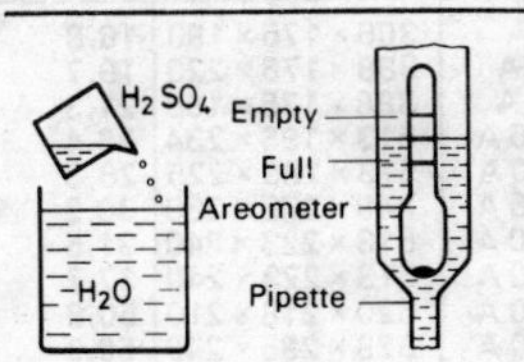

Charge condition of the battery	Density at 20°C kg/dm³		Consumption of the acid mixture: Water percentage volume	H_2SO_4 percentage volume	Freezing point C
Fully charged	Normal	1,28	74	26	−68
	Tropic	1,23	79	21	−40
Half charged	Normal	1,20	82	18	−27
	Tropic	1,16	86	14	−17
Discharged	Normal	1,12	90	10	−11
	Tropic	1,08	94	6	−8

Starter battery (accumulator)

Commissioning Delivery condition of the battery: unfilled, plates preloaded and chemically protected, therefore can be stored.	Closing stopper opened and sulphuric acid of 1.28 kg/dm³ density filled up to acid level mark. The acid level checked after 20 minutes. The battery is now ready for operation, and has 80% of its nominal capacity.	Metallic funnels should not be used for filling the acid. Acid level over the plate top edge: normal about 15 mm for small acid space about 10 mm for motor cycle batteries about 6 mm
Capacity Current quantity (charge) in ampere hours (Ah) which is given out during discharge. The capacity is determined by the quantity of the effective mass.	Nominal capacity K_{20} according to DIN 72311: Discharge capacity in Ah for 20 hours discharge time. Minimum value for discharge close voltage for 6 V batteries 5.25 V, for 12 V batteries 10.5 V. Electrolyte temperature 27°C. Discharge current 1/20 of the capacity value.	While discharging with higher current and at low temperature the battery gives out only a part of its nominal capacity. Example of designation: 12 V 44 Ah 210 A Nominal voltage 12 V Nominal capacity 44 Ah Cold testing current 210 A
Cold testing current The cold testing current gives information on the starting capacity at low temperatures.	In cold testing, the battery is discharged at a temperature of —18°C with the cold testing current. The permissible minimum voltage of each cell is: After 30 seconds discharge time 1.4 V After 180 seconds discharge time 1.0 V	Comparison of two starter batteries: (1) 12 V 110 Ah 490 A (21 plates/cell) (2) 12 V 135 Ah 450 A (19 plates/cell) Battery (1) has in spite of smaller nominal capacity a more favourable starting behaviour in cold than battery (2)
Efficiency For the determination of the efficiency, a fully charged battery is discharged and again charged.	Current quantity (charge) efficiency: $\eta_{Ah} = \dfrac{\text{Current quantity in Ah given out}}{\text{Current quantity in Ah fed in}}$ For 20 hour discharge and 27°C, ηAh amounts to about 90%.	Energy efficiency $\eta_{Wh} = \dfrac{\text{Electrical work in Wh given out}}{\text{Electrical work in Wh fed in}}$ For 20 hour discharge and 27°C, η_{Wh} amounts to about 75%.
Self discharge Charged batteries lose their charge after a particular time even if the loads are not connected.	Due to metallic impurities of the acid and antimony sediments of hard lead, small local elements are formed on the effective mass which lead to self discharge. The self discharge increases at higher temperatures.	The self discharge amounts to 0.2 to 1% of the nominal capacity per day depending upon the age of the battery. Batteries taken out of operation are maintained with a retaining charge (continuous charging with small current) at the fully charged condition.
Break down Discharge performance too small, the voltage falls steeply with load.	Short circuit inside a cell: acid density too low, cells discharge Leakage between two neighbouring cells: acid density too low in both cells, both cells discharge. Acid density too low in all cells in spite of charging: sulphating of the plates. Pole terminals oxidised or loose, blooming of white salts: load current falls. Battery old, positive plates washed out; plate closing by deficient mass, battery capacity small.	In case of load haulage vehicle batteries, the defective plate block or the unsealed block case can be exchanged. In case of passenger vehicles, there is no possibility of repair. White blooms on the connecting poles removed, battery and poles washed, treated with neutralising fluids. Pole terminals are tightened, corroded terminals are exchanged, poles and pole terminals greased with acid protective grease. Life of a starter battery in the normal case is about three years. Defective cells give out gas during high-current loads. Checking in the vehicle by switching on the starter with positioned gear fully braked vehicle.
Sulphating Grey-white lining on the positive and negative plates. Discharge capacity and acid density too low, charging voltage high, strong heating while charging.	Coarse crystalline lead sulphate is formed from the fine crystalline lead sulphate on the plate surface which can be transformed back only with a lot of difficulty. Sulphating of the battery occurs due to longer standing in the uncharged condition.	Battery loaded with low charging current so that the lining slowly transforms back. Charging current about 1 A per 100 Ah nominal capacity. If after many chargings and dischargings, no back transformation of the lining is achieved, the battery cannot be used.

Types of battery

(Selection)

DIN Type-number	Designation of battery	Dimensions $l \times b \times h$, mm	Weight kg	DIN Type-number	Designation of battery	Dimensions $l \times b \times h$, mm	Weight kg
005 16	6 V 4,5 Ah	95 × 85 × 123	1,2	566 18	12 V 66Ah 300 A	306 × 175 × 190	16,8
012 14	6 V 12 Ah 45 A	123 × 61 × 137	1,7	584 11	12 V 84 Ah 280 A	388 × 175 × 220	19,7
066 12	6 V 66 Ah 270 A	178 × 175 × 188	6,3	588 11	12 V 88 Ah 395 A	388 × 175 × 190	21,3
515 11	12 V 15Ah 110 A	196 × 82 × 172	3,9	605 11	12 V 105 Ah 350 A	513 × 196 × 234	26,4
527 12	12 V 27 Ah 130 A	210 × 175 × 175	8,0	610 11	12 V 110 Ah 490 A	513 × 189 × 225	28,3
536 24	12 V 36 Ah 175 A	210 × 175 × 175	8,3	625 11	12 V 125 Ah 435 A	349 × 175 × 289	30,2
540 86	12 V 40 Ah 150 A	263 × 137 × 233	12,0	635 11	12 V 135 Ah 450 A	513 × 223 × 240	31,5
544 81	12 V 44 Ah 210 A	210 × 219 × 223	12,0	643 11	12 V 143 Ah 630 A	513 × 223 × 240	37,7
556 11	12 V 56 Ah 190 A	262 × 182 × 220	13,8	670 15	12 V 170 Ah 470 A	520 × 218 × 210	50,9
560 49	12 V 60 Ah 200 A	272 × 173 × 225	17,1	680 11	12 V 180 Ah 600 A	523 × 295 × 249	50,9

ypes, Functions

. System with bright-dark-limit (European continent)
 a) with symmetric dim light
 b) with asymmetric dim light
2. Additional head light
 a) halogen bright (high beam) head light
 b) fog head light
 c) wide beam head light (not permitted in Germany)
3. System without bright-dark limit (America, Great Britain
 a) sealed beam system
 b) sealed beam four head-light system

1. Largest possible range. Intensity of light at least 1 lux at a distance of 100 m in case of bright light (high beam light), at a distance of 25 m in case of dim light. At 1 lux a dark-clothed person is not clearly visible.
2. Uniform lighting of the road. With non-uniform lighting fast straining of the eye due to bright-dark adaption.
3. Lateral dispersion (journey along a curve).

Main parts: Housing with adjusting device. Mirror (reflector) paraboloid in shape, drawn from steel sheet. Ground inside, lacquered, aluminium vapour coated in vacuum and coated with quartz.

Two wire lamp (bilux lamp) for bright and dim lights.

Dispersion disc with pressed-in cylindrical lenses (ribs) or prisms.

ath of beam

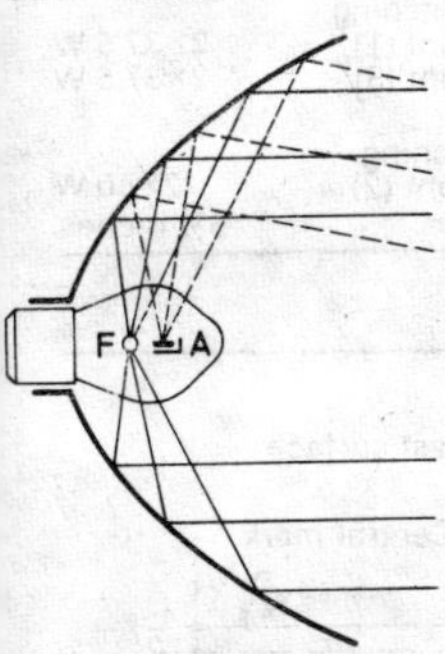

Bright light (high beam)

The bright lighting element (F) of the two-wire lamp lies at the focus of the mirror. All light beams are reflected parallel to the mirror axis (beam formation), therefore large range. The light distribution is uniform due to the dispersion disc.

Dim light

The dim lighting element (A) lies in front of the focus, therefore bending of the generated light beams. The dim light cap does not allow any light into the lower half of the mirror so that only the upwards directed light appears from the upper half of the mirror.

Low parabola mirror conceives large quantity of light.

Focal length	20 to 40 mm
Reflection	about 90 %
Amplification of the light intensity by the reflector	about 1000 times

Two-wire lamp

Temperature of the lighting element (tungsten alloy) about 2000°C

Blackening of the lamp occurs due to vapourised tungsten.

Incandescent lamps should not be held with bare fingers since the residues vapourise or burn.

A voltage loss of 10 % at the lamp gives a light intensity loss of about 30 %.

Symmetric dim

Incandescent lamp　　Dispersion disc

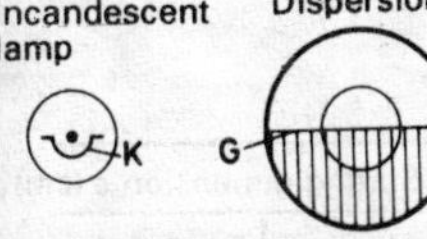

The dim light cap (K) of the incandescent lamp limits the light exactly to the horizontal axis whereby the bright-dark limit (G) occurs.

The light distribution on the road is symmetric to the longitudinal axis of the vehicle.

Incandescent lamp power bright light (high beam)/dim light. 6 V 35/35 W, 12 V 35/35 W, 24 V 45/40 W

Application for single track automobiles, for tractors and automobiles with permanent dim light.

Asymmetric dim light

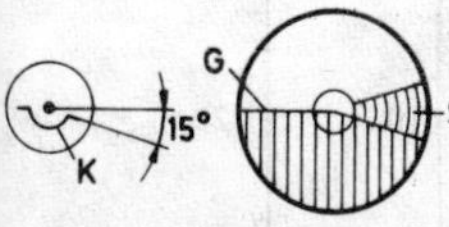

The dim light cap (K) is inclined to one side by 15° whereby light falls on the lower half of the mirror. The resulting light therefore reaches the right side of the road further in the front. The light distribution is asymmetric.

Incandescent lamp power bright light (high beam)/dim light. 6 V 45/40 W, 12 V 45/40 W, 24 V 55/50 W

Permitted only for multitrack automobiles. The sector (S) in the dispersion disc serves for the asymmetric light distribution.

Halogen filled incandescent lamps

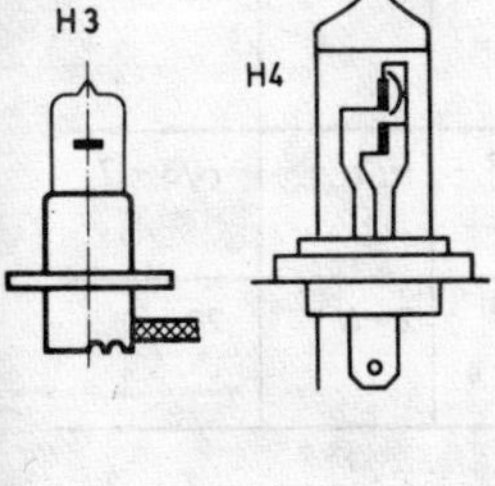

Halogen cyclic process

The iodine mixed to the filling gas combines at the bulb wall with vapourised tungsten to form tungsten iodide and deposits as a result of thermal current on the hot lighting element where it splits again into tungsten and iodine. The blackening of the bulb which must be hot at least 250°C is thereby prevented.

Halogen bright head light permissible for bright light.
1. simultaneously with the main head lights
2. simultaneously with dim light
3. only halogen bright head light.

Incandenscent lamp power:

Type H 1 and H 3	6 v, 12 V, 55 W 24 V, 70 W
Type H 2	12 V 55 W
Type H 4	12 V, 60/55 W 24 V, 75/70 W
Lighting element temperature	up to 3000°C
Gas pressure (operating pressure)	up to 2 bar
Light output compared with a normal incandescent lamp	150 %

Types H 1 and H 2 single wire lamp with lighting element in the lamp axis.
Type H 3 single wire lamp with lighting element perpendicular to lamp axis.
Type H 4 two wire lamp four bright light and dim light.

Head light

Sealed beam system (USA)		

Sealed beam system (USA)

Bright light

The lighting element (F) lies at the focus of the mirror. All light beams are reflected parallel to the axis of the head light.

Dim light

The lighting element (A) lies above the focus and is laterally shifted, therefore, the side of the road is illuminated brighter. Besides the upward directed light, light from the boundary zone of the mirror directed upwards appears.

Sealed beam because incandescent lamp, mirror and dispersion disc form a unit.

Plane, parabolic mirror,

Focal length	about 30 mm
Mirror diameter	7 inches
Bright light/dim light	50/45 W

The dispersion disc is divided into rectangular prisms. It effects the lateral dispersion and the breaking of the upward directed light whereby the blend effect is moderated.

Sealed beam four head light system

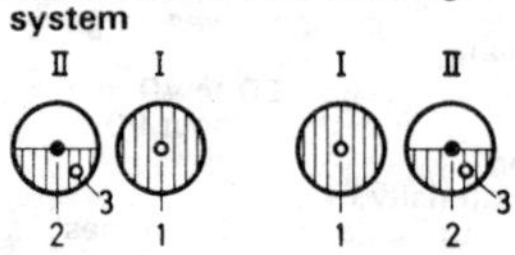

Head light type I
Lighting element (1) at the focus produces sharp beam bright light

Head light type II
Lighting element (2) produces asymmetric dim light
Lighting element (3) produces dispersed additional light for the bright light.

Bright light switching	
Lighting element (1)	2×37,5 W
Lighting element (3)	2×37,5 W
Dim light switching	
Lighting element (2)	2×50 W
Mirror diameter	5¾ inches.

Setting of head light

Test surface (dimensons in cm)

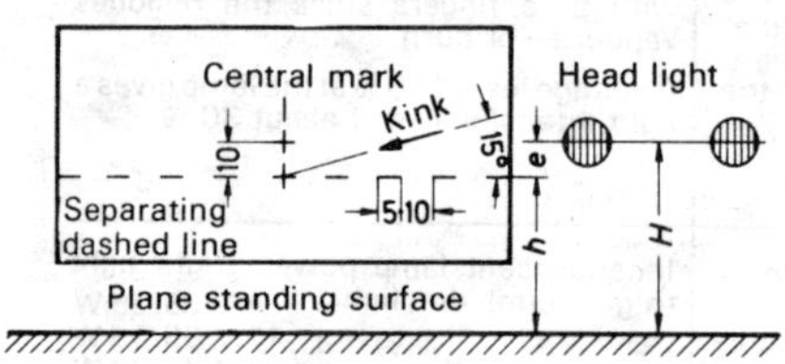

Set up of the test surface

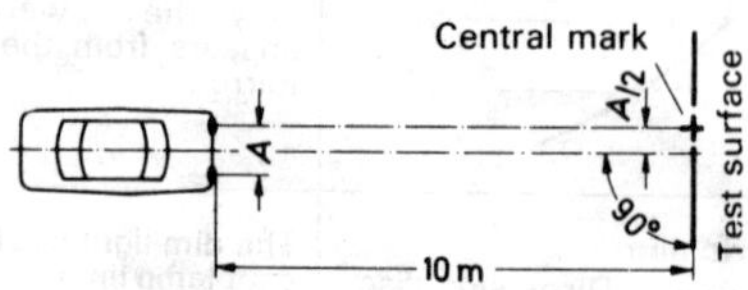

Type of vehicle	Setting load	Setting dimension *e* (cm)	
		Head light	Fog head light
1. Vehicles with upper mirror edge of head light up to 135 cm above the standing surface a) Passenger vehicles b) Automobiles with level-controlled suspension or automatic inclination balance of light beam c) Multi-axle trailor or processing machines d) Single-track vehicles e) Lord haulage vehicles with front loading surface	Passenger vehicles One person or 70 kg on the rear seat. In passenger vehicles without rear seat, one person or 70 kg on the driver's seat. Emergency seats are not considered as rear seats. Load haulage vehicles and other multi-track vehicles: Setting in the unloaded condition	10	20
f) Load haulage vehicles with rear loading surface g) Saddle trailors, Omni buses h) Passenger vehicles that are taken as combination vehicles, however vehicles according to 1b	As above	30	40
2. Vehicles in which the upper mirror edge of the head light lies higher than 135 cm above the standing surface	Like automobiles under 1 and 3	$H/3$	$H/3 + 7$
3. Single-axle trailor or processing machines with permanently dimmed head lights on which the required inclination of the light beam is specified.	Single-track and single-axle trailors or processing machines: 1 person or 70 kg on the driver's seat	$2 \times N$	20

Every head light individually set. The bright-dark light must touch the separating dashed line. In case of asymmetric head lights, the kink must lie on the vertical through the central mark. The light beam middle of the bright light must lie on the central mark. N=dimension in cm by which the light beam shall be inclined at 5 m distance.

Types, Functions

1. Battery-fed equipment
   a) Coil ignition
   b) Transistor or ignistor ignition (TSZ)
   c) High-voltage condensor ignition (HKZ)
2. Magneto ignition equipment
   a) Magneto ignition
   b) Flywheel magneto ignition
   c) High voltage magneto condensor ignition (MHKZ)

1. Generation of a high-voltage impulse (about 30000 V) by transforming the battery voltage.
2. Distribution of the high voltage to the spark plugs.
3. Generation of a spark across the plug electrodes for inflaming the fuel-air mixture.
4. In magneto ignition, the primary energy is produced by a rotating magneto system.

Fields of application:
TSZ: Engines with high speed
HKZ: In case of unfavourable sooting resistance of the plugs, e.g. rotary piston engines
MHKZ: Small engines with high speed, e.g. engine saws

Ignition voltage requirement. High ignition voltages in case of:
large electrode gap
high compression thin fuel-air mixture
low electrode temperature.

Coil ignition

Circuit

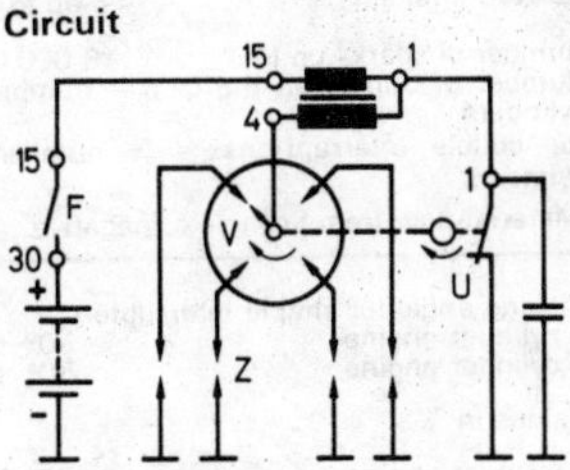

Primary current circuit
Positive of battery, travel switch F, primary winding terminal 15, terminal 1, interruptor U, earth, negative of battery.

Secondary current circuit
Secondary winding terminal 4, distributor V, spark plug Z, earth, negative of battery, travel switch F, terminal 15, secondary winding.

Primary winding
Lies outside, good heat drain.

Number of coils	150 … 200
Diameter of wire	0,4 … 0,6 mm
Steady current	up to 5 A

Secondary winding

Number of coils	about 20000
Diameter of wire	0,06 … 0,08 mm
Voltage, unloaded	20 … 35 kV
Voltage, loaded	6 … 15 kV

Ignition coil

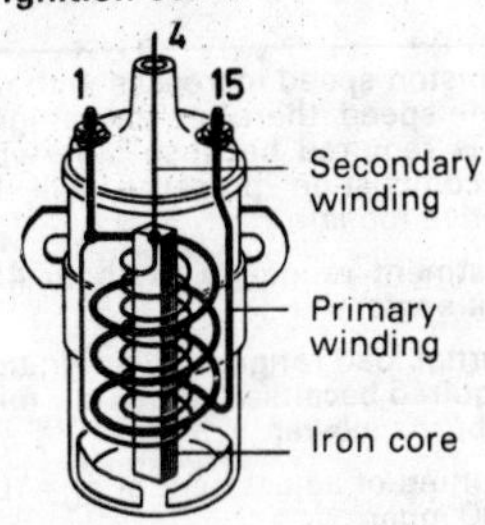

Generation of ignition voltage

The primary current produces a strong magnetic flux in the iron core of the ignition coil.

With the opening of the interruptor, the magnetic flux is suddenly eliminated.

Due to the sudden elimination of flux, a high voltage impulse is induced in the secondary winding.

Ignition coil capacity
Maximum number of sparks per minute

6 V coil	13000
12 V coil	18000

Wattage consumption for 1000 revolutions per minute of the engine

6 V coil	15 … 20 W
12 V coil	20 … 30 W

Transformation ratio primary to secondary

Normal coil	1:100
High performance coil	upto 1:200
Operating temp.	max. 90°C

Preliminary resistance

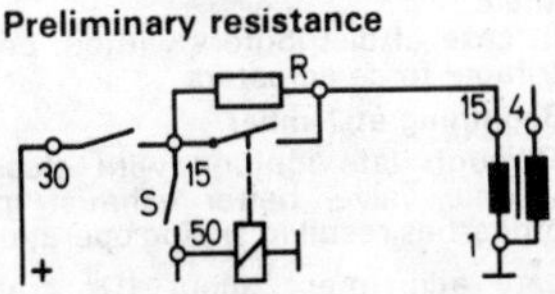

Starting accentuation

The ignition coil has its normal power with the series-connected resistance R. While starting (switch S), the preliminary resistance is bridged by the relay whereby the primary current in the ignition coil increases.

1. Increase of power while starting.
2. Thermal loading of the coil with steady current small because heat is produced outside in the preliminary resistance.

Values of preliminary resistance:
0.4, 0.6, 0.8, 1.2 ohm.

Condensor

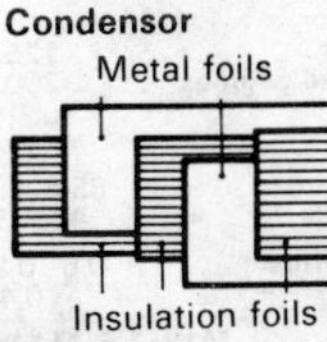

When the interrupter is opened, the condensor is charged by the self-induction voltage and thus prevents contact fire.

It is discharged again immediately through the primary winding which leads to rapid field elimination.

The ignition condensor consists of a winding of two mutually insulated metal foils from which one is connected to the aluminium housing (earth) and the other to the supply line.

Capacity (μF microfarad)	0.2—0.3 μF

Primary current

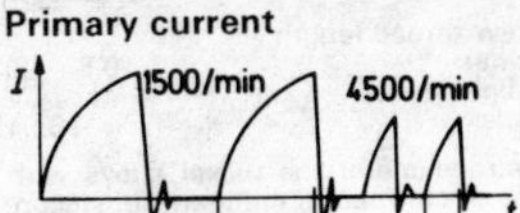

With increasing speed of the engine, the closing time of the interrupter becomes shorter, therefore the primary current reduces.

The ignition capacity drops with increasing speed.

Closing time for a simple interrupter and an engine speed of 4000 1/mm.

four—cylinder engine	4ms
six—cylinder engine	3ms

Ignition equipment

Ignition distributor

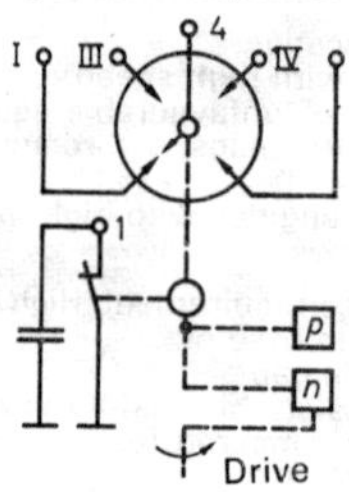

The ignition distributor distributes the ignition voltage to the individual cylinders corresponding to the ignition sequence. The ignition voltage is fed from terminal 4 through sliding carbon to the distributor rotor, jumps over the distributor electrodes and comes to the spark plugs through the ignition cable. Due to the formation of nitrous oxide and ozone through the sparks, the distributor is ventilated. Interrupter (*U*), centrifugal force adjuster (*n*) and negative gauge pressure adjuster (*p*) are built into the distributor

Ignition sequence

4 cyl. in-line engine	1-3-4-2 1-2-4-3
4 cyl. opposed cyl. engine	1-4-3-2
6 cyl. in-line engine	1-5-3-6-2-4 1-2-4-6-5-3 1-4-2-6-3-5
Air gap between the distributor electrodes	0,3...0,7 mm

Speed of the distributor shaft for

4 stroke engine	½ speed of crank shaft
2 stroke engine	speed of the crank shaft

Interrupter and ignition adjuster

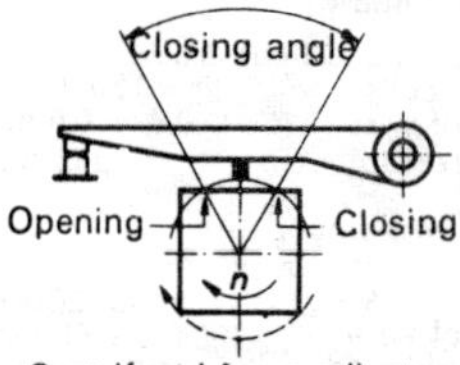

Interrupter

The contact opening begins when the cam of the fibre piece of the insulated supported interruptor lever makes contact.

The ignition point of time is adjusted by changing the contact distance, e.g. increase effects early ignition.

Contact opening	0,3...0,4 mm
Contact force	4...6,5 N
Tungsten support	0,5...0,8 mm
Current intensity	up to 5 A
Number of sparks up to	18 000/min

Number of humps on the cam = number of cylinders,
for double interruptions = ½ number of cylinders
Material flow from positive to negative

Closing angle

Rotation angle of the cam while the contact is closed. Greater contact distance gives smaller closing angle and vice-versa.
Setting of the contact by measuring the closing angle.

Closing angle for simple interrupters:

4 cylinder engine	50° ± 3°
6 cylinder engine	50° ± 3°

Values in %

4 cylinder engine	52...59%
6 cylinder engine	62...71%

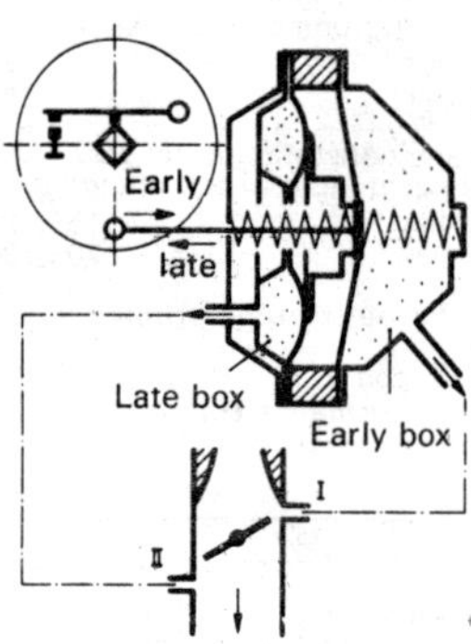

Centrifugal force adjustment
Matching the ignition point of time to the engine speed
Centrifugal weights spread out and rotate the rotatably supported cams in the direction of rotation, more early ignition results thereby.

Negative gauge pressure adjustment
Matching of the ignition point of time to the load on the engine.
Partial load (I):
Negative gauge pressure maximum with partially opened throttle valve. The carrier plate of the interrupter is rotated against the direction of rotation of the cam; ignition early.
Without load (II):
Negative gauge pressure maximum with closed throttle valve: adjustment in the direction late.

The piston speed increases with the engine speed, therefore earlier ignition is required because otherwise the combustion pressure will be effective too late.

Adjustment range up to about 45° Crank angle.

In partial load range, earlier ignition is required because the fuel-air mixture burns slower.

Beginning of adjustment at p_u=100 to 130 mbar
Adjustment range up to about 20° crank angle.
In case of distributors without centrifugal force adjusters
Beginning at 1 mbar
Through late ignition with closed throttle valve, better exhaust gas properties result in sliding operation.

Late adjustment about 10° crank angle.

Spark plug

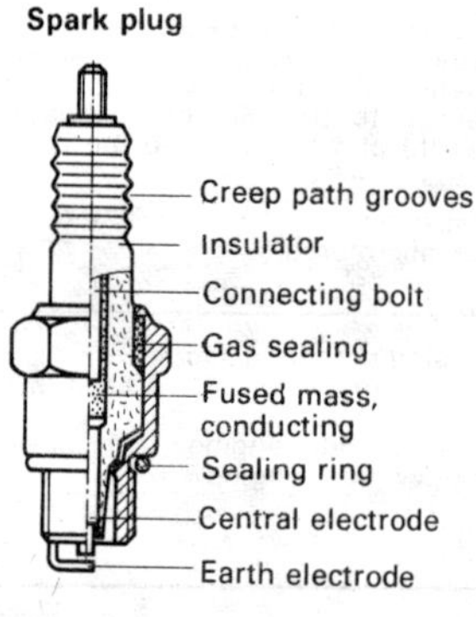

The thermal loadability of a spark plug is specified by the thermal value (comparative value, not a temperature value). Higher the thermal value, higher is the thermal loadability of the spark plug.

Engines with high combustion chamber temperatures at full load require, e.g. spark plugs with high thermal value.

With too low thermal value of the spark plugs, electrodes and insulator are overheated; incandescent ignition comes in. With too high thermal value, the free-burning temperature is not attained; dirt collects on the spark plugs.

For sparking disturbances, there are spark plugs with built-in suppressor resistances of 5000 ohm.

Thermal values for

standard engines	125...175
highly compressed engines	200...280
race engines	up to 500
Spark plug temp.	850...950 °C
Free burning temp.	450...500 °
Electrode gap, normal	0,6...0,8 mm
special cases	0,4...0,5
Screw thread	M 18×1,5, M 14×1,25
rarely used	M 12×1,25, M 10×1

Screw thread length,

normal	9,5...12,7 mm
for light metal heads	18...19 mm

Electrode materials: nickel alloys with additions of chromium, manganese, silicon
Insulator from sintered corundum, 95 % aluminium oxide.

Transistor-coil ignition, contact regulated

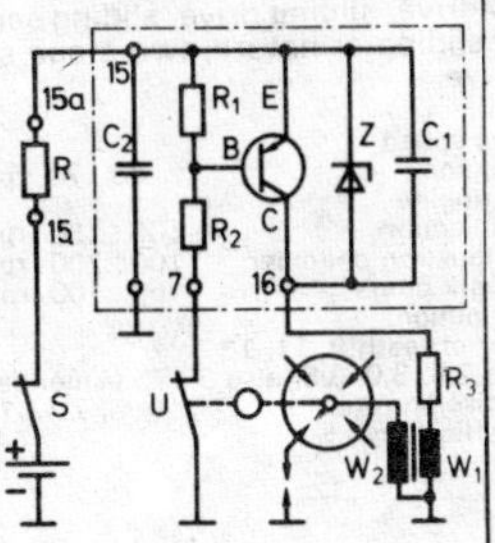

On closing the interruptor, a control current flows from the positive of the battery to the earth through travel switch S, preliminary resistace R, emitter E, base B, resistance R_2, interruptor U.

The control current effects the through-switching of the transistor from emitter E to collector C so that primary current flows through R_3 to ignition coil (15), primary winding W_1, earth. The magnetic field in the ignition coil is built-up.

At the ignition point of time, the control current is interrupted, thereby the transistor suddenly blocks the through flow of the primary current from E to C, hence elimination of field and generation of high voltage.

The condensor C_1 and the Zener diode Z protect the transistor from inductive over-voltages.

Control current	0,8...1,0 A
Primary current (steady current)	8...10 A
Secondary voltage	25...30 kV
Number of sparks, max.	21 000/min
Power for n= 1000/mm	approximately 70 W

Significantly higher ignition voltages and energy-richer ignition sparks in the upper speed range than with traditional coil ignition. Starting behaviour, idle running and rotation of the engine are favourably influenced. Interrupter contacts are loaded only by the small control current and not inductively, therefore no opening spark and only a little wear. All electronic construction elements are assembled protected in the switching device.

Precaution: Ignition to be switched off before working on it!

Transistor-coil ignition, contactless

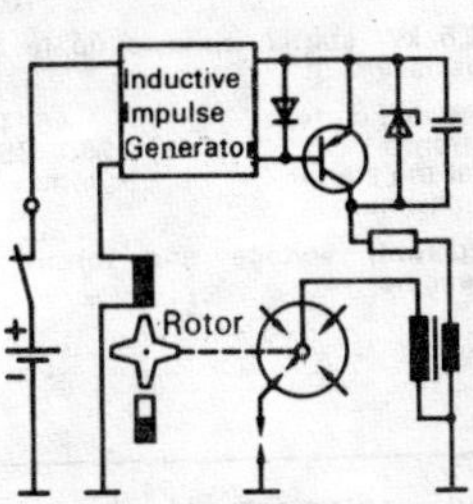

In contactless controlled transistor ignition the interrupter is replaced in the distributor by an inductive impulse generator. A toothed disc (rotor) made from soft-magnetic steel produces magnetic flux changes in a permanent magnet during its rotation. The flux changes effect voltage impulses in the induction coil surrounding the magnet. These control impulses which are modified and amplified electronically actuate the switching transistor and thus produce the ignition voltages.

Besides the inductive generators, sound generators are also used. The voltage impulses are produced in a current through-flowing semiconductor platelet by a speed-dependent magnetic field.

Primary current	up to 10 A
Ignition voltage	25...30 kV
Number of sparks	up to 40000/min

Ingition impulse generators require no maintenance since they work without wear. The ignition point of time is exactly governed with switched on ignition and standstill engine, no steady current flows in the ignition coil.

The number of teeth (spikes) of the rotor is equal to the number of cylinders of the engine.

In race engines, the ignition is controlled by special impulse generators which are built on the fly wheel.

Electronic ignition equipment produce dangerous voltages; they are therefore dangerous equipment according to VDE guidelines.

Magneto transistor ignition

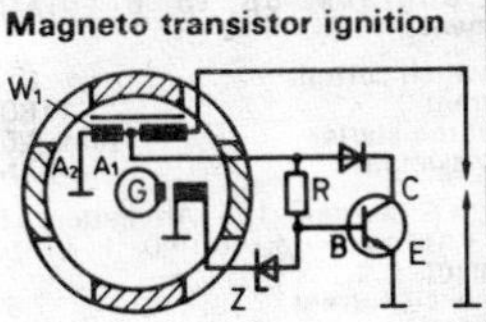

When the magnetic flux of the cogwheel in the primary winding W_1 of the ignition armature produces a positive voltage, the transistor is positive controlled through the resistance R so that it is connected through from C to E. Primary current flows. At the ignition point of time, a negative impulse comes from generator G through the Zener diode Z to B. The transistor interrupts the primary current.

As soon as the primary current is interrupted a rapid magnetic flux change is produced in the ignition armature and hence a high voltage in the secondary winding.

There are magneto transistor ignitions with inherent triggering. They do not need any generator. For the control of the switching device, the primary winding (connection A 1 to A 2) is used.

High-voltage condensor ignition

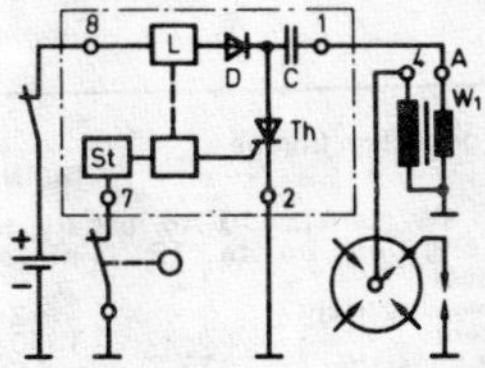

On switching-on the ignition, the battery voltage is transformed to about 400 V load direct voltage in the loaded part through a direct voltage connector and thus the storage condensor C is charged through the diode D. At the ignition point of time, the thyristor Th is controlled by the interrupter or contactless through the control part whereby the thyristor becomes conducting. Now the storage condensor C discharges through the thyristor Th and the primary winding W_1 of the ignition transformer. High voltage is induced in the secondary winding W_2.

Primary voltage	400...450 V
Secondary voltage	about 30 kV
Voltage rise	8000 V per μs
Duration of spark	about 300 μs

Due to the very rapid voltage rise (coil ignition only 400 V per μs), the condensor ignition is insensitive against parallel connections in the ignition circuit, e.g. in case of spark plugs on which dirt has deposited. The short spark duration can however give rise to unfavourable exhaust gas characteristics.

Precaution: Ignition switched-off before working on it!

Magneto high-voltage condensor ignition

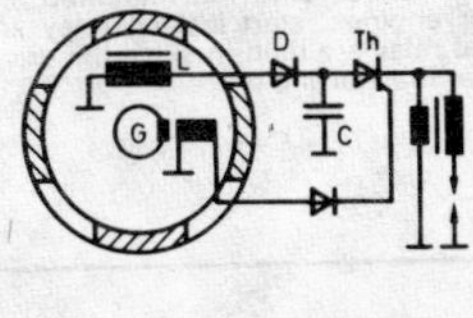

The rotating cogwheel produces in the winding of the load armature an alternating voltage which is rectified through the diode D and charges the condensor C.

A the ignition point of time a voltage impulse comes from the impulse generator to the thyristor Th which thereby becomes conducting. The condensor discharges through the thyristor and the ignition transformer.

A high voltage impulse is produced in the secondary winding of the ignition transformer due to the discharge current of the condensor.

The magneto condensor ignition is insensitive against neighbouring connections, e.g. due to moisture in the ignition circuit. The wear is small, therefore it needs only a little care.

Starter

Method of working

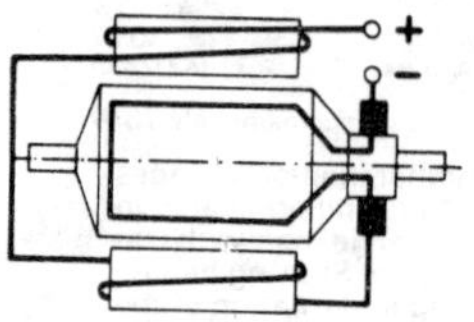

1. The starter is a series connected motor. Excitation and armature windings are connected in series.
2. Full current flows through the excitation winding, therefore large torque.
3. At the beginning of starting (maximum current intensity), the torque is greatest, the motor is actuated
4. After starting the engine, the flow of force is disconnected through roller free wheel or plate clutch.

Differentiation according to the type of engagement:
screw drive, sliding drive, sliding screw drive, sliding armature, two-stage sliding drive.

Starting speed	
Otto engine	40...70 rpm
Diesel engine	
direct injection	100...150 rpm
precombustion chamber	100...200 rpm
with spark plugs	60...100 rpm
Starter pinion:	
Number of teeth 9, 11, 13.	
Module 2,5; 3,0; 3,5; also 3.175 (American).	
Transmission ratio	7:1...12:1
Mean efficiency	45%

Sliding screw drive starter

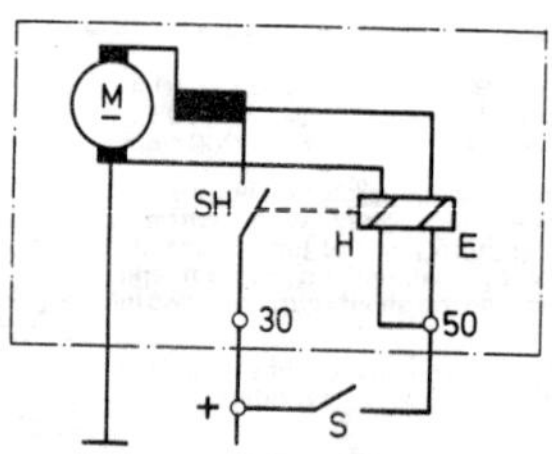

1. With switching-on (S), a current flows through the draw-in winding (E) and holding winding (H).
2. The pinion is moved until meshing, the main current switch (SH) is closed, the draw-in winding is short circuited.
3. The holding winding holds the pinion in mesh until switched-off.
4. After starting the engine, the pinion runs free. The flow of force between pinion and armature is disconnected by the roller free wheel.

6 V, 0.4 kW starter, suitable up to 1.5 l piston displacement:

Ignition switch current	70/13 A
Main current	440...500 A
Voltage at the starter	min. 3,5 V
Torque, maximum	10 Nm

12 V, 0.6 kW starter, suitable up to 25 l piston displacement:

Ignition switch current	46/11 A
Main current	280...350 A
Voltage at the starter	min. 7 V
Torque, maximum	15 Nm

Main current, voltage and torque for blocked engine.

Sliding armature starter

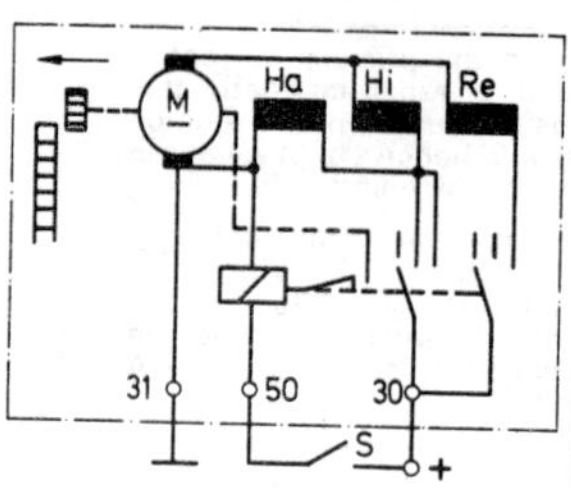

1. Switching stage 1. With switching-on (S) current flows thhrough the auxiliary winding (Hi) and holding winding (Ha).
2. The armature is drawn in axially in the coil field by rotation so that the pinion meshes.
3. The armature movement disconnects the blocking kink, switching state II switches the series connected winding (Re), the armature develops its full torque.
4. The holding winding holds the armature until switching off
5. A plate clutch effects the free running of the pinion after starting.

Overload protection of the plate clutch. 140 Nm

12 V, 3 kW starter, 135 Ah battery, for Diesel engines up to 8l piston displacement

Ignition switch current	7,5 A
Main current	1180 A
Voltage at the starter	min. 4,5 V
Torque, maximum	40 Nm

24 V, 3 kW starter, 135 Ah battery, for Diesel engines up to 10 l piston displacement:

Ignition switch current	7,5 A
Main current	930 A
Voltage at the starter	min. 12 V
Torque, maximum	68 Nm

Main current, voltage, torque for blocked engine.

Sliding drive starter

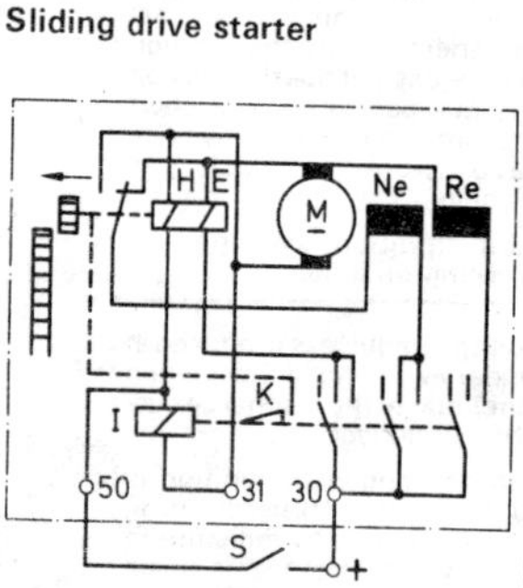

1. Switching stage I. With switching on (S) the control relay (I) draws in. Thereby current flows through the parallely connected winding (Ne) and the draw-in winding (E).
2. The trigger magnet brings the self-rotating pinion into mesh with the switch rod.
3. The blocking kink (K) at the control relay is disconnected, stage II is connected.
4. The series-connected winding (Re) receives current, the armature develops its full torque.
5. Pinion free-running through plate clutch.

Overload protection (torque limiting) 140 Nm

24 V, 4.5 kW starter, 110 Ah, battery, for Diesel engines up to 12 l piston displacement

Ignition switch current	12 A
Main current	1350 A
Voltage at the starter	min. 8,5 V
Torque, maximum	85 Nm

Main current, voltage and torque for blocked engine.

In case of vehicles with rear-mounted and under-floor engines, start locking relay and start repeat relay are built-in since the driver cannot hear the starting process.

Direct current generator

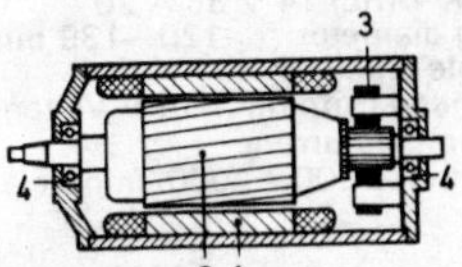

Shunt-wound generator with self excitation:

1. Pole housing with pole shoes and excitation winding.
2. Armature (dynamo lamination pack) with armature winding and commutator.
3. Carbon brushes with holders.
4. Drive bearing and cummutator bearing (ball bearing)

Type of designation according to Bosch.

Example: G (R) 14 V 30 A 25
G Housing diameter 100 109 mm
R Right hand running, 14 V Voltage
30 A maximum current
Speed 20 times 100 = 2500/minute at 2/3 I_{max}

Maximum speed	about 8000/min
Efficiency	50...70%
Bearing load (drive)	300 N

Circuit

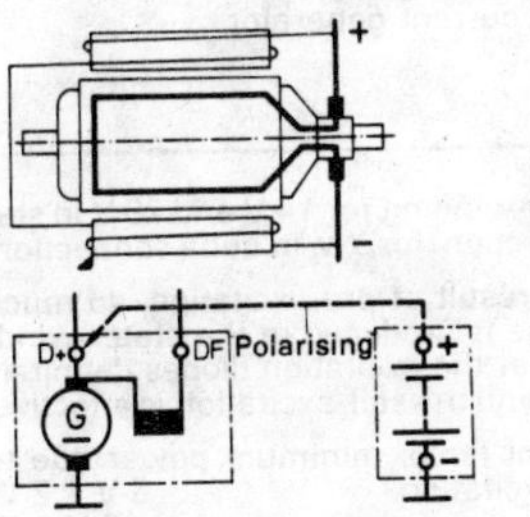

1. The residual magnetism in the housing and in the pole shoes produces a weak magnetic flux in the armature.
2. In the rotating armature winding, a voltage is therefore generated which effects an excitation current in the parallely connected excitation winding.
3. The magnetic flux is amplified by the excitation current and the voltage increases with increasing speed until the regulation sets in.

Polarising
Cable B+ at the regulator is disconnected and connected short to terminal D +.
Direct current generators must be polarised during the initial installation and after commissioning because with false poles the regulator can be destroyed.

Direction of rotation

In false direction of rotation, the generator is not excited. While changing the direction of rotation, the connections of the excitation winding must be interchanged.

Three-element regulator circuit

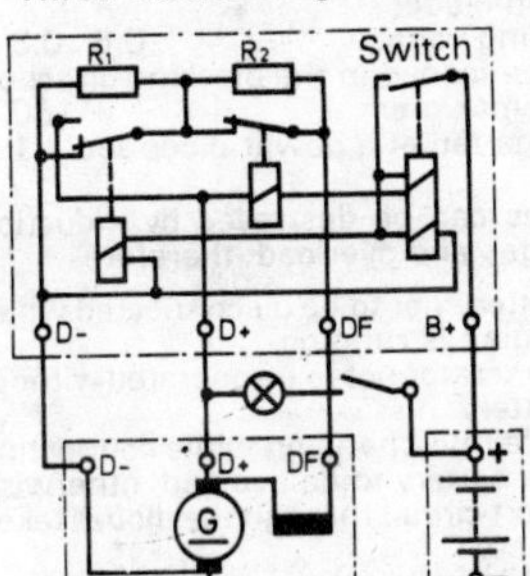

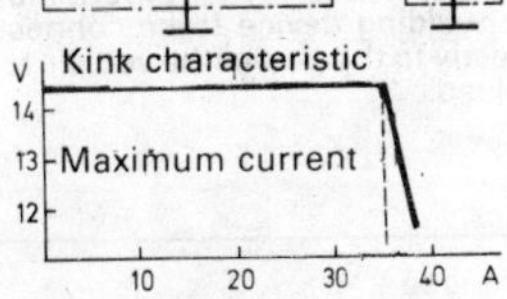

Switching on

When the switching-on voltage is attained, the switch contacts closed and the generator connected to the regulator.

Regulation, lower range.

R_1 is connected and disconnected in quick succession through the regulator contacts so that the excitation current is weakened and the generator voltage is limited.

Regulation, upper range.

The excitation current circuit is short-circuited for a short while in quick succession.

Current limitation

During high current, R_2 is connected into the excitation current circuit through the current regulator.

Switching off

With reducing generator voltage, reverse current flows from the battery to the generator whereby the switch opens.

Switching-on voltage.

For measuring, the voltmeter is connected at the regulator terminal 61 and D+. Engine speed increases slowly. Switching on current while closing the switch contacts (the indicator goes back) read
Values 6,5 V, 13 V, 26 V

Regulation voltage
Measured without battery and load; voltmeter connected at the regulator terminal B + and D-

Values 7...7,5 V
14...14,5 V
28...28,5 V

Kink characteristic
The voltage remains constant in spite of load until the maximum current is attained, and then falls in a kink. Regulators with kink characteristic guarantee best possible charging of the battery and ensure overload protection of the generator.

Variode regulator

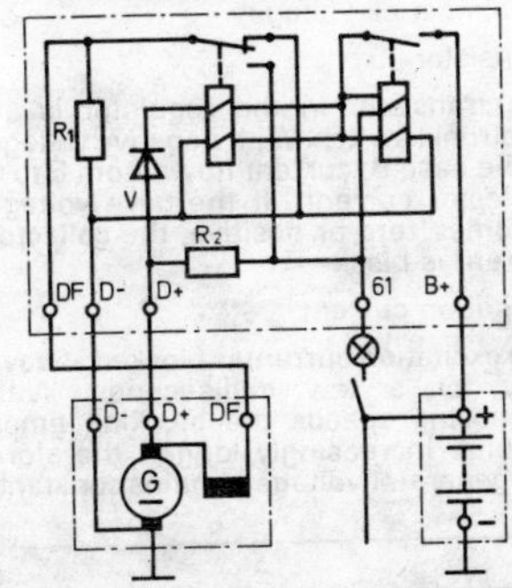

Regulation

As in three-element regulator by resistance R_1 in the lower speed range and by short-circuiting of excitation current in the upper speed range.

Current limitation

The variode V lies parallel to the control resistance R_2 of the load current circuit.

In case of maximum current, a voltage decrease of about 0.3 V takes place at R_2. The variode becomes thereby conducting whereby it allows a control current to the voltage regulator. I_{max} is limited by regulation.

Variode

The variode is a semiconductor element that becomes conducting with a voltage of 0.2 to 0.3 V.

Temperature dependence

At low temperature, the opening voltage of the variode is higher, therefore the regulation sets in only at larger current intensities. The generator is utilised better in short travels and in cold.

Control resistance

In 7 V regulators, the control resistance consists of an elongated conductor set. A change influences the regulation.

Three-phase current generator, Regulator

Claw-pole stationary-field synchronous generator

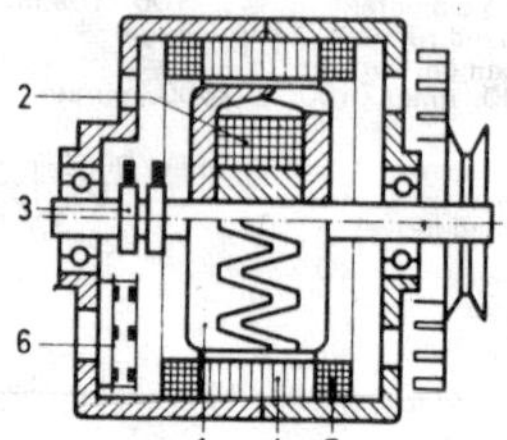

Three-phase current generator with self-excitation:

1. Claw-pole rotor, 12 pole
2. Excitation winding
3. Slip ring for the supply of the excitation current
4. Stator pack
5. Stator winding, 3 phase
6. Rectifier set with 3 excitation and 6 power diodes.

The rotor generates in the stator winding a three-phase alternating current (three-phase current) which is directly fed to the battery through the rectifier.

Type designation according to Bosch:

Example: K 1 (RL) 14 V 35 A 20
K Housing diameter 120—139 mm
1 Claw-pole type
RL Right hand running 14 V Voltage
35 A maximum current
Speed 20 times 100 = 2000/minute at 2/3 I_{max}

Maximum speed up to 12000 rpm
Transmission ratio engine/generator about 1:2, hence power output only in case of engine idle running.

Efficiency 50 to 70 %
Weight to power ratio half as large as in direct current generator

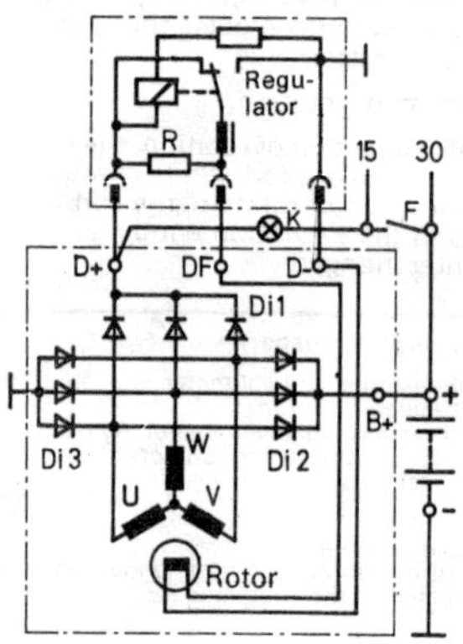

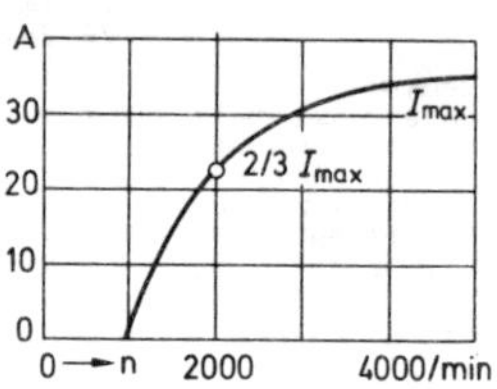

1. The rotor is preexcited from the battery through travel switch F, control lamp K, regulator, excitation winding.
2. With the rotation of the rotor, a three-phase alternating current is produced in the stator winding U, V, W.
3. From U, V, W, rectified by the excitation diode Di 1, excitation current flows through D+ regulator contacts, DF, excitation winding D– and back through the negative diode Di 3.
4. The charging current flows from U V, W, through the positive diode Di 2 to B+, battery, earth and through the negative diode Di 3 back to U, V, W.
5. With increasing voltage, the regulator contacts operate. The excitation current is weakened for a short while (through R) or short circuited and thereby the voltage is maintained constant.
6. A current limitation is not required since the machine does not give out more than I_{max}.
7. In the blocking direction, the diodes do not allow any current, therefore there is no reverse current.

Stator winding for 14 V and 28 V in star connection, for 7 V in delta connection.

As a result of pre-excitation, so much voltage is produced in the stator winding that the excitation diodes definitely open and the self-excitation is effective.

Control lamp, minimum power due to pre-excitation.
6 V 1,2 W
12 V 2 W
24 V 3 W

Silicon diodes
Opening voltage 0,6...0,8 V
Can be loaded in the blocking direction up to maximum 100 V
Voltage fall at a power diode about 1 V

Diodes can be destroyed by induction voltages and overload, therefore

1. Battery not to be disconnected when engine is running.
2. Generator not to be operated without battery.
3. Pole interchanging while connecting the battery to be avoided, otherwise short circuit through the diodes takes place.
4. While welding, the earth terminal of the welding device to be connected directly to the part of the vehicle to be welded.

Transistor regulator

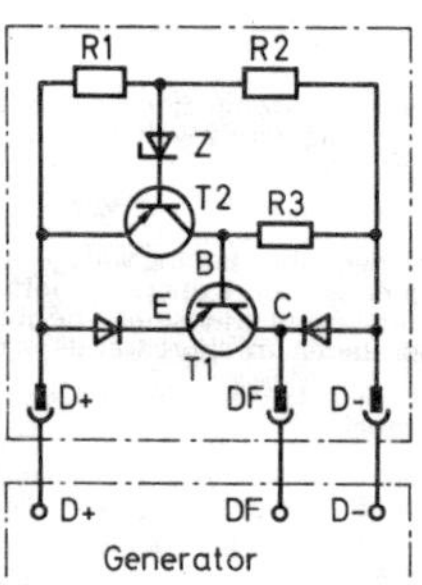

Full excitation

The main transistor is conducting since there is a negative voltage at its base B through the resistance R_3. The excitation current flows from D+ through the emitter E to the collector C of the transistor and to DF.

Regulation

1. With 28 V generator voltage, 14 V comes to the Zener diode Z as a result of voltage division through R_1 and R_2.
2. The Zener diode becomes conducting at 14 V and controls the base of the preliminary transistor T_2, which becomes likewise conducting.
3. A positive voltage comes through T_2 to the base B of the transistor T_1 whereby the excitation current is blocked.

Zener diode

The Zener diode becomes conducting at its break down voltage (e.g. 14 V) and blocks immediately when this voltage becomes lower. There are Zener diodes for almost all voltages.

Transistor

The transistor in the regulator is an electronic switch. With negative voltage at the base B, current flows from E to C (collector current). If the base voltage becomes zero or positive, the collector current is blocked.

Excitation current

The excitation current is blocked always only for a few milliseconds. With increasing speeds the blcoking times become increasingly longer, therefore the generator voltage remains constant.

Electrical wiring in automobiles

DIN 72551

Nominal cross section mm^2	Individual wires Number off	Individual wires Diameter mm	Conductor mm	Cable diameter mm	Resistance at 20°C mΩ/m	Remarks
0,5	16	0,21	1,0	2,3	37,1	Conductor material: Wires from electrolytic copper, soft annealed, no tin plating.
0,75	24	0,21	1,2	2,5	24,7	
1	31	0,21	1,4	2,7	18,5	
1,5	30	0,26	1,6	3,0	12,7	Insulating sheath: Thermoplastic, Polyvinylchloride (PVC) containing softener
2,5	50	0,26	2,1	3,7	7,60	
4	56	0,31	2,7	4,5	4,71	
6	19	0,64	3,3	5,1	2,93	
10	19	0,80	4,1	6,4	1,90	Requirements: Electrical insulation strength up to 3 KV, heat resistant, cold resistant, fuel resistant and lubricant resistant
16	37	0,75	5,4	7,7	1,10	
25	37	0,90	6,4	9,2	0,77	
35	37	1,10	7,8	10,6	0,516	Example of designation. XYZ 500 m cable B 2.5 DIN 72551
50	61	1,00	9,1	11,5	0,380	
70	61	1,20	11,0	14,1	0,263	
95	61	1,40	12,8	16,4	0,193	
120	61	1,60	14,6	18,2	0,146	

Permissible voltage drop

(Reference values)

Cable	Permissible voltage drop in out-going cable for nominal voltage		Permissible voltage drop in the whole circuit for nominal voltage		Remarks
Lighting cable from switch (30) up to the light maximum 15 W or up to trailor socket	0,1 V	0,1 V	0,6 V	0,6 V	Current for nominal voltage and nominal power
Lighting cable from switch (30) up to light above 15 W or up to trailor socket	0,5 V	0,5 V	0,9 V	0,9 V	
Lighting cable from switch (30) upto head light	0,3 V	0,3 V	0,6 V	0,6 V	
Charging cable from generator B+ up to battery	0,4 V	0,8 V	–	–	In insulated return cable, the whole length is effective
Starter main cable, battery to starter,	0,5 V	1,0 V	–	–	Permissible current density upto $30 A/mm^2$
Starter control cable from switch to starter (50), starting relay with one coil.	1,4 V	2,0 V	1,7 V	2,5 V	
Starter control cable from switch to starter (50), starting relay with two coils.	2,4 V	2,8 V	2,8 V	3,5 V	
Other control cables from switch to relay up to wiper, up to horn, etc.	0,5 V	1,0 V	1,5 V	2,0 V	Current for nominal voltage and nominal power

Specific resistance, Conductance (mean values)

Conductor material	Specific resistance ρ_{20} $\frac{\Omega \cdot mm^2}{m}$	Electrical conductance $\varkappa_{20}$ $\frac{m}{\Omega \cdot mm^2}$	Resistance material	Specific resistance ρ_{20} $\frac{\Omega \cdot mm^2}{m}$	Electrical conductance $\varkappa_{20}$ $\frac{m}{\Omega \cdot mm^2}$
Silver	0,0167	60	CuMn12Ni	0,43	2,3
Copper	0,0178	56	CuNi44	0,50	2,0
Aluminium	0,0278	36	CuNi30Mn	0,40	2,5
Tungsten	0,055	18,2	NiCr80 20	1,12	0,9
Brass	0,07	14,3	CrAl20 5	1,37	0,73
Iron	0,13	7,7	Carbon (hard)	40 … 60	0,025 … 0,017

Power consumption of electrical loads

Dim light	each 55 W	Tail disc heating	120 W	Transistor radio	10 … 15 W
Boundary light	each 4 W	Horn, Fanfares	25 .. 40 W	Disc wiper	90 W
Blink lights	each 21 W	Internal lights	each 5 W	Stop lights	each 5 W
Brake lights	each 18 W	Instrument lights	each 2 W	Coil ignition	20 W
Bright head lights	each 60 W	Number plate lights	each 10 W	Starter for passenger vehicles	0,8 … 3 kW
Blower motor	50 … 80 W	Fog head lights	each 55 W	Starter for load-haulage vehicles	2 … 12 kW
Spark plugs	each 60 W	Parking lights	3 … 5 W	Transistor ignition	70 W
Pin type spark plugs	each 100 W	Reverse travel head lights	each 25 W	Cigarette lighter	100 W

Ignition equipment

1	Ignition coil, ignition distributor, ignition device (low voltage) for HKZ
1 a	Ignition distributor with two separate
1 b	Current circuits (low voltage)
2	Short circuit terminal in magneto ignition
4	Ignition coil, ignition distributor (high voltage)
4 a	Ignition distributor with two separate current
4 b	circuits (high voltage)
7	Base resistances from or to ignition distributor (control contact)
7 a	First base resistance for TSZ and HKZ
7 b	Second base resistance for TSZ
7 f	Charging contact for HKZ
15	Exit travel switch
15 a	Entry ignition device HKZ, switching device TSZ and ignition coil preliminary resistance

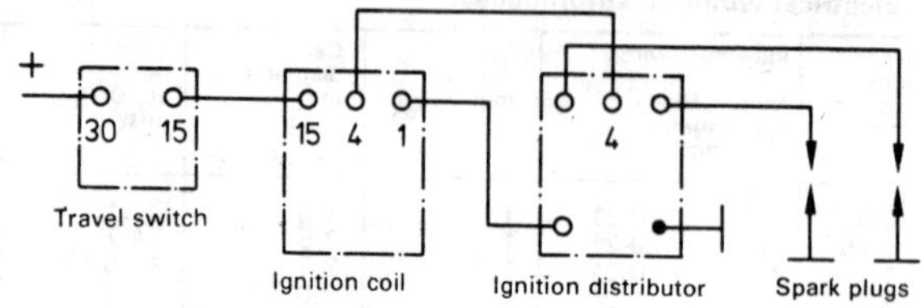

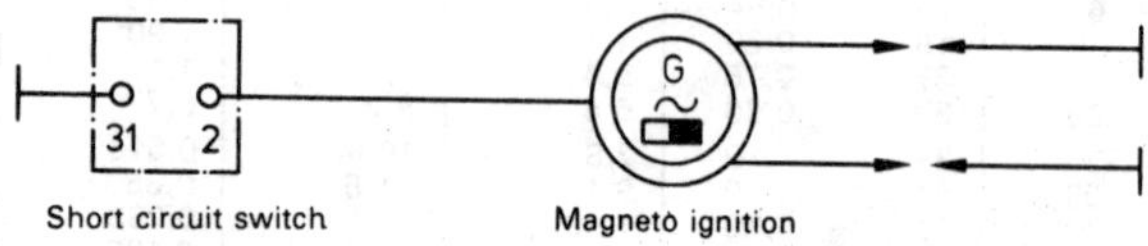

Pre-ignition equipment

15	Ignition starting switch-entry
17	Ignition starting switch Stage II starting
19	Ignition starting switch Stage I preignition
50	Ignition starting switch Starter control

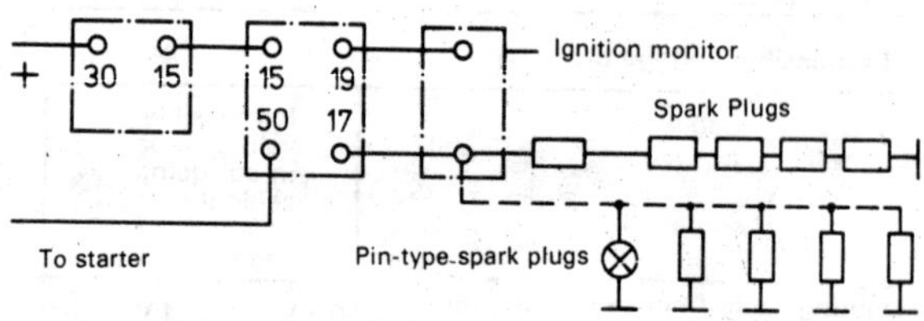

General application

15	Battery plus through switch
30	Battery plus directly from the battery
31	Return line directly to the battery negative, earth
31 b	Return line to the battery negative or earth through switch or relay

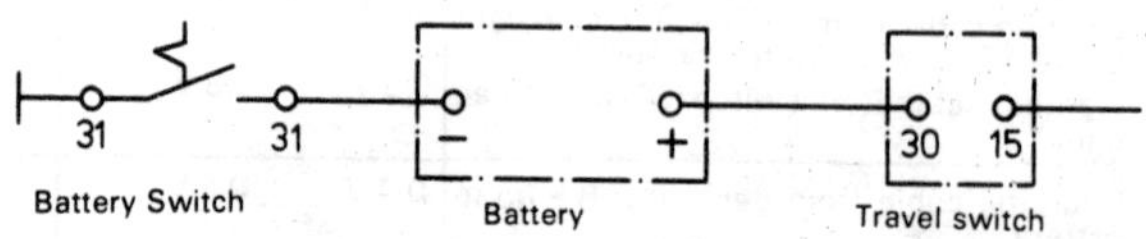

Electric motors

30	Entry directly from battery positive
32	Return line
33	Main connection
33 a	End disconnecting switch
33 b	Shunt-wound field
33 f	For second smaller speed stage
33 L	Direction of rotation left
33 R	Direction of rotation right
86	Relay entry (winding beginning)

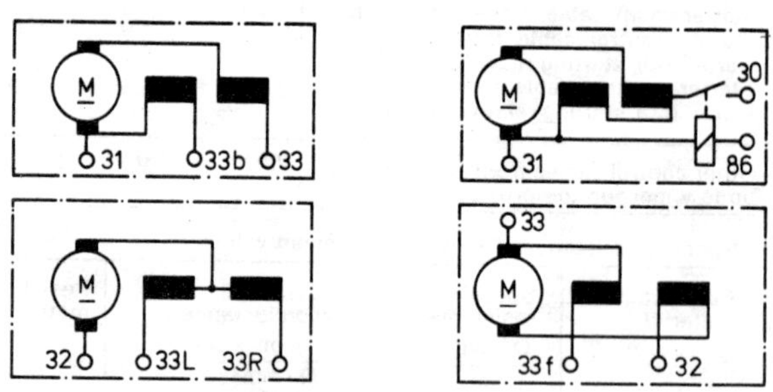

Starter equipment

30	Entry directly from battery positive
30 a	Battery change-over relay, entry from battery II
31	Return line directly to battery negative, earth
31 a	Battery change-over relay, return line to battery II negative
31 c	Battery change-over relay, return line to battery I negative
48	Starting repetition (relay)
50	Starter control direct
50 a	Starter control indirect (battery change-over relay)
50 e	Starting lock relay entry
50 f	Starting lock relay exit
50 g	Starting repeat relay entry
50 h	Starting repeat relay exit
86	Relay entry (winding beginning)

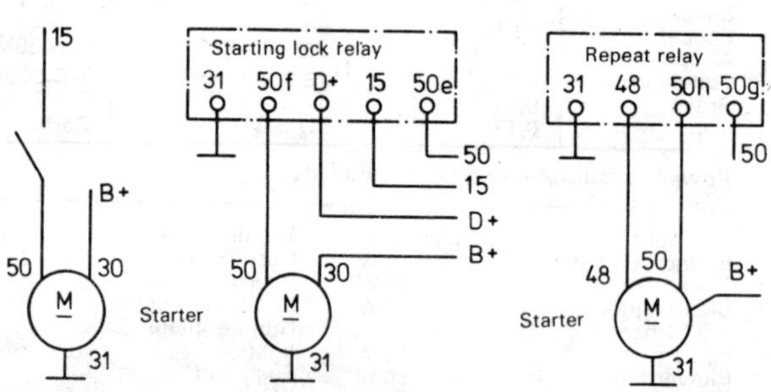

Generators, Regulators

44 Voltage balance at regulators in parallel operation of two generators
51 Direct voltage at the rectifier in case of alternating current generators
51 e Like 51, however with impedance coil for day journey in case of alternating current generators
59 Alternating voltage, exit alternating current generator; entry light switch and rectifier
59 a Load armature
59 b Stop light armature
59 c Brake light armature
61 Load indicator light at generator and regulator
B+ Battery positive
B− Battery negative
D+ Dynamo positive
D− Dynamo negative
DF Dynamo field
DF 1 Dynamo field 1
DF 2 Dynamo field 2
Three-phase current generator with separate rectifiers
J Excitation winding positive
K Excitation winding negative
M p Central terminal

Lighting equipment

54 Brake light
55 Fog head light
56 Head light
56 a Bright head light and bright head light indicator
56 b Dim light
56 d By-pass light contact
57 Stand light for motor cycle head light
57 a Parking light
57 L Parking light left
57 R Parking light right
58 Limiting, stop and number plate lights, instrument lighting
58 b Stop light change-over single axle trailor
58 c Trailor plug connection for stop light laid single wired, separately secured in the trailor
58 d Controllable instrument light
58 L Stop and limiting lights, left
58 R Stop and limiting lights, right

Travel direction indication

49 Blink producer entry (plus)
49 a Blink producer exit (impulse)
49 b Blink producer exit, second blink circuit
49 c Blink producer exit, third blink circuit
C First indication lamp
C 2 Second indication lamp
C 3 Third indication lamp (Two trailer mode)
L Blink light, left
R Blink light, right
L b Two-circuit blinker switch, left
L r Two-circuit blinker switch, right

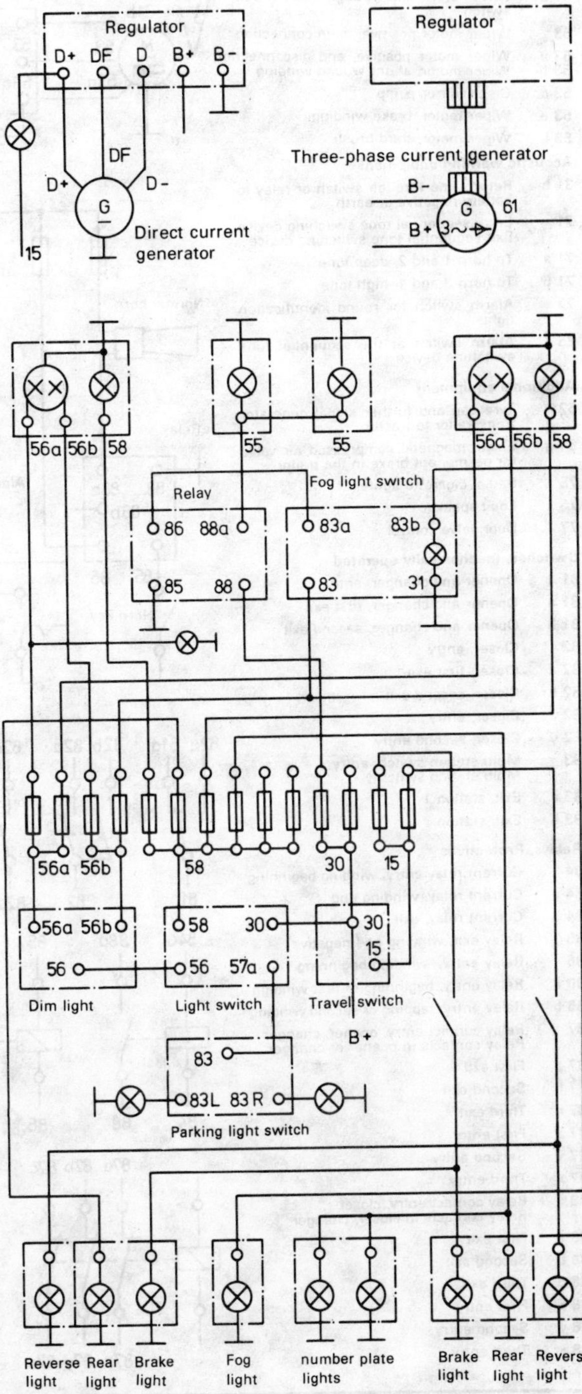

Windscreen wiper and washer

- 31 b Earth return line through short circuit switch
- 53 Wiper motor positive, main connection
- 53 a Wiper motor positive, and disconnection
- 53 b Wiper motor, shunt-wound winding
- 53 c Disc scourer pump
- 53 e Wiper motor, brake winding
- 53 i Wiper motor, third brush

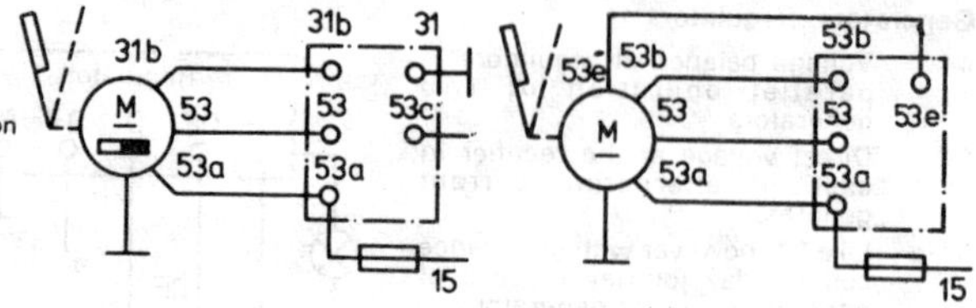

Acoustic warning equipment

- 31 b Return line through switch or relay to battery negative or earth
- 71 Entry sequential tone switching device Exit sequential tone switching device
- 71 a To horn 1 and 2, deep tone
- 71 b To horn 3 and 4, high tone
- 72 Alarm switch for round identification lights
- 85 c Alarm switch at the sequential tone switching device

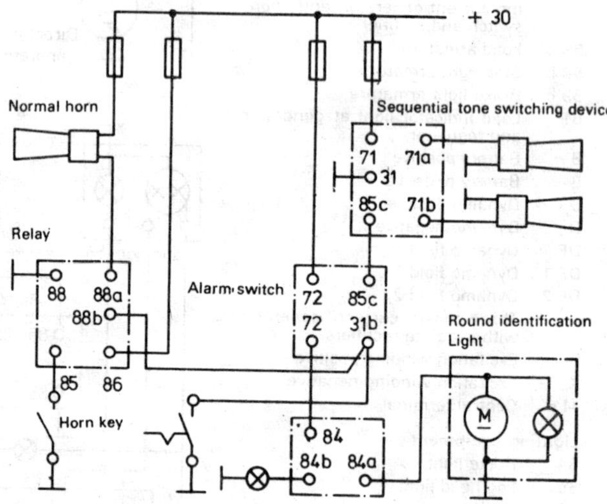

Additional equipment

- 52 Tyre cap and further signal generator from trailor to tractor
- 54 g Electromagnetic compressed air valve for permanent brake in the trailor
- 75 Radio, cigarette lighter
- 76 Loud speaker
- 77 Door valve control

Switches, mechanically operated

- 81 Opener and changer, entry
- 81 a Opener and changer, first exit
- 81 b Opener and changer, second exit
- 82 Closer, entry
- 82 a Closer, first exit
- 82 b Closer, second exit
- 82 z Closer, entry
- 82 y Closer, second entry
- 83 Multi station switch, entry Multi station switch
- 83 a Exit, station 1
- 83 b Exit, station 2

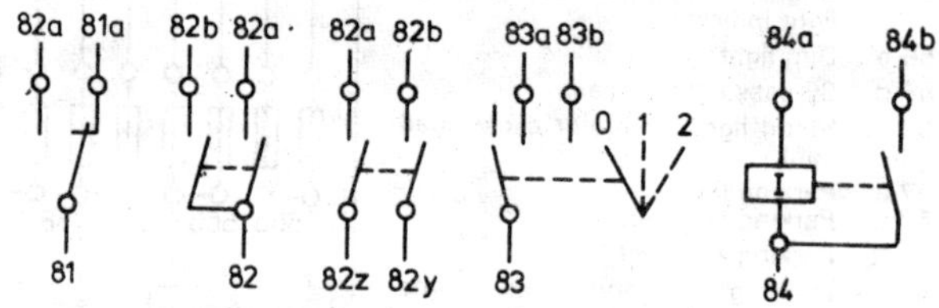

Relays, Protectives

- 84 Current relay entry, winding beginning
- 84 a Current relay, winding end
- 84 b Current relay, exit
- 85 Relay exit, winding end negative
- 86 Relay entry, winding beginning
- 86 a Relay entry, beginning of first winding
- 86 b Relay entry, tapping or second winding
- 87 Relay contact entry, opener, changer Relay contacts in opener or changer
- 87 a First exit
- 87 b Second exit
- 87 c Third exit
- 87 z First entry
- 87 y Second entry
- 87 x Third entry
- 88 Relay contact entry, closer Relay contacts in closer, changer
- 88 a First exit
- 88 b Second exit
- 88 c Third exit
- 88 z First entry
- 88 y Second entry
- 88 x Third entry

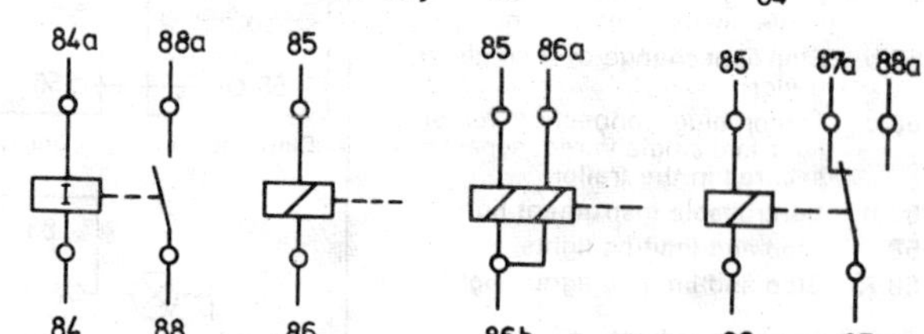

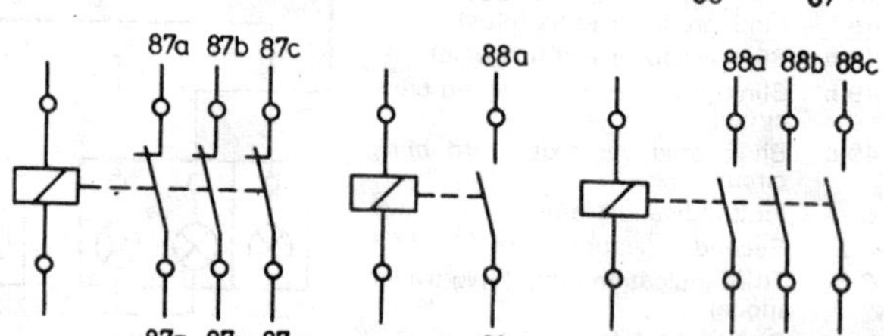

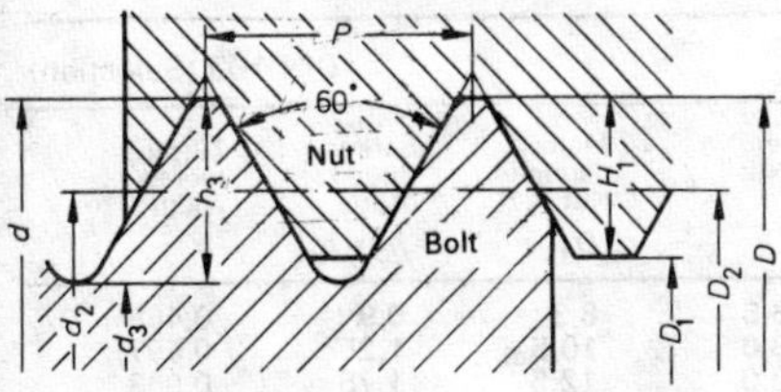

Nominal diameter	$d = D$
Pitch	P
Flank angle	60°
Tap drill diameter	$d - P$
Stress cross-sectional area	$A_s = \left(\frac{d_2 + d_3}{2}\right)^2 \frac{\pi}{4}$

Designation examples:
Coarse-pitch screw thread
Fine-pitch screw thread

d = 20 mm, P = 2,5 mm : **M 20**
d = 20 mm, P = 1,0 mm : **M 20×1**

The screw threads of Series 1 are to be preferably used

Coarse-pitch screw threads Dimensions in mm

Nominal diameter of screw thread $d = D$				Screw thread depth		Minor diameter		Stress cross-sectional area	Tap drill diameter	Through hole diameter	Width across flats of hexagonal spanner
				Bolt	Nut	Bolt	Nut				
Series 1	Series 2	P	$d_2 = D_2$	d_3	D_1	h_3	H_1	A_s (mm²)			
M 3		0,5	2,675	2,387	2,459	0,307	0,271	5,03	2,5	3,4	5,5
	M 3,5	0,6	3,110	2,764	2,850	0,368	0,325	6,77	2,9	3,9	6
M 4		0,7	3,545	3,141	3,242	0,429	0,379	8,78	3,3	4,5	7
M 5		0,8	4,480	4,019	4,134	0,491	0,433	14,2	4,2	5,5	8
M 6		1	5,350	4,773	4,917	0,613	0,541	20,1	5,0	6,6	10
M 8		1,25	7,188	6,466	6,647	0,767	0,677	36,6	6,8	9	13
M 10		1,5	9,026	8,160	8,376	0,920	0,812	58,0	8,5	11	17
M 12		1,75	10,863	9,853	10,106	1,074	0,947	84,3	10,2	14	19
	M 14	2	12,701	11,546	11,835	1,227	1,083	115	12	16	22
M 16		2	14,701	13,546	13,835	1,227	1,083	157	14	18	24
	M 18	2,5	16,376	14,933	15,294	1,534	1,353	192	15,5	20	27
M 20		2,5	18,376	16,933	17,249	1,534	1,353	245	17,5	22	30
	M 22	2,5	20,376	18,933	19,294	1,534	1,353	303	19,5	24	32
M 24		3	22,051	20,319	20,752	1,840	1,624	353	21	26	36
	M 27	3	25,051	23,319	23,752	1,840	1,624	459	24	30	41
M 30		3,5	27,727	25,706	26,211	2,147	1,894	561	26,5	33	46
	M 33	3,5	30,727	28,706	29,211	2,147	1,894	693	29,5	36	50
M 36		4	33,402	31,093	31,670	2,454	2,165	817	32	39	55
	M 39	4	36,402	34,093	34,670	2,454	2,165	975	35	42	60
M 42		4,5	39,077	36,479	37,129	2,760	2,436	1120	37,5	45	65
	M 45	4,5	42,077	39,479	40,129	2,760	2,436	1306	40,5	48	70
M 48		5	44,752	41,866	42,587	3,067	2,706	1470	43	52	75
M 56		5,5	52,428	49,252	50,046	3,374	2,977	2030	50,5	62	85
M 64		6	60,103	56,639	57,505	3,681	3,248	2680	58	70	95

Fine-pitch screw threads Dimensions in mm DIN 103 (Selection)

Screw thread designation	Pitch diameter	Minor diameter		Screw thread designation	Pitch diameter	Minor diameter		Screw thread designation	Pitch diameter	Minor diameter	
		Bolt	Nut			Bolt	Nut			Bolt	Nut
$d \times P$	$d_2 = D_2$	d_3	D_1	$d \times P$	$d_2 = D_2$	d_3	D_1	$d \times P$	$d_2 = D_2$	d_3	D_1
M 3×0,35	2,773	2,571	2,621	M 24×1	23,350	22,773	22,917	M 48×4	45,402	43,093	43,670
M 4×0,5	3,675	3,387	3,459	M 24×1,5	23,026	22,160	22,376	M 56×1,5	55,026	54,160	54,376
M 5×0,5	4,675	4,387	4,459	M 24×2	22,701	21,546	21,835	M 56×2	54,701	53,546	53,835
M 6×0,75	5,513	5,080	5,188	M 30×1,5	29,026	28,160	28,376	M 56×3	54,051	52,319	52,752
M 8×0,75	7,513	7,080	7,188	M 30×2	28,701	27,546	27,835	M 56×4	53,402	51,093	51,670
M 8×1	7,350	6,773	6,917	M 30×3	28,051	26,319	26,752	M 64×2	62,701	61,546	61,835
M 10×0,75	9,513	9,080	9,188	M 36×1,5	35,026	34,160	34,376	M 64×3	62,051	60,319	60,752
M 10×1	9,350	8,773	8,917	M 36×2	34,701	33,546	33,835	M 64×4	61,402	59,093	59,670
M 12×1	11,350	10,773	10,917	M 36×3	34,051	32,319	32,752	M 72×3	70,051	68,319	68,752
M 12×1,25	11,188	10,466	10,647	M 42×1,5	41,026	40,160	40,376	M 72×4	69,402	67,093	67,670
M 14×1	13,350	12,773	12,917	M 42×2	40,701	39,546	39,835	M 80×3	78,051	76,319	76,752
M 14×1,5	13,026	12,160	12,376	M 42×3	40,051	38,319	38,752	M 90×4	87,402	85,093	85,670
M 16×1	15,350	14,773	14,917	M 42×4	39,402	37,093	37,670	M 100×4	97,402	95,093	95,670
M 16×1,5	15,026	14,160	14,376	M 48×1,5	47,026	46,160	46,376	M 125×4	122,40	120,09	120,67
M 20×1	19,350	18,773	18,917	M 48×2	46,701	45,546	45,835	M 140×6	136,10	132,64	133,50
M 20×1,5	19,026	18,160	18,376	M 48×3	46,051	44,319	44,752	M 150×6	146,10	142,64	143,50

Screw threads

ISO trapezoidal screw threads Dimensions in mm

DIN 103 (Selection)

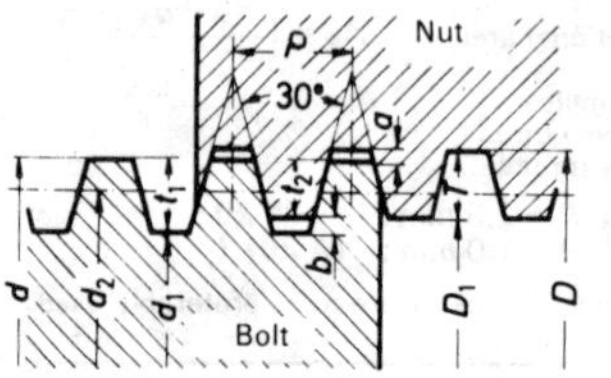

Nominal diameter of screw thread d
Screw thread pitch P
Flank angle 30°

Designation example
screw thread diameter 20 mm, pitch 4 mm:
Tr 20×4

Screw thread designation $d \times P$	Minor diameter Bolt d_3	Minor diameter Nut D_1	Major diameter Nut D_4	Screw thread depth $h_3 = h_4$	Turning tool width b
Tr 8 × 1,5	6,2	6,5	8,3	0,9	0,468
Tr 10 × 2	7,5	8,0	10,5	1,25	0,597
Tr 12 × 3	8,5	9,0	12,5	1,75	0,963
Tr 14 × 3	10,5	11,0	14,5	1,75	0,963
Tr 16 × 4	11,5	12,0	16,5	2,25	1,329
Tr 18 × 4	13,5	14,0	18,5	2,25	1,329
Tr 20 × 4	15,5	16,0	20,5	2,25	1,329
Tr 24 × 5	18,5	19,0	24,5	2,75	1,695
Tr 28 × 5	22,5	23,0	28,5	2,75	1,695
Tr 32 × 6	25,0	26,0	33,0	3,5	1,926
Tr 36 × 6	29,0	30,0	37,0	3,5	1,926
Tr 40 × 7	32,0	33,0	41,0	3,5	2,292
Tr 44 × 7	36,0	37,0	45,0	4,5	2,292
Tr 48 × 8	39,0	40,0	49,0	4,5	2,658
Tr 52 × 8	43,0	44,0	53,0	4,5	2,658
Tr 60 × 9	50,0	51,0	61,0	5,0	3,024

Saw-tooth screw threads Dimensions in mm

DIN 513, 514 (Selection)

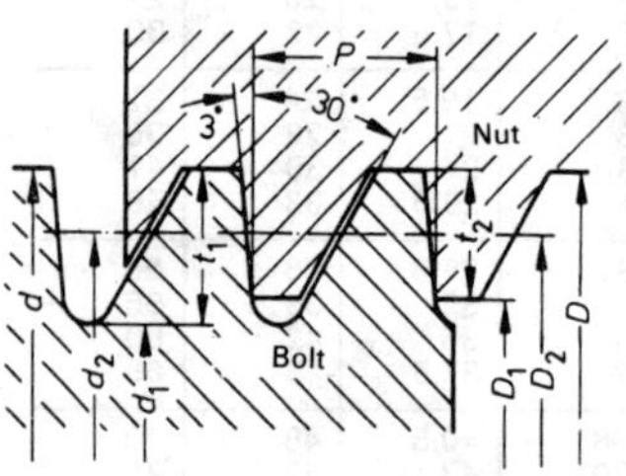

Nominal diameter of screw thread d
Pitch of screw thread P
Flank angle 30° + 3° = 33°

Example of designation
screw thread diameter 24 mm, pitch 3 mm;
S 24×3

Designation of screw thread $d \times P$	Minor diameter Bolt d_1	Screw thread depth Bolt t_1	Minor diameter Nut D_1	Screw thread depth Nut t_2	Pitch diameter $d_2 = D_2$
S 12 × 2	8,528	1,736	9,0	1,5	10,636
S 16 × 2	12,528	1,736	13,0	1,5	14,636
S 20 × 2	16,528	1,736	17,0	1,5	18,636
S 24 × 3	18,794	2,603	19,5	2,25	21,954
S 24 × 5	15,322	4,339	16,5	3,75	20,590
S 30 × 3	24,794	2,603	25,5	2,25	27,954
S 30 × 6	19,586	5,207	21,0	4,5	25,909
S 36 × 3	30,794	2,603	31,5	2,25	33,954
S 36 × 6	25,586	5,207	27,0	4,5	31,909
S 40 × 3	34,794	2,603	35,5	2,25	37,954
S 40 × 7	27,852	6,074	29,5	5,25	35,227
S 48 × 3	42,794	2,603	43,5	2,25	45,954
S 48 × 8	34,116	6,942	36,0	6,0	42,545
S 55 × 9	39,380	7,810	41,5	6,75	48,863
S 60 × 9	44,380	7,810	46,5	6,75	53,863
S 70 × 10	52,644	8,678	55,0	7,5	63,181

Round screw threads

DIN 405 (Selection)

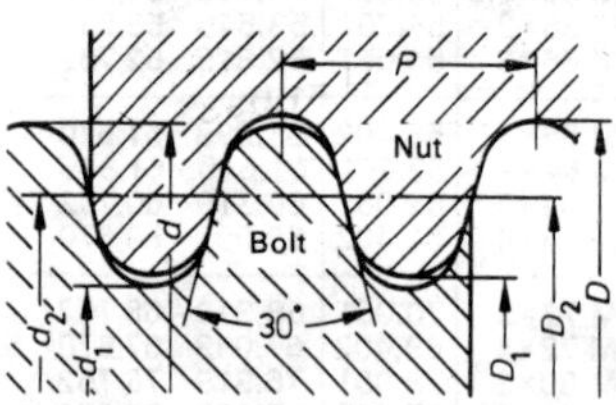

Nominal diameter of screw thread d
Pitch of screw thread P
Flank angle 30°

Example of designation.
screw thread diameter 24 mm, pitch ⅛":
Rd 24×⅛"

Designation of screw thread $d \times P$	Minor diameter Bolt d_1 (mm)	Major diameter Nut D (mm)	Minor diameter Nut D_1 (mm)	Pitch diameter d_2 (mm)	Number of threads per inch z
Rd 8 × 1/10"	5,460	8,254	5,714	6,730	10
Rd 10 × 1/10"	7,460	10,254	7,714	8,730	10
Rd 12 × 1/10"	9,460	12,254	9,714	10,730	10
Rd 16 × 1/8"	12,825	16,318	13,142	14,412	8
Rd 20 × 1/8"	16,825	20,318	17,142	18,412	8
Rd 22 × 1/8"	18,825	22,318	19,142	20,412	8
Rd 24 × 1/8"	20,825	24,318	21,142	22,412	8
Rd 30 × 1/8"	26,825	30,318	27,142	28,412	8
Rd 36 × 1/8"	32,825	36,318	33,142	34,412	8
Rd 40 × 1/6"	35,767	40,423	36,190	37,883	6
Rd 48 × 1/6"	43,767	48,423	44,190	45,883	6
Rd 52 × 1/6"	47,767	52,423	48,190	49,883	6

Whitworth screw threads

British screw threads

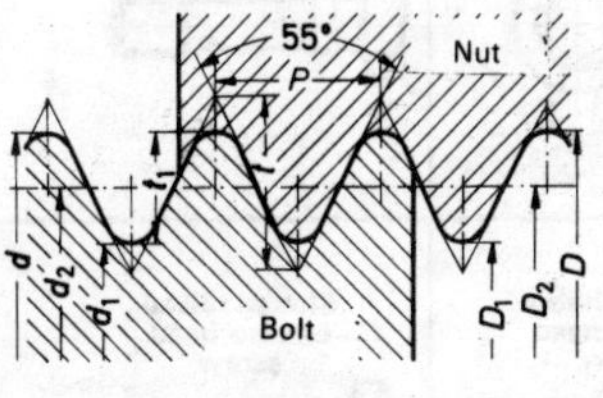

Nominal diameter of screw thread $d = D$

Pitch of screw thread P

Flank angle 55°

Example of designation:
screw threads ½",
number of threads 12: ½"

Designation of screw thread Inch	Major diameter $d = D$ mm	Minor diameter $d_1 = D_1$ mm	Number of threads per inch z	Pitch P mm	Tap drill diameter mm	Width across flats of hexagonal spanner mm
1/4"	6,35	4,72	20	1,270	5,1	11
5/16"	7,94	6,13	18	1,411	6,5	13
3/8"	9,53	7,49	16	1,588	8,0	17
1/2"	12,70	9,99	12	2,117	10,5	22
5/8"	15,88	12,92	11	2,309	13,5	27
3/4"	19,05	15,80	10	2,540	16,5	32
7/8"	22,22	18,61	9	2,822	19,5	36
1"	25,40	21,34	8	3,175	22	41
1 1/4"	31,75	27,10	7	3,629	28	50
1 1/2"	38,10	32,68	6	4,233	33,5	60
1 3/4"	44,45	37,95	5	5,080	39	70
2"	50,80	43,57	4,5	5,645	45	80
2 1/4"	57,15	49,02	4	6,350	50	85
2 1/2"	63,50	55,37	4	6,350	57	95
2 3/4"	69,85	60,56	3,5	7,257	62	105
3"	76,20	66,91	3,5	7,257	68	110

Whitworth pipe screw threads

DIN 259

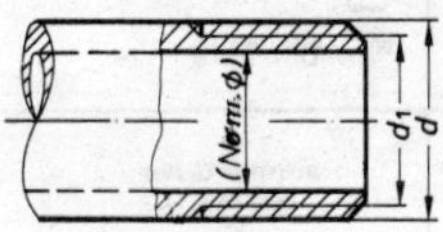

Nominal diameter $d = D$

Nominal diameter of screw thread P

Hank angle 55°

Example of designation-
screw threads ½", number of threads 14: R½"

Designation of screw thread inch	Major diameter $d = D$ mm	Minor diameter $d_1 = D_1$ mm	Pitch P mm	Number of threads per inch z	Depth of screw thread H_1 mm
R 1/8"	9,73	8,57	0,907	26	0,581
R 1/4"	13,16	11,45	1,337	19	0,856
R 3/8"	16,66	14,95	1,337	19	0,856
R 1/2"	20,96	18,63	1,814	14	1,162
R 5/8"	22,91	20,59	1,814	14	1,162
R 3/4"	26,44	24,12	1,814	14	1,162
R 7/8"	30,20	27,88	1,814	14	1,162
R 1"	33,25	30,28	2,309	11	1,479
R 1 1/4"	41,91	38,95	2,309	11	1,479
R 1 1/2"	47,80	44,85	2,309	11	1,479
R 1 3/4"	53,75	50,79	2,309	11	1,479
R 2"	59,62	56,66	2,309	11	1,479

ISO inch screw threads

UNF screw threads (fine-pitch screw threads)

Flank angle 60°

Example of designation:
screw threads ¼", 28 threads: ¼"–28 UNF

Nominal diameter of screw thread inch	Major diameter mm	Number of threads per inch	Minor diameter Bolt mm	Minor diameter Nut mm	Core cross-sectional area mm²
1/4"	6,350	28	5,237	5,367	21,0
5/16"	7,938	24	6,640	6,792	33,8
3/8"	9,525	24	8,227	8,397	52,2
7/16"	11,112	20	9,555	9,738	70,3
1/2"	12,700	20	11,143	11,326	95,9
9/16"	14,288	18	12,555	12,761	122
5/8"	15,875	18	14,143	14,348	155
3/4"	19,050	16	17,102	17,330	227
7/8"	22,225	14	20,0	20,262	310
1"	25,400	12	22,804	23,109	403

UNC screw threads (coarse pitch screw threads)

Flank angle 60°

Example of designation:
screw threads ¼", 20 threads: ¼"–20 UNC

Nominal diameter of screw thread inch	Major diameter mm	Number of threads per inch	Minor diameter Bolt mm	Minor diameter Nut mm	Core cross-sectional area mm²
1/4"	6,350	20	4,793	4,976	17,4
5/16"	7,938	18	6,205	6,411	29,3
3/8"	9,525	16	7,577	7,805	43,7
7/16"	11,112	14	8,887	9,149	60,2
1/2"	12,700	13	10,302	10,548	81,1
9/16"	14,288	12	11,692	11,996	105
5/8"	15,875	11	13,043	13,376	130
3/4"	19,050	10	15,933	16,299	195
7/8"	22,225	9	18,763	19,169	271
1"	25,400	8	21,504	21,963	355

Screws, Nuts, Locking devices

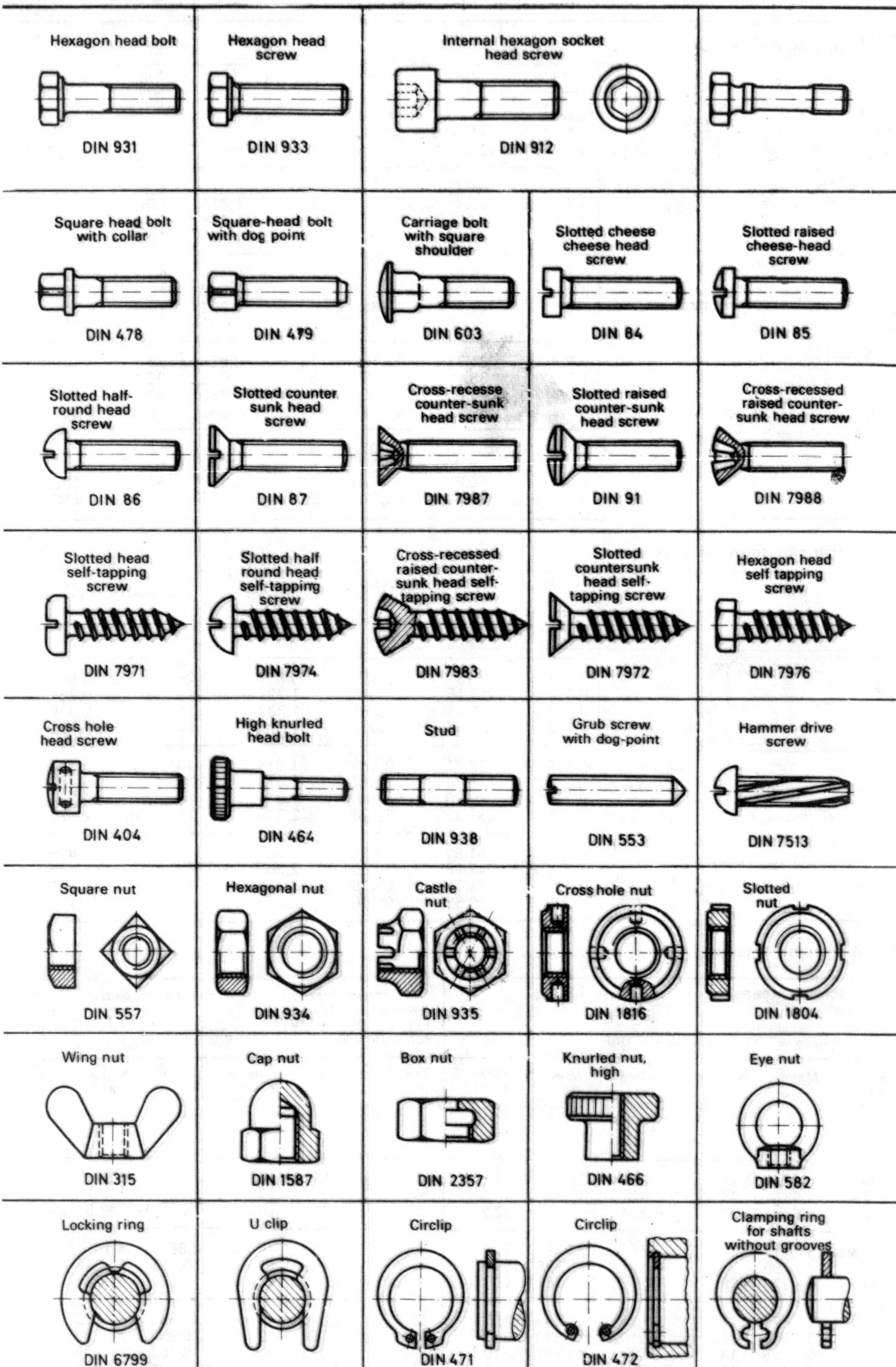

Spring washer

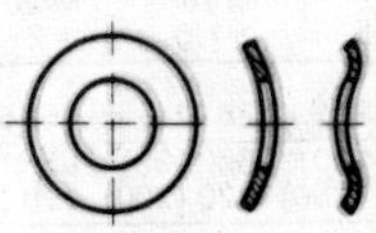

DIN 137

Toothed washer

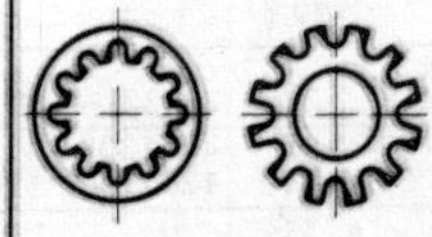

DIN 6797

Spring washer

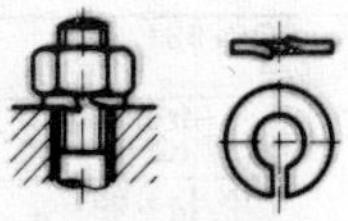

DIN 127

Locking plate with lug

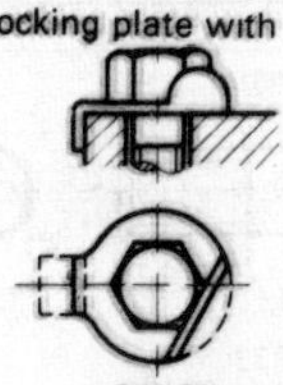

DIN 93

Locking plate with lug

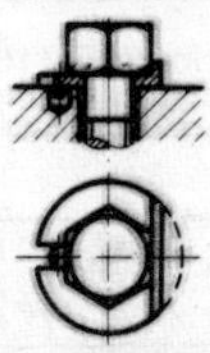

DIN 432

Self locking nut

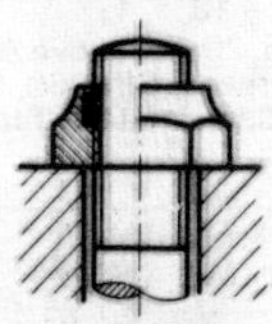

DIN 985

Lock nut

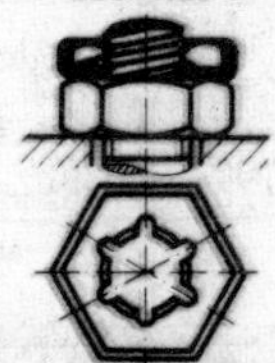

DIN 7967

Locking with hooked snap ring

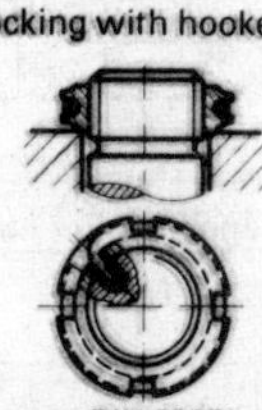

DIN 70851

Lock nut

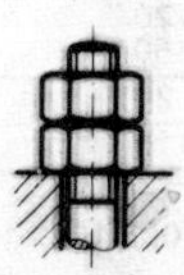

Castle nut with cotter pin

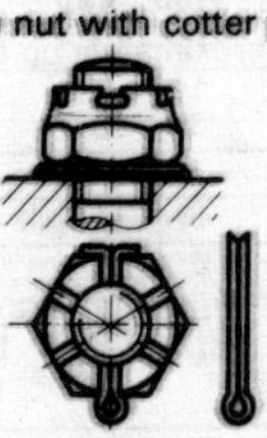

Dubo locking

Wire locking

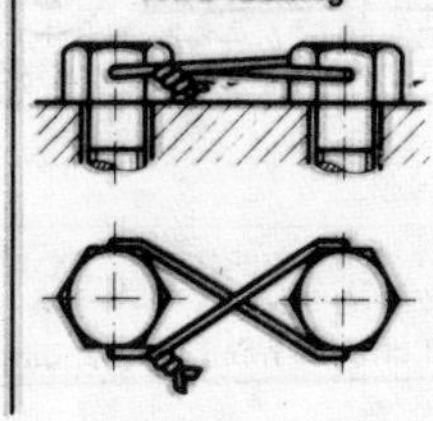

Strength of screws DIN 267

Property class	3.6	4.6	4.8	5.6	5.8	6.6	6.8	6.9	8.8	10.9	12.9	14.9
Minimum tensile strength σ_B (R_m) in N/mm²	340	400	400	500	500	600	600	600	800	1 000	1 200	1400
Minimum elongation at fracture δ_5 (A) in %	25	25	14	20	10	16	8	12	12	9	8	7
Brinell hardness HB minimum	90	110	110	140	140	170	170	170	225	280	330	390

Explanation of the property class	The first number specifies 1/100 of the minimum tensile strength e.g. 400 N/mm²≙4 The second number specifies 10 times the yield limit ratio e.g. for $\sigma_s = 320$ N/mm² and σ_B (Rm) = 400 N/mm² it is equal to {(320N/mm²) / (400N/mm²)} 10 = 8 The product of the two numbers 4·8 = 32 corresponds to 1/10 of the minimum yield strength of 320 N/mm². Note: The yield limit is the tensile stress at which the material begins to flow whereby a permanent strain sets in.
Designation of screws, examples	1. Designation of a hexagon head screw with M 10 screw thread, length $l = 50$ mm and a property class 8.8: Hexagon head screw M 10 × 50 DIN 931—8.8. 2. Designation of an internal hexagon socked head screw, screw thread M 8, length $l = 40$ mm and a property class 6.9: Internal hexagon socket head screw M 8 × 40 DIN 912—6.9.

Types of screws and nuts

Abbreviation (m) medium	Roughness depth $R_t = 25\mu$m for bearing surfaces, screw thread flanks of screws and nuts, shank and root of screw thread of screws. $R_t = 100$ for cups and spanner faces. Any R_t for screw thread major diameter of screws and screw thread minor diameter of nuts.
Abbreviation mg (medium coarse)	$R_t = 25$ for bearing surfaces, screw thread flanks of screws and nuts, shank and screw thread minor diameter of screws. Any R_t for other surfaces.
Abbreviation g (coarse)	$R_t = 40$ for screw thread flanks of screws and nuts and screw thread minor diameter of screws. Any R_t for all other surfaces.

Hexagon head screws, dimensions in mm DIN 931, DIN 933, DIN 960, DIN 961

DIN 931

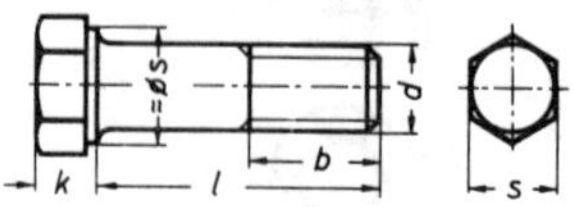

DIN 933

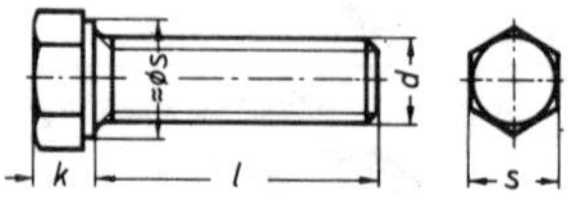

d		M 5	M 6	M 8 M 8 × 1	M 10 M 10 × 1,25	M 12 M 12 × 1,5	M 16 M 16 × 1,5	M 20 M 20 × 2
DIN 931, 960								
b		16	18	22	26 (32)	30 (36)	38 (44)	46 (52)
l	from	30	30	35	40 (130)	45 (130)	55 (130)	65 (130)
	up to	80	90	110	120 (150)	120 (180)	120 (200)	120 (220)
DIN 933, 961								
b		Screw thread approximately up to the screw head						
l	from	6	6	8	8	10	12	16
	up to	80	90	110	150	150	150	200
k		3,5	4	5,5	7	8	10	13
s		8	10	13	17	19	24	30

Steps of length: 10, 12, 16, 20, 25; above 80 mm length in steps of 10 mm. DIN 931, 933 coarse-pitch threads, DIN 960, 961 fine-pitch threads. ( ) for fine-pitch threads. Property class (material) from M 12 screw threads: 5.6, 8.8, 10.9

Internal hexagon socket head screws, dimensions in mm DIN 912

DIN 912

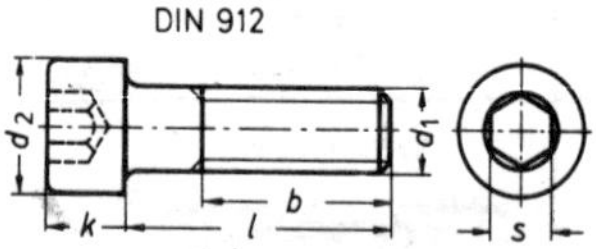

d_1		M 5	M 6	M 8 M 8 × 1	M 10 M 10 × 1,25	M 12 M 12 × 1,5	M 16 M 16 × 1,5	M 20 M 20 × 2
b		16	18	22	26	30	38 (44)	46 (52)
l	from	30	30	35	40	45	60 (140)	70 (140)
	up to	60	60	100	120	120	120	120 (180)
b		Screw thread approximately up to the screw head						
l	from	10	10	16	16	20	20	40
	up to	25	25	30	35	40	50	60
d_2		8,5	10	13	16	18	24	30
k		5	6	8	10	12	16	20
s		4	5	6	8	10	14	17

Steps of length: 10, 12, 16, 20, 25, 30 ... 50, 60 .. 100, 120, 140 ... 180.
Property class (material): . 6.9 8.8 10.9 12.9

Slotted cheese head screw, dimensions in mm DIN 84

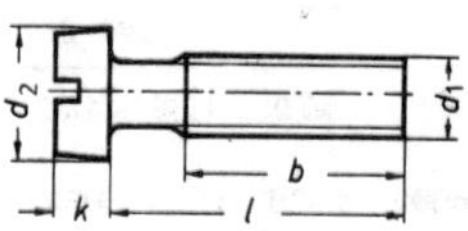

DIN 84

d_1		M 2,5	M 3	M 4	M 5	M 6	M 8	M 10
b		Screw thread approximately up to the screw head						
l	from	3	3	4	6	8	10	12
	up to	20	20	25	25	35	40	45
b		18	19	22	25	28	34	40
l	from	20	25	30	30	40	45	50
	up to	30	40	50	50	50	55	60
d_2		4,5	5,5	7	8,5	10	13	16
k		1,6	2	2,6	3,3	3,9	5	6

Slotted countersunk head screws, Cross-recessed countersunk head screws, dimensions in mm DIN 963, 965

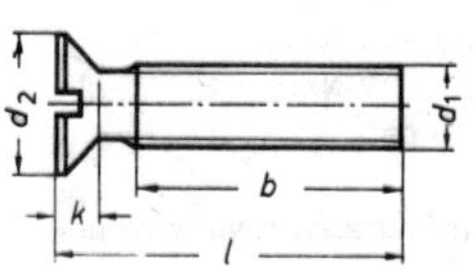

DIN 963

DIN 965 with cross recess

d_1		M 2,5	M 3	M 4	M 5	M 6	M 8	M 10
b		Screw threads approximately up to screw head						
l	from	3	4	5	6	8	10	12
	up to	20	20	25	30	35	40	50
b		18	19	22	25	28	34	40
l	from	25	25	30	35	40	45	55
	up to	25	30	40	50	50	55	60
d_2		4,7	5,6	7,5	9,2	11	14,5	18
k		1,5	1,65	2,2	2,5	3	4	5

Slotted raised countersunk head screws, Cross-recessed raised countersunk head screws, dimensions in mm DIN 964, 966

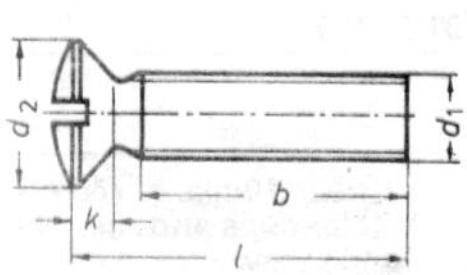

DIN 964

DIN 966 with cross recess

		M 2,5	M 3	M 4	M 5	M 6	M 8	M 10
b		Screw threads approximately upto screw head						
l	from	3	4	5	6	8	10	12
	up to	20	20	25	30	35	40	50
b		18	19	22	25	28	34	40
l	from	25	25	30	35	40	45	55
	up to	25	30	40	50	50	55	60
d_2		4,7	5,6	7,5	9,2	11	14,5	18
k		1,5	1,65	2,2	2,5	3	4	5

Steps of length for DIN 84, DIN 963, 965, DIN 964, 966 actually: 3, 4, 5, 6, 8, 10, 12, 16, 20, 25, 30 ... 60.

Slotted self-tapping screws, Dimensions in mm DIN 7971, DIN 7972, DIN 7973

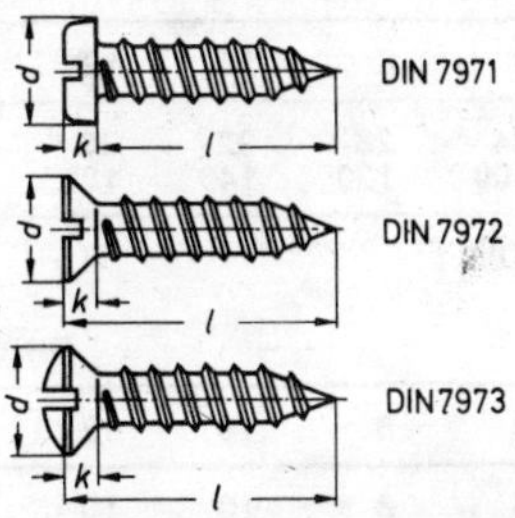

Nom. dia.	2,2	2,9	3,5	(3,9)	4,2	4,8	(5,5)	6,3
ISO No.	2	4	6	7	8	10	12	14
l from	4,5	6,5	9,5	9,5	9,5	9,5	13	13
up to	16	19	25	25	32	38	38	38
Slotted cheese head screws 7971								
d	4,2	5,6	6,9	7,5	8,2	9,5	10,8	12,5
k	1,3	1,7	2,1	2,2	2,4	2,8	3,2	3,6
Slotted countersunk head screws								
d	4,3	5,5	6,8	7,5	8,1	9,5	10,8	12,4
k	1,3	1,7	2,1	2,3	2,5	3,0	3,4	3,8
Slotted raised countersunk head screws								
d	4,3	5,5	6,8	7,5	8,1	9,5	10,8	12,4
k	1,3	1,7	2,1	3,3	2,5	3,0	3,4	3,8

Cross-recessed head self-tapping screws, Dimensions in mm DIN 7981, DIN 7982, DIN 7983

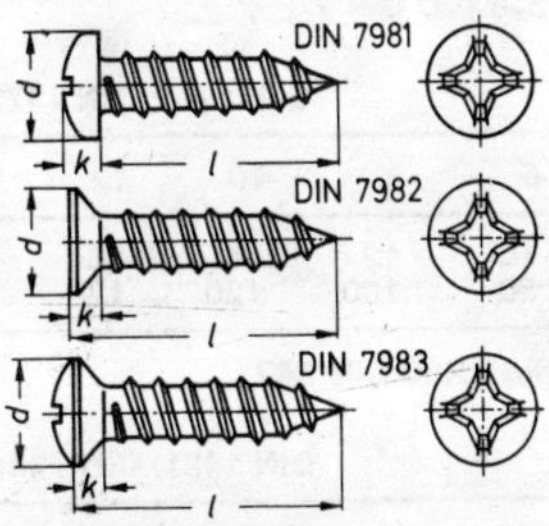

Nom. dia.	2,2	2,9	3,5	(3,9)	4,2	4,8	(5,5)	6,3
ISO No.	2	4	6	7	8	10	12	14
l from	4,5	6,5	9,5	9,5	9,5	9,5	13	13
up to	6	19	25	25	32	38	38	38
Cross-recessed raised cheese head self-tapping screws								
d	4,2	5,6	6,9	7,5	8,2	9,5	10,8	12,5
k	1,8	2,2	2,6	2,8	3,0	3,5	3,9	4,5
Cross-recessed countersunk head self-tapping screws								
d	4,3	5,5	6,8	7,5	8,1	9,5	10,8	12,4
k	1,3	1,7	2,1	2,3	2,5	3,0	3,4	3,8
Cross-recessed raised countersunk head self-tapping screws								
d	4,3	5,5	6,8	7,5	8,1	9,5	10,8	12,4
k	1,3	1,7	2,1	2,3	2,5	3,0	3,4	3,8

Hexagonal head self-tapping screws, Dimensions in mm DIN 7976

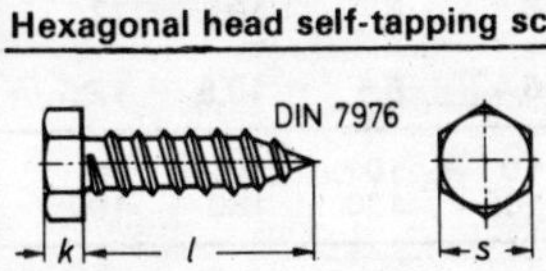

Nom. dia.	2,2	2,9	3,5	(3,9)	4,2	4,8	(5,5)	63
ISO No.	2	4	6	7	8	10	12	14
l from	4,5	6,5	9,5	9,5	9,5	9,5	13	13
up to	13	19	25	25	32	50	50	50
s	3,2	5	5,5	7	7	8	8	10
k	1,3	1,5	2,3	2,3	2,8	3	4	4,8

Countersunks and counterbores for screws (medium type), Dimensions in mm DIN 75

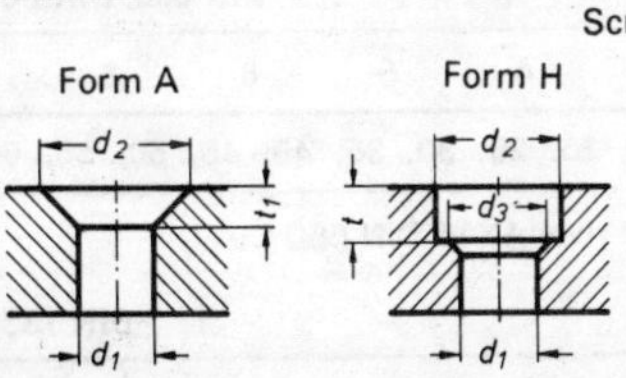

Screw thread	M 4	M 5	M 6	M 8	M 10	M 12	M 16	M 20
Form A								
d_1	4,5	5,5	6,6	9	11	14	18	22
d_2	8,6	10,4	12,4	16,5	20,4	24,4	32,4	40,4
t_1	2,1	2,5	2,9	3,7	4,7	5,2	7,2	9,2
Form H								
d_1	4,5	5,5	6,6	9	11	14	18	22
d_2	8	10	11	15	18	20	26	33
d_3	–	–	–	–	–	16	20	24
t	3,2	4	4,7	6	7	8	10,5	12,5

Tap drill diameters for self-tapping screws DIN 7975

Nominal diameter	ISO number	Sheet metal thickness above up to	Tap drill diameter Steel sheet[1])	Aluminium sheet	Nominal diameter	ISO number	Sheet metal thickness above up to	Tap drill diameter Steel sheet[1])	Aluminium sheet
2,2	2	0,56 ... 0,75	1,7	1,6	4,2	8	0,50 ... 1,13	3,2	3,0
		0,75 ... 0,88	1,8	1,6			1,13 ... 1,38	3,3	3,2
		0,88 ... 1,13	1,85	1,6			1,38 ... 2,5	3,5	3,5
2,9	4	0,56 ... 0,75	2,2	2,2	4,8	10	0,50 ... 1,13	3,7	3,7
		0,75 ... 1,38	2,4	2,2			1,13 ... 1,75	3,9	3,7
		1,38 ... 1,75	2,5	2,25			1,75 ... 3,0	4,0	3,8
3,5	6	0,56 ... 0,88	2,7	2,65	5,5	12	1,13 ... 1,50	4,3	4,1
		1,0 ... 1,38	2,8	2,65			1,50 ... 1,75	4,5	4,2
		1,38 ... 1,75	2,9	2,75			1,75 ... 3,0	4,6	4,5
3,9	7	0,50 ... 1,13	2,95	2,9	6,3	14	1,38 ... 1,75	5,0	5,0
		1,13 ... 1,38	3,0	2,95			1,75 ... 2,0	5,2	5,0
		1,38 ... 2,0	3,2	3,0			2,0 ... 3,0	5,3	5,2

[1]) These values are applicable for sheets of brass, copper and nickel also.

Pins, Clamping sleeves, Rivets, Cotter pins

Taper pins

DIN 1

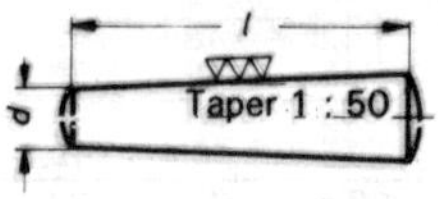

d		3	4	5	6	8	10	12
l	from up to	14 50	16 60	20 70	24 100	28 120	32 140	36 165

Example of designation: Taper pin 5×20 DIN 1

Cylindrical pins

DIN 7

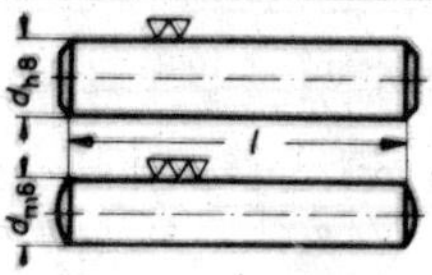

d		3	4	5	6	8	10	12
l	from up to	4 32	5 40	5 50	6 60	8 80	10 100	10 120

Example of designation: Cylindrical pin 5_m6×20 DIN 7

Grooved taper pins, Grooved cylindrical pins

DIN 1471, DIN 1473

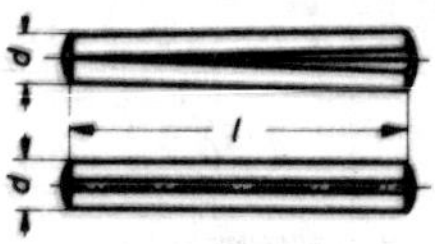

d		3	4	5	6	8	10	12
l	from up to	6 40	6 60	8 60	10 80	12 100	14 120	16 120

Example of designation: Grooved taper pin 6×50 DIN 1471

Clamping sleeves

DIN 1481, DIN 7346

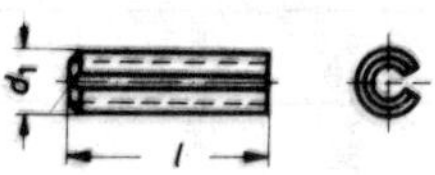

Nominal diameter		3	4	5	6	8	10	12
d_1		3,3	4,4	5,4	6,4	8,5	10,5	12,5
l	from up to	4 40	4 50	5 80	10 100	10 120	10 160	10 180

Example of designation: Clamping sleeve 8×50 DIN 1481

Half round rivets, Countersunk rivets

DIN 660, DIN 661

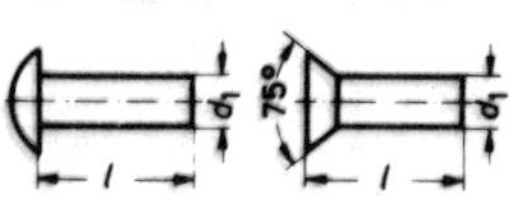

d_1	2	2,6	3	4	5	6	8
l	3, 4, 6, 8, 10, 15, 20, 25, 30, 35, 40, 45, 50, 55, 60						

Example of designation: Half round rivet 4×15 DIN 660 Cu

Rivets for brake linings and coupling linings

DIN 7338

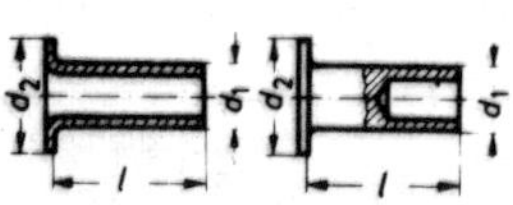

d_1		3	4	5	6	8
d_2		5,5	7,5	9,5	11,5	15,5
l	from up to	5 10	6 15	6 18	8 40	10 40

Example of designation: Rivet 45×20 DIN 7338 Al

Split cotter pins

DIN 94

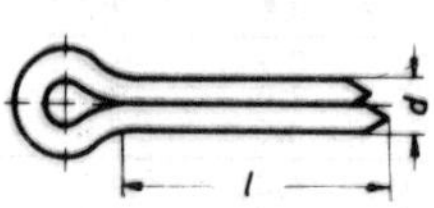

Ø of bolt above...up to	3 … 4	4 … 6	6 … 8	8 … 11	11 … 17
Ø of screw above...up to	4 … 5	5,5 … 7	8 … 10	11 … 14	16 … 20
Ø of split cotter	1	1,5	2	3	4
l	6 … 15	8 … 30	10 … 40	15 … 60	20 … 70

Example of designation: Split cotter pin 3×20 DIN 94 St

Woodruff keys

DIN 6888

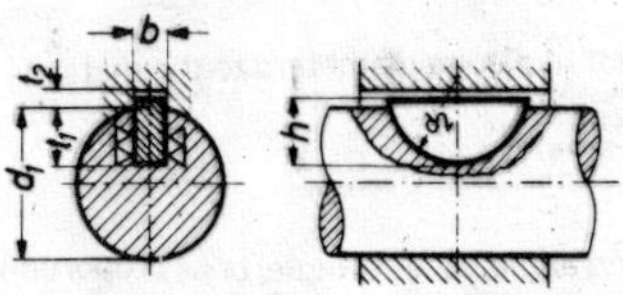

Diameter of shaft	Woodruff key			Depth of keyway in shaft	Depth of keyway in hub
	b_{h9}	h_{h12}	d_2		
> 8 up to 10	3	3,7/5	10/13/16	2,5/3,8/5,3	1,4
>12 up to 17	5	6,5/7,5	16/19/22	4,5/5,5/7,0	2,2
>17 up to 22	6	7,5/9	19/22/28	5,1/6,6/8,6	2,6
>22 up to 30	8	9/11	22/28/32	6,2/8,2/10	3,0
>30 up to 38	10	11/13	28/32/45	7,8/9,8/13	3,4

Parallel keys

DIN 6885

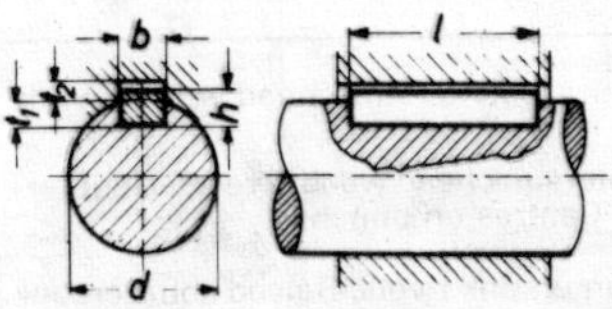

Diameter of shaft	Parallel key (high form)			Depth of keyway in shaft	Depth of keyway in hub with clearance
	b	*h*	*l*	t_1	
> 12 up to 17	5	5	10 ... 45	2,9	2,2
> 17 up to 22	6	6	12 ... 56	3,5	2,6
> 22 up to 30	8	7	16 ... 70	4,1	3
> 30 up to 38	10	8	20 ... 90	4,7	3,4

Round-end taper keys, Flat-end taper keys

DIN 6886

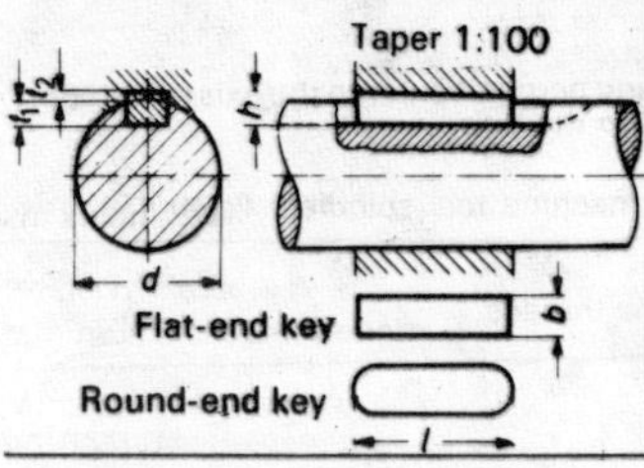

Diameter of shaft	Taper key (taper 1:100)			Depth of keyway in shaft	Depth of keyway in hub
	b	*h*	*l*		
>12 up to 17	5	5	12 ... 56	2,9	1,8
>17 up to 22	6	6	16 ... 70	3,5	2,1
>22 up to 30	8	7	20 ... 90	4,1	2,4
>30 up to 38	10	8	25 ... 110	4,7	2,8
>38 up to 44	12	8	32 ... 140	4,9	2,6

Flat taper keys

DIN 6883

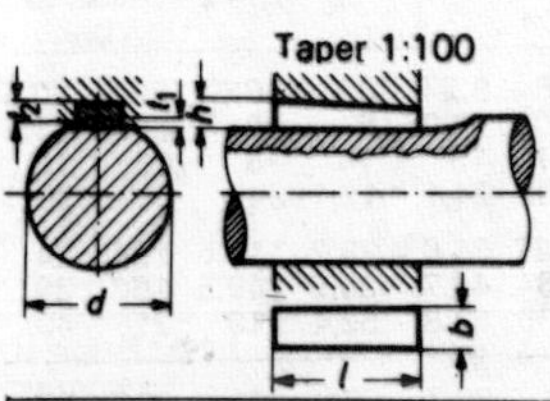

Diameter of shaft	Taper key (taper 1:100			Depth of keyway in shaft	Depth of keyway in hub
	b	*h*	*l*		
> 22 up to 30	8	5	20 ... 70	1,3	3,2
> 30 up to 38	10	6	25 ... 90	1,8	3,7
> 38 up to 44	12	6	32 ... 125	1,8	3,7
> 44 up to 50	14	6	36 ... 140	1,4	4

Saddle keys

DIN 6881

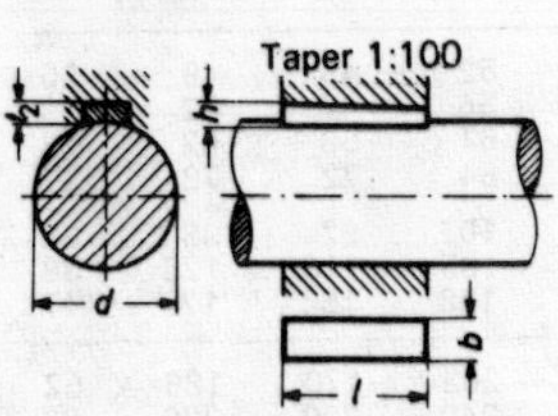

Diameter of shaft	Taper key (taper 1:100)				Depth of keyway in hub
	b	*h*	*l*		
> 22 up to 30	8	3,5	20 ... 90		3,2
> 30 up to 38	10	4	25 ... 110		3,7
> 38 up to 44	12	4	32 ... 140		3,7
> 44 up to 50	14	4,5	40 ... 160		4
> 50 up to 58	16	5	45 ... 180		4,5

Definitions and functional values

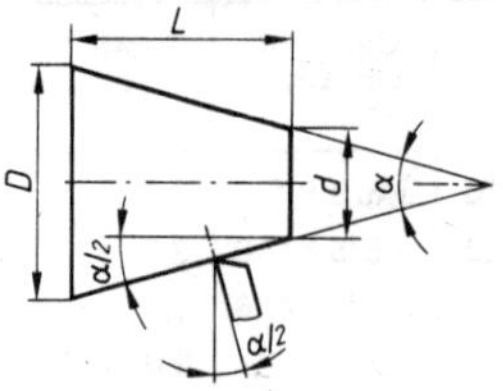

Cone Taper (DIN ISO 3040)

The cone taper is the ratio of the difference of two cone diameters to the distance between them, therefore:

$$\text{Cone taper} = \frac{\text{large diameter-small diameter}}{\text{taper length}} \text{ or } C = \frac{D-d}{L}$$

The cone taper can be specified by the included angle α (cone angle) or as proportionate value, e.g. 1:5 or 0.2:1 or 20%

The setting angle on the processing machine is equal to half the cone angle.

Preferred values

DIN 254

Taper C 1:x	Cone angle α	Setting angle $\alpha/2$	Fields of application
1:0,289	120°	60°	Protecting countersinks of centre holes, Black countersunk head screws with square collars.
1:0,500	90°	45°	Valve tapers, Countersunk head screws, Countersunk head wood screws, Countersunk rivets, Locking screws (Screwed plugs), Centres on punches.
1:0,866	60°	30°	Centre holes, Countersunk head screws, Countersunk rivets, Raised countersunk rivets, Centre punches, Sealing tapers for light pipe fittings.
1:5	11°25′	5°42′	Tapered tongues on shafts and taper holes in hubs for easily detachable machine parts (V belt pulleys) Friction couplings, Hose connectors for compressed air.
1:6	9°32′	4°46′	Sealing tapers tor cocks (friction cocks)
1:10	5°43′	2°52′	Taper ends of shafts, Machine parts for loading perpendicular to the axis on rotation and adjustable bearing bushes longitudinal to the axis.
1:20	2°52′	1°26′	Metric tool tapers and mounting tapers of machine tool spindles, Taper holes in steering wheel hubs.
1:50	1°9′	34′	Taper pins, Taper reamers, Taper pipe screw threads.

Tool tapers

DIN 228

Taper shank

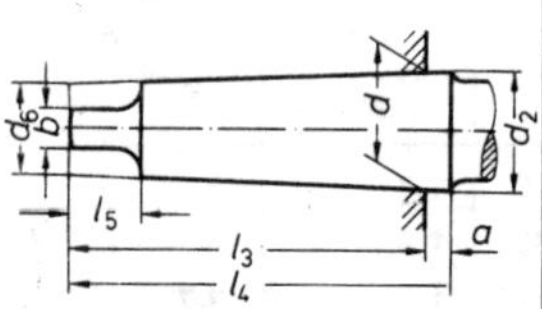

Taper	Number	Taper ratio	Setting angle	a	b	d_1	d_2	d_6	l_3	l_4	l_5
Morse taper	0	1:19,212	1°29′27″	3	3,9	9,05	9,2	6,1	56,5	59,5	10,5
	1	1:20,047	1°25′43″	3,5	5,2	12,06	12,2	9	62	65,5	13,5
	2	1:20,020	1°25′50″	5	6,3	17,78	18	14	75	80	16
	3	1:19,922	1°26′16″	5	7,9	23,82	24,1	19,1	94	99	20
	4	1:19,254	1°29′15″	6,5	11,9	31,27	31,6	25,2	117,5	124	24
	5	1:19,002	1°30′26″	6,5	15,9	44,34	44,7	36,5	149,5	156	29
	6	1:19,180	1°29′36″	8	19	63,35	63,8	52,4	210	218	40
Metric taper	80	1:20	1°25′56″	8	26	80	80,4	69	220	228	48
	100	1:20	1°25′56″	10	32	100	100,5	87	260	270	58

Taper socket

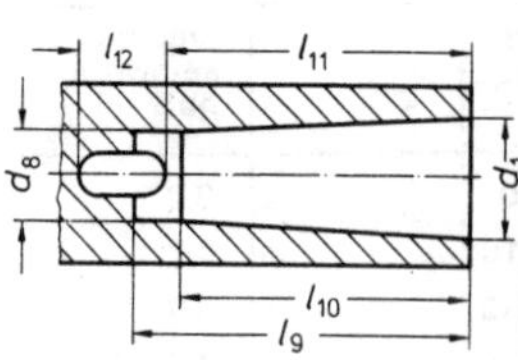

Taper	Number	Taper ratio	Setting angle	d_1	d_8	l_9	l_{10}	l_{11}	l_{12}
Morse taper	0	1:19,212	1°29′27″	9,054	6,70	52	45	49	15
	1	1:20,047	1°25′43″	12,065	9,7	56	47	52	19
	2	1:20,020	1°25′50″	17,780	14,9	67	58	62	22
	3	1:19,922	1°26′16″	23,825	20,2	84	72	78	27
	4	1:19,254	1°29′15″	31,267	26,5	107	92	98	32
	5	1:19,002	1°30′26″	44,399	38,2	135	118	125	38
	6	1:19,180	1°29′36″	63,348	54,8	188	164	177	47
Metric taper	80	1:20	1°25′56″	80	71,5	202	170	186	52
	100	1:20	1°25′56″	100	90	240	200	220	60

Angles on turning tool cutting edge

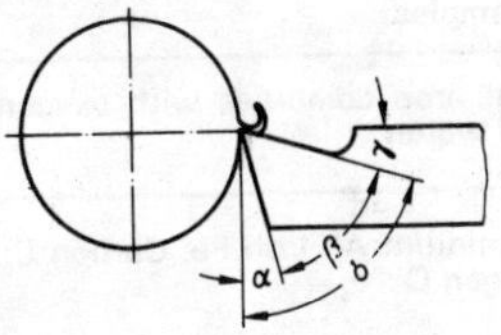

Clearance angle $\alpha = 3° ... 10°$

Angle between the cutting wedge and the perpendicular to the machining surface. When the clearance angle is 0°, the cutting edge is not relieved and the clearance face lies on the machined surface and rubs.

With too small clearance angle, there are large cutting force, strong heating of the cutting edge, rough machined surface, small tool life (tool life=span of working time between grinding and regrinding of the tool)

Wedge angle $\beta = 40° ... 95°$

Angle of the cutting wedge which penetrates into the material.

Effects of wedge angle: Smaller the wedge angle, easier the cutting wedge penetrates into the material, smaller is the required cutting force. Small wedge angles are suitable for soft materials. With too small angles there is the danger of breakage of the tool cutting edge.

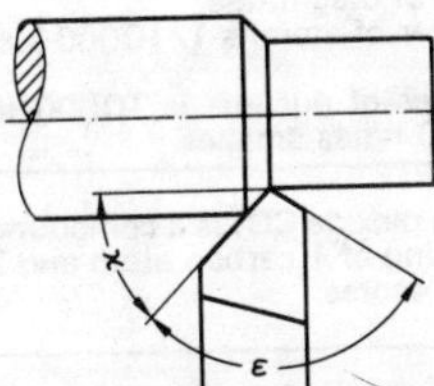

Rank angle $\gamma = 0° ... 45°$

Angle between the rake face and the horizontal plane. Effects of rake angle: A large rake angle effects easy chip formation, smooth machined surface, small cutting force, continuous chip flow. Too large rake angle gives too weak cutting cross-section, thereby heat accumulation and tool life. In general, a medium rake angle is chosen in order to obtain shorter chips.

Side rake angle $\lambda = 3° ... 5°$

Angle by which the rake face is inclined. By the inclination of the rake face a better chip disposal is effected. A drawing cut and hence an easy peeling off of the chip is obtained thereby. With interrupted cut, the cutting tip is loaded less strongly.

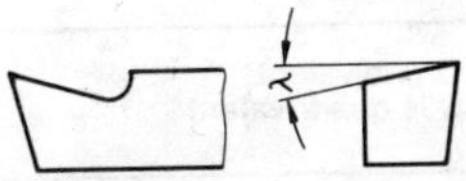

Cutting angle $\delta = \alpha + \beta$

Cutting angled = clearance angle + wedge angle

The cutting angle is generally smaller than 90°. With negative rake angles, it can be greater than 90°

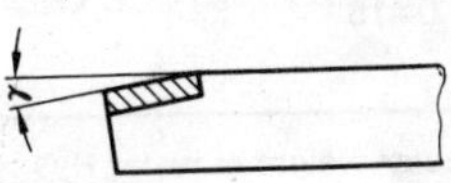

Tip angle $\varepsilon = 60° ... 120°$

Angle at the cutting tip.

A large tip angle effects a smaller thermal loading of the cutting tip and hence a higher tool life at high cutting speeds.

Setting angle $\varkappa = 40° ... 50°$

Angle between the main cutting edge and the machined surface.

A large setting angle effects a short cutting length, therefore small cutting force but high thermal loading of the cutting edge. A large setting angle is favourable against chatter.

Reference values for processed material	High-speed steel turning tool Clearance angle ∢ α	Wedge angle ∢ β	Rake angle ∢ γ	Hard metal turning tool Clearance angle ∢ α	Wedge angle ∢ β	Rake angle ∢ γ
Steel, Cast steel up to 600 N/mm²	8°	60° ... 67°	15° ... 20°	4° ... 6°	65° ... 72°	14° ... 18°
Steel, Cast steel above 600 N/mm²	8°	65° ... 68°	14° ... 18°	4° ... 6°	70° ... 76°	10° ... 12°
Cast iron soft	6° ... 8°	68° ... 70°	14°	5°	73° ... 75°	10° ... 12°
Cu-Zu alloys	6° ... 10°	75° ... 85°	0° ... 5°	6° ... 8°	77° ... 85°	0° ... 5°
Aluminium, Aluminium alloys	10°	35° ... 40°	30° ... 40°	8°	48° ... 52°	30° ... 35°
Synthetic resin casting materials	6° ... 8°	50° ... 65°	20° ... 30°	6° ... 8°	55° ... 70°	15° ... 25°

Twist drills

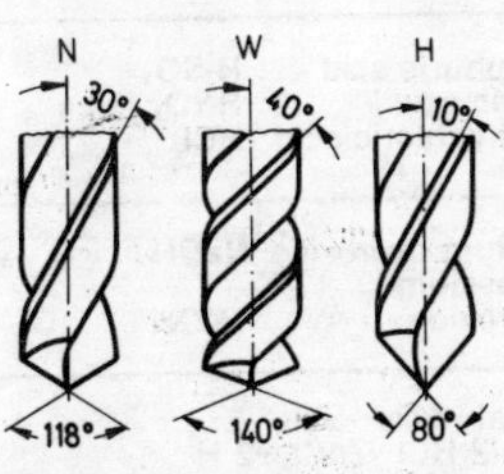

Tool type	Point angle	Material to be processed
N	118°	Normal materials, Steel and Cast steel up to 700 N/mm², Cast iron, Malleable iron, Brass from CuZn 40
H	80° 118°	Hard materials, Synthetic resin casting materials, Marble, Brass up to CuZn 40
W	140°	Soft materials, Copper, Aluminium, Aluminium alloys.

Smaller the spiral angle, greater is the spiral pitch, smaller is the rake angle of the main cutting edge.

With unequally long cutting edges of the main cutting edge the hole becomes too large.

Information on materials

Basic concepts in chemistry

Concept	Explanation	Examples
Chemical reactions	Reactions in which new materials with new properties are produced by decomposition or composition	Rust; iron combines with oxygen and water.
Basic materials or elements	Materials which cannot be split further chemically into simpler materials (materials with exactly equal atoms)	Aluminium Al, Iron Fe, Carbon C, Oxygen O
Atom	Smallest particle of an element which cannot be furtner split chemically. It consists of a nucleus consisting of protons and neutrons and an envelope of electrons.	Orders of magnitude: Diameter of atom is 1/10000000 mm Diameter of nucleus is 10000 to 100000 times smaller
Chemical compound	Material which is obtained by the composition of two or more elements. The compound has properties other than of its elements	Carbon dioxide CO_2 is a compound consisting of 1 carbon atom and 2 oxygen atoms.
Molecule	Smallest particle of a chemical compound consisting of at least 2 atoms. Gaseous elements (rare gases excepted) occur freely like this in molecules.	Carbon dioxide molecule CO_2 Oxygen molecule O_2 Ozone molecule O_3
Valency	Number which specifies how many hydrogen atoms (univalent) an atom can combine with or replace	H_2O; O is bivalent CO_2; C is quadrivalent
Atomic weight	Number which specifies how many times an atom is heavier than a hydrogen atom. Accurate atomic weights are referred to the carbon atom (atomic weight 12).	Atomic weight of hydrogen is 1.0 Atomic weight of oxygen $16 \times 1.0 = 16$
Molecular weight	Sum of the atomic weights of all atoms of a molecule	Molecular weight of water H_2O is $2 \times 1 + 16 = 18$.
Analysis	Decomposition of a material or material mixture into elements for the determination of the composition.	Decomposition of water with the help of an electric current. $H_2O \rightarrow 2H + O$
Synthesis	Preparation of a chemical compound from elements or simply prepared compounds	56 g of iron and 32 g of sulphur combine to form 88 g of iron sulphide $Fe + S \rightarrow FeS$
Oxidation	Combination of material with oxygen whereby an oxide is produced. Any combustion is an oxidation.	Burning magnesium, carbon $Mg + O \rightarrow MgO$ (magnesium oxide) $C + 2O \rightarrow CO_2$ (carbon dioxide)
Reduction	Removal of oxygen from an oxide. Oppsite of oxidation application in extraction of metals.	Mercuric oxide is reduced to mercury by heating $2\ HgO \rightarrow 2\ Hg + O_2$
Acid	Chemical compound of hydrogen with a non-metal oxide (acid residue) or non-metal. Acids taste sour, turn litmus red.	Sulphuric acid H_2SO_4 Nitric acid HNO_3 Hydrochloric acid HCl
Alkali or base	Chemical compound of metal oxides with water. Alkalis consist of a metal and the OH group; they turn litmus blue.	Sodium hydroxide NaOH Potassium hydroxide KOH
Salt	Compound of a metal with an acid. The hydrogen of the acid is thereby replaced by a metal.	Metal+Acid→Salt $Zn + 2\ HCl \rightarrow ZnCl_2 + 2\ H$

Concept	Explanation	Examples
Physical processes	Processes in which energies are effective (heat, mechani cal work) but no material changes take place. State, shape and position change.	Change of state: gasification of petrol, Water becoming steam Change of shape: bending of sheet Change of position: lifting of vehicle
Types of state	The types of state or aggregate states are solid state, liquid state and gaseous state. Solids have a definite shape and a definite volume, liquids have a definite volume but no definite shape, gases have no definite shape and no definite volume.	Solid: iron, aluminium, glass, plastic Liquid: water, gasoline Diesel fuel, mercury Gas: air, oxygen
Change of state	The state of a material is changed by: heating: liquids become vapours, solids melt. cooling: liquids become solids. compressing: gases become liquids decompressing: liquids become gases.	Water becomes steam, water becomes ice ice melts to water, Propane gas becomes liquid under pressure in a flask and with releasing becomes gas again.
Melting point, Freezing point, Solidifying point	Temperature at which a solid becomes liquid or a liquid becomes solid. Freezing point for materials which are liquids at room temperature.	Melting and solidifying points of lead amounts to 327°C, the freezing point of water 0°C.
Boiling point, Condensation point	Temperature at which a liquid becomes gas or a gas becomes liquid. Liquids change partly even below the boiling point gradually into gas (evaporation)	Water boils at 100°C and changes into steam with bubble formation. At room temperature water evaporates gradually.
Quantity of heat	The quantity of heat which is contained in a material or which a material can give out is specified in the SI unit joule (J) or kilojoule (kJ)	The quantity of heat 1 J is equal to 1 wattsecond (Ws) and equal to 1 newtonmeter (Nm)
Specific heat capacity	Quantity of heat in joule (J) or kilojoule (kJ) which is required to heat 1 kg of a material by 1 K (kelvin) or 1°C (Celsius)	To heat 1 kg of water by 1°C, 4,1868 kJ are required.
Specific heat of evaporation (latent heat)	Quantity of heat in J or kJ which is required to vapourise 1 kg of a material at the boiling point. The heat of evaporation is released when the vapourized material becomes liquid again (heat of condensation)	To convert 1 kg water at 100°C into steam at 100°C, 2 257 kJ are required.
Specific heat of melting (fusion heat)	Quantity of heat in J or kJ which is required to melt 1 kg of a material at the melting point. The heat of melting is released when the liquid solidifies.	To convert 1 kg ice at 0°C into water at 0°C, 335 kJ are required.
Thermal conductivity	Quantity of heat in kJ which flows in 1 hour through a wall of 1 m thickness and 1 m^2 area when the two surface temperatures differ by 1 K (1°C).	All metals are good heat conductors. Silver and copper stand at the top.
Mass	The dimensional unit for the mass is 1 kg. The mass of 1 kg is fixed by the international kilogram prototype (platinum cylinder)	1 kg is approximately the mass of 1 dm^3 water at 4°C and an air pressure of 1 013 mbar.
Density	Density of a material is the quotient obtained by dividing the mass by the volume. It is understood as the mass in kg which is contained in 1 dm^3 of a material.	Density of steel is 7.86 kg/dm^3. density of gasoline 0.74 kg/dm^3.
Force	The SI unit for the force is 1 newton (N). 1 N is the force which imparts to a body of 1 kg mass the acceleration of 1 m/s^2.	$1N = 1\ kg \cdot 1\ m/s^2 = 1\ kg\ m/s^2$
Cohesion	Connecting force with which the molecules of a material are mutually attracted. The forces of attraction are effective only over very small distances.	Solids have great cohesion, liquids only a little cohesion.
Adhesion	Force of attraction between the molecules of different materials (complementary force). Nearer the bodies are brought together, greater is the adhesion.	Chalk sticks to the black-board, oil sticks to a steel shaft.

Chemical elements

Selection

Element	Chemical symbol	Valency	Atomic Weight
Metals			
Aluminium	Al	3	26,98
Antimony	Sb	3, 5	121,75
Barium	Ba	2	137,34
Beryllium	Be	2	9,01
Lead	Pb	2, 4	207,19
Calcium	Ca	2	40,08
Chromium	Cr	2, 3, 6	52,00
Iron	Fe	2, 3, 6	55,85
Germanium	Ge	2, 4	72,59
Gold	Au	1, 3	196,97
Indium	In	3	114,82
Iridium	Ir	3, 4	192,20
Cadmium	Cd	2	112,40
Potassium	K	1	39,10
Cobalt	Co	2, 3	58,93
Copper	Cu	1, 2	63,54
Lithium	Li	1	6,94
Magnesium	Mg	2	24,31
Manganese	Mn	2, 3, 4, 6, 7	54,94
Molybdenum	Mo	3, 4, 6	95,94
Sodium	Na	1	22,99
Nickel	Ni	2, 3	58,71
Osmium	Os	2, 3, 4, 8	190,20
Palladium	Pd	2, 4	106,40
Platinum	Pt	2, 4	195,10
Mercury	Hg	1, 2	200,59
Radium	Ra	2	226,00

Element	Chemical symbol	Valency	Atomic weight
Metals			
Silver	Ag	1, 2	107,87
Tantalum	Ta	5	180,95
Titanium	Ti	2, 3, 4	47,90
Uranium	U	3, 4, 5, 6	238,03
Vanadium	V	2, 3, 4, 5	50,94
Bismuth	Bi	3, 5	208,90
Tungsten	W	2, 3, 4, 5, 6	183,85
Zinc	Zn	2	65,37
Tin	Sn	2, 4	118,69
Non-metals			
Arsenic	As	3,5	74,92
Barium	B	3	10,81
Bromium	Br	1, 5	79,91
Iodine	J	1, 3, 5, 7	126,90
Carbon	C	2, 4	12,01
Phosphorus	P	3, 5	30,97
Sulphur	S	2, 4, 6	32,06
Selenium	Se	2, 4, 6	78,96
Silicon	Si	4	28,09
Gases			
Argon	Ar	0	39,95
Chlorine	Cl	1, 3, 5, 7	35,45
Fluorine	F	1	19,00
Helium	He	0	4,00
Crypton	Kr	0	83,80
Neon	Ne	0	20,18
Oxygen	O	2	16,00
Nitrogen	N	2, 3, 5	14,01
Hydroger	H	1	1,00

Chemical compounds

Selection

Industrial and chemical term	Formula
Acetone	$(CH_3)_2 \cdot CO$
Acetylene	C_2H_2
Alcohol, Ethanol	C_2H_5OH
Benzene	C_6H_6
Litharge, Lead monoxide	PbO
Borax, Sodium tetra borate	$Na_2B_4O_7$
Pyrolusite, Manganese dioxide	MnO_2
Chile saltpeter, Sodium nitrate	$NaNO_3$
Chromic acid, Chromium oxide	CrO_3
Rust, Ferric oxide hydrate	Fe_2O_3
Green vitriol, Ferrous sulphate	$FeSO_4$
Acetic acid	CH_3COOH
Hydrofluoric acid	HF
Gypsum, Calcium sulphate	$CaSO_4$
Glauber's salt, Sodium sulphate	Na_2SO_4
Glycerin	$C_3H_5(OH)_3$
Lunar caustic, Silver nitrate	$AgNO_3$
Potash lye, Potassium hydroxide	KOH
Saltpeter, Potassium nitrate	KNO_3
Caustic lime (quick lime), Calcium oxide	CaO
Slaked lime (hydrated lime), Calcium	$Ca(OH)_2$
Calcite (limestone), Calcium carbonate	$CaCO_3$
Carbide, Calcium carbide	CaC_2
Silica (silicic acid), Silicon dioxide	SiO_2
Common salt, Sodium chloride	$NaCl$
Carbonic acid gas, Carbon dioxide	CO_2
Carbon monoxide	CO

Industrial and chemical term	Formula
Corundum, Aluminium oxide	Al_2O_3
Chalk, Calcium carbonate	$CaCO_3$
Blue vitriol, Copper sulphate	$CuSO_4$
Killed spirit (soldering fluid), Zinc chloride solution	$ZnCl_2$
Marble, Calcium carbonate	$CaCO_3$
Minium, Lead oxide	Pb_3O_4
Methyl alcohol, Methanol	CH_3OH
Molybdenum disulphide	MoS_2
Soda lye (caustic soda, soda liquor), Sodium hydroxide	$NaOH$
Sodium saltpeter, Sodium nitrate	$NaNO_3$
Polishing red (calcother, crocus, rouge), Ferric oxide	Fe_2O_3
Potash (pearl ash), Potassium carbonate	K_2CO_3
Sal ammoniac, Ammonium chloride	NH_4Cl
Aqueous ammonia, Ammonia water	NH_4OH
Saltpeter, Sodium nitrate	$NaNO_3$
Nitric acid	HNO_3
Hydrochloric acid	HCl
Sulphuric acid	H_2SO_4
Silicon carbide	SiC
Soda, Sodium carbonate	Na_2CO_3
Carbon tetrachloride	CCl_4
Trichloroethylene	C_2HCl_3
Potassium permanganate	$KMnO_4$
Hydrogen peroxide	H_2O_2
Zinc chloride	$ZnCl_2$
Zinc white, Zinc oxide	ZnO
Potassium cyanide	KCN

Properties of materials

Material	Density kg/dm³	Melting temperature °C	Boiling temperature °C	Specific heat kJ/kgK	Melting heat kJ/kg	Coefficient of linear expansion 1/K	Coefficient of thermal Conductivity kJ/mhK
Metals, pure							
Aluminium	2,7	658	2060	0,904	356	0,000024	757,8
Antimony	6,68	630	1440	0,209	163	0,000012	62,8
Barium	3,5	701	1537	0,293	—	0,000019	—
Beryllium	1,85	1279	2970	1,879	—	0,000012	602,8
Lead	11,3	327	1717	0,054	24,3	0,000029	125,6
Calcium	1,55	850	1270	0,630	328	—	—
Chromium	7,1	1890	2400	0,542	134	0,000006	—
Iron	7,86	1539	3000	0,456	270	0,000012	242,8
Germanium	5,32	936	2700	0,310	409	0,000006	211
Gold	19,3	1063	2677	0,054	66,5	0,000014	1109
Cadmium	8,6	321	767	0,234	64,4	0,000031	330,7
Cobalt	8,9	1490	3168	0,435	242	0,000013	251,2
Copper	8,93	1083	2300	0,389	204	0,000017	1339
Magnesium	1,74	650	1103	0,837	379	0,000025	565,2
Manganese	7,3	1260	1960	0,502	251	0,000023	—
Molybdenum	10,2	2600	3560	0,272	286	0,000003	494
Sodium	0,97	97,5	889	1,26	113	—	481
Nickel	8,85	1452	3000	0,460	305	0,000013	209,3
Platinum	21,5	1770	4400	0,134	112	0,000009	251,2
Mercury	13,6	−39	357	0,138	11,7	—	29,3
Silver	10,5	960	1950	0,234	108	0,000020	1507
Tantalum	16,6	3030	—	0,138	171	0,000007	196,7
Titanium	4,54	1670	3000	0,583	88	0,000008	58,6
Vanadium	6,1	1730	3000	0,502	343	0,000009	—
Tungsten	19,3	3370	5900	0,142	54	0,000005	711,7
Zinc	7,14	420	907	0,205	114	0,000029	397,7
Tin	7,28	231	2275	0,226	59,5	0,000027	230,2
Alloys							
Bronze (CuSn)	7,5 ... 8,9	950	2400	0,376	—	0,000017	167,4
Chrom nickel	7,5	1430	2300	0,502	—	—	188,4
Cast iron	7,25	1200	2500	0,536	125	0,000010	175,8
Constantin	8,9	1260	2400	0,418	—	0,000015	83,7
Mg-alloy	1,8	650	1500	1,204	—	0,000025	375
Brass	8,5	950	2400	0,376	167	0,000019	355
Gun metal	8,8	1000	2400	0,376	—	—	251
Steel, alloyed	7,85	1500	2500	0,502	209	0,000012	188
White metal	7,5 ... 10	300 ... 400	2100	0,146	—	—	209
Non-metals							
Asbestos	2,2 ... 2,8	1150 ... 1550	—	—	—	—	—
Asphalt	1,2 ... 1,5	80 ... 100	300	0,921	—	—	2,51
Concrete	1,7 ... 2,5	—	—	0,879	—	—	4,18
Borax	1,72	741	—	1,204	—	—	—
Diamond	3,51	3550	4000	0,502	—	0,000001	—
Ice	0,92	0	100	2,0	333	0,000051	8,37
Fat, mineral	0,92 ... 0,94	30 ... 175	300	0,6 ... 0,8	—	—	0,75
Gypsum	2,3	1200	—	1,088	—	—	1,45
Glass	2,4 ... 2,7	700	—	0,837	—	0,000007	2,93
Mica	2,6 ... 3,2	1300	—	0,837	—	—	2,93
Graphite	2,26	3550	4350	0,795	—	0,000008	350
Hard rubber (ebonite, vulcanite)	1,2 ... 1,8	—	—	1,423	—	0,000070	0,62
Hard paper	1,3 ... 1,4	2050	2700	0,962	—	0,000015	62
Lime stone	2,7	—	—	0,921	—	—	7,9
Common salt	2,15	802	1440	0,879	—	—	—
Carbon	3,51	3540	4000	0,50	—	0,00000118	—
Cork	0,1 ... 0,3	—	—	1,6 ... 2	—	—	0,15
Corundum	3,9 ... 4	2050	2700	0,962	—	—	62
Paraffin	0,9	52	300	3,27	—	—	0,75
Plexglass	1,18	—	—	—	—	—	4,44
Porcelane	2,4	1600	—	0,80	—	0,0000045	3,50
Quartz	2,5 ... 2,8	1485	2230	0,80	—	—	35,5
Sulphur	2,07	114	445	0,753	46	—	0,96
Pit coal	1,2 ... 1,3	—	—	1,26	—	—	0,96
Silicon	2,33	1420	2600	0,753	—	0,000004	301
Wax	0,97	60	—	3,433	176	—	0,30
Cement	0,9 ... 1,9	—	—	1,130	—	—	3,76

Liquids	Density kg/km³	Boiling Temperature °C	Freezing Temperature °C	Specific heat kJ/kgK	Heat of evaporation (latent heat) kJ/kg	Coefficient of volumetric expansion 1/K	Coefficient of thermal conductivity kJ/mhK
Ether	0,72	35	− 116	2,26	376	0,0016	0,50
Ethyl alcohol	0,80	78,5	− 114	2,42	904	0,0011	0,71
Gasoline	0,72	40 … 200	− 40	2,09	—	0,001	0,58
Benzene, pure	0,88	80	5,4	1,67	393	0,0013	0,50
Diesel fuel	0,86	210 … 380	− 30	—	—	0,001	0,54
Glycerine	1,26	290	− 40	2,42	—	0,0005	1,04
Fuel oil	0,9	170 … 300	− 5	—	—	0,0001	0,41
Machine oil	0,92	370 … 410	− 5	1,88	—	0,0008	0,46
Methyl alcohol	0,80	65	− 98	2,51	1109	—	0,75
Kerosine	0,82	150 … 300	− 70	2,09	—	0,0001	0,58
Nitric acid	1,56	86	− 42	1,67	—	—	1,88
Hydrochloric acid 10%	1,05	102	− 14	3,14	—	—	1,80
Sulphuric acid	1,84	338	—	1,38	—	0,00055	1,67
Silicon oil	0,94	—	—	1,04	—	—	0,79
Spirit	0,81	78	− 90	2,42	1004	0,0012	0,58
Terpentine	0,87	160	− 10	1,92	293	0,001	0,37
Trichloroethylene	1,47	87	− 86	1,29	—	0,0012	0,58
Water	1,00	100	0	4,18	2256	0,00018	2,09

Gases	Density in the normal condition kg/m³	Boiling temperature °C	Freezing temperature °C	Specific heat at constant volume kJ/kgK	Specific heat at constant pressure kJ/kgK	Coefficient of thermal conductivity at 20°C J/mhK	Coefficient of thermal conductivity Air = 1
Acetylene	1,17	− 81	− 84	1,67	1,34	71	0,77
Ammonia	0,77	− 34	− 78	1,67	2,21	83	0,90
Etnylene	1,26	− 102	− 169	1,25	1,54	138	1,5
Argon	1,78	− 186	− 190	0,33	0,54	65	0,7
Butane	2,7	1	− 135	—	—	—	—
Helium	0,18	− 269	− 272	3,18	5,23	543	5,9
Carbon monoxide	1,25	− 191	− 205	0,75	1,05	83	0,9
Carbon dioxide	1,98	− 78,5	− 57	0,67	0,88	55	0,6
Lighting gas	0,58	− 210	− 230	1,59	2,14	230	2,5
Oxygen	1,29	− 192	− 220	0,71	1,00	92	1,0
Nitrogen	0,72	− 162	− 184	1,67	2,22	117	1,27
Hydrogen	2,02	− 45	− 190	—	—	—	—
Air	1,43	− 183	− 218	0,67	0,92	92	1,0
Methane	1,25	− 196	− 210	0,75	1,05	92	1,0
Propane	0,09	− 253	− 257	10,0	14,28	129	1,4

Calorific values

Solids	kJ/kg	Liquids	kJ/kg	Gases	kJ/kg
Anthracite	31 400 … 33 500	Basoline	41 000 … 43 900	Acetylene	56 940
Brown coal	18 800 … 20 100	Benzene, pure	40 190	Butane gas	117 230
Charcoal	27 200 … 31 400	Diesel fuel	41 000 … 44 380	Natural gas	29 300 … 37 860
Wood	14 650 … 15 900	Fuel oil	41 000 … 44 380	Lighting gas	15 900 … 18 840
Coke	25 100 … 29 300	Kerosine	41 800	Propane gas	92 100
Peat	12 500 … 14 600	Tar oil	36 400 … 38 500	Hydrogen	10 760

Coefficients of friction (mean values)

Material	Static friction μ_H dry	Static friction μ_H lubricated	Sliding friction μ_G dry	Sliding friction μ_G lubricated	Rolling friction μ_R dry	Journal bearing friction lubricated
Cast iron on cast iron	0,15…0,20	0,10…0,12	0,10…0,15	0,05…0,10	–	–
Light metal on cast iron	–	–	0,10…0,15	0,01…0,06	–	0,01…0,06
Steel on cast iron	0,18…0,33	0,10…0,15	0,10…0,20	0,01…0,05	–	0,01…0,06
Steel on copper-tin alloy or on lead-tin alloy	0,18…0,20	0,08…0,15	0,10…0,20	0,01…0,05	–	0,01…0,05
Steel on steel	0,12…0,33	0,08…0,15	0,10…0,25	0,01…0,05	0,005…0,008	0,001…0,002 rolling bearing
Brake lining on steel	0,60…0,70	0,25…0,35	0,50…0,60	0,20…0,35	–	–
Vehicle tyre on asphalt[1]	0,70…0,85	0,35…0,50	0,55…0,65	0,30…0,40	0,01…0,02	–

[1] Further coefficients of friction and coefficients of rolling resistance for vehicle tyres see p. 64/65

Iron ores, iron content		Chemical formula, colour	Main location of find
Magnetite	60 ... 80 %	Fe_3O_4 Eerric oxide, iron black	Sweden, Norway, North America, Brazil, Africa, Russia
Red haemetite	40 ... 60 %	Fe_2O_3 Ferrous oxide, reddish brown	Germany, England, North America, Spain, Canada
Brown haemetite	30 ... 50 %	$Fe_2O_3 + H_2O$ Ferrous oxide with water, yellowish brown	Germany, Luxembourg, Lorraine
Siderite	30 ... 45 %	$FeCO_3$ Ferrous carbonate, grey	Germany, Austria

Preparation of iron ores

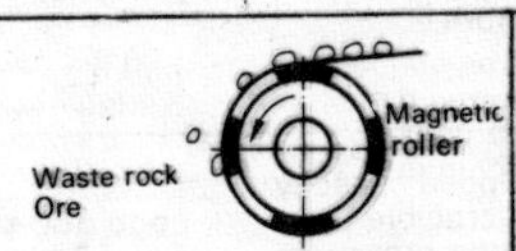

The ore is pulverized and separated from waste rock by magnetic rollers. By roasting of the ores, water, sulphur and carbine dioxide are expelled. By sintering, the fine-grained ore is baked together into briquette form.	Reduction in weight by roasting up to 30 % By the addition of lime during sintering, self-fluxing sinter is obrained which does not require any further additions in the blast furnace.

Smelting of the ores in the blast furnace

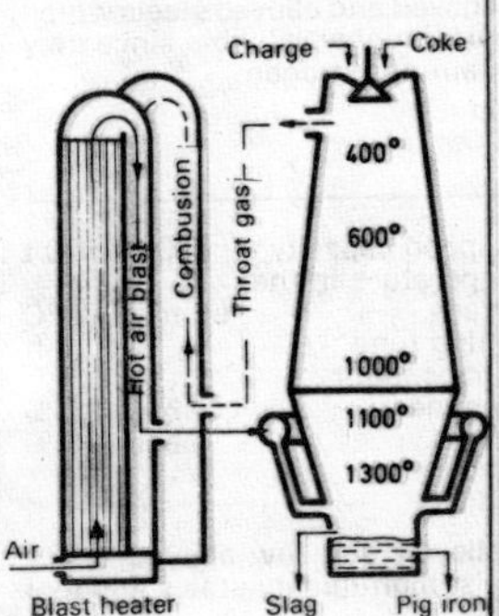

The blast furnace is alternately fed with coke and charge (ore with lime and dolomite additions). Fresh air is heated by the blast heater and blowed into the blast furnace. The coke burns whereby the carbon monoxide required for the reduction and the necessary heat are generated.

Preheating

In the upper part of the blast furnace, water and sulphur are expelled by the heating of the charge.

Reduction

Oxygen is taken away from the hot iron ore by the gases containing carbon monoxide. The ferrous oxide is reduced to iron.

Carbonization

The temperature of the free iron is now so high that it absorbs carbon from the coke.

Melting

By the addition of carbon, the melting point is lowered, the iron becomes liquid and descends into the hearth. The slag liquified by the lime floats on the iron and led out through a slag channel. The pig iron is cast into pigs or brought to steel factories in the liquid condition.

Blast furnace of medium size.	
Height	30 to 40 m
Diameter	6 to 12 m
Pig iron tapping	every 3 h
Daily capacity	
Ore input	2000 t
Pig iron	1000 t
Coke consumption	1000 t
Lime additions	800 t
Cooling water consumption	20 000 t
Temperatures	
Preheat zone	400° C
Reduction zone	up to 800°C
Carbonization zone	up to 1000°C
Melting zone	up to 1700°C

Blast heater

Steel cylinder with grid-type brick lining, up to four units per blast furnace. The blast heaters are alternately heated with throat gas and operated with fresh air which is heated in the grid work and fed to the blast furnace as hot air blast.

Height	30 to 40 m
Diameter	6 to 8 m
Hot blast temperature	upto 900°C

Products of the blast furnace

Grey pig iron	Grey, granular fracture surface. The carbon separates as graphite flakes as a result of the high silicon content. Grey pig iron is used in the form of pigs for the manufacture of cast iron.	Impurities Carbon 3 to 5 % Silicon up to 3 % Manganese, sulphur, phosphorus in small quantities.
White pig iron	White, needle shaped fracture surface. The carbon is combined with iron as iron carbide (Fe_3C) as a result of high manganese content. Use predominantly in the liquid state for the production of steel.	Impurities Carbon 3 to 4 % Manganese 2 to 6 % Silicon, sulphur, phosphorus in small quantities.
Byproducts	Throat gas, combustible gas, for the operation of blast heaters, gas engines, coking plants. Slag for the manufacture of slag stones, slag wool, portland cement.	Components of the throat gas. Carbon monoxide 22 to 25 % as well as hydrogen, carbondioxide and nitrogen.

Production of steel

Thomas process

Liquid pig iron is taken from the pig iron mixer and filled into the tilting converter (pear). Compressed air is then blown in through the bottom jets and the pear is set upright. The air flowing through the melt burns first the silicon, manganese and sulphur and then carbon. Phosphorus is combined due to the basic pear feed and lime addition. At the end of the cleaning process, carbon is again added in the form of spiegel-eisen.

In the latest method, acid-oriented air is partly blown in.

The Bessemer process differs from the Thomas process by its acidic feed. It is hardly used in Germany.

Equipped capacity of a pear	20 to 100 t
Compressed air	1.5 to 2.5 bar
Blast duration	15 to 20 min.
Temperature	up to 1600°C

Scrap metal is required at higher temperature for cooling.

Products

Thomas steel with 0.05 to 0.5 % carbon for rolling mill products lke sheets, profiles, bars, steel blocks.

Thomas slag pulverized to Thomas powder is a valuable phosphoric manure.

Oxygen blast method

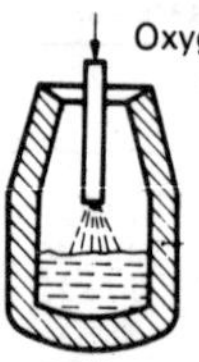

Also known as LD method (Linz-Donawitz method). In the LD crucible, oxygen is blown over the liquid pig iron through a water-cooled lance (pipe). The carbon and the harmful ingredients are thereby burnt. Lime dust is blown in with oxygen in case of pig iron containing phosphorus. As a result of combustion with oxygen, the temperatures rise very high, therefore scrap metal is added for cooling.

Equipped capacity of a crucible	up to 350 t
Oxygen pressure	4 to 12 bar
Duration of blowing	15 min.
Temperature	up to 2000°C

Products

Unalloyed and alloyed steels which are very highly valuable since they contain no nitrogen.

Siemens—Martin Process

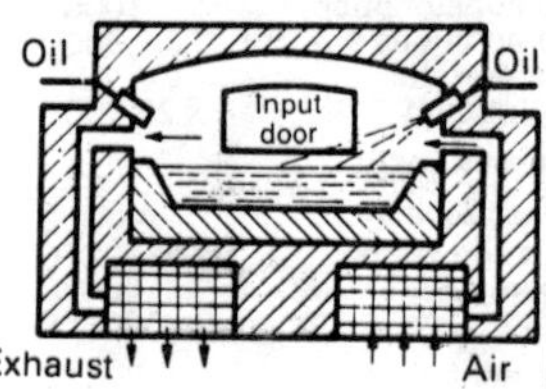

Flame furnace with gas or oil firing and air pre-heating through heat chambers which are alternately heated by the fire gases. In order to achieve high temperatures, the preheated air is further mixed with oxygen. Scrap metal or pig iron with scrap metal is used. The carbon and the harmful ingredients are oxidised by the oxygen of the burning air and by the oxygen addition of the input material. Low alloy steels can be produced by the addition of chromium, nickel, vanadium at the end of the process.

Equipped capacity	100 to 500 t
Temperature in the furnace	up to 2000°C
Melting time	5 to 6 h
Furnace input	
scrap metal	about 85 %
pig iron	about 15 %

Products

Unalloyed and low-alloyed utility steels, high-quality steels and tool steels.
The economic significance of the process lies in the utilisation of scrap metal.

Production of high-quality steel

Electric arc process

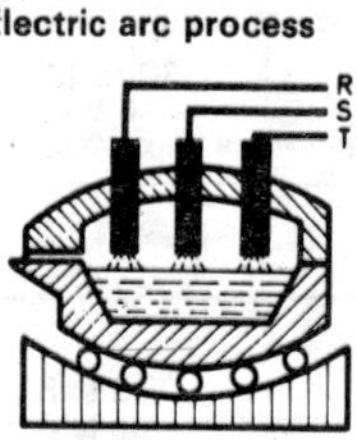

Electro-steel process for the smelting of specially pure high-quality steel. In the electric arc furnace, an electric arc is produced between the melt material and the graphite electrodes. The input material is freed from the harmful ingredients and alloyed according to requirement in the melting process. Because of the high furnace temperature, the process is specially suitable for alloying with difficult-to-melt components like tungsten and molybdenum.

Equipped capacity	up to 100 t
Voltage	50 to 100 V
Current	about 4000 A
Temperature	up to 3500°C

Furnace input

Mild steel, high quality scrap metal

Products

Alloy and high-alloy tool steels and high-quality structural steels.

Induction process

Melt material

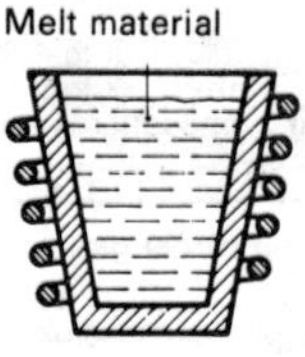

The melting crucible is surrounded by a water-cooled copper coil through which high-frequency alternating current flows. Due to the alternating magnetic field, eddy currents are produced (induced) in the input material which melt the steel. Due to the remelting, the harmful ingredients are separated in the form of flakes and hence the steel becomes of high-quality. Besides this, alloying elements can also be added.

Equipped capacity	up to 50 t
Alternating current frequency for large equipment	up to 500 Hz

Furnace input

Mild steel, high-quality scrap metal

Products

Alloy and high-alloy steels with special high-quality characteristics.

Cast iron with graphite flakes (Grey Iron)

Graphite flakes

Cast iron is smelted from grey pig iron, broken castings and steel scrap together with lime as slag medium in the cupola furnace or electric furnace. With slow cooling in the sand mould, the carbon separates out as graphite in the form of flakes. The flakes separate the iron crystals and thereby reduce the strength and elongation.

Cannot be forged, can be cast and machined well, sensitive against changes of form, compression resistant, low elongation, good running characteristics, vibration resistant, corrosion and fire resistant.

Density	7,25 kg/dm³
Melting point	1150...1250 °C
Tensile strength	100...400 N/mm²
Elongation	quite low
Carbon	2,5...3,5%

Cast iron with spheroidal graphite (SG Iron)

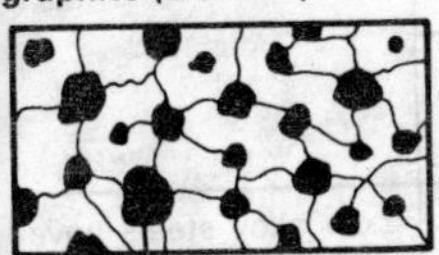

Graphite spheres

An "injection material", e.g. magnesium in the form of an alloy is added to the molten cast iron overheated to about 1400°C. While cooling, the graphite separates in the form of spheres. The graphite spheres do not have any notch effect, therefore the tensile strength and elongation are significantly higher than in case of cast iron with graphite flakes.

Better impact strength and toughness than GG, good wear strength, can be machined and surface hardened well.

Density	7,1...7,3 kg/dm³
Melting point	1400 °C
Tensile strength	380...700 N/mm²
Elongation	17...2%
Carbon	3,5...3,8%

White heart malleable iron

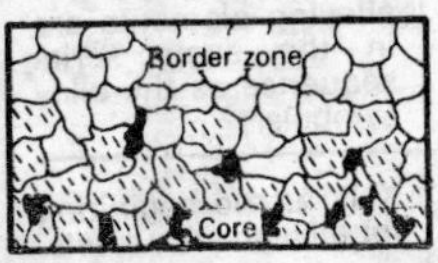

Border zone, Core Ferrite, Graphite flakes, Pearlite

Malleable iron parts are cast from white solidifying malleable hard iron (graphite free). The malleable hard iron parts are packed in cast iron crucibles together with red haematite and annealed for 2 to 5 days. The oxygen released from red haematite removes the carbon from cast parts.

In the later processes, the hard iron parts are annealed without red haematite in a mixture of carbon dioxide and carbon monoxide.

White fractured surface, mechanical properties similar to steel, hardenable

Annealing temperature	900...1050 °C
Density	7,4 kg/dm³
Tensile strength	350...650 N/mm²
Elongation	15...2%
Wall thickness	up to 10 mm
Carbon (tempered)	0,5%

Black heart malleable iron

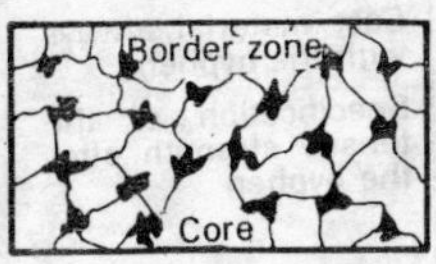

The malleable hard iron parts are packed into the malleable crucible with quartz sand and annealed for 2 days. There is no decarburisation here but a structure change. The iron carbide decomposes into iron and graphite flakes. The fractured surface is therefore black.

Black fractured surface, mechanical properties similar to steel, hardenable

Annealing temperature	900...950 °C
Density	7,4 kg/dm³
Tensile strength	350...700 N/mm²
Elongation	12...3%
Wall thickness	any
Carbon (tempered)	1,8%

Centrifugally cast iron

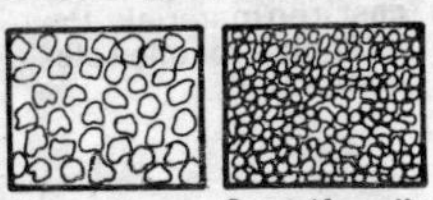

Normal cast iron — Centrifugally cast iron

GGZ centrifugally cast iron, GSZ centrifugally cast steel (centrifugal casting).

The casting mould, e.g. for a cylindrical bush, rotates fast about its axes. Due to the centrifugal force the liquid casting material is pressed on to the mould wall.

Dense, non-porous and shrink-free castings, greater purity level.

Values differ depending upon the casting material used.

Cast steel (steel mould casting)

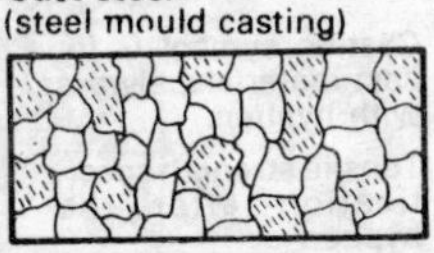

In moulds of cast unalloyed or alloyed steel. Because of the strong shrinkage, stresses are introduced; therefore cast steel must be annealed after casting for about 10 hours at 900°C.

Steel properties

Density	7,85 kg/dm³
Tensile strength	380 ... 800 N/mm²
Elongation	22 ... 8%

Shrinkage values, colours of casting models

	GG	GGG	GT	GS	Heavy metal	Light metal
Shrinkage in %	1	1,2	1 ... 2	2	1,5	1,25
Model scratch, base colour	Red	Lilac	Grey	Blue	Yellow	Green
Surfaces to be machined	Yellow	Yellow	Yellow	Yellow	Red	Yellow
Core marks	Black					

Iron and Steel, Material Terminology

According to DIN 17006

Abbreviation	Material	Explanation of the short designation	Remarks
St 34 – 1	General structural steel, unalloyed, with 340 N/mm² minimum tensile strength	St Abbreviation for steel 34 Minimum tensile strength 340 N/mm² 1 Quality group 1	Abbreviation St for unalloyed structural steels which are not meant for heat treatment.
C 15	Unalloyed utility steel with 0.15% carbon	C Symbol for carbon 15 Nominal number of carbon, 15/100 %	Abbreviation C for unalloyed steels which are meant for heat treatment.
Ck 22	Unalloyed heat treable steel with 0.22 %carbon	C Symbol for carbon 22 Nominal number of carbon, 22/100 % k Small phosphorus and sulphur content	Identification according to composition
C 110 W 1	Unalloyed tool steel with 1.1% carbon quality class 1	C Symbol for carbon 110 Nominal number of carbon, 110/100 % W 1 Tool Steel quality class 1	
15 Cr 3	Low alloy utility steel with 0.15% C and 0.75 % Cr.	15 Nominal number of carbon, 15/100 % Cr Alloy symbol for chromium 3 Code for chromium content, 4%	Low alloy steels have about upto 5 % aloying additions. Without carbon symbol Code numbers for alloying elements are in the same serial sequence as the alloy symbols.
24 CrMo 5 4	Low alloy heat treable with 0.24% C, 1.25 % Cr and 0.4% Mo	Cr, Mo Alloy symbols for chromium and molybdenum 5 Code for chromium content, 5/4% 4 Code for Mo content, 4/10%	
X12CrNi 18 8	High alloy steel with 0.12% C, 18 % Cr and 8 % Ni, stainless, acid resistant	X Letter code for high alloying 12 Nom. number of carbon 12/100% Cr Ni Alloy Symbols for chromium and nickel 18 Code no. for Cr content, 18% 8 Code no. for nickel content 8%	X is at the beginning for all high alloy steels. Multiplier 1 for alloying elements.
GG – 15	Cast iron with graphite flakes (grey iron) with 150 N/mm² minimum tensile strength	GG Cast iron with graphite flakes 15 Minimum tensile strength 150 N/mm²	Cast material always with the hyphen Specification of the tensile strength after the hyphen
GGG – 40	Cast iron with graphite spheroids with 400 N/mm² minimum tensile strength	GGG Cast iron with spheroidized graphite (globular) 40 Minimum tensile strength 400 N/mm²	
GGL – NiCuCr 15 6 2	Austenitic cast iron with graphite flakes, high alloyed, with 15% Ni, 6 % Cu, 2 % Cr	GGL Austenitic cast iron with graphite flakes Ni, Cu, Cr Alloy symbols for nickel, copper, chromium 15,6,2 code numbers for nickel, copper and chromium contents in%	In case of high alloy cast iron materials, the multiplier for alloy elements is 1
GTW – 40	White-heart malleable iron with 400 N/mm² minimum tensile strength	GTW Malleable iron, white heart 40 Minimum tensile strength 400 N/mm²	Casting symbol G for cast materials, always with hyphen. Tensile strength specification after the hyphen
GTS – 45	Black-heart melleable iron with 450 N/mm² minimum tensile strength	GTS Melleable iron, black-heart 45 Minimum tensile strength 450 N/mm²	
GS – 60	Cast steel with 600 N/mm² minimum tensile strength	GS Cast steel 60 Minimum tensile strength 600 N/mm²	

Multiplication factors and alloy symbols

Code numbers for alloying elements in the short designation = % content × multiplication factor
While solving the short designation, the code numbers must therefore be divided by the corresponding multiplication factor.

Multiplication factor	4		10		100	
Alloying additions, other than carbon, for high-alloy steels	Cr Co Mn Ni Si W	Chromium Cobalt Manganese Nickel Silicon Tungsten	Al Be Pb Cu Mo V	Aluminium Beryllium Lead Copper Molybdenum Vanadium	C P S N	Carbon Phosphorus Sulphur Nitrogen

General Structural Steel

According to DIN 17100

Abbreviation	Tensile strength σ_B (R_m) N/mm²	Elongation $\delta(A)$ %	Carbon %	Properties, Application
St 33 – 1	330 … 350	18	—	Suitable for subordinate purposes
St 34 – 1 St 37 – 1	340 … 420 370 … 450	25 … 27 23 … 25	0,17 0,20	Can be fusion welded and resistance welded, tough, not cold and hot brittle; for simple parts, screws, pins, levers, rods.
St 42 – 1	420 … 500	20 … 22	0,25	For higher stresses, e.g., shafts, bolts, cranks
St 50 – 1 St 60 – 1	500 … 600 600 … 720	18 … 20 15	0,30 0,40	Same as St 42 but for higher stresses, e.g., gears, shafts, keys, pins, dies

Sheet Steel (Sheet thickness below 3 mm)

DIN 1623

Abbreviation	Tensile strength σ_B (R_m) N/mm²	Elongation $\delta(A)$ %	Carbon %	Properties, Application
St 10	280 … 500	—	0,15	For parts with good surface, suitable for galvanising
USt 12 RSt 13	280 … 420 280 … 400	24 27	0,10 0,10	Suitable for deep drawing, can be spray painted; for simple deep-drawn parts
RRSt 14	280 … 380	30	0,10	Can be deep drawn very well; for parts of automobile bodies, covering sheets

Utility Steel

According to DIN 17210

Abbreviation	Tensile strength σ_B (R_m)N/mm²	Elongation $\delta(A)$ %	Carbon %	Alloying elements %	Properties, Application
C 10 Ck 15	420 … 520 500 … 650	19 16	0,10 0,16	— Low phosphorus and sulphur content	Piston rods for small engines, pins, levers, hinges
15 Cr 3 15 CrNi 6	600 … 850 900 … 1 200	13 9	0,15 0,15	0,65 Cr 1,5 Cr; 1,5 Ni	For higher stressed parts, piston rods, cam shafts, drive shafts, gears
16 MnCr 5 18 CrNi 8	800 … 1 100 1 200 … 1 450	10 7	0,16 0,18	1,25 Mn; 1 Cr 2 Cr; 2 Ni	For highly stressed transmission gears, bevel gears, axle drive gears of trucks

Heat Treatable Steels

According to DIN 17200

Abbreviation	Tensile strength σ_B (R_m)N/mm²	Elongation $\delta(A)$ %	Carbon %	Alloying elements %	Properties, Application
C 22	500 … 600	22	0,22	—	For parts with small heat-treatment cross sections
Ck 45 Ck 60	650 … 800 750 … 900	16 14	0,45 0,60	Low phosphorus and sulphur content	For parts in the automobile industry with normal stress, axles, shafts, screws, nuts, levers
34 CrMo 4 37 MnSi 5	800 … 950 900 … 1 050	14 12	0,34 0,37	1 Cr; 0,2 Mo 1,25 Si; 1,25 Mn	For parts with high fatigue strength like connecting rods, crank shats, front axles
42 MnV 7 42 CrMo 4	1 000 … 1 200 1 000 … 1 200	11 11	0,42 0,42	1,75 Mn; 0,1 V 1 Cr; 0,2 Mo	For higher stressed parts in automobile industry like ball pins, steering parts, spring bolts
30 CrMoV 9 36 CrNiMo 4	1 250 … 1 450 1 000 … 1 200	9 11	0,30 0,36	2,5 Cr; 0,15 Mo, V 1 Cr; 1 Ni; 0,2 Mo	For very highly stressed parts like coupling rods, drive shafts

Nitriding Steel

According to DIN 17211

Abbreviation	Tensile strength σ_B (R_m)N/mm²	Elongation $\delta(A)$ %	Carbon %	Alloying elements %	Properties, Application
34 CrAl 6 34 CrAlMo 5 34 CrAlNi 7	750 … 950 800 … 1 000 800 … 1 000	14 12 14	0,34 0,34 0,34	1,4 Cr; 1 Al 1,2 Cr; 1 Al; 0,2 Mo 1,7 Cr; 1 Al, Ni	Wear resistant, very hard surface even at temperature above 450°C. For piston rods, transmission gears, coupling rods, measuring tools (gauges)

Steel

Spring steel According to DIN 17220 to 17225

Abbreviation	Tensile strength $\sigma_B (R_m)$ N/mm²	Elongation $\delta (A)$ %	Carbon %	Albying elements %	Properties, Application
55 Si 7 65 Si 7	1300 ... 1500 1300 ... 1500	6 6	0,55 0,65	1,7 Si; 0,7 Mn 1,7 Si; 0,7 Mn	For normally loaded Beleville springs, taper springs and leaf springs, spring washers, spring plates for screw locks
67 SiCr 5 50 CrV 4	1500 ... 1700 1350 ... 1700	5 6	0,67 0,50	1,3 Si; 0,5 Cr 1 Cr; 0,1 V	For highly stressed leaf springs, helical springs, torsion springs, Beleville springs and valve springs

Valve steel

37 MnSi 5	800 ... 950	14	0,37	1,25 Mn; 1,25 Si	For normally stressed inlet valves
X 45 SiCr 4	900 ... 1050	14	0,45	4 Si; 2,6 Cr	For highly stressed in let valves and normally stressed exhaust valves
X 45 CrSi 9 X 45 CrNiW 18 9	900 ... 1050 800 ... 1000	14 25	0,45 0,45	9 Cr; 3 Si 18 Cr; 9 Ni; 1 W	For exhaust valves with very high thermal stress and high mechanical stress
X 210 Cr 12	800 ... 950	8	2,10	11,5 Cr	For valve seating rings

Free-cutting steel

9 S 20 9 S Pb 23	370 370	25 25	0,09 0,09	0,2 S; 0,7 Mn 0,2 S; 0,23 Pb	Soft steel; for parts with low stress
22 S 20	420	20	0,22	0,2 S; 0,7 Mn	Heat-treatable free-cutting steel; for parts of higher stress

Tool steel, unalloyed [1]) W = Water O = Oil, L = Air as quenching medium

Abbreviation	Carbon %	Alloying elements %	Heat treatment °C		Properties, Application
			Soft annealing	Hardening	
C 110 W 1	1,10	P u. S 0,025	680 ... 710	750 ... 780 W	Quality class 1, chisel hard; for screw thread cutting tools, reamers, sections
C 130 W 2	1,30	P. u. S 0,03	680 ... 710	750 ... 780 W	Quality class 2, hard; for knives, files, screw thread, taps
C 70 W 2	0,70	P u. S 0,03	680 ... 710	780 ... 810 W	Quality class 2, tough, for hammers, chisl. els, dies
C 90 W 3	0,90	P u. S 0,035	680 ... 710	780 ... 810 Ö	Quality class 3, tough hard; for collets, drawing bars
C 55 WS	0,55	P u. S 0,03	680 ... 710	790 ... 820 W	Special quality, very tough; for hand saws, axes, hammers, tongs

Tool steel, alloyed

31 Cr V 3	0,31	0,7 Cr; 0,1 V	680 ... 720	830 ... 860 W	Very tough; for spanners and wrenches of all types
120 W V 4	1,20	1 W; 0,1 V	710 ... 750	810 ... 840 Ö	Cold working steel; for twist drills, machine cutters
105 W Cr 6	1,05	1,2 V; 1 Cr	710 ... 750	800 ... 830 Ö	Hard tough, for reamers, drills, milling cutters, gauges
X 85 WV 12 3	0,85	12 W; 3 V; 4,2 Cr	770 ... 820	1220 ... 1260 L	Universal high speed steel, for turning tools, milling cutters, drills
X 120 WCrV 7 4	1,20	7 W; 4 Cr; 3 V	770 ... 820	1220 ... 1260 L	High power high speed steel; for turning tools
X 90 CrMoV 18	0,90	18 Cr; 1,2 Mo 1 V	800 ... 850	1075 ... 1100 Ö	Rolling bearing steel, stainless
X 210 CrW 12	2,10	12 Cr; 0,7 W	800 ... 840	950 ... 980 Ö	High wear resistant; for cutting tools, broaches

Cast iron with graphite flakes (grey iron) According to DIN 1691

Abbreviation	Tensile strength σ_B (R_m) N/mm²	Elongation $\delta(A)$ %	Carbon %	Alloying elements %	Properties, Application
GG-10 GG-15	100 150	— —	3,3 ... 3,6 3,3 ... 3,6	1,5 ... 2,5 Si 1,8 ... 2,4 Si	Can be machined well; for cast parts with normal stress
GG-20 GG-25	200 250	— —	3,2 ... 3,4 2,8 ... 3,2	1,6 ... 2 Si 1,2 ... 1,8 Si	For cast parts with higher stress, cylinder blocks, exhaust manifolds
GG-30 GG-40	300 400	— —	2,6 ... 3,0 2,6 ... 3,0	1,2 ... 1,6 Si 1,2 ... 1,6 Si	High performance cast iron, can be surface hardened, for parts with very high stress, cam shafts

Cast iron with spheroidal graphite (SG iron) According to DIN 1693

Abbreviation	Tensile strength σ_B (R_m) N/mm²	Elongation $\delta(A)$ %	Carbon %	Alloying elements %	Properties, Application
GGG-38 GGG-42	380 420	17 12	3,8 3,8	— —	Can be machined and welded well, flame and induction hardening possible, ferritic structure
GGG-50	500	7	3,8	—	Can be surface hardened well, ferritic and pearlitic structure, gears, crank shafts, cam shafts
GGG-60 GGG-70	600 700	2 2	3,5 3,5	— —	High wear resistance, can be surface hardened well, fernitic and pearlitic structure, gears, crank shafts, cam shafts

Austentic cast iron (high alloy cast iron) According to DIN 1694

Abbreviation	Tensile strength σ_B (R_m) N/mm²	Elongation $\delta(A)$ %	Carbon %	Alloying elements %	Properties, Application
GGL-NiCuCr 15 6 2 GGL-NiCuCr 15 6 3	150 180	2 1 ... 2	3,0 3,0	15 Ni; 6 Cu; 2 Cr 15 Ni; 6 Cu; 3 Cr	With graphite flakes, corrosion and heat resistant, good sliding properties; cylinder bushes, piston ring carriers
GGL-NiCr 30 3	170	7 ... 18	2,6	30 Ni; 3 Cr	With graphite flakes, heat-resistant up to 800°C, exhaust manifolds, exhaust-driven turbo-superchargers
GGG-NiCr 20 2 GGG-NiCr 20 3	380 400	8 6	3,0 3,0	20 Ni; 2 Cr 20 Ni; 3 Cr	With spheroidal graphite, cor rosion and heat resistant, good sliding properties, cylinder running bushes, exhaust manifolds; exhaust-driven turbo-superchargers, compressor housings

Malleable iron According to DIN 1692

Abbreviation	Tensile strength σ_B (R_m) N/mm²	Elongation $\delta(A)$ %	Carbon %	Alloying elements %	Properties, Application
GTW-35 GTW-40 GTW-65	350 400 650	4 5 3	— — —	— — —	White heart malleable iron; for cast parts up to 10 mm wall thickness; levers, fittings keys (spanners, wrenches), brake drums, brake shoes
GTS-35 GTS-45 GTS-70	350 450 700	12 7 2	— — —	— — —	Black-heart malleable iron can be machined well, for heavy, thick-walled cast parts

Cast steel, unalloyed According to DIN 1681

Abbreviation	Tensile strength σ_B (R_m) N/mm²	Elongation $\delta(A)$ %	Carbon %	Alloying elements %	Properties, Application
GS-38 GS-45 GS-60	380 450 600	20 16 8	0,15 0,25 0,45	— — —	For normally stressed cast steel parts; rigid trailor couplings, rear axle funnel, wheel body or wheel spider for heavy duty trucks

Cast steel, alloyed According to DIN 17245

Abbreviation	Tensile strength σ_B (R_m) N/mm²	Elongation $\delta(A)$ %	Carbon %	Alloying elements %	Properties, Application
GS-C 25 GS-22 Mo 4 GS-22 CrMo 5 4	450 ... 600 450 ... 600 530 ... 700	22 22 20	0,25 0,22 0,22	0,3 Cr 0,3 Cr; 0,4 Mo 1 Cr; 0,4 Mo	For cast steel parts which must have an appropriate strength even at high temperatures (up to 540°C)

Heat Treatment of Steel

Annealing:

Process	Temperature°C	Details of process	Effects, Purpose
annealing	680 ... 730	Workpieces slowly heated and depending upon the specification annealed up to 2 hrs. then slowly cooled again	Hardened or strain-hardened steel (e.g. by deep drawing) becomes soft and machinable
Normal annealing	750 ... 900	Workpieces held only for a short time at the annealing temperature, then cooled again	Refinement or normalisation of the structure of forged, welded or carburised work-pieces before hardening
Stress relieving	500 ... 600	Workpieces annealed for 1 to 2 hours and then cooled very slowly	Removal of stresses in forged or welded workpieces

Hardening of unalloyed tool steel

Heating	750 ... 820	First slowly, then brought to hardening temperature quickly; annealing colour is dark cherry red to light cherry red	Transformation of the steel structure, iron carbide Fe3C disintegrates, the carbon goes into the solution.
Quenching	—	Immersed in water (long workpieces possibly vertical), stirring to form steam bubbles to be avoided	Reformation of the structure is avoided, the steel is glass hard, brittle, has strains.
Tempering	220 ... 320	Cleaned and reheated to the desired temperature (tempering colours observed) Only partly quenched workpieces can be tempered with their own heat	Hardening stresses are removed, toughness is increased, hardness reduced. The workpiece has its working hardness now.

Hardening of alloyed tool steel

Heating	low-alloyed 780 ... 850 high-alloyed 950 ... 1 250	Heated slowly at first, then heated to the hardening temperature quite fast; temperatures maintained accurately according to the specification of the supplier, e.g. for high-alloy steels	Transformation of the structure depending upon the alloying elements only at higher temperatures
Quenching		Low alloyed: mostly in oil, in case of hot bath hardening in salt solution of 200°C, then air cooling High alloyed in oil, compressed air or in salt bath, according to specification of supplier	Slow re-formation of the structure as a result of the alloying elements, therefore good through hardness, less danger of cracks, no distortion.
Tempering	low-alloyed 220 ... 320	Cleaned and reheated till the tempering temperature (tempering colour) is attained	Removal of hardening stresses, reduction of hardness, increase of toughness
	high-alloyed 500 ... 600	Tempered in salt or lead bath for 30 minutes to several hours, twice in special cases	Increase of hardness

Tempering of structural steel

Heating	800 ... 950	The workpieces are annealed longer than in hardening, annealing duration 0.5 to 2 hours	Purpose of tempering is the refinement of structure for increasing the strength.
Quenching		Unalloyed heat treable steel cooled in water or oil, alloyed steel in oil	Unalloyed steel up to 1000 N/mm² Alloyed steel up to 1000 N/mm²
Tempering	530 ... 670	Tempering duration up to several hours; high tempering temperature effects toughness	Front axles, parts for wheel suspension and steering

Annealing and tempering colours

Annealing colour	Temp. °C	Annealing colour	Temp. °C	Tempering colour	Temp. °C	Tempering colour	Temp. °C
Dark cherry red	740	Yellowish red	950	Straw yellow	220	Violet	280
Cherry red	780	Yellow	1 100	Dark yellow	240	Dark blue	290
Light cherry red	810	Light yellow	1 200	Brown red	260	Light blue	310
Light red	850	Yellowish white	1 300	Purple red	270	Grey	320

Surface hardening by carburising

Process	Media	Details of process	Explanation, Application
Burning process	Hardening powder containing carbon Potassium ferro cyanide	1. Workpieces heated to light red heat. 2. Hardening powder spread on the surface and heated again, possibly repeated. 3. Quenched in water.	Annealing temperature 850 to 900°C Hardening depth 0.05 to 0.1 mm For smaller parts of structural steel up to 0.2 % carbon, e.g., screw heads, nuts, thumb screws
Pack hardening	Packing powder of charcoal, bone coal, charred leather, horn chips, barium carbonate	1. Workpieces packed in packing powder in sheet boxes which can be locked, layer thickness 20 mm. Places not to be carburised, e.g., screw threads, are covered with paste or copper plated. 2. Packing boxes annealed for 2 to 8 hours. 3. Workpieces from the packs quenched (water or oil) or first normalised and then quenched	Annealing temperature 850 to 900°C Hardening per hour 0.1 mm Total hardening up to 2 mm While annealing, the carbon enters into the surface in the form of gas (CO) and forms iron carbide (Fe_3C) Piston rods, drive gears made from unalloyed and alloyed utility steel
Bath used	Cyanide salt bath (highly poisonous)	1. Preheated, dry workpieces are brought into the cyanide salt solution (water sticking to the workpiece causes explosions). 2. The parts are quenched in water or oil straight from the bath.	Bath temperature 850 to 950°C Hardening per hour 0.5 mm Drive gears, bevel and crown gears, shifting forks, drive shaft, Cyanide salts are highly poisonous for the stomach, therefore special safety precautions.
Gas used	Lighting gas Propane gas Butane gas Acetylene	1. The workpieces are heated to light red heat in densely packable furnaces, then gas is let in for carburisation. 2. Workpieces are cooled after carburisation slowly in the beginning, then reheated to hardening temperature and quenched in oil.	Carburisation temperature 900 to 950°C Hardening in 6 hours 1 mm Uniform carburisation, burn free surface For mass production of pack hardened parts

Surface hardening by nitriding

Process	Media	Details of process	Explanation, Application
Gas nitriding	Ammonia gas (NH_3)	1. The workpieces are set out to an ammonia gas current in the electrically heated furnace. 2. The nitrogen enters into the surface and produces a very hard boundary layer. No quenching required.	Nitriding temperature 500 to 520°C Nitriding duration 0.5 to 4 days Hardening in 10 hours 0.1 mm Very hard, wear resistant, burn free surface, no distortion of the workpieces. Drive gears, piston rods, exhaust valves
Bath nitriding (soft nitriding)	Nitriding salt bath containing cyanide	1. Dry workpieces are brought into the salt bath. C and N form the "coupound layer" on the surface and the nitriding layer 1 mm deep. 2. The parts are cooled in salt water, cast workpieces in the air	Bath temperature 550 to 570°C Nitriding duration 90 to 120 min Compound layer 0.02 mm High wear and fatigue strengths, better running properties and corrosion resistance. Crank shafts, cam shafts, drive parts, axle swivels

Flame and induction hardening

Process	Media	Details of process	Explanation, Application
Flame or burn hardening	—	1. The workpiece surface is heated very quickly to the hardening temperature with acetylene burners. 2. Before the heat has entered into the inner layers, it is quenched with water or oil; core remains soft	Hardening temperature 800 to 900°C Hardening depth 2 to 3 mm. Parts of unalloyed and alloyed heat treatable steel with 0.3 to 0.7% C, e.g., crank shafts, cam shafts
Induction hardening	—	1. The workpieces are brought into an induction coil in which high-frequency alternating current flows. The workpiece surface is heated quickly to red heat as a result of the induction current 2. Quenching by water or oil sprinkler.	Hardening temperature 800 to 900°C Hardening depth at frequencies 0.5 to 20 kHz above 15 mm 40 to 600 kHz below 1.5 mm Crankshafts, cam shafts, rear axle shafts, gear box parts

Sintered Materials, Hard Metals

Production of pressed parts	Liquid metals are powdered by means of compressed air or steam and the powder grains are collected in a water basin. Metal carbides and metal oxides are powdered or pulverised by chemical or electrolytic process. The powder mixture is pressed in press tools on hydraulic presses to formed parts. For obtaining uniform density, high press pressure is required	**Advantages of sintering technology** Combination of materials with very different melting points and properties, Shaping without waste, Highly porous and dense materials can be produced
Sintering	The pressed parts are heated in induction furnaces in protective gas until the workpiece is doughy. The soft powder grains are thus combined strongly with one another as a result of the cohesion, the volume reduces. After sintering, the workpieces are reformed in a calibration tool	Sintering temperatures Bronze 600...800°C Sinter iron 1100...1300°C Hard metal 1400...1600°C Sintering duration 30...120 min
Sintered materials for filters and sliding bearings	Sintered bronze Sint-A Raw material tin bronze; used for filters, e.g. air filters and oil filters for automobile engines, hot oil filters. Sintered bronze Sint-B, Sintered iron Sint-B Raw material tin bronze or steel; used for bearing material for oil impregnation Sintered bronze Sint-B for solid lubrication Raw material tin bronze with graphite, molybrenum disulphide.	Sint-A, highly porous Pore volume 50 to 60% Density 4 to 6 kg/dm³ Sint-B, porous Pore volume 18 to 30% Density 5 to 7 kg/dm³
Sintered material for precision parts	Sintered steel Sint-C, Sint-D Raw material steel powder with additions of copper, graphite and lead. Sintered bronze or brass Sint-C, Sint-D Raw material bronze powder or brass powder	Sint-C pore volume 20% for precision parts with good adhesion capacity of lubricants Sint-D pore volume 15% for precision parts of high strength
Oxide ceramic materials	Sintered corundum Raw material aluminium oxide powder with small addition of other metal oxides. Application for ceramic oxide cutting tips (99.7% Al_2O_3) because of the high hardness and for electrical insulators, e.g. for spark plugs	Density 3,8...4,1 kg/dm³ Tensile strength 250 N/mm² Compressive strength 3000 N/mm² Hardness (Mohs) 9,9 Softening temperature 1800°C high electrical maintenance
Hard metals	Raw materials are powdered tungsten carbide (WC), Titanium carbide (TiC), tantalum carbide (TaC) and cobalt as binding medium. Because of the great hardness and wear resistance, hard metal is used for cutting tips for mounting on tool edges.	Tungsten carbide 30...90% Ti- and Ta-carbide 2...65% Cobalt 4...18% Compressive strength up to 5700 N/mm² Hardness (Mohr) 9,9 Density 7,5...15 kg/dm³

Hard metals, Main machining groups and application groups DIN 4990 (Selection)

Main groups of machining	Machining—Application group, working conditions
Identification letter P for long-chip producing materials Identification colour blue	P 10 For the machining of steel and cast steel with high cutting speed and small to medium feed P 20 For the machining of steel and cast steel with medium cutting speed and medium feed P 30 Tough hard metal type for turning, milling and planing of steel and cast steel at low cutting speed and large feed, also interrupted cut P 50 Toughest hard metal type for difficult rough machining at low cutting speed and large chip cross-section
Identification letter M for long and short chip-producing materials Identification colour yellow	M 10 Wear resistant multipurpose type for turning at high cutting speed and small to medium feed M 20 Tough multipurpose type for turning and milling of steel, cast iron and malleable iron at medium cutting speed M 30 For the machining of steel, non-ferrous metals and light metals at low to medium cutting speed and medium feed
Identification letter K for short chip-producing materials Identification colour red	K 10 For the machining of plastics, hard paper, hard cast iron, glass, porcelane, for the cutting portion of drills, counter sinks, counter bores and reamers. K 20 For the machining of cast iron and non-ferrous metals as well as steel of low strength, large rake angle possible.

Material	Occurrence, Extraction	Properties, application
Antimony Sb	Antimony ores, sulphur-antimony compounds; Algeria, Bolivia, Japan, China Roasting of the ores, reduction in red-hot coke purification by remelting	Density 6,68 kg/dm³ Melting point 630 °C Silver white, very brittle, used only as an alloying element. Lead, white metal become hard by antimony additions, e.g., hard lead for battery erminals, bearing metal
Lead Pb	Galena (lead sulphide) Pbs, 85% Pb, Canada, USA Mexico, Russia, Spain, Roasting of the ores for transformation into lead oxide, then reduction in the pit furnace to raw lead and purification by remelting or electrolysis	Density 11,3 kg/dm³ Melting point 327 °C Tensile strength ≈ 20 N/mm² Acid resistant, corrosion resistant, shields X-rays and radio active rays, poisonous Accumulators, cable covers; lead pipes, alloys, corrosion protection, whitelead, red lead, lead glass
Chromium Cr	Chromite; South Africa, Turkey, Russia By the reduction of the ore, ferrochrom is obtained, an iron-chromium alloy, the most common commercial form	Density 7,1 kg/dm³ Melting point 1 890 °C Good corrosion resistance, Chromium plating of workpieces, automobile parts, hard chromium plating of tools, alloying element for steel
Cadmium Cd	Mostly occurs together with zinc, extracted as a by-product in zinc production	Density 8,6 kg/dm³ Melting point 321 °C For low-melting alloys, galvanic coatings on steel
Cobalt Co	Mostly in combination with nickel; Separation from nickel by chemical process	Density 8,9 kg/dm³ Melting point 1 490 °C Alloying addition for steel magnetic material, hard metals
Copper Cu	Copper glance, copper pyrites, coloured copper ore, copper slate, 40 to 80 % Cu, Chile, USA, Canada, Congo, in Germany only Mansfeld, copper slate 2% Cu. Desulphurisation of the ores in the roasting furnace, reduction in the pit furnace, burning of the impurities in the converter. Pure copper is obtained by remelting (refining) of the convertor copper, purest cathode electrolytic copper, KE-Cu 99.9%, by electrolysis	Density 8,9 kg/dm³ Melting point 1 070 ... 1 090 °C Tensile strength 200 ... 370 N/mm² Elongation 30 ... 38% Elongation, drawn 2 ... 8% Good electrical conductor, good heat conductor, can be alloyed well, corrosion and fire resistant, tough, can be forged in hot and cold condition well, casting and machining poor, forms a green patina layer in moist air, green chip (poisonous) by acetic acid. Electrotechnology, Heat Engineering, alloys, roof covers.
Manganese Mn	Manganite MnO_2, manganese-containing iron ore Extraction in the blast furnace while smelting iron ore, e.g., as ferromanganese or spiegeleisen	Density 7,3 kg/dm³ Melting point 1 260 °C Hard, brittle, only as an alloying addition for steel, cast iron, copper alloys
Molybdenum Mo	Molybdenum glance, MoS_2; Australia, USA, Russia, Sweden. Transformation of the ore into molybdenum oxide by roasting, then reduction with hydrogen.	Density 10,2 kg/dm³ Melting point 2 600 °C Alloying addition for high quality steel, contact material
Nickel Ni	Magnetic pyrites (nickel-iron-sulphur compund); Canada, USA, Mexico, Russia. Nickel oxide is extracted from the ore by roasting and nickel by reduction using coal	Density 8,85 kg/dm³ Melting point 1 452 °C Tough, can be polished well corrosion resistant, magnetic Galvanic coatings, alloys, alloying addition for steel and cast iron

Non-ferrous heavy metals

Material	Occurrence, Extraction	Properties, Application
Titanium Ti	Raw materials ilmenite or titaniferous iron ($FeTiO_3$) and rutile or titanium dioxide (TiO_2) occur in many minerals. Extraction by costly chemical processes, therefore cost of the pure metal very high.	Density 4,5 kg/dm³ Melting point 1670 °C Silver-white metal, high tensile strength, corrosion resistant, small heat elongation, titanium alloys with Al, Fe, Cr, Ni, above 1000 N/mm², for connecting rod, crank shafts of race engines, for aircraft, rockets.
Mercury Hg	Cinnabar, mercury-sulphur compound HgS; Spain, Mexico, Russia, USA. The sulphur burns when the cinnabar is roasted, the mercury collects on the cool surfaces.	Density 13,6 kg/dm³ Melting point −39 °C The only liquid metal at room temperature, poisonous, forms compounds with almost all metals, amalgams (dental fillings), thermometers, mercury switches.
Vanadium V	Contained in iron ores, vanadium ores are rare. Extraction of vanadium oxide first, then reduction with aluminium powder.	Density 6,1 kg/dm³ Melting point 1 730 °C Alloying addition for structural and tool steels in the form of ferrovanadium.
Tungsten W	Tungsten ore wolframite, sheelite; China, Australia, Canada, India, Russia. Chemical transformation of the ore using tungsten acid into tungsten oxide and reduction with hydrogen to tungsten powder.	Density 19,3 kg/dm³ Melting point 3370 °C Highest melting point of all metals, corrosion resistant, acid resistant, hard, brittle. Lighting elements for incandescent lamps, contacts, alloying addition for steel, tungsten carbide for hard metals.
Zinc Zn	Zinc blende, ZnS, calamine or zinc spat $ZnCO_3$, 50 to 65% Zn; USA, Australia, Russia, Congo, small occurrence in Rheinland Harz, Erzgebirge. Transformation of the ore into zinc oxide by roasting, then reduction through coal whereby zinc is obtained in vapour form.	Density 7,1 kg/dm³ Melting point 420 °C Tensile strength ≈ 150 N/mm² At room temperature and above 200°C brittle, coarse-crystalline fracture, large heat elongation, small resistance against acids, alkalies, salts, but good corrosion resistance to air. Zinc diecasting, alloys, corrosion protection of steel by hot galvanising and galvanic coating.
Tin Sn	Tin stone SnO_2, tin pyrite; Asia, Australia, England, South America. Roasting of the ore for purification and concentration, then reduction through coal; purification of the raw tin by remelting or electrolysis.	Density 7,3 kg/dm³ Melting point 231 °C Tensile strength ≈ 40 N/mm² Corrosion resistance, against organic acids, "tin cry" while bending, disintegration to grey powder "tin pest" at −20°C. Tin coated steel sheet (white sheet), tin solders, alloys.
Silver Ag	Silver ores are silver glance (silver-sulphur compound) and silver-antimony glance; North and South America, Mexico, South Africa, Australia, Russia. Smelting of the ore in pit furnaces; pure silver is obtained by remelting (refining). Silver of the highest purity is extracted by electrolysis.	Density 10,5 kg/dm³ Melting point 960 °C Best electrical conductivity of all metals, good heat conductor, not affected by alkalies and weak acids. Sulphur compounds blacken silver. For jewellery, cutlery, mirrors, contacts.
Gold Au	Gold occurs almost only in pure form as powder or in the form of grain in rock; South Africa, North and South America, Russia, Mexico. Gold is washed out from the pulverised stone. If the gold content is small, it is extracted using mercury by the formation of gold amalgam.	Density 19,3 kg/dm³ Melting point 1 060 °C Specification of gold content 1000 parts = 24 carat 585 parts = 14 carat Resistant against chemical effects, dissolved only by aqua-regia (nitric acid+hydrochloric acid)
Platinum Pt	Platinum occurs in pure form and as alloy with palladium and indium; Russia, North and South America, South Africa. Separated from its companion metals using chemical processes.	Density 21,5 kg/dm³ Melting point 1770 °C Extraordinarily resistant against chemicals, not oxidised even when red hot, attacked only by aqua-regia.

Abbreviation	Tensile $\delta_B (R_m)$ N/mm²	Elongation $\delta (A)$ %	Composition % ≈	Properties, Application
Copper-zinc wrought alloys (brass)				According to DIN 17660
CuZn 40 Pb 3	390 ... 670	35 ... 0	57 Cu, 40 Zn, 3 Pb	Main alloy for chip removal machining processes; for screws, profile-turned parts
CuZn 40	340 ... 590	30 ... 3	60 Cu, 40 Zn	Forged brass, suitable for bending, upsetting, flanging, rivetting, deep drawing, die forging.
CuZn 37	300 ... 620	50 ... 0	63 Cu, 37 Zn	Can be cold-formed well, for deep drawing, pressing, rolling, curling, for tubes, for radiators.
CuZn 30	280 ... 530	50 ... 0	70 Cu, 30 Zn	Can be cold-formed very well, for deep drawing, rolling, curling; for tubes and structural parts.
CuZn 31 Si	360 ... 600	32 ... 10	68 Cu, 31 Zn, 1 Si	Special brass for sliding stress even at high loads; bearing bushes, sliding parts.
CuZn 40 Al 2	550 ... 650	18 ... 10	57 Cu, 40 Zn, 2 Al, 1 Mn	Special brass with high strength; for increased demands of sliding loads.
Copper-zinc wrought alloys (tin bronze)				According to DIN 17662
CuSn 6	350 ... 750	55 ... 5	94 Cu, 6 Sn	Tin bronze of high strength and elasticity; for springs of all types and sliding parts
CuSn 8	400 ... 700	60 ... 5	92 Cu, 8 Sn	High strength and great wear resistance; for wires, tubes, gears for small drives
Copper-zinc casting alloys (casting brass)				According to DIN 1709
G-CuZn 33 Pb	180	12	65 Cu, 33 Zn, 2 Pb	Casting brass for sand cast parts such as housings, instrument parts, fittings
GK-CuZn 37 Pb	280	12	60 Cu, 37 Zn, 2 Pb, 1 Ni	Ingot casting and pressure casting with metallic bright surface for machine building, electro technology
G-CuZn 25 Al 5	750	8	65 Cu, 25 Zn, 5 Al, 5 Mg	Casting brass for highly stressed parts for sliding load; bearing bushes
Copper casting alloys with Sn, Zn, Pb, Al				According to DIN 1705, DIN 1714, DIN 1716
G-CuSn 14	250	5	86 Cu, 14 Sn	Wear-resistant tin bronze for sliding bearing shells, worm gears, fittings
G-CuSn 10	280	20	90 Cu, 10 Sn	Corrosion resistant tin bronze for high lead; fittings, gears, valve seating rings
G-CuSn 7 ZnPb	300	20	83 Cu, 7 Sn, 4 Zn, 6 Pb	Gun metal with high wear resistance and good emergency running properties; sliding bearings
G-CuPb 5 Sn	300	12	85 Cu, 5 Pb, 10 Sn	Lead-tin bronze with good sliding properties and high wear resistance; connecting rod bearings
G-CuPb 25	–	–	70 Cu, 25 Pb, 2 Ni, 3 Zn	Lead bronze for compound casting; connecting rod bearings and crank shaft bearings for internal combustion engines.
G-CuPb 20 Sn	200	10	70 Cu, 20 Pb, 5 Sn, 3 Zn, 2 Ni	Lead-tin bronze with very good sliding properties; sliding bearings with high surface pressure.
G-CuAl 9	450	25	90 Cu, 9 Al, 1 Ni	Copper-aluminium bronze with good corrosion resistance; fittings for chemical industry
G-CuAl 11 Ni	750	8	77 Cu, 11 Al, 6 Fe, 6 Ni	Copper aluminium bronze for the highest load; gears, sliding bearings, hydraulic parts
Fine zinc casting alloys				According to DIN 1743
GD-ZnAl 4	250	1,5	95 Zn, 4 Al, 0,6 Cu, 0,4 Mg	Pressure casting alloys with goal dimensional stability, attainable accuracy ± 0.02 mm for workpieces upto 15 mm wall thickness. For housings of carburettors and fuel pumps.
GD-ZnAl 4 Cu 1	270	2	94,6 Zn, 4 Al, 1 Cu, 0,4 Mg	
Gk-ZnAl 4 Cu 3	260	2	93 Zn, 4 Al, 3 Cu	Ingot casting alloys; properties similar to pressure casting alloys.

Non-ferrous heavy metal alloys

Abbreviation	Term	Composition in %	Application
Lead and tin-based bearing metals			According to DIN 1703
Lg-Pb-Sb 12	Hard bearing lead	12 Sb; 0.9 Cu; up to 0.3 Ni; up to 1 5 As; rest Pb	For normal load in general machine building and automobile engineering
Lg-Pb-Sn 5	White metals	5 Sn; 1 Cu; 15.5 Sb; rest Pb	For higher demands on sliding properties and load
Lg-Pb-Sn 10	White metal 10	10 Sn; 1 Cu; 15.5 Sb; rest Pb	For higher demands, electric motors, gear boxes, machine tools
Lg-Sn 80	White metal 80	80 Sn; 6 Cu; 12 Sb; 2 Pb	Sliding bearings for high speeds, for shock and impact load

Solders

Abbreviation	Melting range °C solid liquid	Composition in %	Application
Soft solders for heavy metals			According to DIN 1707
L-PbSn 20 Sb	186 ... 270	20 Sn; 0,2 ... 1,2 Sb; Rest Pb	Lead-tin-soft solder 20, lubricating solder, for flame and immersion soldering; carriage building, radiator building
L-PbSn 40 Sb	186 ... 225	40 Sn; 0,5 ... 2,4 Sb; Rest Pb	Lead-tin soft solder 40, sheet metal solder, for flame, immersion and piston soldering
L-Sn 50 Pb (Sb)	183 ... 215	50 Sn; 0,12 ... 0,5 Sb; Rest Pb	Tin-lead soft solder 50, no antimony, for flame, immersion and piston soldering, fine soldering, galvanising, electro technical purposes
L-Sn 60 Pb	183 ... 190	60 Sn; Rest Pb	Tin-lead soft solder 60, antimony-free for flame, immersion and piston soldering, fine soldering
L-Sn 63 Pb	183 ... 183	63 Sn; Rest Pb	Tin-lead soft solder 60, antimony-free, sump solder, goes over to liquid state without becoming pasty
Hard solders for heavy metals (copper and silver solders)			According to DIN 8513
L-Ms 60	890 ... 900	60 Cu; up to 0.3 Si and Mn up to 0.5 Sn, rest Zn	For steel, malleable iron, copper and copper alloys, nickel and nickel alloys; small demands on strength
L-Ms 54	880 ... 890	54 Cu; Rest Zn	
L-Ms 42	835 ... 845	42 Cu; Rest Zn	Preferably for nickel and copper alloys
L-SoMs	870 ... 890	56 ... 62 Cu; Rest Zn	Special brass solder for joining solders of high strength
L-Ag 12	800 ... 830	12 Ag; 48 Cu; Rest Zn	Silver solder for steel, malleable iron, copper, copper alloys, nickel, nickel alloys
L-Ag 15 P	650 ... 810	15 Ag; 5 P; Rest Cu	Silver solder for copper, brass, bronze, gunmetal
L-Ag 49	625 ... 705	49 Ag; 8 Mn; 5 Ni; 16 Cu; Rest Zn	Silver solder for hard metal on steel
L-Ag 72	779 ... 779	72 Ag; Rest Cu	Silver solder for copper, nickel and their alloys
Hard and soft solders for aluminium materials			According to DIN 8512
L-AlSi 12	575 ... 590	12 Si; Rest Al	Hard solders of pure aluminium and aluminium alloys, joining solders and application in case of casting alloys
L-AlSiSn	520 ... 580	10 ... 12 Si; 8 ... 12 Sn + Cd; 2 ... 4 Cu + Ni; 72 Al	
L-SnZn 40	200 ... 330	50 ... 70 Sn; Rest Zn	Soft solder, friction solder suitable for soldering with flux also
L-CdZn 20	265 ... 280	17 ... 25 Zn; Rest Cd	Corrosion resistant soft solder, suitable for soldering with flux

Aluminium	Occurrence, Extraction	Propertıes, Application
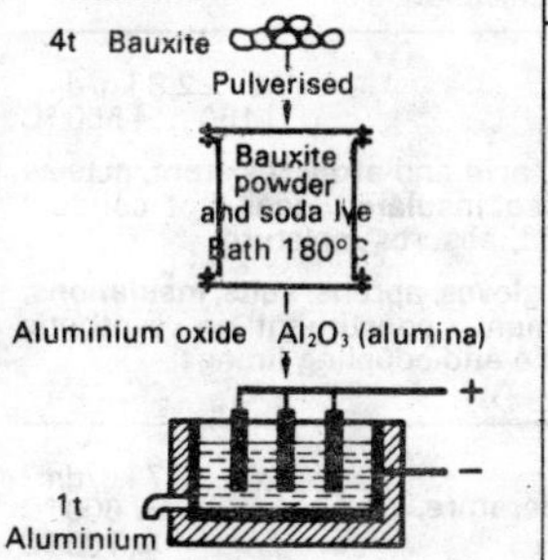	Raw material bauxite with 50 to 60 % aluminium oxide (alumina) Al_2O_3 France, Yugoslavia, Hungary, USA, Russia, India Pure, white aluminium oxide is extracted from bauxite by means of sodium lye at 8 bar pressure and 180°C. Reduction of the aluminium oxide in electrolysis furnace, carbon electrodes positive pole, lining nagative pole The melting point of aluminium oxide is reduced from 2000°C to 950°C by cryolite addition	2t aluminium oxide is obtained from 4t bauxite, 1t aluminium from it. Energy consumption for 1t Al 18000 kWh; Voltage 5 to 6 V, current 3000 to 100000 A Density 2,7 kg/dm³ Melting point 658°C Tensile strength 40...140 N/mm² Elongation 32...3% Good conductivity for electricity and heat, can be cast and machined well, corrosion resistant. High purity aluminium Al 99.98 for search light mirrors, pure aluminium Al 99 to 99.9 for harness, alloys.
Magnesium	Raw material magnesite, dolonite, karnalite, even sea water. By chemical means, magnesium chloride is extracted first, then magnesium by electrolysis	Density 1,74 kg/dm³ Melting point 650°C Small strength, therefore Mg is only used for alloying; Magnesium burns

Aluminium wrought and casting alloys — According to DIN 1725, 1745

Abbreviation	Identification colour	Tensile strength N/mm²	Elongation %	Composition in %	Properties Application
AlCuMg 1	Red	380 ... 400	12 ... 10	4 Cu; 1 Mg; 0,5 Si; 0,5 Mn; Rest Al	Very strength wrought alloy, sheets, tubes rods, profiles; machine building automobile building
AlCuSiMn	Red-Violet	370	6	4,5 Cu; 0,8 Si; 0,8 Mn; 0,5 Mg; Rest Al	Wrought alloy, specially well suitable for forged parts
AlMg 5	Green black	240 ... 320	14 ... 4	5 Mg; Rest Al	Sea water resistant wrought alloy, can be welded well; ship building, automobile building
G-AlSi 12	—	170 ... 220	8 ... 4	12 Si; Rest Al	Casting alloy for sand casting shock resistant suitable for thin walled parts
G-AlSi 10 Mg	—	180 ... 240	5 ... 2	10 Si; 0,3 Mg; Rest Al	Vibration-resistant cast parts, engine housings, gear boxes, cylinders
GD-AlSi 12	—	200 ... 280	3 ... 1	12 Si; Rest Al	Cast parts with complicated shape and good chemical resistance

Magnesium wrought and casting alloys — According to DIN 1729

Abbreviation	Identification colour	Tensile strength N/mm²	Elongation %	Composition in %	Properties Application
MgMn 2	Yellow-black-red	200 ... 230	10 ... 1,5	2 Mn; Rest Mg	Wrought alloy, can be welded, corrosion resistant, can be formed well
MgAl 6 Zn	Yellow-black-white	260 ... 280	10 ... 6	6 Al; 1 Zn; 0,3 Mn; Rest Mg	Vibration-resistant cast parts for high load
G-MgAl 6 Zn 3	Yellow-white	160 ... 200	6 ... 3	6 Al; 3 Zn; 0,2 Mn; Rest Mg	Casting alloy for sand casting, normally loaded cast parts
GK-MgAl 9 Zn 1	Yellow-black	240 ... 280	10 ... 6	9 Al; 1 Zn; 0,3 Mn; Rest Mg	Ingot casting for cast parts of normal load.

Piston alloys (not standardised)

Abbreviation	Identification colour	Tensile strength N/mm²	Elongation %	Composition in %	Properties Application
AlSi 12 CuNi	—	200 ... 370	—	12 Si; 1 Cu; 1 Ni; 1 Mg; Rest Al	For ingot casting and pressed pistons, normal thermal expansion
AlSi 18 CuNi	—	180 ... 300	—	18 Si; 1 Cu; 1 Ni; 1 Mg; Rest Al	Piston alloy with small thermal expansion $\alpha = 0{,}000019$
AlSi 25 CuNi	—	140 ... 200	—	25 Si; 1 Cu; 1 Ni; 1 Mg; Rest Al	Thermal expansion extremely small $\alpha = 0{,}0000175$

Non-metals

Material	Occurrence, Extraction	Properties, application
Asbestos	Mineral fibre from asbestos stone hornblende asbestos (magnesium-calcium silicate) or serperatine asbestos (magnesium-aluminium silicate) Canada, USA, Russia, Africa The fibres are extracted by pounding of the stone; long fibres for asbestos fabric, short fibres for asbestos sheets	Density 2,2 ... 2,8 kg/dm³ Melting point 1 150 ... 1 550 °C Fire resistant, acid and alkali resistant, acts as heat and noise insulator, does not conduct electric current, absorbs moisture; Fire-resistant gloves, aprons, suits, insulations, asbestos cement, constituent in synthetic materials brake and coupling linings.
Glass	Raw materials are quartz sand, soda and chalk. The materials are finely powdered and melted to glass mass at 1400 to 1500°C Soda serves as flux, chalk effects hardness, brightness and strength. Metal oxides are mixed for colours, potash, borax and Glauber's salt for special types of glass The ductile liquid glass mass is further processed at 1200° C by blowing, pressing, rolling or drawing	Density 2,4 ... 2,7 kg/dm³ Forming temperature ≈ 600 °C Melting point ≈ 700 °C Hard, transparent, brittle, sensitive against temperature changes, bad heat and electricity conductor. Hollow glass is blown, sheet glass rolled or drawn, mirror glass cast, ground and polished, pressed glass is pressed in moulds, glass fibre serves for heat and noise protection.
Safety glass	Prestressed glass The finished glass sheet is heated and suddenly cooled. It retains thereby internal stressles and cannot be further machined now. Multilayer glass or compound glass: Two or more thin glass sheets are joined with one another by transparent, tough intermediate layer of acrylic glass. In the case of breakage of multi-layer glass, there is less visibility hindrance than in the case of prestressed glass	Prestressed glass can become opaque suddenly through shock or impact due to innumerable fine cracks. In case of breakage it crumbles without sharp edges whereby the danger of injuries due to cuts is reduced. In the case of breakage of multi layer glass, the danger of splinters is reduced due to the elastic intermediate layers. Safety glass is prescribed for automobiles by law.
Mica	Stone or mineral from potassium silicate, sodium silicate or magnesium silicate which can be split into transparent thin sheets. Mica occurs mainly in volcanic rocks and in marine deposits. Main locations are in India, North America and Austria (Kaernten)	Density 2,6 ... 3,2 kg/dm³ Does not burn, temperature resistant up to 600°C, transparent, elastic high electrical resistance. Mica does not absorb water, therefore it is well-suited as electrical insulator, e.g., in case of heating devices.
Porcelain	White alumina (kaolin), quartz and feldspar are raw materials. Powdered quartz and feldspar are mixed with kaolin to a pulp, from which water is partly removed in filter presses. After giving the required shape, the parts are burnt at about 800°C and after applying the gloss they are burnt again at about 1400°C.	Density 2,3 ... 2,5 kg/dm³ Fire resistant up to 1 650 °C High electrical resistance, does not conduct heat, not attacked by acids (except hydrofluoric acid) and solvents, neutral in taste. Electrical insulators utensils, laboratory equipment.
Electro-carbon	Carbon powder from soot, powdered coke, petrol coke and graphite is mixed with binders (asphalt, tar) and pressed. The pressed parts are heated to 1200°C to 1400°C in the absence of air. The binder is thereby charred and the powder grains are conglomorated. The volume shrinks, the strength of the pressed part is increased.	Hard carbon consists mainly of coke and soot, soft carbon contains graphite. By the addition of copper, brass or bronze powder, metallic carbon is obtained, e.g. carbon brushes of starters. Hard carbons for carbon brushes of small motors, sliding contacts, resistances.
Cork	Plantation fabric on the bark of stems and roots of cork oak. The stems are peeled of every 10 to 15 years. Commercial forms; cork pieces, cork shots, cork powder. Producing countries are Algeria, Portugal, Spain	Density 0,2 kg/dm³ Bad conductor for heat, sound, vibrations, almost unpenetratable for gases and water, elastic, does not rot, insensitive against most solvents, almost fire resistant. Heat and sound insulators, linokeum (cork shots), motor cycle coupling linings, packings.

Material	Occurrence, Extraction	Properties, application
Paper	Finely powdered or ground wood is transformed by means of alkalies and sulphur compounds into cellulose by boiling for a long time. The cellulose is processed with water and glue to paper pulp in paper grinders. The paper pulp is predried on sieves and then pressed on hot rollers (calanders) to paper rolls.	Simple papers, e.g., news print, are produced directly from wood grind. For wood-tree paper only pure cellulose is used; depending upon the type, cotton and fabric residue are mixed. For the manufacture of card-boards several paper layers are rolled on one another.
Leather	Animal hide is depilated and tanned with plantation tans (oak bark, bark) or chemically with chromium salts. Plantation tanned leather becomes hard at 60°C whereas chrome leather temperatures amount up to about 90°C. For driving belts, core leather from the rear part of cattle hide is used.	Density 0,85 ... 1,2 kg/dm³ Leather is degreased by cleaning with soap and solvents and hence brittle. Maintenance medium must contain plantation or animal fat, mineral greases make the leather hard and breakage-prone. Chrome leather is used for packings, seals and stuffing boxes in automobiles.
Natural rubber	Raw material for natural rubber is the milky juice (latex) from the bark of the rubber tree. 1. Latex is coagulated by acetic acid or formic acid 2. The white coagulated natural rubber mass is rolled to sheets 3. Sheets made durable by smoking are brown and are called smoked sheets. Air dried, white rubber is called crepe. Main producing countries are Indonesia, Malaya, Srilanka, Africa.	Natural rubber is plastic, non-resistant to form and temperature, becomes soft and sticky, brittle and hard in cold. Natural rubber dissolves in mineral oils and grease. Rubber solution is natural rubber dissolved in gasoline or benzene. Main consumer of cured natural rubber is the tyre industry. Crepe is used for the manufacture of light coloured rubber.
Rubber Mixture	The manufacture of rubber is divided into two parts: 1. Manufacture of rubber mixture 2. Vulcanisation Rubber mixture The natural rubber is masticated in rolling mills (calanders) so that it is able to absorb the additive materials which must be finely distributed in the mixture. Example of rubber mixture for tyres, composition by weight 1. Natural rubber 100 2. Softener (oils, fatty acids) 7 3. Strengthening filler materials: Gas soot 45 Zinc white 5 4. Vulcanising medium, sulphur 3 5. Vulcanising accelator 1 6. Aging protective material 1,5	Up to 90 % synthetic rubber is used for passenger vehicle tyres. Softeners improve the plasticisability and machinability of the natural rubber. Gas soot is the most important additive material for a tyre rubber mixture; it effects wear resistance and cutting strength. Zinc white is used instead of soot for light coloured rubber. Sulphur effects the chemical change during vulcanisation; hard rubber contains 20 to 30 % sulphur. Vulcanisation accelerator reduces the vulcanisation time. Aging protective materials retard the effects of the acids and light by which the rubber becomes brittle and cracky.
Rubber	Vulcanisation The plastic rubber mixture is heated and simultaneously pressed in the heated form by means of water pressure, steam pressure or air pressure. A thick-walled heat hose is introduced in the cover in the tyre. By heating the mixture, the sulphur molecules combine with the fibre-shaped natural rubber molecules; the plastic rubber thereby becomes elastic rubber.	Density soft robber 1,1 ... 1,2 kg/dm³ hard rubber 1,2 ... 1,8 kg/dm³ Elongation of soft rubber up to bis 600 % Mean vulcanisation temperature 150 °C Vulcanisation pressure. 5 ..15 kp/cm² Rubber is elastic, dense, form resistant, temperature resistant up to about 120°C, non-conductor of electricity, resistant against acids and alkalies. Natural rubber is attacked by mineral oils and greases; it swells and breaks.

Plastics

General

"Plastics" are synthetically produced materials. They are macro-molecular organic materials. Their manufacture is either by combination of small molecules (monomers) to macromolecules (polymers) or by modification of macromolecular natural materials. Raw materials for plastics are petroleum, natural gas, coal, chalk, mineral salt, water, air or similar materials. Raw materials for the modified natural materials are protein, cellulose, natural rubber and natural resins. Synthetic materials are not low-cost replacements but in many cases lie above the natural materials.

Chemical structure

Thermoplastics 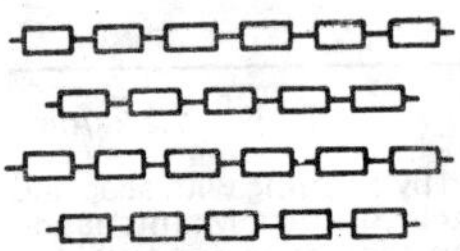	Thermoplastics become soft and formable on heating to 80 to 160°C. After cooling they retain their form. On further heating they transform into pasty condition. Thermoplastics consist of fibre-shaped giant molecules. Many thousand C-atoms are in series with one another in the form of a chain. The binding forces of the individual molecular fibres with each other are small at higher temperatures, so that the fibres slide by on one another.	Polyethylene Polypropylene Polystyrol Polyvinyl chloride Polymethacrylate Polyacrynitrile Polyamide Saturated polyester Polycarbonate Linear polyurethane
Elastomers 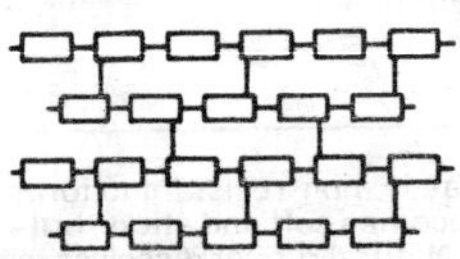	Elastomers are rubber-like, elastic and cannot be formed or only slightly formed by heat. The chain-shaped macromolecules are connected loosely with one another by lateral combinations. The snorled chains can side slip on one another and extend. But they are still held by the molecular lattice structure. With the subsiding of the external force, the molecules take back their original position.	Buna and mixed polymerisates of butadine with styrol or acrylic nitrile Polyisoprene Butyl rubber Methyl rubber Polychloroprene Polyurethanes Silicone rubber
Duroplastics 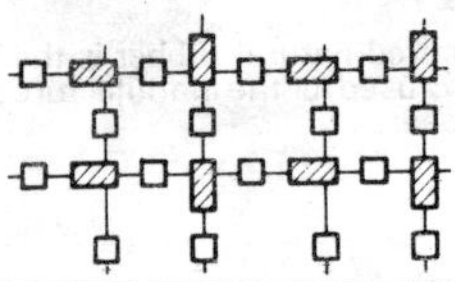	Duroplastics are converted once for all into the plastic state by heating and then harden. After hardening they can not be further formed by heat. The large molecules combine on heating in all directions to rigid formations. The binding forces between the individual molecular fibers are greater, therefore duroplastics are hard and brittle.	Phenoplastics Aminoplastics Unsaturated Polyester resins Epoxy resins Ethoxylin resins Combined

Manufacture (formation reactions)

Polymerisation poly = many, meros = parts) 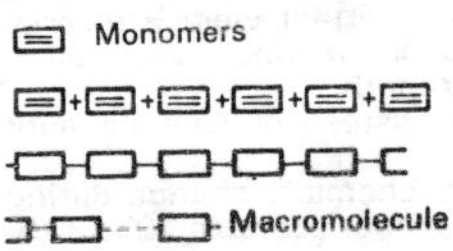	Polymerisation is the linkage of small, equal molecules to a giant molecule. The individual molecules contain two or more C-atoms which are twice or many times linked to one another. These double bonds are opened up under pressure and heat or with a catalyst. Thereby each molecule can hang with the liberated connecting arm from the neighbouring molecule.	Polyethylene Polypropylene Polystyrol Polyvinylchloride Polyisobutylene Polymethacrylate Polyacrylicnitrile Polytetrafluoroethylene
Polycondensation 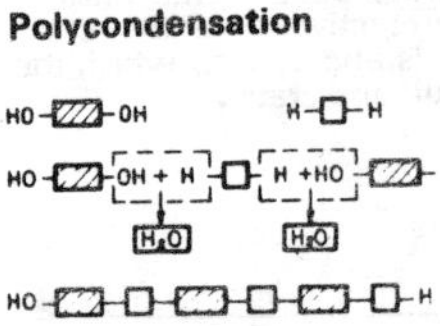	Polycondensation is the linkage of same or different types of molecules by the splitting of a material. The connection results in which every molecule liberates one or more atoms and in that place joins with another to a chain. The molecules given out together form the split product. Molecules with 2 linkage points form chain type polycondensates, those with 3 points specially connected polycondensates.	Phenol resin Aminoplastic Alkyd resin Polyester resin Silicone Polyamide Polycarbonate Anilin resin Melamine resin Carbamide resin
Polyaddition 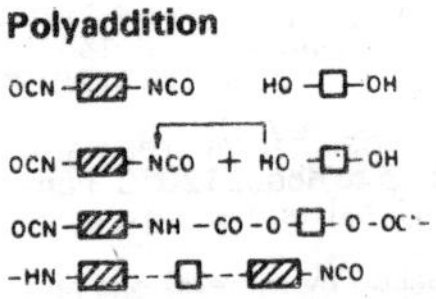	Polyaddition is the connection of same or different types of molecules without the splitting of a material. In this one atom of the first molecule moves to the second molecule and combines there. Thereby a connecting arm with the first molecule becomes free and produces the connection to the second molecule. Chain type or lattice-structured products can be formed. No split products are produced.	Epoxy resins Lattice structured and linear Polyurethane

Duroplastics		Extracts
Phenoplastics a) Casting resins without filler material Trolon Decorite Precious resin b) Pressed mass Bakelite Resinol Trolitan c) Layer press material Hard fabric Ferrocell Novotex Biratex Hard paper Pertinax Trolitan Birax Hardwood Lignofol Durofol PAG	Phenol and formaldehyde are the raw materials. By the heating of the two materials in the presence ot a catalyst pure phenol resin is obtained as low molecular, meltable and soluble pre-condensate (Resole). The pre-condensates are brought to the market with or without filler material as hardenable intermediate stage (Resitol). The complete hardening takes place during the shaping by pressing and simultaneous heating to 100 to 180°C. The resin thereby becomes soft initially, then a lattice structure of the molecules is formed through further condensation (Resit). Hardened resin is insoluble and not meltable. Resin powder, cellulose, textile pulp, paper and stone powder serve as filler material for press mass, whole rolls of paper, textile fabric or plywood serve as filler material for moulded laminated plastics. The filler shall stretch the resin, form the resin carrier and improve the properties of the raw resin.	Good electrical insulator Bad heat conductor Resistant against weak acids, alkalies, water, alcohol, gasoline, mineral oils, acetone and chlorhydro carbons. Non-resistant against strong acids and alkalies. Cannot be melted, softened or welded, insoluble in the hardened condition. Density 1,3...1,4 kg/dm³ Heat resistant ≈ 100...150°C Phenol resins carbonise above 300°C, Insensitive against moisture. Strength, hardness and brittleness dependent upon filler. No adhesiveness on smooth surface. Becomes dark, therefore coloured yellow brown or black
Aminoplastics Urea resins Carbamide resins Melamin resins Thicarbamide resins Anilin resins belong to this type a) Moulded plastics Pollopas Bakelite b) Moulded laminated plastics Formica Ultrapas c) Foams Iporka	Formaldehyde and amino compounds (urea, melamin, aniline, proteins) are the raw materials. The manufacture of the resins takes place by condensation of the amino compounds alone or in mixtures with aqueous formaldehyde under very accurately maintained re-action conditions. The resins harden over intermediate materials under the action of heat to insoluble and unmeltable products. It is therefore moulded at 140°C. The hardening process can be accelerated and controlled by catalysts. The amino resin can be processed pure or with fillers and textile cuttings.	Can be manufactured colourless. Colour-fast, therefore mostly light coloured. Good electrical insulator High creeping current strength. Odourless and tasteless. More fracture-resistant than porcelain. Chemical resistance like phenol resins Heat resistant ≈ 130 ... 150°C Melamin resins are resistant against water and heat; boil proof and less crack prone than urea resins.
Unsaturated polyester resins a) Casting resins Palatol Polyleit Leguval Vestopal Alpolit b) Finished parts Lamilux Scobalit Spimalit Tronex Filon Marcolit	Esters are compounds of acids and alcohols. The unsaturated polyesters are mixed codensates of saturated and unsaturated dicarbonic acids (Malein acids) with bivalent alcohols. The esters so obtained are dissolved in polymerisable vinyl compound, mostly styrol. The finished duroplastic is then lattice-structured by polycondensation and in the second stage polymerised by styrol. The reaction temperatures lie between 80 and 100°C. By the addition of a hardener, it can be hardened at 20°C. The hardening times vary between 2 minutes and a few hours. The pot time amounts to 2/3 of the hardening time. Polyester-styrol resin can attain very high strength with glass fibres as filler. A compound material is obtained due to the high adhesiveness of of the resin on the glass fibre.	Hard and brittle plastics. Can be manufactured transparent or in covered colours, does not colour on it own. Heat resistance ≈150...200°C. High flammability. Polyester resins reinforced by glass fibres have very high strength. In case of hard blow, polyester springs out in the form of a star at the point of blow, bad accident behaviour. Corrosion resistant. Very good electrical insulator. High creep flow strength. Resistant against water, weak acids, alkalis and salt solutions. Soluble in methyl chloride and acetone. Density 1.5 to 25 kg/dm³ depending upon glass fibre content. Finished parts can be easily repaired. Strong shrinkage of 6 to 8 % Manufacture of carriage parts uneconomical till today.

Thermoplastics and Elastomers Extract

Polyethylene Lupolene Hostalene Vestolene Supralene Trolene	Raw material: Ethylene gas from coal and lime, petroleum, natural gas or alcohol. The polymerisation takes place at 1000 to 2000 bar and 200°C in the presence of a catalyst. The molten materials is pressed into strands and cooled. The cold strands are cut to suitable grains.	Elastic, non-fragile, milky transparent in natural colour. Can be coloured to any colour, paraffin-like, replenishes greasy density 0.92 to 0.96 kg/dm^3, heat resistant 60 to 90°C, melting point 115 to 140°C. Large heat expansion, gas and air transparent, resistant against water, acids, alkalies, solvents, sensitive against oils and greases, good electrical insulator.
Polystyrol Polystyrol Styron Trolitul Vestyron as foam Poresta Styropor	Raw material: Styrol (vinyl benzene) from benzene and ethylene. Polymerisation according to various methods. Styrol is quite readily polymerised, it polymerises even at room temperature under the effect of light. Mixed polymerisation with other monomers permits the modification of properties.	Hard, brittle, breaks easily, beautiful surface gloss, clear as glass, can be manufactured in all colours. Density 1.05 kg/dm^3, good electrical insulator, softens at 70°C, not boil resistant, melting point 140 to 160°C, resistant against water, weak acids and alkalies, alcohol and oils; sensitive against ether, benzene, gasoline and acetone.
Polyvinyl chloride (PVC) Astraton Misolam Supradur Trovidur Hostalite Vestolite Vinoflex	Raw material: Vinyl chloride from acetylene and hydrochloric acid. Vinylchloride polymenses easily in the presence of a suitable catalyst at 50 to 80°C and 10 bar. The polymerisate comes out as milky emulsion whose solid parts are extracted by spray or roller drying. Softeners, lubricants, colours and stretchers can be added to the PVC powder before processing.	Hard PVC tough and hard, soft PVC soft and elastic, becomes soft at 80 to 100°C, strength reduces 40°C onwards, brittle in cold condition, large heat expansion, corrosion resistant, density, 1.2 to 1.6 kg/dm^3, hard PVC resistant against water, weak acids and alkalies, alcohol, gasoline and oils; sensitive against ether, benzene and acetone; soft PVC resistant only against weak acids and alkalies.
Acrylic glass (polymethyl methacrylate) Plexiglass Plexigum Plextol Resartglass	Raw material: Acrylic compounds from acetylene and hydrocyanic acid. Acrylic resins are extracted from acrylic compounds by polymerisation by using catalysts in a complicated process. A modification of the properties is possible by mixed polymerisation with acrylic nitrile and PVC.	Light and fracture resistant, clear as glass, high transparency even in colours, aging and weathering resistant, heat resistant 70 to 90°C, thermoplastic at 130°C. Cold resistant, large thermal expansion, small thermal conductivity good electrical insulator, density 1.2 kg/dm^3, resistant against water, weak acids, alkalies and salt solutions, gasoline; sensitive against benzene, alcohol, acetone.
Polyamides Nylon Perlon Supramide Ultramide Vestamide	Raw material: Adipin acid for nylon, caprolactam for perlon. Basic material for both is phenol. Polyamides are obtained by poly condensation of dicarbonic acids with diamines or polyaddition of lactams. There are many types in polaymides. The linear polyurethanes are very near to polyamides.	Tough hard and wear resistant, elastic and non-fragile, abrasion and scour resistant, whitish-milky when colourless, possible in all colours, stretched polyamides have high tensile strength, heat resistant 100°C, melting point 150 to 200°C, water absorbent, dimensionally unstable, no good electrical insulator, density 1.13 kg/dm^3 stable against oils, alcohols, gasoline.
Polybutadines (elastomers) Buna 32 Butadine-styrol Buna S Stadine acrylic nitrile Buna N Perbunan Butadine chlor Neoprene Perbunan C	Raw material: Butadines from acetylene, carbohydrates, petroleum, natural gas, coal. The butadines polymerise in the presence of sodium, therefore the name Buna (butadine natrium). The molecules resulting from the polymerisation are reduced in size by breaking up and vulcanised according to the forming. The chain molecules are arranged into a lattice by vulcanisation whereby the material obtains the rubber-elastic properties.	Rubber like, expandable, elastic, covers natural rubber in many properties, abrasion resistant, stable against heat, aging, gasoline and oils. The synthetic rubber can be introduced sweepingly to all consumer desires by mixed polymerisation.

From natural materials			Extract
Modification of cellulose 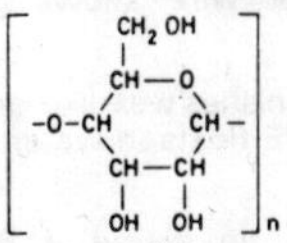 $(C_6H_{10}O_5)_n$ Glucose	The most important natural material that comes for improvement is cellulose. It is a macromolecular compound which consists of a large number of glucose building blocks connected by oxygen bridges. By chemical interferences, the cellulose can be modified to new materials. Cellular material which is produced from wood or cotton linter serves as raw material. The macromolecules of proteins are likewise suited for the manufacture of plastics.		Cellulose hydrate Viscose Parchment paper Vulcan fibre Cellulose ester Cellulose nitrate Cellulose acetate Cellulose aceto butyrate Cellulose ether Synthetic horn
Cellulose hydrate Cellophane Heliocell Transparit Viscose	Cellulose is dissolved with sodium hydroxide and carbon dissulphide to a viscose The plastic is formed when the viscose is run into a chemical bath.	Clear as glass, expandable, glossy and non-sticky foils, odourless and tasteless, unbreakable, air tight, non-inflammable, stable against greases, gasoline	Foils for groceries, cigarette packing, sausage covers, insulation tapes for electric cables
Vulcan fibre Dynos Hornex Lederstein	Zellulosebahnen werden mit Zinkchlorid getrankt, geschichtet, ausgewaschen und gepreßt.	Tough, hard, flexible, leather like, splinter resistant, density 1.1 to 1.45 kg/dm³, absorbs water, stable against fuels, alcohols, oils, weak acids and alkalies	Boxes, knobs, protective labels, packings, gears, brake linings, coupling discs, bearings
Cellulose nitrate Celluloid Cell horn	Cellulose treated with nitric acid gives nitrocellulose (collodium wool). This is kneaded with alcohol (swell medium) and camphor (softener)	Strong, springy becomes soft at 60 to 70°C, absorbs water, burns violently, density 1.28 kg/dm³, stable in gasoline, weak acids and alkalies, soluble in acetone, esters.	Spectacle frames, combs, toilet articles, dolls, knobs, grips, door protector, drawing equipment, table tennis balls, sports material, nitrolack, earlier photofilms.
Cellulose acetate Cellidor Trolit W Cellon Supraphan Triafol	Cellulose treated with acetic acid	Becomes soft at 60 to 70°C, stable against light, swells in moist conditions, density 1.3 kg/dm³, slightly inflammable, stable in gasoline, soluble in acetone and esters.	Safety films, toilet articles, spectacle frames, knobs, sports goods, packing foils, document envelopes, grips, hardwa re, safety glass, lamp shades
Synthetic horn Galalith	Proteins are raw material. Milk protein of toned milk (casein) is hardened in formaldehyde	Absorbs water, can be hot formed at 70°C, can be polished and coloured well, density 1.4 kg/dm³, stable in oil, alcohol, gasoline, acids, sensitive against water and alkalies	Knobs, grips buckles, sports marks, pencils
Silicones			
Silicones Silastic Siloprene D C silicone	Quartzsand and coal are raw materials. Silicones contain the inorganic element silicon. The macromolecules are formed by the continuous arrangement into a lattice structure of silicon and oxygen atoms.	Silicone oil is colourless, odourless, it is replenished by grease. All materials are rubber-like, extreme temperature resistant −70 to +200°C, water repellant, costly, stable in all common solvents, corrosion resistant, good electrical insulator.	Hydraulic and coolant impregnation medium, water resistant polishes, cutting oil for casting moulds. Silicone rubber for hoses, packings, cable insulators, pressed parts with rock powder or glass fibre, stove, enamels, silicone resins, antifoamant, silicone paste, silicone greases.

Identification of plastics

Designation of plastic	Burn test			Other identifications
	Flame	Smell	Specific characteristics	
Polyvinyl-chloride (PVC)	Yellowish, light green, brillant	Pungent like hydrochloric acid	Chars in flame, extinguishes outside the flame	Flame test with red hot copper wire shows green flame
Polyethy-lene (PE)	Shines with blue core	Like ex-tinguished candle flame	Melts, burning drops fall down	Replenishes wax-like, greasy; soft PE floats in water
Polystyrol (PS)	Shines, yellow, very sooty	Uncomfortably sweetish	Suspended soot flakes, becomes fluidy, threads can be drawn	Light, tin sound if PS is allowed to drop on table
Acrylic glass (PMMA)	Shines, yellowish, crackling	Sweetish fruitlike	Vapourises and burns com pletely, no drops	Sounds hollow when thrown lightly on table
Polyamide (PA)	Bluish, yellow border	like burnt horn	Burns with difficulty, drops in bubbles and threads and can be drawn like sea-ling wax	Thin components made from PA can be bent without being broken
Polyvinyl-denchloride	Yellowish-greenish	Pungent like hydrochloric acid	Extinguishes outside the flame	Coarse, extraordinarily tear-resistant bristles and strings in all colours
Polyisobuty-lene	Yellowish	Like tar	Burns slowly with breakage	Primarily as thick black, rubber-like foils for buil-ding protection
Polypropy lene (PP)	Bluish	Pungent, resin like	Melts, drops burn again	Not fully to the greasy, wax-like grip of polyethylene
Polyacrilic nitrile (PAN)	Shining, yellowish, a little sooty	Sharp, acidic precaution, hydro-cyanic acid (poison)	Vapourises, burns fast, forms hard black ash	Added to wool as textile fibre, difficult to identify
Phenol resin (PF)	Burns in the flame light, sooty	Pungent like phenol	Can hardly be fired, chars, the organic filler burns the highest	Typical phenol smell when machined, no continuous chip
Amino plastics	—	Repulsive like ammonia	Can hardly be fired, it chars, mostly white edges	Aminoplastics are fast to light, therefore mostly co-loured light
Polyester unsaturated (UP)	Shining, light yellow, highly sooty	Uncomfortably sweet	Suspended soot flakes (due to styrol content)	The glass fibre mat serving for reinforcement can be recognised well
Polyurethane lattice struc-tured (PUR)	Shining, yellowish	Uncomfortably pungent (isocyanate)	Burns after firing, melts (vulkollan)	Elastic, light foams whole material is rubber-like, brownish
Vulcan fibre	Light bluish	Like burnt paper	Burns with difficulty and slowly	Hard, leather-like, horny material, splinter resistant
Cellulose hydrate (cellulose glass)	Light like paper, yellowish	Like burnt paper	Burns after firing conti-nuously, chars, does not melt	packing foils for groceries (packing which keeps fresh)
Cellulose nitrate (CN)	Light, yellowish brown vapurs	Like comphor with celluloid	Burns violently and fast, burns almost com-pletely	Celluloid comfortable in grip, can be coloured into beauti-ful patterns
Cellulose acetate (CA)	Yellowish green with sparkle, drips	Like accetate and burnt paper	Black ash, burns heavier than celluloid, melts, drops burn	Very beautiful patterns, e.g., bead pattern, good surface gloss
Cellulose acetobut yrate (CAB)	Yellow shining with bluish border	Like rancid butter burnt paper	Melts while burning, drips, burns again after firing	Hornlike, good gripping, beautiful surface gloss
Synthetic horn	Shining, some times dull	Like burnt hair and horn	Burns in the flame, extin guished outside	Look and sound in vory-like, horn-type plastic

Abbreviations

According to DIN 7728

ABS	Acrylicnitrile-Butadien-Styrol	PC	Polycarbonate	PVAL	Polvvinylalcohol
AMMA	Acrylicnitrile-Methylmethacrylate	PE	Polyethylene	PVB	Polyvinvlbutvral
CA	Celluloseacetate	PETP	Polyethyleneterephthalate	PVC	Polyvinylchloride
CAB	Celluloseacetobutyrate	PF	Phenol formaldehyde	PVCC	Chlorated polyvinylchloride
CF	Cresolformaldehyde	PIB	Polyisobutylene	PVF	Polyvinylfluoride
CN	Cellulose nitrate	PMMA	Polymethylmethacrylate	PVFM	Polyvinyl formal
CS	Casein	POM	Polyoxymethylene	SAN	Styrol-Acrylicnitrile
EC	Ethyl cellulose	PP	Polypropylene	SB	Styrol-Butadien
EP	Epoxide	PS	Polystyrol	SI	Silicone
FEP	Perfluoro ethylene propylene	PTFE	Polytetrafluoroethylene	SMS	Styrol α Methylstyrol
MF	Melamin formaldehyde	PUR	Polyurethane	UF	Urea formaldehyde
PA	Polyamide	PVAC	Polyvinylalcohol	UP	Unsaturated polyester

Abbreviations for reinforced plastics

According to DIN 7728

FK Fibre-reinforced plastic

The type of fibre or whisker is specified by the letter placed before the actual abbreviation FK or WK

GFK	Glassfibre reinforced plastic
AFK	Asbestos-fibre reinforced plastic
BFK	Boronfibre-reinforced plastic
CFK	Carbon fibre-reinforced plastic
MFK	Metal fibre reinforced plastic
SFK	Synthetic fibre reinforced plastic
MWK	Metal whisker reinforced plastic

WK Whisker reinforced plastic

The type of metallic or synthetic fibre is supplemented by prefixing the correct designation

Cu-MFK	Copper-fibre-reinforced plastic
PA6-SFK	Polyamide-fibre-reinforced plastic

The type of plastic to be reinforced is specified after the abbreviation by deleting the letter K and adding a hyphen:

SF-PF	Synthetic-fibre reinforced phenol resin
GF-UP	Glass fibre reinforced unsaturated polyester
GF-PP	Glass fibre reinforced polypropylene

Physical properties

According to DIN

Plastic DIN Type		PC 7744	PE 7740	PMMA 7745	PS 7741	PVC 7748 3	PF 7708 11	UF 7708 155	EP 16946 1000-6	UP 16946 1130
Raw material density	g/cm³	1,20	0,94	1,18	1,05	1,40	1,8	2,0	1,8	1,2
Mechanical properties										
Modulus of elasticity	N/mm²	–	–	–	3300	3000	10000	10000	–	3500
Bending strength	N/mm²	75	–	110	100	–	50	40	110	60
Compressive strength	N/mm²	80	–	130	100	80	120	140	200	–
Tensile strength	N/mm²	65	–	74	50	40	15	15	40	20
Ball indentation hardness	N/mm²	106	40	170	110	–	310	380	–	–
Thermal properties										
Form stability in heat	°C	100	–	90	80	75	150	130	100	90
Coefficient of linear expansion	$\times 10^{-6}$ 1/K	60	200	70	70	80	25	20	40	–
Thermal conductivity	kJ/m h K	0,71	1,47	0,67	0,62	0,59	2,52	2,52	–	–
Specific heat	kJ/kg K	1,18	2,58	1,47	1,30	–	–	–	–	–
Electrical properties										
Surface resistance	Ω	10^{15}	10^{14}	10^{15}	10^{14}	–	–	–	10^{13}	–
Specific resistivity	Ω cm	$4 \cdot 10^{15}$	10^{17}	10^{15}	10^{16}	10^{15}	10^{11}	10^{9}	10^{14}	10^{14}
Dielectric dissipation factor	tan δ	0,01	0,0004	0,02	0,001	0,0001	0,3	0,4	0,02	0,02
Dielectric constant	ε_r	3,0	2,3	2,6	2,5	4,0	13	7,5	4,0	4,0
Dielectric strength	kV/cm	250	800	300	400	500	115	100	500	500
Optical properties										
Refractive index		1,587	–	1,492	1,592	–	–	–	–	–
Transparency	%	87	–	92	90	–	–	–	–	–
Other properties										
Water absorption	mg	10	–	70	6	–	45	200	15	–

Plastics in automobiles

Fan, air filter body	PA, PP
Suction pipe, fuel pump	POM
Carburettor parts, float	PA
Fuel line, hydraulic line	PA
Cleaning tank, fuel tank	PE
Shock damper protective tubes	PE
Ball and socket joint of the tie rod	PTFE
Balancing tank, jets	PE, PP
Knobs, switches, housings	PF, UF
Distribution cap, plug connector	PC, PP
Binder for brake lining	PF
Upholstery material, protective covers	PVC-PUR
Roof, inside covering	PVC-PUR
Heating hoses, heating body	PP
Fresh air boxes, wheel caps	PP
Bumper, dashboard	PUR
Radiator grill, decorative rings	ABS, PP
Canisters, first-aid box	PE
Warning triangle, decorative inserts	PVC-POM
Towing cable, safety belts	PA, PETP
Steering wheel, horn ring, grips	PP, POM
Mud guard, parts of the body	GF-UP, PA
Floor covers, keder	PVC
Light housings, battery housings	SB, PP, PS
Light covers, discs	PMMA
Cable covers, identification	PVC
Lacquer, smoothener	UP, EP, CN
Foils for connecting discs	PVB
Gears, sliding bearing	PA, PETP
Tyres, cord fabric	SK, PETP

Composition of Petroleum

Petroleum

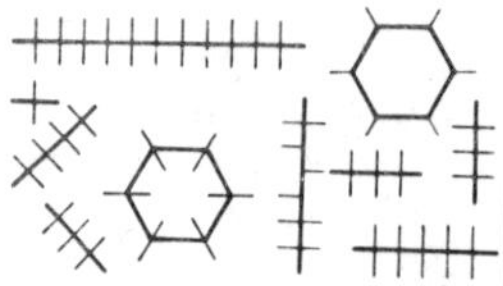

Petroleum is a mixture of many different types of hydrocarbon compounds. Light components (gasoline) difficult to boil components (gas oil, lubricating oil) and asphalt type components (bitumen) come under this.

The chemical and physical properties of the hydrocarbons depend upon

1. The size of the molecule, i.e., the number of the C and H atoms
2. The structure or form of the molecule i.e., the arrangement of the C and H atoms

Paraffins

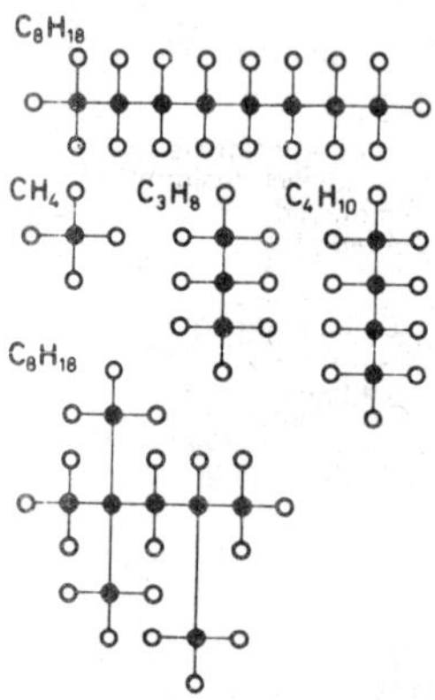

Hydrocarbons with chain structure are called paraffins. In compounds up to 4 C-atoms they are gaseous, up to 16 C-atoms liquid and above 16 C-atoms, solid.

Paraffins are the main components of petroleum. Besides a few mentioned here, there is a large number of paraffins which are contained in petroleum or formed while processing petroleum.

Methane CH_4, main component of natural gas, is the simplest hydrocarbon molecule. The C-atom is quadrivalent, H atom is monovalent. 1 C-atom can combine with 4 H-atoms.

Several C-atoms can be combined with each other in the form of a chain, e.g., propane C_3H_8 or butane C_4H_{10}. Propane and butane are the main components of the fuel gas. They become liquid at 20°C and a gauge pressure of about 3 bar.

Hydrocarbon molecules of the same composition which have, besides the straight line chain, side branches also are called isomers and therefore contain the prefix 'Iso'. They have other properties and are knockproof.

Octane C_8H_{18} (without branching), boiling point 126°C, less knockproof.

Iso-octane C_8H_{18} (3 side branches), boiling point 99°C, highly knock proof, used as fuel for the determination of the octane number.

Olefins

C_4H_6

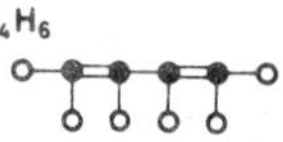

Olefins are unsaturated hydrocarbons. They have double bonds between the C-atoms instead of H-atoms. The example shown beside shows the structure of the gaseous butadiene C_4H_6 which is used for the manufacture of synthetic rubber. Olefins occur only in small quantities in petroleum, they however result from cracking. Because of their anti-knocking property, they are valuable component of the cracked gasoline.

Naphthenes

C_6H_{12}

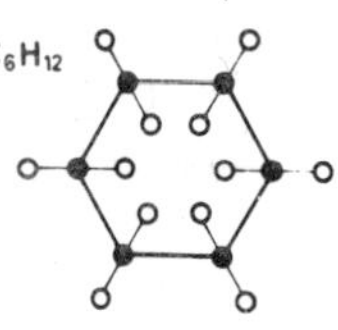

Naphthenes are ring-shaped, saturated hydrocarbons with 5 or 6 C-atoms in the ring (rarely 7 and 8) with or without paraffinic side branches. The paraffin chain is closed to a ring by eliminating both the outer H-atoms. Napthenes are contained in all gasolines and petroleums. Russian and Venezuelan petroleums are specially rich in naphthenes. The example shown beside shows the structure of cyclohexane C_6H_{12}.

Aromatics

C_6H_6

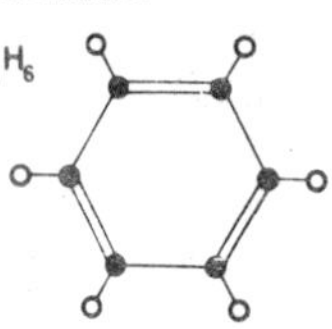

Aromatics are chemical compounds, from carbon and hydrogen which contain the benzene nucleus as the simplest building block. Benzene C_6H_6 is a ring-shaped unsaturated hydrocarbon with 6 C-atoms and 3 double bonds. All general aromatic hydrocarbons consist of benzene rings with one or more side chains. Pure benzene is alone not well-suited as fuel (firing point+5°C), therefore it is mixed with the aromatics toluene and xylene. The mixture is called engine benzene.

Petroleum contains only small quantities of aromatics. But since they impart valuable properties to the gasoline (high performance and antiknock property), the distillate gasolines in the refining process are partly converted into aromatic hydrocarbons.

Occurrence of petroleum	The main petroleum producing countries are 1. U.S.A.: Gulf of Mexico, Texas, Luisiana, Oklahoma, Pennsylvania 2. Near East: Arabia, Iraq, Kuwait, Iran 3. Latin America: Venezuela 4. East Europe: Between Ural and Volga, Caucasus, Rumania, Galicia 5. Africa: Libya, in the Sahara Less significant for the world production but significant for the covering of own consumption are the oil fields in the Federal Republic of Germany. Schleswig-Holstein, between Elle, Weser and Ems, in Emsland and Oberrheintal, in Bavaria.	1973 production in million tonnes North America 613,1 Latin America 272,1 Near East 1061,0 East Europe 493,7 West Europe 15,5 Africa 277,2 Far East 109,5 World 2842,1 Federal Republic of Germany Production 6,6 Consumption 136,6
Petroleum exploration	Methods: Surface exploration, Gravity Measurement, Magnetic measurement, Siesmographic exploration. In the siesmic method, an explosive charge is electrically detonated in a 15 to 50 m deep drilled hole. The shock waves emanating from the explosion are echoed in different ways from the different types of rock. The geophones distributed on the earth surface receive the reflected vibrations. They convert the waves into current pulses which a siesmograph records on a strip. These figures enable conclusions on the type and position of the rock inside the earth. The knowledge whether the rocks contain petroleum must be obtained by a test drilled hole.	Propagation velocity of the waves in m/s. Loose earth 600 Water 1435 Sandstone 2500 Salt 5000 Limestone 6000 Deposits: Salt dome trap Stratigraphic trap Thrust trap Saddle trap Oil bearing rock Large-pored sandstone
Petroleum drilling	Methods: Percussion drilling or rotation drilling. In rotation drilling there are rotary drilling, turbine drilling and electrical drilling In rotary drilling, the roller type tool (diamond tool or fish tail tool) scraps rotating into the earth with its hard teeth. It is held perpendicular by the drill stem and driven with engine force by the drill bit from a receiver bit and a rotary table. The drill slurry is pumped downwards through the hollow drill stem to cool the drill bit, to flush the drill chips to the earth surface, to seal the drill hole and to avoid an abrupt eruptive fracture of the oil storage. The hole is telescopically filled with tubes in the course of sinking. In turbine and electrical drilling the drill bit is not driven on the earth but directly at the tool.	Drilled hole diameter 20 ... 65 cm Drilled hole depth up to 7000 m Drilling tower: Height 40 ... 60 m Weight ≈ 45 t Carrying load ≈ 400 t Feed of the tool Soft rock up to 60 m/h Hard rock 10 ... 20 cm/h Drilling cost: 1 m depth 100 ... 1000 DM Time for drill change for 2700 m depth about 7 h Power consumption: Rotary table ≈ 150 kW Lifting work ≈ 500 kW Flushing pump ≈ 600 kW
Petroleum conveying	Methods: Primary method by eruptive conveyance with gas or water drive or by conveyance with pumps. Secondary conveyance by gas lifting or water flooding. If the oil is under sufficiently strong pressure, it flows on its own to the probe, rises up in the drilled hole and freely flows out. Longer this pressure maintains better the storage deposit can be exploited. The outflow is regulated by valves and nozzles so that a premature gas extraction is avoided. If the storage deposit pressure reduces, it is investigated to obtain it by pumped in natural gas or water. If the petroleum cannot be brought to the earth surface due to dissolved gas, the gas cap or the boundary water, deep pumps must be used. They however convey only so much oil that can flow through the rock fissures or the rock pores.	Pressure increase per 10 m depth ≈ 1 bar Temperature increase per 100 m depth ≈ 3 °C Piston pump (rarely centrifugal pump) Number of strokes 5 to 20 per min. Stroke length up to 3m capacity up to 50 m³/day Exploitation of the storage deposit in the Primary method 10 ... 30% Secondary method up to 80%

Refining of Petroleum

Fuels

Atmospheric distillation

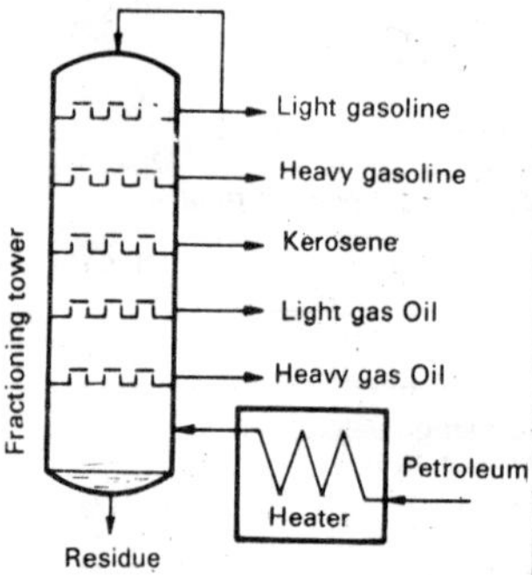

While distilling the petroleum is split into various groups (fractions) of hydrocarbons.

The petroleum is heated in the heater to 360°C and fed to the fractioning tower. The greatest part of the petroleum vapourises at this temperature. The parts remaining as liquid collect at the bottom of the tower. The petroleum vapours rise through the bell shaped bottom. A return flow of cold gasoline cares for gradual decrease of temperature over the whole height of the column. Thereby, the rising vapours cool and again condense to a liquid. Depending upon the boiling point the vapours of the individual fractions rise to different heights and hence can be collected at the individual levels and separated from each other.

Input product: petroleum

Furnace temperature	360 °C

Exploits:

Light distillate (4 to 20 % gasoline)	45 %
Heavy distillate	30 %
Distillation residue	25 %

Boiling range of the extracted products in °C

Gas	below 30
Light gasoline	30 ... 150
Heavy gasoline	150 ... 200
Kerosene	200 ... 250
Gas oil	
Light	250 ... 280
Medium	280 ... 360
Residue	above 360

Cracking TCC
(Thermofor Catalytic Cracking)

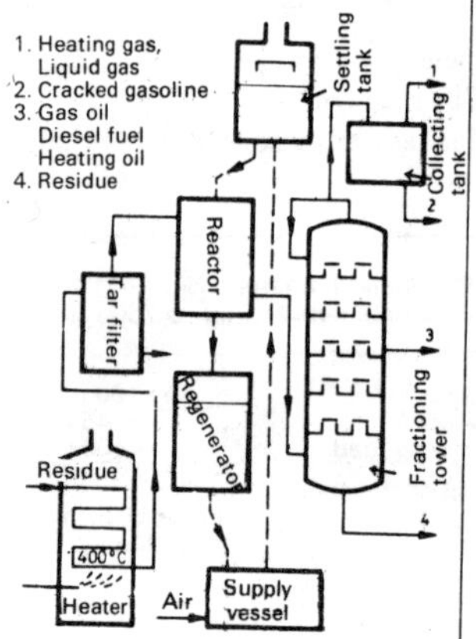

Larger hard-to-boil oil molecules are converted to smaller ones through breaking in cracking.

The input product is treated to 400°C in a heater. The resulting oil vapours are fed to a reactor and brought into contact with a catalyst there which consists of small pearls and has a temperature of 550°C. The input product are thereby cracked. The cracked vapours are split into various products in a connected fractioning tower.

In the cracking process C-atoms become free which are deposited on the catalyst whereby it becomes ineffective. It is therefore blown into a regenerator. The carbon burns here at a temperature of 650°C by the oxygen of the infed air. The cleaned catalyst is fed back in to the reactor.

Input product:

Heavy tractions of petroleum

Catalyst:

Aluminium silicate

Temperatures in °C

Tube heater	400
Reactor	550
Regenerator	650

Finished products:
Heating gas
Liquid gas
Cracked gasoline
Gas oil
Diesel fuel
Heating oil
Residue

Gasoline exploits 50...60%

Reforming
(Platinum reforming-equipment)

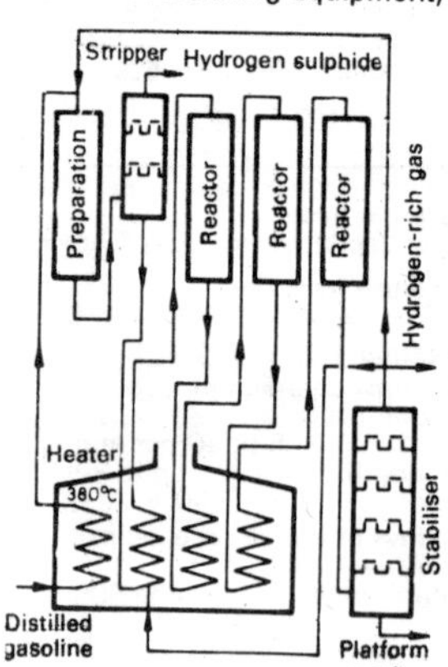

Distillate gosoline has a small octane number. By reforming, the chain-shaped hydrocarbon molecules (paraffins) of the distillate gasoline are converted into ring-shaped and branched compounds (naphthenes and aromatics). The antiknock property of these molecules is very high.

The input product is heated in a heater to 380°C, is mixed with hydrogen-rich gas and fed to the preparation reactor filled with cobalt molybdenum catalyst. At about 380°C and 30 bar the hydrocarbons are freed from sulphur.

The gasoline preheated for the protection of the valuable platinum catalyst is heated to 500°C in the heater and fed through 3 mutually connected reactors with fixed platinum catalyst bed. The paraffins are converted into aromatics in the reactors. The hydrogen thereby becomes free which is brought into the desulphurisation equipment.

Input product:

Distillate gasoline with low octane number and higher boiling range of 80-200°C (heavy gasoline)

Catalysts:

Cobalt, molybdenum and platinum

Temperatures in °C

Preparation	380
Reforming	500

Finished products:

Heating gas
Liquid gas
High-octane gasoline (platform gasoline)
Sulphur

Lubricants

Vacuum distillattion

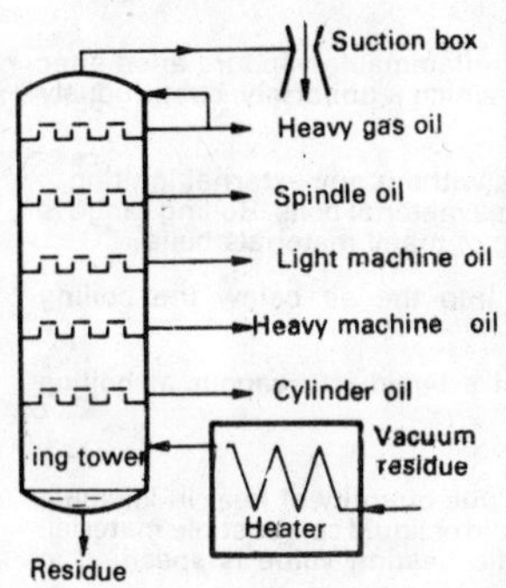

Hydrocarbons are not temperature stable. They split (crack) at temperatures above 400°C. In order to avoid this, they are distilled for lubricating oils under reduced air pressure which reduces the distillation temperature by more than 100°C.

The residue collecting at the bottom of the atmospheric column goes as input material into the vacuum distillation. After heating the residues in an oil heater, the vapours go into a distillation column which is under vacuum. The fractioned condensation takes place in the same way as in atmospheric distillation into the various height positions of the columns.

Input material:

Residue from atmospheric distillation.

Temperature about 360°C
Pressure 10 to 80 mbar. Boiling temperature reduces by about 150°C at 10 mbar

End products
1. Heavy gas oil
2. Spindle oil
   12 to 70 cSt at 20°C
3. Machine oil
   20 to 90° cSt at 50°C
4. Cylinder oil
   12 to 75 cSt at 100°C
5. Vacuum residue
   e.g. bitumen

Double-solvent refining

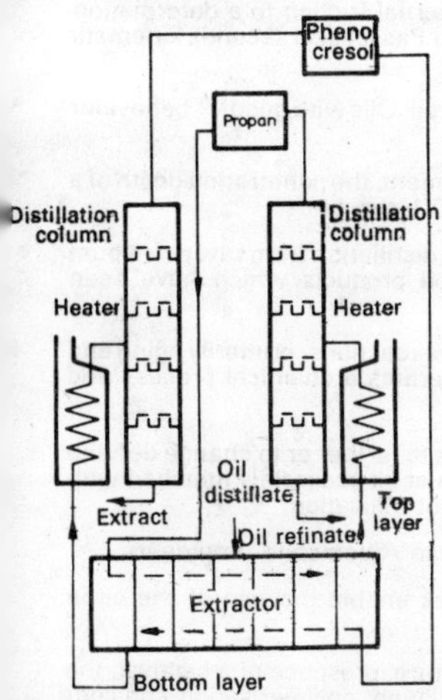

In the double solvent refining, the lubricating oil distillates are freed from undesirable components (aromatics). The separation takes place with two selective (chosen) solvents, e.g., propane and phenol-cresol mixture. The propane drains the desired paraffinic lubricating oils from the distillate, the phenol cresol dissolves the undesired components.

The separation takes place in cylindrical pressure-vessels (extractors). At one end of the vessel propane is fed in, at the other end phenol-cresol and in the middle lubricating oil distillate. The 3 components are mixed in the extractor. With the ensuing disposition of the mixture, a bottom layer with the undesired components and a top layer with paraffinic lubricating oil are formed due to the different weights.

Both layers are fed to the oil heaters and distillation columns. From the columns, the solvents can be re-extracted as well as the pure lubricating oil refinate and the byproducts (extracts) can be taken off.

Input material:

Lubricating oil distillate from vacuum distillation

Solvents:

Propane and phenol-cresol mixture

Separation into
1. Aromatics-rich layer with phenol-cresol (bottom)
2. Aromatics-less extracted lubricating oil refinate with propane (top)

End products:
1. Paraffinic lubricating oil refinate
2. Byproducts (extract) e.g. aromatics

Deparaffinisation equipment

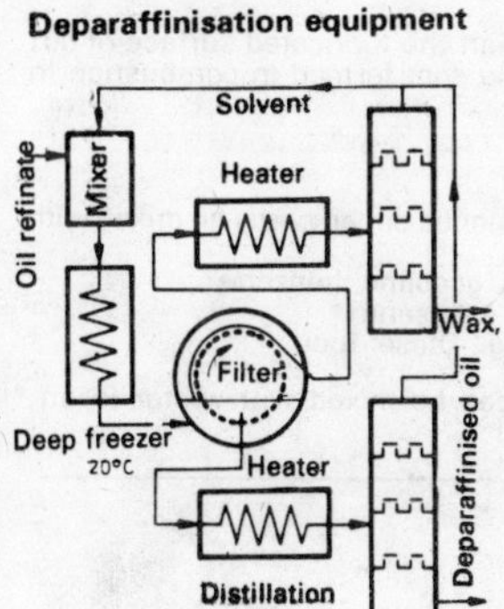

In deparaffinisation, the solid paraffins are separated from the lubricating oil since they unfavourably influence the flow of the oil at low temperatures.

The oil is mixed with solvents e.g. with toluene and butanon and cooled in deep freezers to about −20°C. The solid paraffins thereby freeze into white crystals (wax). The wax is separated from the dissolved oil in a rotating filter drum. An extractor roller extracts the wax cakes from the filter. The solvents are re-extracted after heating in two separate distillation equipment and used afresh for dissolving the oil.

Input material:

Lubricating oil refinate from double solvent refining.

Solvents:

Toluene and butanon

Cooling temperature: - 20°C

End products:
1. Deparaffinised lubricating oil
2. Wax for candle and floor polishing industries

Technological Terms of Fuel and Lubricating

Flow point, Drop point	The flow point is the temperature at which a grease begins to flow after heating under fixed conditions. Drop point is the temperature at which the first drop of a melting material drops from the glass rod.
Pour point	The pour point is the temperature at which a liquid flows no more under the influence of gravity.
Flame point, fire point	The flame point is the temperature at which the first inflammable vapours arise after heating under specified conditions. The temperature at which a uniformly, continuously burning flame is formed is designated as fire point.
Auto ignition point	Temperature at which an ignitable material self ignites without any external ignition.
Boiling point and boiling range	Boiling point is the temperature at which a homogeneous material boils. Boiling range is the temperature range within which a liquid consisting of many materials boils.
Evaporation	Vapourisation of a liquid through the free surface into the air below the boiling temperature.
Specific heat of Evaporation (latent heat	Quantity of heat which is required to convert 1 kg of a liquid into vapour at boiling temperature.
Specific heating value (calorific value)	The specific heating value (calorific value) Hu is the same quantity of heat in kilojoule which is released by complete combustion of 1 kg of a solid or liquid combustible material. In case of gaseous combustible materials, the specific heating value is specified in kilojoules per m^3 standard volume of the gas.
Octane number	Scale for the antiknock property of a fuel. Fuels with high octane number are knock proof.
Cetane number	Scale for the ignition quality of a fuel. Higher the cetane number, better the ignition quality of the fuel.
Viscosity	Viscosity is the property of a liquid or gas to oppose internal friction to a deformation. Dimensional units: dynamic viscosity in poise (P) or SI unit Pas (pascal second), kinematic viscosity in stokes (St) or SI unit m^2/s.
Viscosity index and polar altitude	Dimensions for the viscosity temperature behaviour of an oil. Oils with good VT behaviour have a high index and a low polar altitude.
Penetration	Dimension for the consistency of a grease. In the measurement, the penetration depth of a metal ball into the grease probe at 25°C is specified in 0.1 mm units.
Distillates and refinates	Distillates are mineral oil fractions which are obtained by distillation from raw petroleum without further improvement. Refinates are mineral oil products which have been improved by a refining process.
Inhibitor, catalyst	Inhibitor is an effective material which decelerates definite reactions, primarily aging and corrosion processes. An effective material which accelerates a chemical process and controls in its direction is called catalyst.
Additives (effective materials)	Additive materials to mineral oils and mineral oil products to further or to change definite properties or to achieve entirely new properties. They must be accurately matched with one another, with the base oil used and to the purpose of utilisation.
SAE	Society of Automotive Engineers, Association of American Automobile Engineers.
HD oil	Heavy Duty oils, engine lubricating oils whose additives enable the use of the oil in engines under severe working conditions.
Aging of oil	Under the influence of oxygen of air, increased temperature, presence of a catalyst, the lubricant changes. The most important changes are oxidation, polymerisation and condensation. In all cases, materials with changed properties result so that a precipitation or turbidity of the oil sets in.
Detergents and dispersants	Detergents give the lubricating oils the capacity to clean the lubricated surface of dirt particles and together with the dispersants to keep the soot formed in combustion in suspension.
Danger classes	Classification of combustible liquids. Group A: Liquids and mixtures or solutions which cannot be or can partly be mixed with water: Class I with a flame point below 21°C, e.g. gasoline, benzene Class II with a flame point 21 to 55°C, e.g. kerosene Class III with a flame point 55 to 100°C, e.g., Diesel fuel Group B: Liquids and mixtures or solutions which can be mixed with water in an proportion and have a flame point below 21°C.

Boiling curve and boiling coefficient
(DIN 51751)

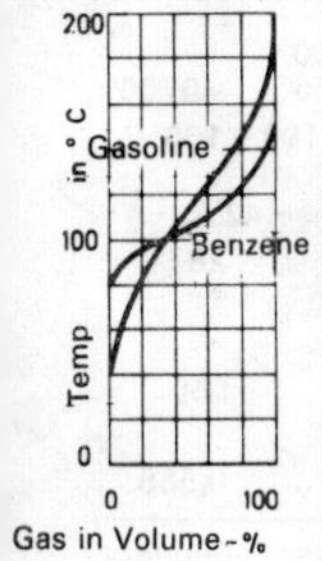

Fuels are a mixture of hydrocarbons, therefore they have no boiling point but only a boiling range. The dependence of temperature in °C on distillation quantity in percentage volume is specified by the boiling curve. It is determined in a boiling analysis in standardized equipment under fixed conditions (DIN 51751). The mean temperature in °C at which 5, 15, 25 ... 95% are over distilled in the boiling analysis is called boiling coefficient.

The vapourised quantities at 70, 100 and 200°C are of special significance. The beginning of boiling influences the vapour bubble formation, vapour bubble loss and start behaviour, the middle boiling range the transition behaviour in cold engine and carburettor ice formation, the end of boiling influences engine oil thinning and cleanness of the engine.

Portions of the fuel which are not yet gasified at 200°C are called boiling tails. They precipitate at the cylinder wall, wash off the lubricating film there and seep into the crank case.

Boiling range in °C

Gasoline	30 ... 200
Benzene	80 ... 160
Gasoline-benzene mixture	30 ... 200
Diesel fuel	180 ... 360
Engine kerosine	160 ... 300
Alcohol	78

Boiling coefficient in °C

Gasoline	≈ 110
Benzene	≈ 100
Gasoline-Benzene mixture	≈ 105
Diesel fuel	250 ... 280
Engine kerosene	≈ 230
Alcohol	—

Octane number
(DIN 51756)

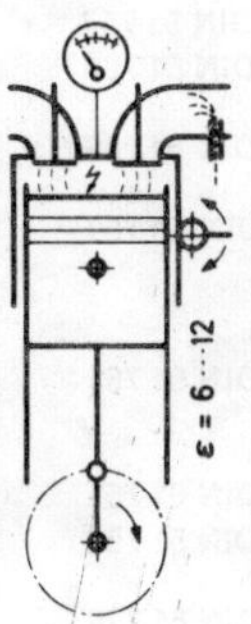

The octane number (OZ) is the dimension for the antiknock property of a carburettor fuel.

For the determination of the OZ, the fuel to be tested is compared with a standard fuel which consists of a mixture of isooctane (100 OZ) and n-heptane (0 OZ) up to 100 OZ and a mixture of isooctane and lead tetra ethyl above 100 OZ. A knocking engine with variable compression and a detonation meter are used for this. There are many test methods which give octane numbers differing from one another. Research method (ROZ) and Engine method (MOZ).

The measured octane number expresses what percentage by volume isooctane is found in the standard mixture which has the same antiknock property as the fuel to be tested as determined according to one of the two methods

The octane number can be increased by

a) refining of gasoline
b) addition of benzene and lead tetraethyl

Octane numbers (ROZ)

Gasoline	91 ... 94
Engine benzene	over 100
Gasoline-benzene mixture	98 ... 100
Diesel fuel	10 ... 30
Engine kerosine	30 ... 50
Alcohol	over 100
Fuel gas	95
Methane gas	100

Operating conditions in the determination of octane numbers in CFR test engine:

	ROZ	MOZ
Engine speed	600	900 rpm
Preignition	13°	17 ... 26°
Temperatures:		
Mixture	—	149 °C
Inlet air	52	38 °C
Coolant	100	100 °C
Oil	57	57 °C

Cetane number
(DIN 51773)

Cloud point
(DIN 51597)

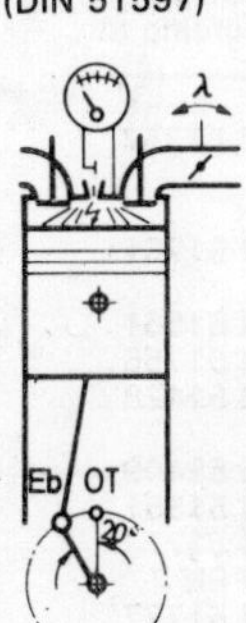

The ignition property of a Diesel fuel is expressed in cetane number (CZ)

The determination of cetane number takes place in the BASF test engine by comparison of the probe with measured mixture of cetane (100 CZ) and alphamethylnaphthaline (0 CZ) with the same operating values for the beginning of spray, fuel quantity and ignition delay. The quantity of suction air necessary for the setting of this ignition delay is a measure for the ignition property of the probe. The cetane number in this case is given by the cetane portion in the measured mixture expressed in percentage volume.

Higher the cetane number, better is the ignition property of the fuel. The antiknock property is in reverse proportion as the ignition property. Higher the antiknock property of the fuel, worse is its ignition property.

Cetane number ≈ 60 — 0.5 × octane number

The cloud point is the temperature at which the probe on cooling begins to become turbid or paraffins or other solids separate in the form of flakes. Filterability and flow can be insufficient, therefore in winter Diesel fuels are in the market with less paraffin. The cetane number is thus reduced.

Cetane numbers (CZ)

In summer	55
In winter	45

Operating conditions in the determination of the cetane numbers in BASF test engine:

Engine speed:	1000 rpm
Compression	18,2:1
Coolant temperature	100 °C
Spray beginning	20°
Ignition delay	20 ° KW
Fuel throughput	8 ml/min
Suction air temperature	20 °C

Beginning of combustion at the top dead centre

Filterability in °C

Summer 0°C, winter −12°C

Cloud Point

Summer	0 °C
Winter	− 15 °C

Properties of Fuels

General

		Normal gasoline	Super gasoline	Diesel fuel	Engine benzene	Ethyl alcohol	Methanol	Fuel gas
Specific calorific value (mean values)	kJ/l kJ/kg	32000 43100	32500 42700	36000 42300	35300 40200	21400 26800	15600 19750	– 40000
Octane (cetane number)	ROZ (CZ)	91 ... 94	98 ... 100	(45 ... 55)	over 100	over 100	over 100	100
Boiling range	°C	25 ... 210	25 ... 210	150 ... 360	80 ... 160	78	65	– 42 ... – 1
Air consumption, theoretical	l/l kg/kg	8500 14,8	8700 14,2	9500 14,5	8900 13	5600 9	3900 6,4	28 15
Flame point	°C	– 40	– 40	55 ... 110	– 10	+ 12	–	–
Self ignition*	°C	480 ... 550	480 ... 700	350	700	450	500	550
Cold stability	°C	under – 50	under – 40	0 ... 15	under – 20	– 114	– 98	–
Density	kg/dm³	0,72 ... 0,75	0,73 ... 0,78	0,81 ... 0,86	0,88	0,80	0,79	0,558

* The self ignition temperature is very strongly influenced by foreign materials (e.g. lead tetraethyl) and also dependent upon the actually prevalent conditions.

Minimum requirements of Otto-engine fuels

DIN 51600

Requirements	Test values for		Normal gasoline	Super gasoline	Testing according to
Density at 15 °C		g/ml	0,715 ... 0,755	0,703 ... 0,780	DIN 51757
Antiknock property	Minimum Minimum	ROZ MOZ	91 82	97,4 87,2	DIN 51756
Lead content	Maximum	gPb/l	0,15	0,15	DIN 51769
Boiling curve, summer gasoline					
Total evaporated quantity up to 70 °C up to 100 °C up to 180 °C	Minimum Minimum Minimum	Vol.-% Vol.-% Vol.-%	15 .. 40 42 ... 65 up to 90	15 ... 40 42 ... 65 up to 90	DIN 51751
Boiling curve, winter gasoline					
Total evaporated quantity up to 70 °C up to 100 °C up to 180 °C	Minimum Minimum Minimum	Vol.-% Vol.-% Vol.-%	20 ... 45 45 ... 70 up to 90	20 ... 45 45 ... 70 up to 90	DIN 51751
Distillation residue	Maximum	Vol.-%	2	2	DIN 51751
Vapour pressure: summer gasoline winter gasoline	Maximum Maximum	bar bar	0,45 ... 0,70 0,60 ... 0,90	0,45 ... 0,70 0,60 ... 0,90	DIN 51754
Evaporation residue	Maximum	mg/l	50	50	DIN 51776
Sulphur content	Maximum	weight - %	0,1	0,1	DIN 51768
Cold stability	up to	°C	– 30	– 30	DIN 51421
Corrosion effect on copper	Degree of corrosion		1 – 50 A 3	1 – 50 A 3	DIN 51759
Impurity	Free from inorganic acids, solid foreign materials and visible water				
Compatibility against elastomers	The compatibility must be ensured				

Minimum requirements of Diesel fuels

DIN 51601

Requirements	Test values			Testing according to
Density at 15°C		g/ml	0,815...0,855	DIN 51757
Boiling curve:				
Total evaporated quantity up to 250°C up to 350°C	Maximum Minimum	Vol.-% Vol.-%	65 85	DIN 51751
Kinematic viscosity at 20°C	Higher than	mm²/s (cSt)	1,8...10	DIN 51561
Flame point according to Abel-Pensky		°C	55	DIN 51755
Limiting value of filterability	in summer maximum in winter maximum	°C °C	0 -12	DIN 51428
Sulphur content	Maximum	weight-%	0.50	DIN 51409
Coke residue according to Conradson	Maximum	weight-%	0,1	DIN 51551
Ignition property (Cetane number)	Minimum	CZ	45	DIN 51773
Ash (oxide ash)	Maximum	weight-%	0,02	DIN EN 7
Water content according to Karl Fischer	Maximum	mg/kg	1000	DIN 51777
Impurity, Turbidity	Free from mineral acids and solid foreign materials and smooth at room temperature (18 to 28°C according to DIN 50014)			

General

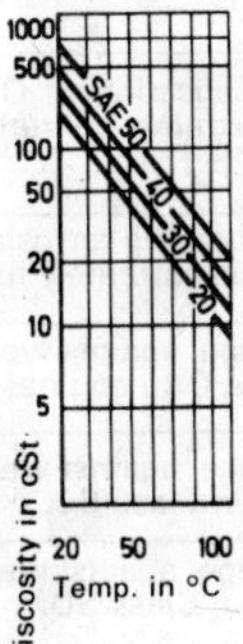

The function of modern lubricating oils in automobiles are:

1. Lubricating capacity
   a) reduction of friction
   b) reduction of wear
   c) pressure bearing capacity of the oil film
2. Purity and cleaning effect
3. Dispersion or distribution capacity of residues obtained from combustion
4. Acidic conversion products resulting from neutralisation
5. Reduction of sludge and to hold sludge in suspension
6. Reduction of aging or oxidation
7. Reduced temperature-dependent viscosity change
8. Reduction of pour point
9. Prevention of foam
10. Compatibility with sealing material
11. Sealing
12. Cooling

These requirements are fulfilled only by the "compounded oils" (HD oils), they are oils with oil-soluble chemical additives. HD is the abbreviation for heavy duty. Depending upon the application, the individual properties are of differing importance. Correspondingly, the individual additives are controlled by the manufacturer (2 to 30 %)

Additives for oil (effective materials, additives, dopes):

Oxidation inhibitors (anti oxidants)
Corrosion protective materials
Purifying additives (detergents)
Sludge carrier additives (dispersants)
High pressure additives (Extreme pressure (EP) additives)
Oiliness improving agents
Viscosity index improving agents (VI agents)
Pour point improving agents
Defoamants

Viscosity

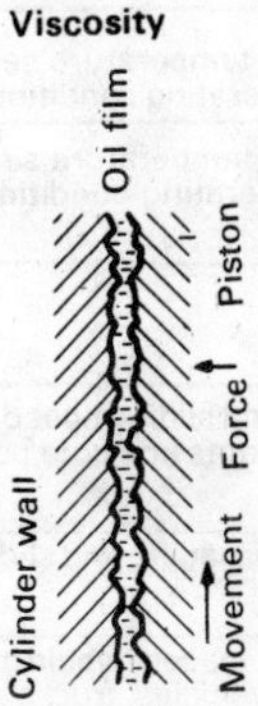

The viscosity is the characteristic for the capacity to flow of an oil. High viscosity signifies thick fluid, low viscosity thin fluid. In parallel flow of a fluid, a resistance (shear stress) proportional to the velocity gradient occurs between two neighbouring layers.

The quotient shear stress by velocity gradient is called dynamic viscosity and has the SI unit pascal second (Pas). The quotient dynamic viscosity by density is the kinematic viscosity. Its SI unit is m^2/s. The units poise (P) for the dynamic viscosity and stokes (St) for the kinematic viscosity are permitted up to 31.12.77 according to DIN 1301.

1 poise = 1 P = 0.1 Pas;   1 stokes = 1 St = 1/10000 m^2/s.

Since these units give mostly too small numerical values, the specification of viscosity in centipoise (cP) and centistokes (cSt) is common.
The American Association of Automobile Engineers (SAE = Society of Automotive Engineers) classifies the viscosity of oils into SAE classes. The SAE classes have been explained for standard in Germany. The viscosity of an oil is dependent upon the pressure and temprature. Higher the temperature thinner the oil becomes. The viscosity temperature index is specified by the viscosity index (VI) Higher the index more temperature stable the viscosity of the oil is.

Single range oils

SAE class	Viscosity in cSt at 50°C
5 W	Below 17
10 W	21,5 ... 29
20 W	29 ... 50
20	25 ... 50
30	50 ... 75
40	75 ... 120
50	120 ... 200
80	Below 65
90	82 ... 155
140	250 ... 520
250	Above 520

Multirange oils

SAE Class	Viscosity in cSt at 50 °C	100 °C
10W/20	25	6
10W/30	40	11
20W/40	70	—

Viscosity index

Uncompounded oils	65 ... 75
Compounded oils	80 ... 95
Multirange Oils	90 ... 110
ATF	130 ... 150

Oil change

Following factors influence the change or deterioration of the oil during use:

1. Design of the engine
2. Operating conditions
3. Fuel quality
4. Oil thinning of fuel
5. Disintegration of water
6. Effectiveness of the oil filter
7. Filling quantity
8. Oil consumption
9. Compounding degree of oil
10. Travel distance
11. Engine loading
12. Period of year

The influence of these factors can be very different. A generally valid answer on the question about time for oil change can hence not be given. It is recommended to adhere to operating instructions without fail. It is warned against so called "quick test".

If the oil has to be changed or supplemented and the oil used till then is not available another brand of the same quality can be filled without being afraid of a damage. Special change of oil regulations are not to be cared for in this case. Without convincing reason, however, frequent change of brands is to be avoided.

Intervals of oil change for passenger vehicles in km;

Otto engine	5000 ... 10000
Diesel engine	2500 ... 5000
Oil bath air filter	5000
Sliding and automatic gear boxes	20000
Rear axle	20000

More frequent in case of severe operating conditions (short distance transport, high dust generation)

Oil change in engine at least twice a year, in automatic gear boxes at least once a year.

Properties of Lubricants

Engine oils, API classification (API = American Petroleum Institute)

New API class	Old API class	Properties, type of application, operating conditions
SA	**ML**	Pure mineral oil, no quality prescriptions. For the simplest application of Otto engines; low piston displacement capacity, low speed, no extreme operating temperatures.
SB	**MM**	Oil with effective materials against galling wear, oil oxidation and bearing corrosion. For the simplest application of Otto engines; moderate speed, no extremely high operating temperatures.
SC	**MS**	Otto engine oil with effective materials against cold and hot sludge, against wear, rust and bearing corrosion. For medium severe application of the Otto engines.
SD	**MS**	Otto engine oil with effective materials against cold and hot sludge, against wear, rust and bearing corrosion. Suitable for higher requirements than Class SC.
SE	—	Otto engine oil with effective materials against cold and hot sludge, against wear, rust and bearing corrosion. Suitable for higher requirements than Class SD.
CA	**DG**	Diesel engine oil with effective materials against high temperature sedimentations and bearing corrosion. For mild to moderate operating conditions with fuels of high quality.
CB	**DM**	Diesel engine oil with effective materials against high temperature sedimentations and bearing corrosion. For moderate operating conditions with fuels of higher sulphur content.
CC	**DM**	Diesel engine oil with effective materials, against high and low temperature sedimentations, against rust and corrosion. For medium to severe operating conditions.
CD	**DS**	Diesel engine oil with effective materials against high and low temperature sedimentations, against rust and bearing corrosion. For severe operating conditions with fuels of differing quality.

Transmission oils, API classification

API-GL-1	For gear boxes which operate under mild conditions with regard to tooth loading end sliding speed, so that uncompounded mineral oil is sufficient. Oxidation and rust inhibitors, defoamants and pour point improving agents are added many times.
API-GL-2, 3	For gear boxes which operate under medium heavy conditions with regard to tooth loading and sliding speed, however, not suited for hypoid gear boxes.
API-GL-4	Multipurpose transmission oil for highly loaded sliding gear boxes and for medium loaded hypoid gear boxes with not too great shift of axis. Sufficient for most gear boxes in passenger vehicles, trucks and construction machinery.
API-GL-5	For operating conditions as in API-GL-4, however under influence of shock loads and for hypoid gear boxes with great shift of axis. Higher content of EP (extreme pressure) additions. Higher requirements are placed on the sealing material.
ATF Suffix A Dexron	ATF = Automatic Transmission Fluids. Thin liquid automatic transmission oil which is oriented more to extreme temperature stability than to load carrying capacity. Other important properties are: small viscosity change, oxidation stability and uniform friction behaviour.

Lubricating greases

Designation	Drop point °C	Penetration $^1/_{10}$ mm	Grease base	Properties	Application
Multipurpose grease	180	220 ... 340	Lithium	Water repellant Heat resistant	All lubricating points
Lubricating grease	Above 95	310 ... 340	Lime	Water repellant Heat sensitive	Nipple lubrication
Rolling bearing grease	Above 160	265 ... 295	Sodium	Water sensitive Heat resistant	Rolling bearings
Water pump grease	Above 140	150 ... 250	Lithium	Water repellant Heat resistant	Water pump
Transmission flowing grease	Above 90	360 ... 380	Sodium	Water sensitive Heat resistant	Transmission with sealing action

Shock absorber oil	Basic material: Mineral oil refinate or a mixture of various liquids. Free from water, solid, foreign materials and hard asphalt, chemical neutral, clear and transparent, stable against aging, good viscosity temperature behaviour, stable in cold, non stick slip (chatter effect).	Density at 15°C 0,89 kg/dm³ Viscosity at 50°C 8,5 … 17 cSt Flame point 160 °C Pour point −50 °C
Brake fluid	Basic material: Multivalent alcohol (glycols). The brake fluid should not come in contact with brake lining. It is corrosive and absorbs lacquers. Brake fluids shall be able to be mixed with one another. Other liquids, e.g., lubricating oils should not be added since they can destroy the sealing cups and thus lead to failure of the brake. Brake fluids are dirt free and water free but absorb dust and water from the air. Therefore, it is recommended to renew them annually.	Density ≈ 1,13 kg/dm³ Viscosity at 20°C 17 cSt at 50°C 5 cSt Cold resistant upto bis −40 °C Boiling point 190 … 260 °C Flame point 90 … 135 °C Corrosion resistant Capable of lubricating, chemically stable Rubber swelling behaviour 0 … 6 %
Antifrost agent	Basic material: Multivalent alcohol (glycols) with corrosion-preventive additives: At temperatures of about or below the fire point, an antifrost agent must be added to the cooling water. The quantities are guided by the temperature and the capacity of the cooling system. Till now, the density of the mixture which can be determined by means of areometer was same for all brands. Today it is different due to the addition of corrosion protective agents. Fire point / Water / Anti frost agent / Density −10 °C / 4 parts / 1 part / 1,040 kg/dm³ −20 °C / 2 parts / 1 part / 1,060 kg/dm³ −30 °C / 5 parts / 4 parts / 1,074 kg/dm³ −40 °C / 8 parts / 9 parts / 1,084 kg/dm³	Ethylene glycol Density at 25°C 1,113 kg/dm³ at 82°C 1,08 kg/dm³ Boiling point 197°C Flame point 110°C Ice flaking point pure −14°C 2:1 below −16°C 1:1 below −33°C Corrosion resistant against cast iron, aluminium, brass and tin solder. Rubber swelling behaviour maximum 6% Soluble in water
Corrosion protective and first operation oil	Basic material: Mineral oil refinate with corrosion protective and cleaning additives. These oils serve for the cleaning action, and corrosion protection for the usual running-in, transport, storage and shut off periods. In running in process, additives of special type convey the surface roughnesses into soft compounds and carry away these gently. In case of a shut down of the vehicle up to 6 weeks it does not require any preservation, between 6 and 26 weeks preservation run of the engine, reduction of the rear axle oil and of the fuel tank and the prescribed battery attention are sufficient. In case of any longer shut down the preservation steps prescribed by the factory are to be observed.	SAE 10, 20 W/20, 30 Characteristics for SAE 20 W/20 Density 0,925 kg/dm³ Viscosity at 50°C 40 cSt Flame point 255 °C Pour point −25 °C
Base protective	Basic material: Wax, bitumen or PVC with additives and solvents. The base protection must form a uniform, durable, corrosion resistant, elastic, best gripping and 85°C temperature resistant film. The lacquer should not sweil or dissolve. The preservative must be able to be sprayed with a spray gun and must be clearly coloured so that it can be tested whether all parts are sufficiently wetted.	Wax or bitumen portion 20 … 40 % pH value 6 … 7 Flow point above 85 °C Flame point above 21 °C Danger class Drying class 60 … 70 min Boiling curve of the solvent 130 … 190 °C
Special additives for lubricants	The additives offered improve only one property of the lubricating oil. This action shows itself clearly in case of uncompounded oils. In comparison with proved compounded oils, further advantages are hardly established. A simple test by the user can give no accurate support on the behaviour in operation. It is therefore warned against an arbitrary filling in of additives not recommended in practice.	

Corrosion

Conception 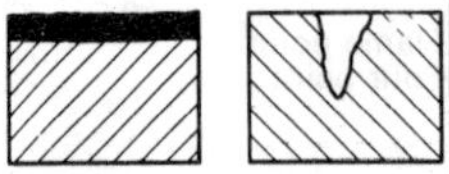	Corrosion is the destruction of a material by chemical actions or by electrochemical processes. Corrosion can be effective on the whole surface of the workpiece (surface corrosion) or only at one place as pitting. In many metals, e.g. copper or aluminium the corrosive layer acts as protective layer and preserves the material from further destruction. Noble metals like platinum, gold and silver are specially resistant.	Corrosion is accelarated by heat, by aggressive materials like acids, alkalies and salts as well as moisture. Air polluted by burnt gases causes corrosion on buildings. Corrosive damages are brought about in vehicles by moisture and by dispersed salt. Losses in multimillions are caused by corrosion every year.
Chemical corrosion 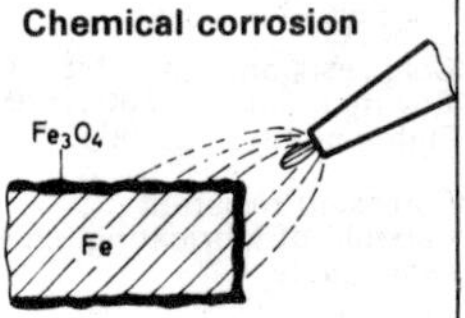	Chemical compound of the material with oxygen, carbonic acid, sulphur dioxide. Disintegration by absorbed liquids like acids, alkalies, salt solutions and solvents as well as aggressive gases and vapours.	While annealing steel, the iron on the surface combines with oxygen, it gives rise to a blue-grey tinder (Fe_3O_4). The oxygen in the air in combination with moisture produces rust. Acetate forms verdigris on copper, carbonic acid a patina layer.
Electrochemical corrosion 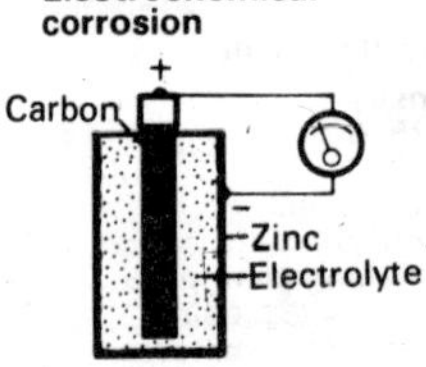	The corrosion processes are similar to those in galvanic element. Such an element consists of two different, electrically conducting materials and a conducting liquid (electrolyte). The material which forms the negative pole (see voltage series) is disintegrated when current flows.	Torch battery consisting of a zinc-carbon element with ammonium chloride solution as electrolyte. The zinc case is the negative pole and hence it is corroded. The voltage amounts to 1.5 volt. The negative pole is attacked to such an extent as the two materials are separated from each other in the voltage series.
Corrosion within a crystal 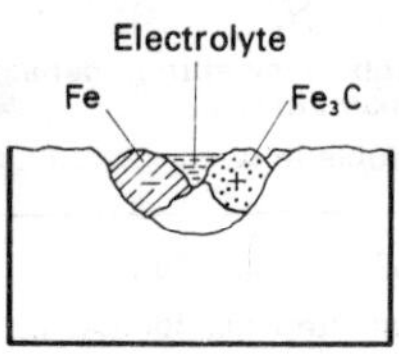	On the surface of a bright metal, a galvanic element can form between the various crystals. Moisture, humidity, hand weld in chamfering act as electrolyte. Weak currents flow in the crystals as a result of which the crystals which form the negative pole are corroded.	On unprotected surface of unalloyed steel, galvanic elements occur between the iron crystals (Fe) and the iron carbide crystals (Fe_3C). The iron crystals form the negative pole; hence they are destroyed.
Contact corrosion 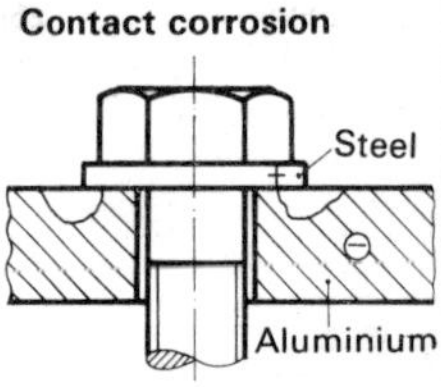	When different metals lying one after the other in the voltage series are in contact, then with the aid of moisture a galvanic element is formed, e.g., steel-copper, steel-brass, steel-aluminium. Current flows between the two materials. The material which forms the negative pole corrodes.	Steel screws in aluminium body; the body is corroded around the screw head. Spokes nipple in the rim of magnesium alloy break due to corroded holes.
Eddy current corrosion 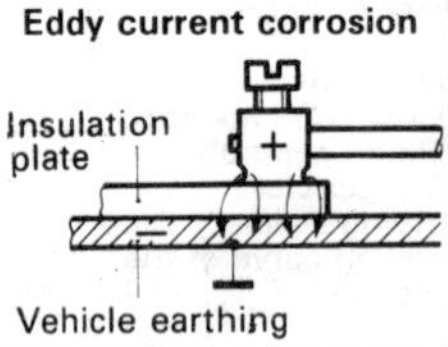	In direct current equipment, a weak current (eddy current) flows, as a result of moisture, from positive pole to negative pole through the insulation. The material which forms the positive electrode is corroded at the exit of the eddy current.	Eddy current corrosion occurs in electrical equipment of vehicles. Higher the voltage, stronger is the action, e.g., in 24 V equipment, the danger through corrosion is greater than in 12 V equipment.
Electrochemical voltage series	Voltage in volt Au + 1,50 · Pb – 0,13 · Fe – 0,44 Pt + 0,86 · Sn – 0,14 · Cr – 0,56 Ag + 0,80 · Ni – 0,23 · Zn – 0,76 Hg + 0,79 · Co – 0,29 · Al – 1,67 Cu + 0,34 · Cd – 0,40 · Mg – 2,40	The elemental voltage is equal to the voltage differences of the two materials, e.g., an element from copper and zinc has a voltage of 1.11 volt.

General	1. Various pure metals, e.g., copper, aluminium, zinc, form with the components of air a dense protective layer which prevents a progress of corrosion. Iron or magnesium are in contrast prone to corrosion since the superficial layer is porous. 2. The corrosion resistance can be improved or worsened by alloying. Bronzes, Al-Si alloys, Cr-Ni steel, etc. are corrosion resistant whereas Al-Cu alloys or Mg alloys are very much prone to corrosion. 3. Corrosion is prevented when no galavanic element can be formed, e.g. by keeping away from moisture, no combination of various metals without intermediate insulation, protective coatings on metal surfaces.	
Oil, grease, wax	Use of acid-free mineral oil or grease (vaseline) or hard wax. Painting dipping, spraying.	Hard wax is more resistant than oil and grease. Process and storage protection.
Chassis protection **Sealing of holes**	Bitumen based or wax-based agents are applied on the cleaned and descaled or best on the still new chassis of the vehicle. To seal holes, protective agents made from oil, dissolved wax and rust inhibitors are sprayed into the bored holes of the vehicle body.	The protective film of the chassis is elastic after drying. It is chemically neutral and does not corrode rubber or plastic. The preservatives displace the sweat water in the holes, transform the rust and build a protective film which prevents further rusting.
Paints, Lacquers	Parts descaled, cleaned (steel brush, blasting) or treated with wash primer. Minimum or wash primer is applied first, then intermediate paint and finally cover paint or lacquer	Wash primer reacts with the metal, prevents chassis rusting. Paint and lacquer films should be dense, elastic and temperature resistant.
Plastic coating	Polyethylene is sprayed in powder form on the preheated parts by means of compressed air. PVC coatings are produced mostly by dipping the parts in liquid solution.	Coating thickness 0.1 mm to several mm; elastic, tear resistant, stable against chemical loading.
Metallic coatings	All metallic protective coatings require a careful pretreatment of the surface, e.g., cleaning, scouring, descaling, pickling, degreasing, grinding, polishing.	Protective coatings of more electropositive metals than the base metal, e.g., nickel or tin on steel protect only so long as they are dense. If the protective coat is lost, an element is formed in which the base metal is negative and hence is destroyed. The corrosion progressed under the protective coating, the protective metal chips off. If the coating metal is less electropositive than the base metal, e.g., zinc on steel, then with element formation the zinc layer is sacrificed, the base metal is protected.
Galvanising	The parts are brought into a metallic salt bath and connected as negative pole to direct current. A plate made from the metal to be deposited forms the positive pole.	
Coating in molten mass	The parts are dipped in the molten coating metal e.g., in tin, lead, zinc (hot galvanizing).	
Hot spraying	The coating metal is melted in wire or powder form in the gun and sprayed with compressed air or protective gas.	
Plating	The coating metal is hot rolled into thin films; the cutting edges are not protected.	
Oxide protection **Anodising**	Electrical oxidation of aluminium and aluminium alloys. The parts are brought as positive pole, a lead plate as negative pole into a sulphuric acid bath and connected to a direct current. An oxide layer is formed on the surface which mainly grows towards inside.	Layer thickness. Porous, strongly adhering, hard, flexible, not electrically conducting, can be painted, higher protective value.
Phosphating **Bonderising**	The steel or zinc parts are dipped in sodium ammonium phosphate solution or sprayed with it by which a phosphate layer is formed.	Grey, porous layer in case of steel, must be oiled or wax coated, otherwise only a little protection, good base for paints.
Blackening **Browning**	Steel parts dipped in oil and burnt in flame. In case of browning, dipped in hot sodium hydroxide containing sodium nitrate.	Brown-black to blue black coating, is oiled, only a small protective value.

Lacquers

Properties	Nitrolacquers	Synthetic resin lacquers
Composition	Solution of nitro-cellulose in volatile solvents with the addition of colouring agents and additives	Solutions of synthetic resins (alkyde resin with urea resins, melamin resins, epoxy resins or phenol resins) in volatile solvents with the addition of colouring agents and additives.
Drying	Thinning of the solvents, air drying in 30 min to 4 h, faster at increased temperatures, e.g. 30 min at 60°C	Thinning of the solvent and chemical transformation of the film builder air drying furnace drying burning
Flowing capacity and covering capacity	Highly viscous types are without solids, give thin, elastic films, more lacquer layers are necessary, bad flowing	Rich in solids, with filling power, dry slower and flow better
Gloss and light fastness	Feeble after drying, polishing therefore required, finer, total gloss, older lacquers turn pale and yellow	High gloss after drying, no further processing required, gloss stable, greasy gloss, only a little yellowing.
Elasticity and hardness	Soft and elastic, hardness in course of time, therefore preservative required.	Hard, porcelain-type surface, elastic film, become slightly softer with time.
Resisting capacity	Sufficient tear, shock and scratch strength, stable against weather with regular care, sensitive against chemical effects	impact, shock, tear and scratch strengths good, very good stability against weathering, has resisting capacity against chemical effects.
Combustibility	Easily combustible (explosion type), flame point below 21°C, danger class AI	Less danger of combustion, flame point 21 to 55°C, danger class AII.
Health hazard	Care while handling and storage, vapours injurious to health	
Solubility	Soluble in same lacquer or solutions after drying, hence can be painted badly, can only be sprayed, delivery prescriptions to be observed	No more soluble after drying, hence can be painted well, base coat of furnace drying lacquer can be dissolved with benzene
Identification	Smooth surface, sharp contours of mirror images, soluble in thinner	Orange type surface, wavy contours of the mirror image, not soluble, swell and pickling seams can occur in furnace drying types.
Maintenance	Requires care, they must be regularly polished and preservative applied.	Requires hardly any care, hard surface is damaged by grinding, hence only polishing or preservative applied.

Lacquer maintenance agents	Composition	Property	Application
High gloss polishing agents Polish Polishing water Wax solution Wax paste	Emulsions of oil or wax with solvents, saponification agents or suspensions, small quantities of soft acting polishing agent (clay, chalk),silicone oil	Highly glossy, renewing and preserving, dissolved dirt and grease, does not corrode lacquer. Silicone gives durable, water resistant high gloss	Polishing, renewing and cleaning. Silicone can form flakes on old lacquer by strong crater effect on unequally absorbing underbase.
Grinding polishing agents a) Polishing paste	Wax solutions or emulsion with soft working grinding agents.	Cleaning, smoothening glossy, corrodes the under base only a little	Cleaning and polishing of smooth lacquers, repolishing not necessary
b) Grinding polishing paste	Wax solution with more and sharp-acting grinding agents.	More smoothening than glossy, corrodes the underbase stronger	Cleaning, polishing, grinding of older lacquers, repolishing recommended.
c) Grinding paste	Wax solutions with larger quantities of sharp-acting grinding agents	Exclusively smoothening, not glossy, corrodes the underbase strongly	Pregrinding of new lacquers, grinding off old lacquers, post polishing necessary

Structure of Original Lacquering and Repair Lacquering of Passenger Vehicles

Parts		Preliminary treatment (adhesive base, passivation, corrosion protection)		Priming (Corrosion protection, adhesive base)		Preliminary lacquering (transition)	Cover lacquering (gloss)
Casing work	Overlappings Concave parts Inaccessible spots	—	Zinc dust paint Thickness : 25 μm	Burnt dip priming Thickness: 25 μm	—	—	—
Outer covering a) Original lacquering	Method I	Sheet freed from dust, moisture, greases, oxides and soldering residues before the beginning of lacquering works	Phosphatising: Alkaline phosphate, zinc phosphate or manganese phosphate. Dipping Thickness 1 to 3 μm	Burnt base Dipping Thickness: 15 to 40 μm, Burning 12 to 30 min, 145 to 190°C Runner separated, Seam ground	Burnt base, Spraying, Thickness 20 to 40 μm, Burning 10 to 30 min 130 to 170°C Seam ground	Burnt filler or preliminary lacquer, Spraying, Thickness 25 μm, Burning 10 to 30 min 130 to 165°C Topping	Burnt cover lacquer, Spraying, Thickness 25 to 35 μm. Burning, 20 to 30 min 120 to 130°C
	Method II	a) Grease and dirt by cleaning agents and solvents	Reaction base (wash primer): alcoholic binding medium with zinc chromate and phosphoric acid. Spraying Thickness 1 to 10 μm	Burnt base Dipping Layer thickness 30 to 50 μm Burning 12 to 30 min 150 to 170°C Seam ground		Burnt filler, Spraying, Thickness 25 to 30 μm, Burning 15 to 30 min 130 to 160°C, Drying, Grinding	Burnt cover lacquer Spraying, Thickness 25 to 30 μm Burning 20 to 30 min 120 to 130°C
b) Repair	Sheet metal parts Spare parts	b) Thick film of rust by descaling agents	Sheet metal parts with reaction base (wash primer). Spare parts with phosphating	Burnt spray base or burnt dipped base, Layer thickness 25 to 35 μm, Spray base required, About 60 min 80°C		Burnt spray base, Thickness: 30 to 35 μm, About 60 min at 80°C	Burnt cover lacquer Thickness: 30 μm, About 60 min at 80 °C
	Original lacquering	Original lacquering mostly by dipping method. Repair by washing off or spraying off	Polishing residues and silicone residues separated with solvents (silicone separators) or by grinding. Weathered lacquer layers ground off till clean layers are met. Old lacquerings can be pickled off with lacquer separators. After about 30 min, the softened lacquering is scraped off with spatula.			Older lacquering: Burnt spray base, Thickness 25 μm About 60 min at 80°C	Burnt cover lacquer Thickness: 30 μm About 60 min at 80°C
Chassis wheel carriers Hatch sheets	Original lacquering		Phosphatisation	Burnt dipped base Thickness: 25 μm 12 to 30 min 150 to 170°C		—	PVC chassis protection Thickness: 800 μm
	Repairs: Bright sheet Spare parts		Smooth sheet with reaction base, spare parts with phosphatisation	Burnt sprayed base or burnt dipped base Layer thickness: 25 μm Sprayed base: 60 min 80°C		—	Natural rubber Chassis protection e.g. Antirombo Thickness: 1000 μm

Abrasives, Grinding wheels

General	Natural and synthetic abrasives are distinguished. The natural abrasives, e.g., quartz (sand stone), corundum and emery are hardly used any more. Even the natural diamond is displaced by synthetic diamond in grinding technology. Synthetic abrasives like electrocorundum (aluminium oxide Al_2O_3) and silicon carbide (S_2C) are almost exclusively used.	Natural abrasives are not uniform in their composition. Their abrasive performance is very less partly due to the blunt abrasive grains. Besides this very low cutting speeds can only be worked with.
Electrocorundum Abbreviation A	Electrocorundum is smelted in electric furnaces at a temperature of about 2000°C from bauxite or pure alumina. Impurities are reduced by adding coke. Large crystal formations called "druse" are obtained.	Density 3,9 ... 4,0 kg/dm^3 Mohs'hardness 9 ... 9,25 Electrocorundum is colour brown and light rose to white. It is suitable for grinding of unhardened and hardened steel, cast steel and malleable iron.
Silicon carbide Abbreviation C	Silicon carbide is smelted in electric furnaces at a temperature of about 2500°C from quartz sand, coke and additives. Equally large crystal formations as in electrocorundum are obtained here also. Silicon carbide is approximately as hard as diamond.	Density 3,17 kg/dm^3 Mohs'hardness 9,5 ... 9,75 Silicon carbide is coloured glossy dark green to black. It is suitable for grinding of cast iron, hard metals, copper alloys and soft metals.
Diamond Abbreviation D	Natural diamonds which are not clear and not suitable for jewellery, or synthetic diamonds are used. These are obtained from carbon by using very high pressures and temperatures.	Density 3,51 kg/dm^3 Mohs'hardness 10,0 Diamond abrasives are suitable for fine grinding, grinding of hard metals and dressing of wheels.
Grain	The large crystals are powdered or reduced in size by means of drop hammers and assorted according to grain size by vibrating screens. Powdery fine grains (lapping media) are obtained in the sedimentation process. The grain sizes are designated with numbers. The number specifies the number of meshes per one inch length of the concerned sieve.	Designation of the grain Coarse 10, 12, 14, 16, 20, 24 Medium 30, 36, 46, 54, 60 Fine 70, 80, 90, 100, 150, 180 Very fine 220, 240, 280, 400, 500, 600 The grain sizes of diamond wheels are specified in μm.
Bond	Bond is the holding together of abrasive particles of a grinding wheel. Different bonds are; 1. Ceramic bond from feldspar, clay and quartz. The abrasive grains are mixed with these materials, pressed to grinding wheels and burnt in furnace at 1500°C. 2. Bond with rubber shellac or with synthetic resin (phenol resin).	Abbreviation for bonds: Caramic V, synthetic resin B, rubber R, shellac E, sintered (diamond wheels) Sint. Ceramic bond wheels are insensitive against high temperature but very brittle and shock and impact sensitive. Rubber or synthetic resin wheels are elastic and therefore suitable for thin grinding wheels.
Hardness	Hardness of a grinding wheel is the bonding force with which the abrasive grains are held together by the binding medium. Easier the abrasive grains break off, softer is the grinding wheel. Soft grinding wheels are used for hard workpieces so that abrasive grains that have become blunt break off at the right time and new grains are exposed.	Designation of the degree of hardness of grinding wheels with capital letters: Very soft D, E, F, G Soft H, I, J, K Medium L, M, N, O Hard P, Q, R, S Very hard T, U, V, W
Structure	Structure of grinding wheel is the magnitude of the pores. In open, large pored structure, the distance of the abrasive grains from one another is greater than in dense structure. To grind soft materials grinding wheels with open structure are used so that the grinding chips are not fixed	Designation of the structure in numbers e.g., dense 0, 1, 2, 3 medium 4, 5, 6, 7 open 8, 9, 10, 11, 12 Grinding wheels with open structure adhere less than those with dense structure, but are consumed faster.
Peripheral speed	Highest permissible peripheral speed for grinding wheels with ceramic bond in manual infeed 25 m/s in automatic infeed 35 m/s	Grinding wheels with permissible peripheral speed above 35 m/s are identified by a coloured band, e.g. blue $<$ 45 m/s, yellow $<$ 60 m/s, red $<$ 80 m/s and green $<$ 100 m/s.

Test method	Execution and result
Tensile testing 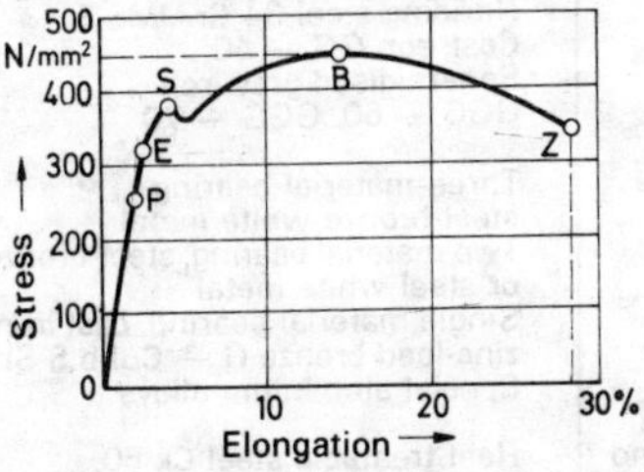	A carefully machined, dimensionally standardized standard bar (e.g. diameter = 20 mm, gauge length = 200 mm) is loaded in a tensile testing machine until fracture Stress and elongation are plotted on a graph. P: Proportionality limit—till then stress and elongation are proportional (linear) E: Elastic limit—till then elastic behaviour, no residual elongation, S: Yield limit—from here onwards, residual deformation B: Fracture limit—highest possible workpiece loading (Rm), after this large elongation and neck formation (in steel) up to the rupture limit Z New designations according to DIN 50145 Tensile strength till now σ_B, new R_m, Unit N/mm² Elongation at fracture till now δ, new, A, Unit %
Brinell hardness testing 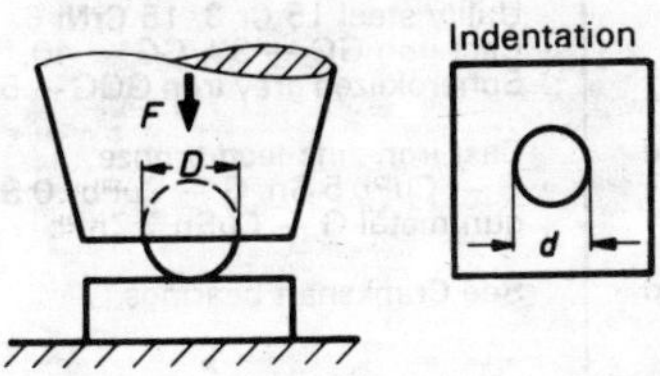	A hardened steel ball (or hard metal ball) is loaded and thereby an indentation is formed in the surface of the workpiece to be tested. In this, the loading of the ball as well as the ball diameter depend upon the type and the thickness of the material of the part to be tested. The ball diameter amounts to 25 mm for probes up to 3 mm thickness, 5 mm for probes from 3 to 6 mm thickness and 10 mm for probes above 6 mm thickness. For the determination of the Brinell hardness value, the diameter of the ball indentation is measured. $\text{Brinell hardness HB} = \frac{1.102 \times \text{Test force in N}}{\text{Surface of the ball indentation in mm}^2}$ $HB = \frac{0{,}102 \cdot F}{A} = \frac{0{,}102 \cdot 2 \cdot F}{D \cdot \pi (D - \sqrt{D^2 - d^2})}$ Example of a hardness specification: 120 HB means: Brinell hardness 120, tested with a ball of 10 mm diameter, test force 3000 kg × 9.81 m/s² = 29430 N From the Brinell hardness, the tensile strength can be approximately determined. For steel $\sigma_B \approx 35$ HB N/mm²
Vickers hardness testing 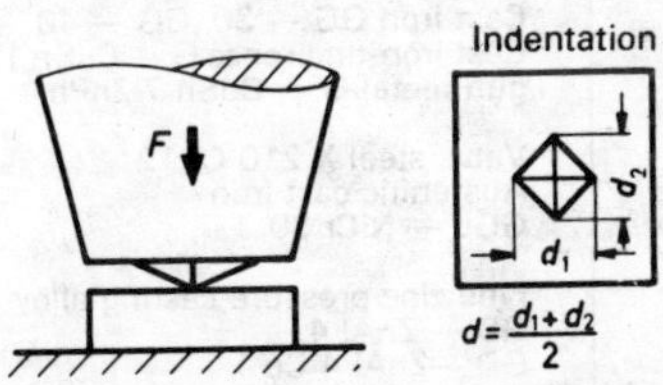	A square diamond pyramid in which the opposite faces form an angle of 136° is indented with a test force which can lie between 49 N and 980 N into the surface of the workpiece to be tested. The lengths of the diagonal of the indentation is very accurately measured and the mean value d is the basis for the calculation. $\text{Vickers hardness} = \frac{0.102 \times \text{Test force in N}}{\text{Surface of indentation in mm}^2}$ $HV = \frac{0{,}102 \cdot F}{A} = \frac{0{,}102 \cdot 1{,}854 \cdot F}{d^2}$ Example of a hardness specification 240 HV 50 means: Vickers hardness 240, tested with a test force of 50 kg × 9.81 m/s² = 490 N The test method according to Vickers is suitable specially for very small, very hard and very thin probes.
Rockwell hardness testing 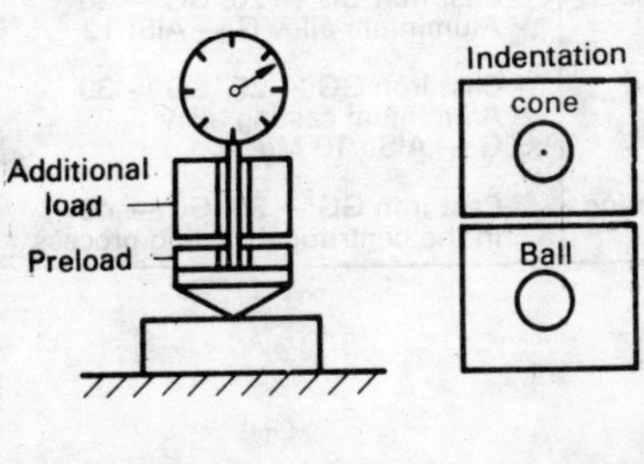	A diamond cone with a cone angle of 120° or a hardened steel ball of 1/16″ diameter is indented first with the use of a preload of 98 N (into the surface of the workpiece to be tested). Now the load is increased by an additional force of 1373 N (for cone) or 883 N (for ball). The resulting indentation depth e after increasing to the additional force is measured. Larger the indentation depth, smaller is the Rockwell hardness. $\text{Rockwell hardness (cone)} = 100 - \frac{\text{resulting indentation depth}}{0.022}$ $HRC = 100 - \frac{e}{0{,}002}$ $\text{Rockwell hardness (ball)} = 130 - \frac{\text{resulting indentation depth}}{0.002}$ $HRB = 130 - \frac{e}{0{,}002}$

Materials of Important Automobile Parts

Parts	Material, abbreviation
Axles	Heat treatable steel C 45, Ck 60, 34 CrMo 4, 37 MnSi 5
Axle webs	Heat treatable steel 34 CrMo 4, 42 MnV 7
Axle web bolts	Utility steel 15 Cr 3, 15 CrNi 6, 18 CrNi 8
Trailer couplings	Cast steel GS — 38, GS — 45, GS — 60
Exhaust valves	Valve steel X 45 CrSi 9, X 45 CrNiW 18 9
Exhaust manifolds	Cast iron GG — 25, GG — 30 Austenitic cast iron GGL — NiCr 30 3
Brake jaws	Steel sheet St 60, St 70; Malleable iron GTS — 35; Aluminium casting alloy G — AlSi 10 Mg
Brake drums	Cast iron GG — 25, GG — 30; Malleable iron GTW — 40 Cast steel GS — 45, GS — 60 Compound casting aluminium cast iron
Inlet valves	Valve steel 37 MnSi 5, X 45 CrSi 4
Springs (leaf, helical, torsion)	Spring steel 55 Si 7, 65 Si 7, 67 SiCr 5, 50 CrV 4
Spring bolts	Utility steel C 15, 15 Cr 3
Drive shafts	Heat treatable steel 42 MnV 7, 30 CrMoV 9
Gear shafts, Gear wheels	Utility steel 15 Cr 3, 15 CrNi 6, 16 MnCr 5, 18 Cr Ni 8 Nitriding steel 34 CrAl 6, 34 CrAlNi 7
Rear axle shafts	Heat treatable steel 30 CrMoV 9, 36 CrNiMo 4
Carriage	Deep drawing sheet steel USt 12, RRSt 14
Pistons	Piston alloys page S. 237
Piston rods	Utility steel C 15, 15 Cr 3, 15 CrNi 6 Nitriding steel 34 CrAl 6, 34 CrAlMo 5
Piston rings	Cast iron GG — 30, GG — 40 in the centrifugal casting process
Fuel tanks	Steel sheet, lead coated
Fuel pump body	Fine zinc pressure casting alloy GD — ZnAl 4
Radiators	Brass CuZn 37, Copper
Crank shafts	Heat treatable steel 34 CrMo 4, 42 CrMo 4 Nitriding steel 34 CrAlMo 5 Cast iron GG — 40 Spheroidised grey iron GGG — 60, GGG — 70
Crank shaft bearings	Three-material bearing, steel-bronze-white metal Two material bearing, steel-bronze or steel-white metal Single material bearing, cast iron, zinc-lead bronze G — CuPb 5 Sn Special aluminium alloys
Longitudinal and transverse links	Heat treatable steel Ck 60, 34 CrMo 4, 37 MnSi 5
Steering rods	Heat-treatable steel 42 CrMo 4, 42 MnV 7
Camshafts	Utility steel 15 Cr 3, 15 CrNi 6 Cast iron GG — 30, GG — 40, Spheroidized grey iron GGG — 50
Connecting rod bushes	Cast iron-zinc-lead bronze G — CuPb 5 Sn, G — CuPb .0 Sn gun metal G — CuSn 7 ZnPb
Connecting rod bearings	See Crankshaft bearings
Connecting rods	Heat treatable steel 34 CrMo 4, 37 MnSi 5
Wheels	Steel sheet St 60, St 70 Cast steel GS — 45, GS — 60
Frames	Structural steel St 50, St 60
Valve springs	Spring steel 67 SiCr 5, 50 CrV 4
Valve guides	Cast iron GG — 30, GG — 40 Cast iron-tin bronze G — CuSn 14 gun metal G — CuSn 7 ZnPb
Valve seating rings	Valve steel X 210 Cr 12 Austenitic cast iron GGL — NiCr 30 3
Carburettor body	Fine zinc pressure casting alloy GD — ZnAl 4, GD —ZnAl 4 Cu 1
Rolling bearing	Rolling bearing steel 100 Cr 6
Cylinder	Cast iron GG — 25, GG — 30 Aluminium casting alloy hard chrome plated Compound casting aluminium cast iron
Cylinder block	Cast iron GG — 20, GG — 25 Aluminium alloy G — AlSi 12
Cylinder head	Cast iron GG — 25, GG — 30 Aluminium casting alloy G — AlSo 10 Mg
Cylinder running bushes	Cast iron GG — 30, GG — 40 in the centrifugal casting process

Square, hexagonal and round steel

Size s, d mm	Weight in kg/m (square, s)	Weight in kg/m (hexagonal, s)	Weight in kg/m (round, d)	Size s,d mm	Weight in kg/m (square, s)	Weight in kg/m (hexagonal, s)	Weight in kg/m (round, d)
10	0,785	0,680	0,617	**20**	3,140	2,719	2,466
11	0,950	0,823	0,746	21	3,462	2,998	2,719
12	1,130	0,979	0,888	22	3,799	3,290	2,984
13	1,327	1,149	1,042	23	4,153	3,596	3,261
14	1,539	1,332	1,208	24	4,522	3,916	3,551
15	1,766	1,530	1,387	**25**	4,906	4,249	3,853
16	2,010	1,740	1,578	26	5,307	4,596	4,168
17	2,269	1,965	1,782	27	5,723	4,956	4,495
18	2,543	2,203	1,998	28	6,154	5,330	4,834
19	2,834	2,454	2,226	29	6,602	5,717	5,185

Plate steel, weight in kg per 1 m length

Width mm	Thickness in mm 3	4	5	6	8	10	12	15	20	25	30	40
20	0,471	0,628	0,785	0,942	1,256	1,570	1,884	2,355	3,140	3,925	4,710	6,280
22	0,518	0,691	0,864	1,036	1,382	1,727	2,072	2,591	3,454	4,318	5,181	6,908
25	0,589	0,785	0,981	1,178	1,570	1,963	2,355	2,944	3,925	4,905	5,888	7,850
28	0,659	0,879	1,099	1,319	1,758	2,198	2,638	3,297	4,396	5,495	6,594	8,792
30	0,705	0,942	1,177	1,413	1,884	2,355	2,826	3,532	4,710	5,888	7,065	9,420
32	0,745	1,005	1,256	1,507	2,010	2,512	3,014	3,768	5,024	6,280	7,536	10,05
35	0,824	1,009	1,374	1,649	2,198	2,748	3,297	4,121	5,495	6,869	8,243	10,99
38	0,895	1,193	1,492	1,790	2,386	2,983	3,580	4,474	5,966	7,458	8,949	11,93
40	0,942	1,256	1,570	1,884	2,512	3,140	3,768	4,710	6,280	7,850	9,420	12,56
45	1,060	1,413	1,766	2,120	2,826	3,533	4,239	5,299	7,065	8,831	10,60	14,13
50	1,177	1,570	1,962	2,355	3,140	3,925	4,710	5,887	7,850	9,813	11,78	15,70
55	1,295	1,727	2,159	2,591	3,454	4,318	5,181	6,476	8,635	10,79	12,95	17,27

Equal angle steel DIN 1028

Profile

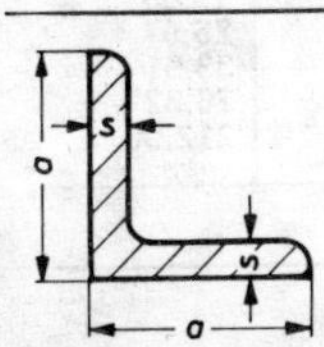

Abbreviation	mm a	mm s	Cross sectional area cm²	Weight kg/m	Abbreviation	mm a	mm s	Cross sectional area cm²	Weight kg/m
L 20 × 3	20	3	1,12	0,88	L 35 × 6	35	6	3,87	3,04
25 × 3	25	3	1,42	1,12	40 × 4	40	4	3,08	2,42
25 × 4	25	4	1,85	1,45	40 × 5	40	5	3,79	2,97
25 × 5	25	5	2,26	1,77	40 × 6	40	6	4,48	3,52
30 × 3	30	3	1,74	1,36	45 × 5	45	5	4,30	3,38
30 × 4	30	4	2,27	1,78	50 × 6	50	6	5,69	4,47
30 × 5	30	5	2,78	2,18	50 × 7	50	7	6,56	5,15
35 × 4	35	4	2,67	2,10	60 × 8	60	8	9,03	7,09
35 × 5	35	5	3,28	2,57	60 × 10	60	10	11,1	8,69

Channel steel DIN 1026

Profile

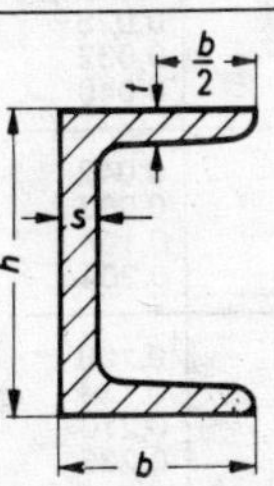

Abbreviation	Dimensions in mm b	h	s	t	Cross sectional area cm²	Weight kg/m
⊏ 40 × 20	20	40	5	5,5	3,66	2,87
40	35	40	5	7	6,21	4,87
50 × 25	25	50	5	6	4,92	3,86
50	38	50	5	7	7,12	5,59
60	30	60	6	6	6,46	5,07
80	45	80	6	8	11,0	8,64
100	50	100	6	8,5	13,5	10,6
120	55	120	7	9	17,0	13,4
140	60	140	7	10	20,4	16,0
160	65	160	7,5	10,5	24,0	18,8
180	70	180	8	11	28,0	22,0
200	75	200	8,5	11,5	32,2	25,3
220	80	220	9	12,5	37,4	29,4

Sheets, wires, tubes

Sheets

Steel sheet DIN 1541, 1542

Thickness in mm	Weight in kg/m²	Supply size width×length in mm
0,18	1,42	530 × 760 500 × 1000 600 × 1200 700 × 1400 800 × 1600
0,2	1,57	
0,22	1,78	
0,24	1,93	
0,28	2,21	
0,32	2,4	
0,38	3	
0,44	3,5	
0,5	4	
0,56	4,5	
0,63	5	530 × 760 800 × 1600 1000 × 2000
0,75	6	
0,88	7	
1,0	8	
1,13	9	
1,25	10	
1,38	11	
1,5	12	800 × 1600 1000 × 2000 1250 × 2500
1,75	14	
2,0	16	

White sheet

Thickness in mm	Weight in kg/m²
0,15	1,27
0,19	1,54
0,22	1,75
0,24	1,91
0,27	2,15
0,28	2,18
0,31	2,36
0,32	2,39
0,36	2,82
0,37	2,96
0,41	3,38
0,46	3,6
0,52	4,2
0,58	4,6
0,64	5,0
0,70	5,5
0,80	6,3
0,90	6,9
1,0	7,9

Copper, brass, aluminium sheet DIN 1751, 1752, 1753

Thickness in mm	Cu Weight in kg/m²	Brass Weight in kg/m²	Al Weight in kg/m²
0,10	0,89	0,85	—
0,15	1,34	1,28	—
0,20	1,78	1,70	0,54
0,25	2,23	2,13	0,68
0,30	2,67	2,55	0,81
0,35	3,12	2,98	0,95
0,40	3,56	3,40	1,08
0,45	4,01	3,83	1,22
0,50	4,45	4,25	1,36
0,55	4,90	4,68	—
0,60	5,34	5,10	1,64
0,65	5,79	5,53	—
0,70	6,23	5,95	1,91
0,75	6,68	6,38	—
0,80	7,12	6,80	2,18
0,85	7,57	7,23	—
0,90	8,01	7,65	2,46
1,0	8,90	8,50	2,73
1,1	9,79	9,35	3,00
1,2	10,70	10,20	3,28

Zinc sheet DIN 9721

Thickness in mm	Weight in kg/m²
0,1	0,7
0,143	1,0
0,15	1,08
0,186	1,3
0,2	1,44
0,228	1,6
0,25	1,75
0,3	2,10
0,35	2,45
0,4	2,80
0,45	3,15
0,5	3,50
0,58	4,06
0,6	4,31
0,66	4,62
0,7	5,03
0,8	5,74
1,0	7,18
1,2	8,62
1,5	10,80

Wires

Steel wire DIN 177				**Copper wire** DIN 1766		**Brass wire** DIN 1757		**Aluminium wire** DIN 1790	
Diameter in mm	Weight in kg/1000 m	Diameter in mm	Weight in kg/1000 m	Diameter in mm	Weight in kg/1000 m	Diameter in mm	Weight in kg/1000 m	Diameter in mm	Weight in kg/1000 m
0,10	0,062	0,80	3,95	0,10	0,070	0,20	0,267	0,5	0,53
0,12	0,089	0,90	4,99	0,15	0,157	0,25	0,417	1,0	2,12
0,14	0,121	1,00	6,17	0,20	0,280	0,30	0,601	1,5	4,77
0,16	0,158	1,20	8,88	0,25	0,437	0,40	1,068	2,0	8,48
0,18	0,200	1,40	12,08	0,30	0,629	0,50	1,670	2,5	13,23
0,20	0,247	1,60	15,78	0,40	1,118	0,60	2,403	3,0	19,06
0,40	0,986	1,80	19,98	0,50	1,747	0,70	3,271	3,5	25,97
0,50	1,540	2,00	24,70	0,80	4,474	0,80	4,273	4,0	33,91
0,60	2,220	2,20	29,80	1,00	6,990	1,00	6,676	6,0	76,32
0,70	3,020	2,50	38,50	1,50	15,727	2,00	26,704	10,0	212,00

Tubes

Precision steel tubes, seamless drawn DIN 2385

Wall thickness in mm	Outside diameter in mm	Inside diameter in mm	Weight in kg/m
0,5	8	7	0,092
	10	9	0,117
0,75	12	10,5	0,208
	15	13,5	0,264
1,0	10	8	0,222
	14	12	0,321
1,5	8	5	0,240
	10	7	0,314
2,0	20	16	0,888
	25	21	1,134
2,5	9	4	0,401
	30	25	1,695

Copper, brass, aluminium-alloy tubes, seamless drawn DIN 1754, 1755, 1795

Wall thickness in mm	Outside diameter in mm	Inside diameter in mm	Cu Weight in kg/m	Brass Weight in kg/m	Al-alloy Weight in kg/m
0,5	6	5	0,08	0,07	0,023
	7	6	0,09	0,09	0,028
	8	7	0,10	0,10	0,032
	10	9	0,13	0,13	0,040
1,0	6	4	0,14	0,13	0,042
	12	10	0,31	0,29	0,093
	18	16	0,48	0,45	0,144
	25	23	0,67	0,64	0,204
1,5	10	7	0,36	0,34	0,108
	12	9	0,44	0,42	0,134
	18	15	0,70	0,66	0,210
	20	17	0,78	0,74	0,235

Drawing parallels

Set square placed on line g

Second set square placed on the first

First set square moved, parallel drawn through A

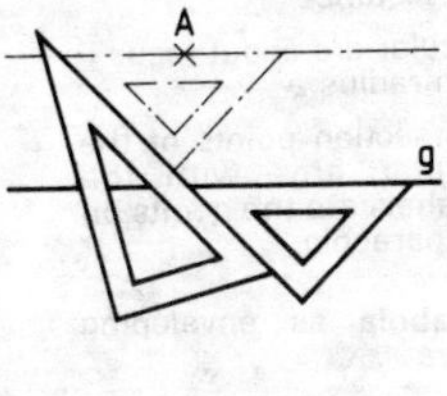

Dividing lines

Ray drawn from A at any angle

Any equal number of parts marked on the line, e.g. 5

End point of part 5 joined to B

Parallels drawn

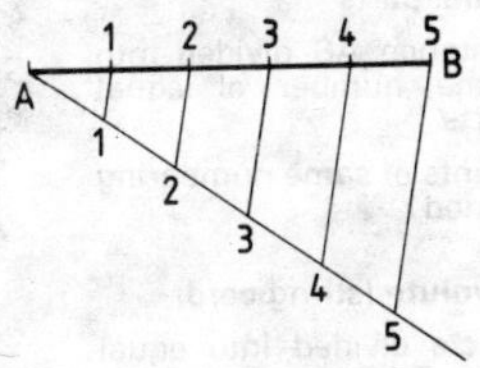

Erecting perpendicular bisector

Circular arcs about A and B with radius r

Connecting line CD is the perpendicular bisector

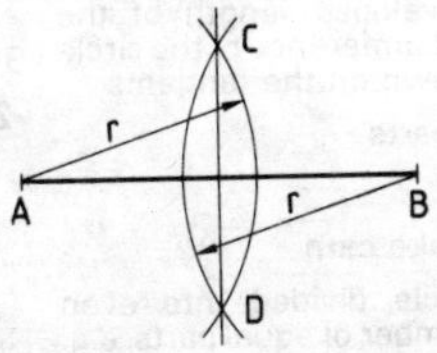

Dropping perpendicular

Circular arc about A with any radius r

Circular arcs about B and C with any radius r_1

Connecting line AD is the perpendicular

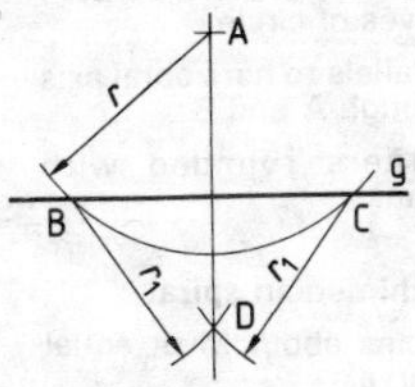

Bisecting angle

Circular arc about S with and radius r

Circular arcs about A and B with any radius r_1

Connecting line SC is the angle bisector

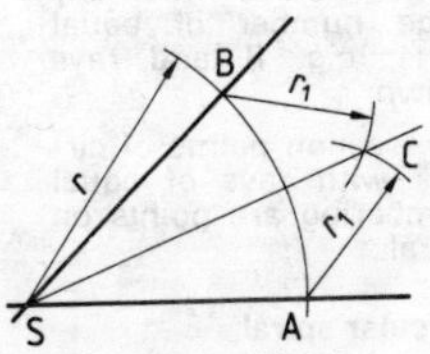

Trisecting right angle

Circular arc about S with any radius r

Circular arcs about A and B with same radius r

Connecting lines SC and SD are trisectors

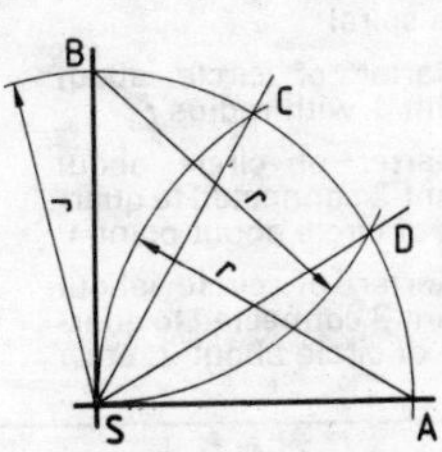

30° angle with circles

Circular arc about S with any radius r

Circular arc about A with radius r

Circular arc about B with radius r

Connecting line SD drawn

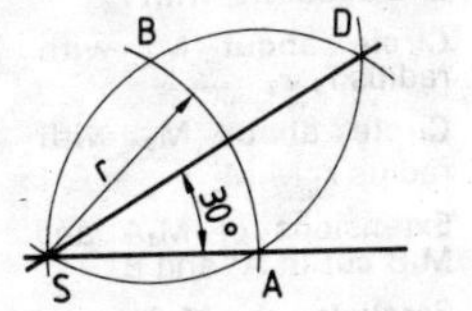

60° angle with circles

Circular arc about S with any radius r

Circular arc about A with radius r

Connecting line SB drawn

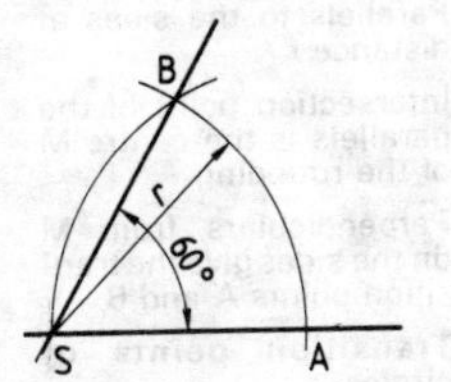

120° angle with circles

Circular arc about S with any radius r

Circular arc about A with radius r

Circular arc about B with radius r

Connecting line SC drawn

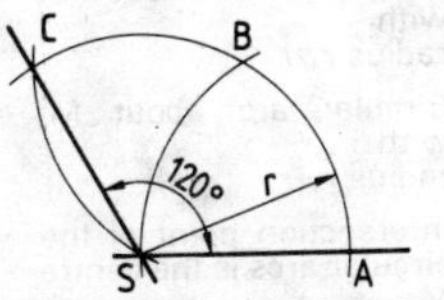

Determining Centre of circle

Any chords AB and CD drawn

Perpendicular bisectors erected on the chords

Intersection point of the perpendicular bisectors is the centre M of the circle

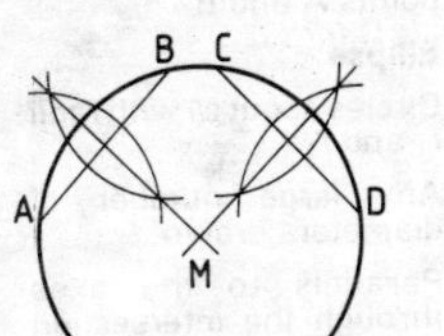

Tangents to circle

Connecting line AM drawn

Line AM bisected

Circle about M' with radius r=AM/2

Connecting lines AB and AC are tangents

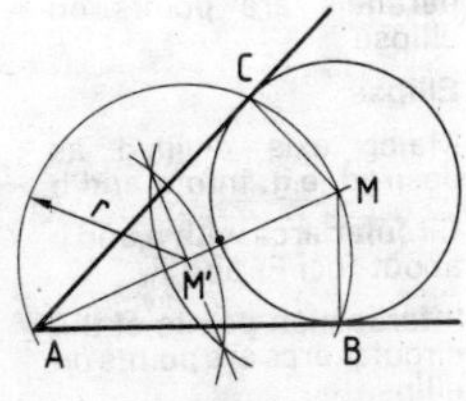

Tangents to two circles

Circle about M_1 with radius $r_1 - r_2$

Tangents M_2A and M_2B drawn

Extensions of M_1A and M_1B give contact points A' and B'

Parallels to M_2A and M_2B through A' and B' drawn

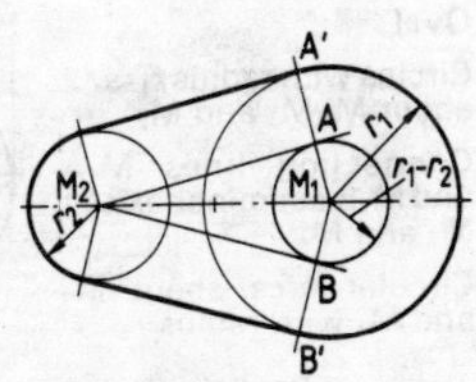

Basic geometric constructions

Cams

Circle about M_1 with r_1 and about M_2 with r_2

Circle about M_1 with radius r_1–r_2

Circle about M_3 with radius r=M_1M_2

Extensions of M_1A and M_1B cut in A^I and B^I

Parallels to M_1A^I and M_1B^I

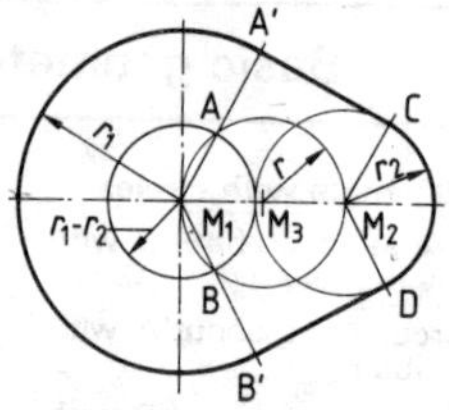

Rounding

Parallels to the sides at distance r

Intersection point of the parallels is the centre M of the rounding

Perpendiculars from M on the sides give the transition points A and B

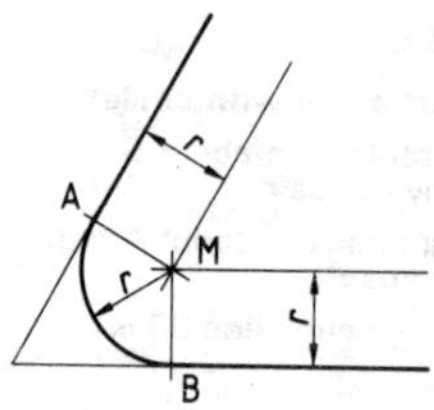

Transition points of circles

Circular arc about M_1 with radius r_1+r

Circular arc about M_2 with radius r_1+r

Intersection point of the circular arcs is the centre M

Connecting lines MM_1 and MM_2 give transition points A and B

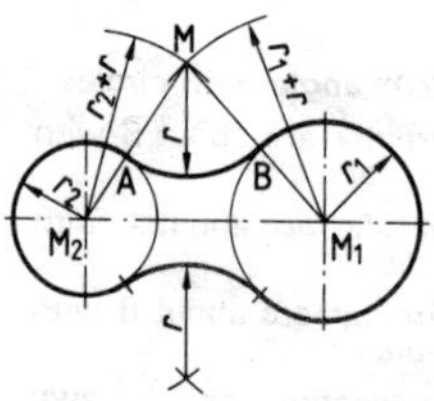

Ellipse

Circles about M with radii r_1 and r_2

Any large number of diameters drawn

Parallels to the axes through the intersection points (diameters with circles) drawn

Intersection points of parallels are points on ellipse

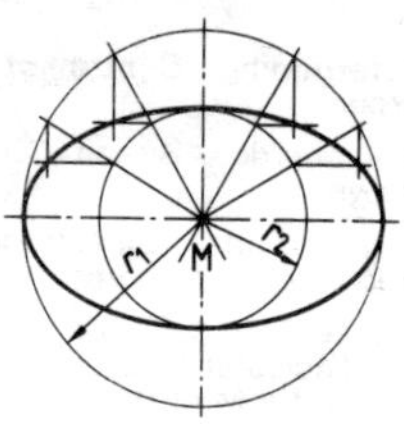

Ellipse

Major axis divided as desired, e.g. into a and b

Circular arcs with a and b about foci F_1 and F_2

Intersection points of the circular arcs are points on ellipse

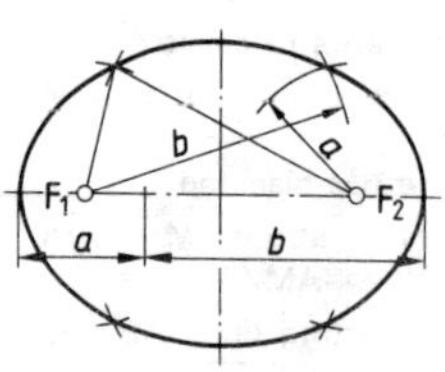

Oval

Circles with radius r_1=a/2 about M, M_1 and M_2

Connecting lines M_1A and M_2B cut minor axis in M_3 and M_4

Circular arcs about M_3 and M_4 with radius r_2

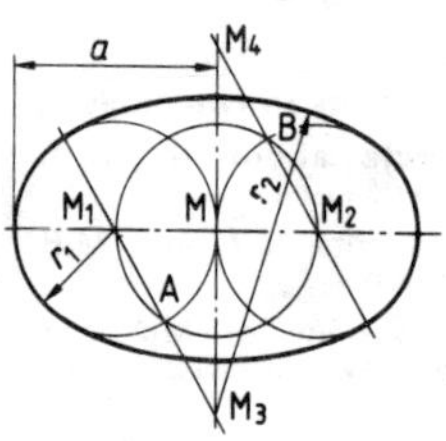

Parabola

Parallels to directrix L at any distance

Circular arc about focus F with radius a

Intersection points of the circular arcs with the parallels are the points on the parabola.

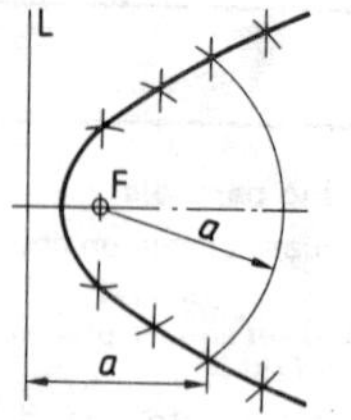

Parabola as enveloping curve

Distance AS divided into equal parts

Distance AB divided into same number of equal parts

Points of same numbering joined

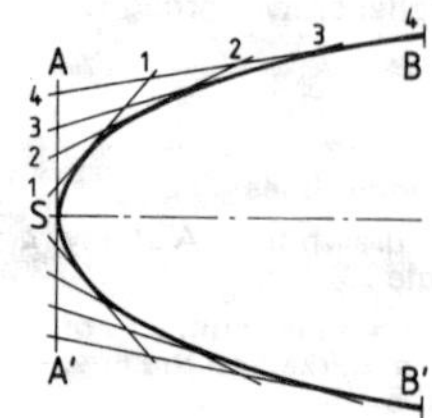

Involute (string cord)

Circle divided into equal parts Tangents to the circle at the division points

Developed length of the circumference of the circle drawn on the tangents

9 parts

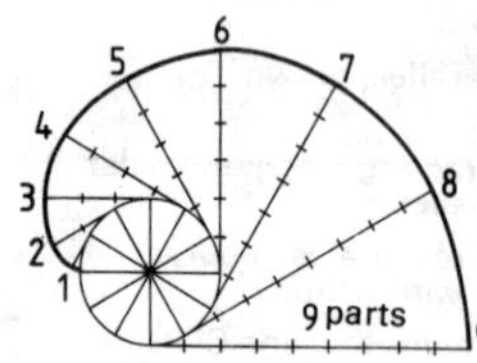

Brake cam

Circle divided into even number of equal parts, e.g. 12

Involutes drawn from both halves of circle

Parallels to horizontal axis through A and B

Corners rounded with radius r

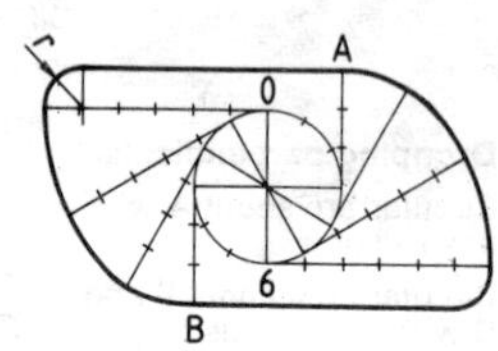

Archimedian spiral

Circles about M at equal intervals

Circle divided into any large number of equal parts, e.g. 8 and rays drawn

Intersection points of circles with rays of equal numbering are points on spiral.

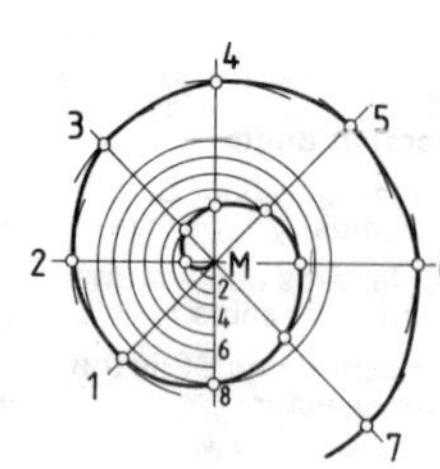

Circular spiral

Quadrants numbered in the direction of rotation of the spiral

Quarter of circle about point 1 with radius r

Quarter of circle about point 2 connected to quarter of circle about point 1

Quarter of circle about point 3 connected to quarter of circle about 2, etc.

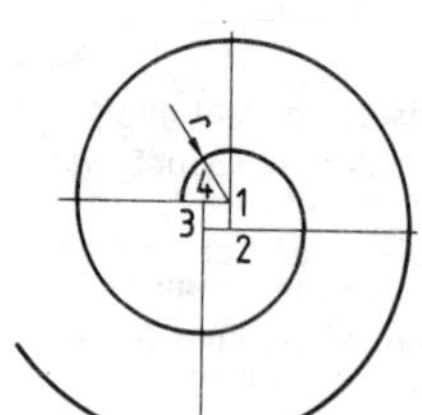

Inscribed circle in a triangle

Desired triangle ABC drawn

Angles of triangle bisected

Intersection point of the angle bisectors is the centre M_i of the inscribed circle

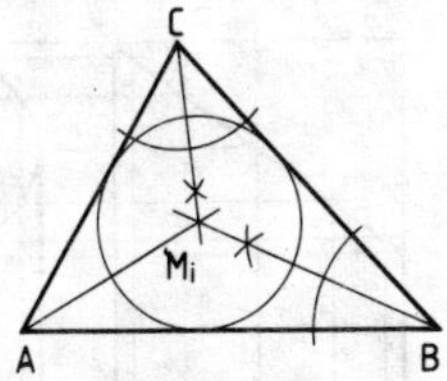

Circumscribed circle and sum of angles

Desired triangle ABC drawn

Perpendicular bisectors erected on the sides of triangle

Intersection point of the perpendicular bisectors is the centre M_u of the circumscribed circle

The sum of the angles in the triangle amounts to 180°

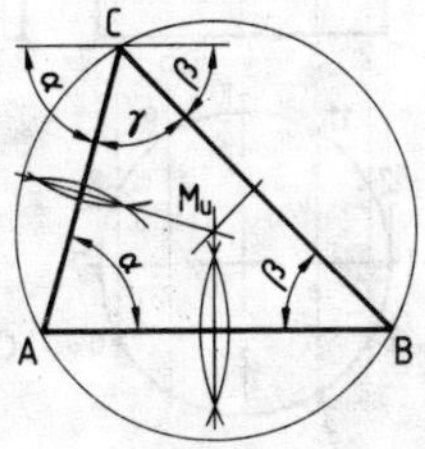

Centre of gravity

Desired triangle ABC drawn

Medians of sides drawn

Intersection point of medians is centre of gravity S

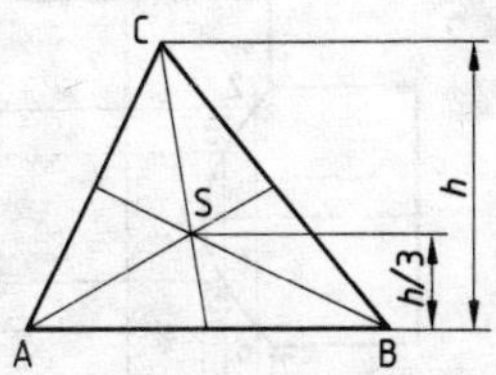

Rectangle

Rectangle ABCD drawn

The diagonals are bisected in M

M is the centre of the circumscribed circle and the centre of gravity

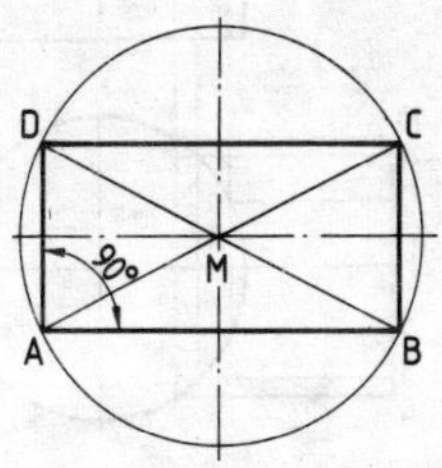

Rhomboid

Rhomboid ABCD drawn

The diagonals are bisected in M

M is the centre of gravity

Opposite angles α, γ and β, δ are equal

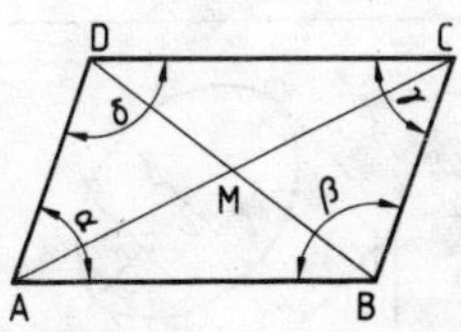

Rhombus

Rhombus ABCD drawn

The diagonals are perpendicular to each other and are bisected in M

M is the centre of the inscribed circle and centre of gravity

Oppsite angles α, γ and β, δ are equal

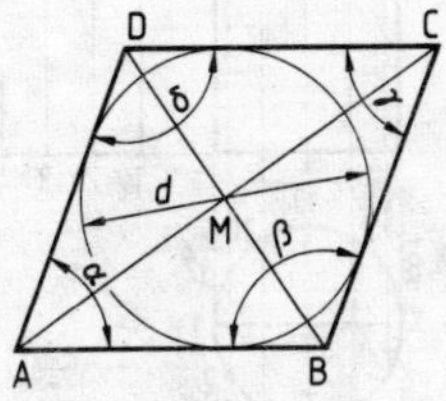

Isosceles trapezium

Parallels to AB and circles about A and B with same radius intersect each other in C and D

Intersection point of the perpendicular bisector is the centre M of the circumscribed circle

Angles $\alpha = \beta$ and $\gamma = \delta$

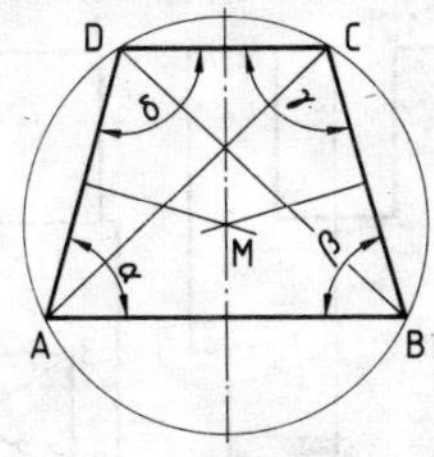

Equilateral triangle

Circles about A and B with radius AB

Angle bisectors, medians of sides, perpendicular bisectors mutually intersect in M

Diameter of inscribed circle $= 1/2$ diameter of circumscribed circle

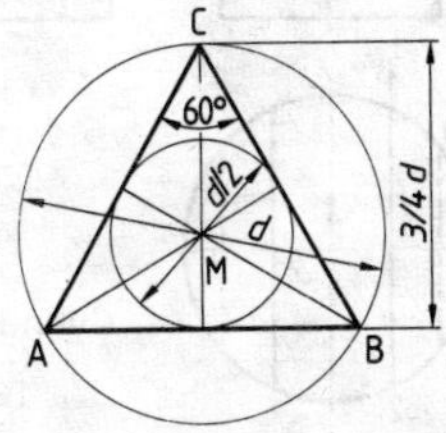

Square, Octagon

Circle about M with radius *r*

Intersection points of the axes with the circumscribed circle connected with one another

For octagon axes drawn at 45°

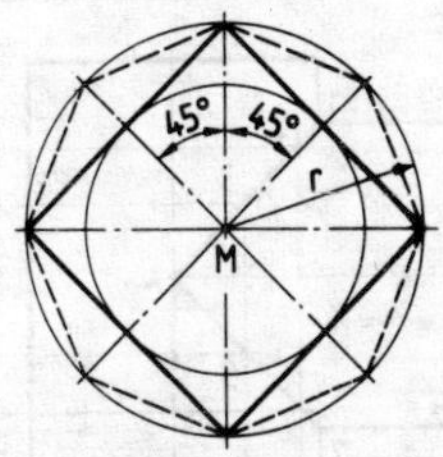

Pentagon, Decagon

Circle about M with radius *r*

Line AM bisected

Circular arc about M_1 with radius M_1B

BC is side of pentagon

Bisecting circular arc BD gives side of decagon

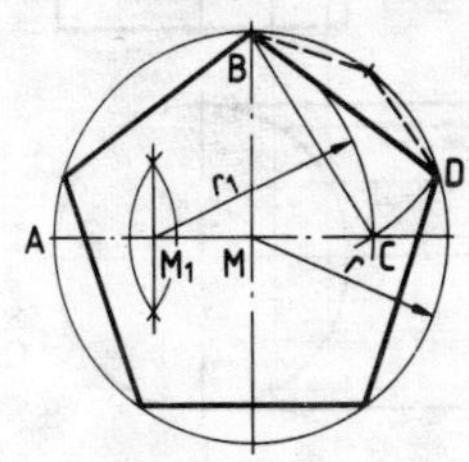

Hexagon, dodecagon

Circle about M with radius *r*

Circular arcs about A and B with radius *r*

For dodecagon circular arcs about C and D with radius *r*

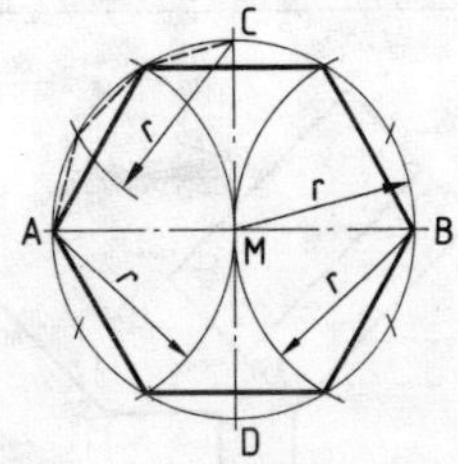

Universal Construction

Circle about M with radius *r*

Line AB divided into as many parts as sides of the polygon, e.g. into 7

Circular arcs about A and B with radius AB

Connecting line of C with every second division point

AD is side of polygon

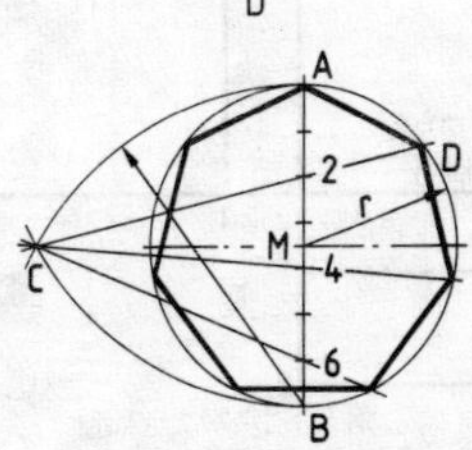

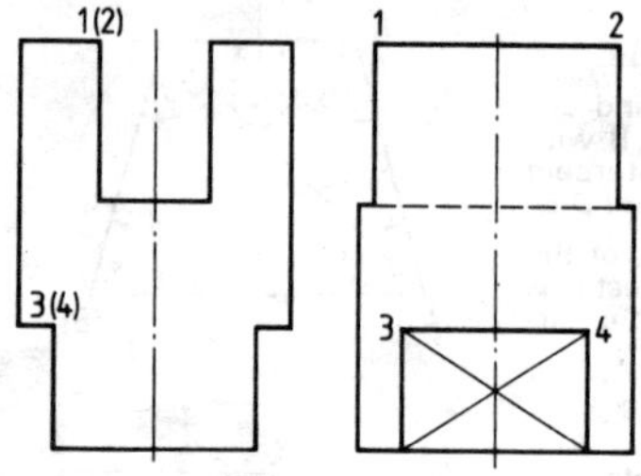

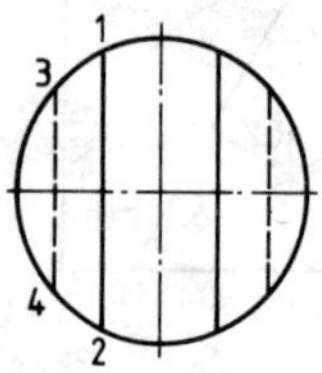

Cylinder section inclined
to the axis

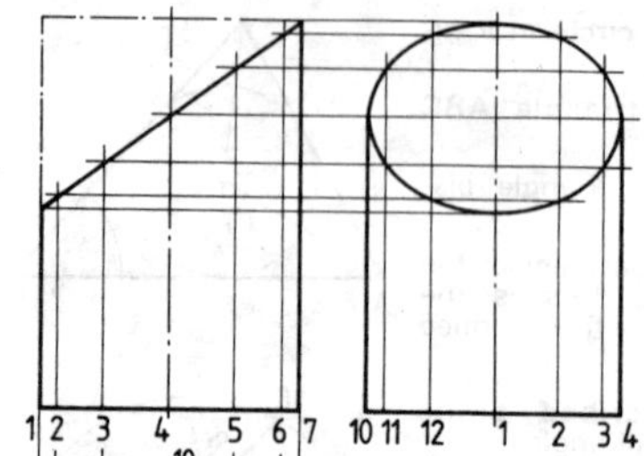

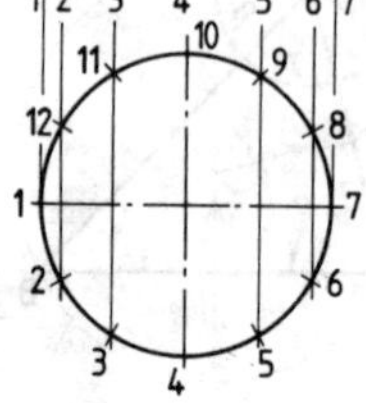

Cylinder section parallel
to the axis

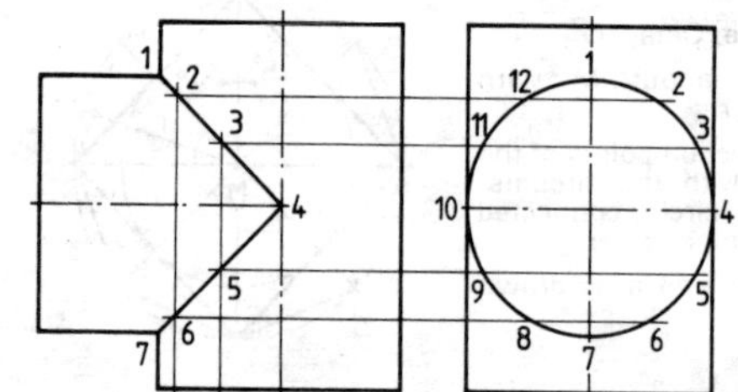

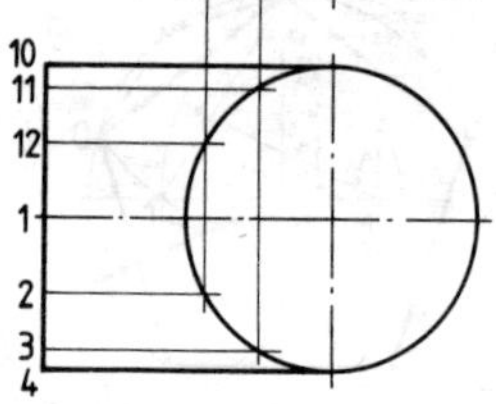

Cylinder penetration

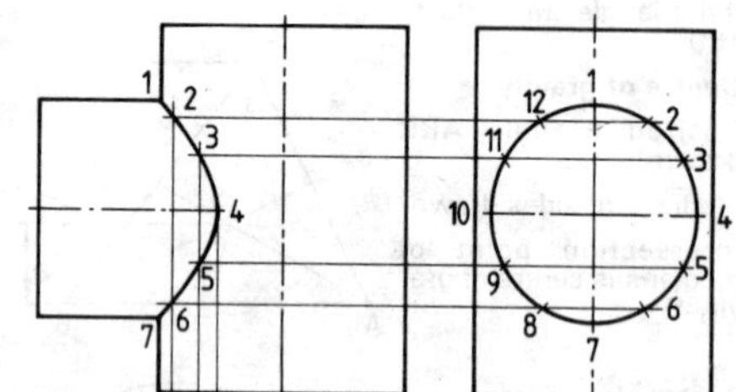

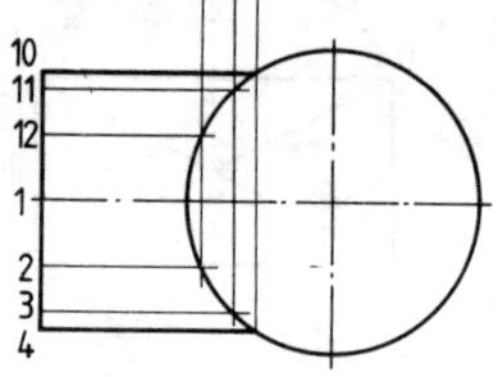

Cylinder penetration

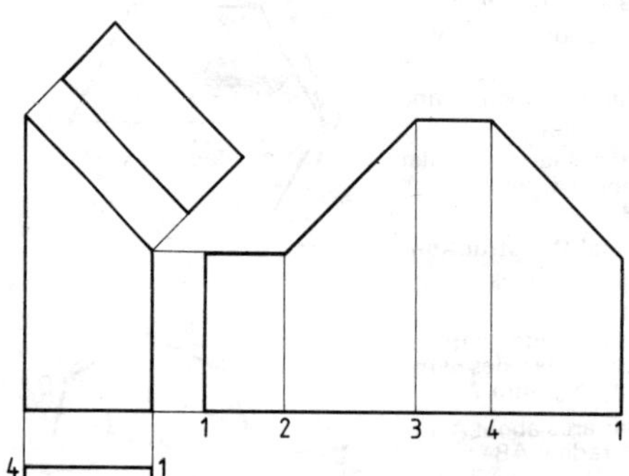

Development

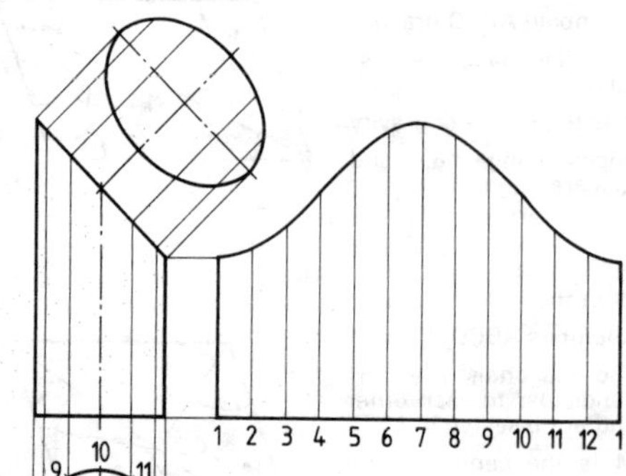

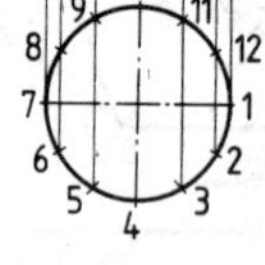

Development

Diagram, Spring characteristics

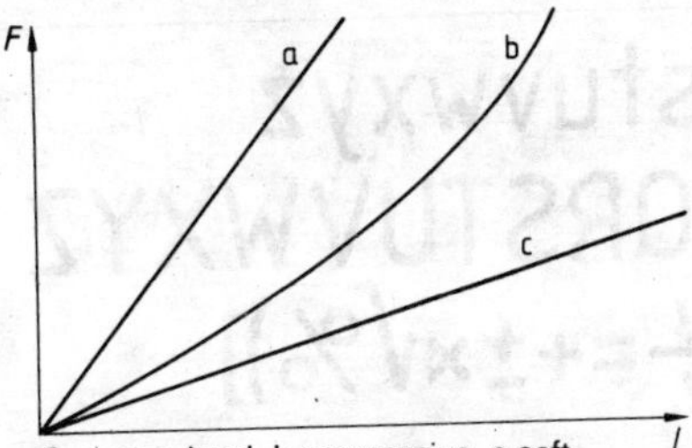

Spring: a hard, b progressive, c soft

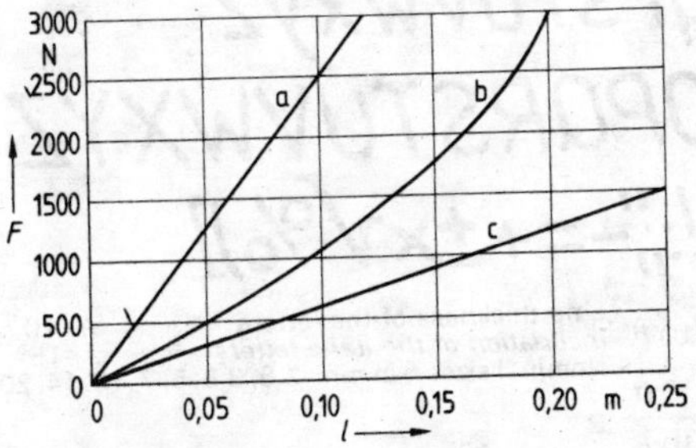

Diagram, Diesel engine characteristics

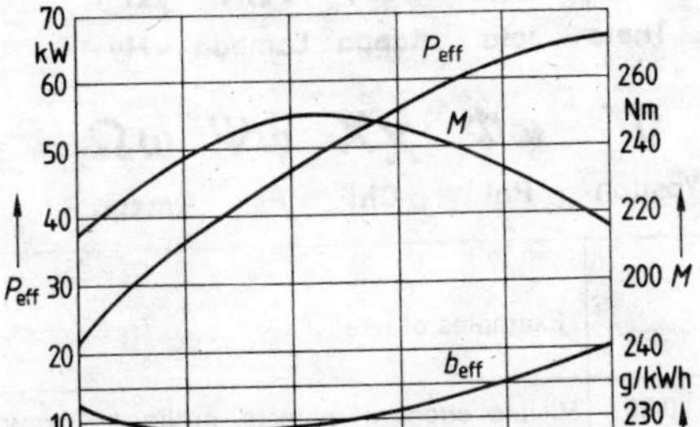

Percentage circle, Petroleum reserves of the world

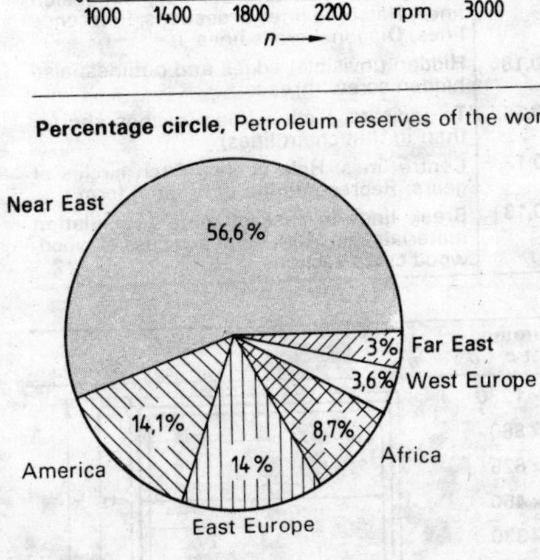

Coordinate system (according to DIN 461)

Figures show in coordinate system the dependence of a quantity upon another quantity, e.g. of spring deflection *l* upon the force *F*. For the plotting of the diagrams, the following rules should be noted.

1. Symbols
They are placed at the left of the arrow head of the vertical axis (abscisca) and below the arrow head of the horizontal axis (ordinate). Arrows may also be parallel to the axes; the symbols or nomenclature are then at the root of the arrow.

2. Set of curves
Identification of the individual curves with reference numbers in italics or with upright reference alphabets (a, b, c)

3. Several dependent variables
If on the same independent variables several dependent variables are to be placed, different types of lines can be used. If only one type of line is used, the designation of the curves with symbols or reference numerals is required.

4. Qualitative representation
Coordinate system without division (survey diagram). However coordinates of prominent points may be specified.

5. Quantitative representation
Numerical scale on each axis; Supplement to the coordinate network usual. The units are placed between the last two numbers. Numerical values and symbols should be readable without rotation of the figure.

6. Line thickness
The line thickness are taken from DIN 15. The ratio is: Network to axes to curves as 1 to 2 to 4

Area representation, Energy distribution in Otto Engine

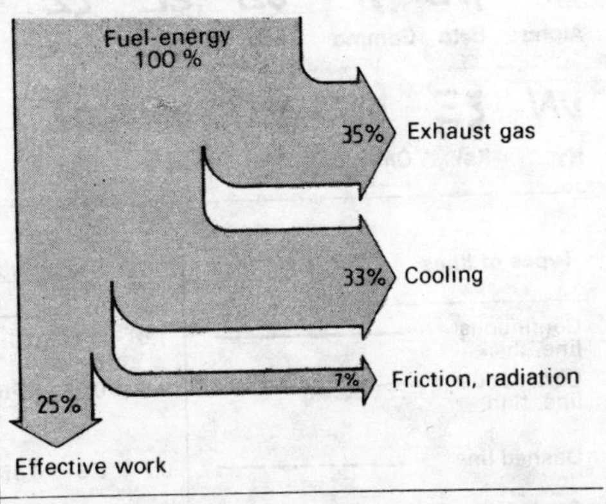

Line representation, Mineral oil consumption per head

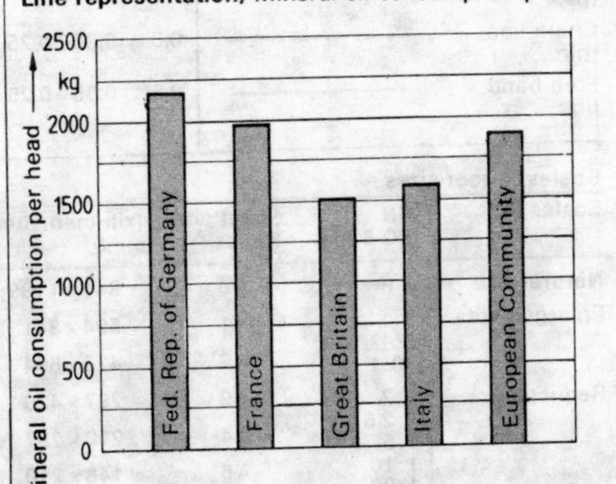

Standard letters. Type B

abcdefghijklmnopqrstuvwxyz
ABCDEFGHIJKLMNOPQRSTUVWXYZ
IVX1234567890[(&?!:;"-=+±×√%)]

abcdefghijklmnopqrstuvwxyz
ABCDEFGHIJKLMNOPQRSTUVWXYZ
IVX123456789[(&?!:;"-=+±×√%)]

Height of the capital letters=nominal size of the letters=*h* =10/10 *h*
Height of the lower case letters (without upper or lower extensions) =7/10 *h*
Minimum distance between the base lines of the letters =16/10 *h*

Line thickness of the letters =1/10 *h*
Inclination of the italic letters =15°
Nominal sizes *h* in mm 2.5, 3.5, 5, 7, 10, 14, 20

Greek letters (inclined letters)

αA	βB	γΓ	δΔ	εE	ζZ	ηH	ϑΘ	ιI	κK	λΛ	μM
Alpha	Beta	Gamma	Delta	Epsilon	Zeta	Eta	Theta	Jota	Kappa	Lambda	My
νN	ξΞ	oO	πΠ	ρP	σΣ	τT	υϒ	φΦ	χX	ψΨ	ωΩ
Ny	Ksi	Omikron	Pi	Rho	Sigma	Tau	Ypsilon	Phi	Chi	Psi	Omega

Types of lines	Line thickness in mm line groups					Examples of use
Continuous line, thick	1,0	0,7	0,5	0,35	0,25	Visible edges of objects, outlines; Screw thread boundaries; Name plate symbols
Continuous line, thin	0,5	0,35	0,25	0,18	0,13	Dimension lines and auxiliary dimension lines; Hatching lines of sections; Reference lines; Diagonal cross lines
Dashed line	0,7	0,5	0,35	0,25	0,18	Hidden (invisible) edges and outlines; also hidden screw threads
Chain line, thick	1,0	0,7	0,5	0,35	0,25	To indicate section lines (dashes shorter than in thin chain lines)
Chain line, thin	0,5	0,35	0,25	0,18	0,13	Centre lines; Hole circles; Pitch circles of gears; Representation of original form
Free hand line	0,5	0,35	0,25	0,18	0,13	Break lines in case of metals, insulation materials, etc.; Break lines in case of wood, wood cross sections

Scales, Sheet sizes

Scales (M)	DIN ISO 5455	Sheet size Series A	Trimmed sheet *a* × *b*	Untrimmed sheet *c* × *d*
Natural size	1:1	A0	841 × 1189	880 × 1230
Enlargements	2:1 5:1 10:1	A1	594 × 841	625 × 880
		A2	420 × 594	450 × 625
Reductions	1:2 1:2,5	A3	297 × 420	330 × 450
	1:5	A4	210 × 297	240 × 330
	1:10 1:20	A5	148 × 210	165 × 240
	1:50	A6	105 × 148	120 × 165

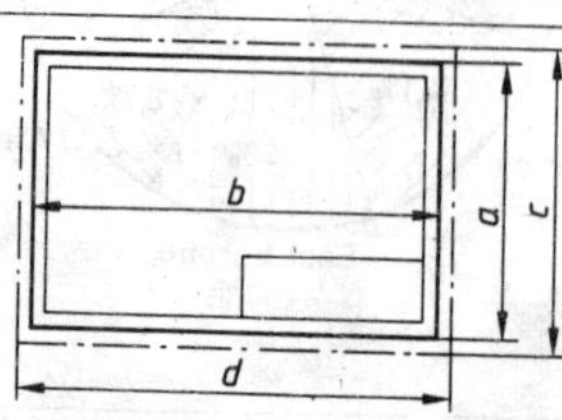

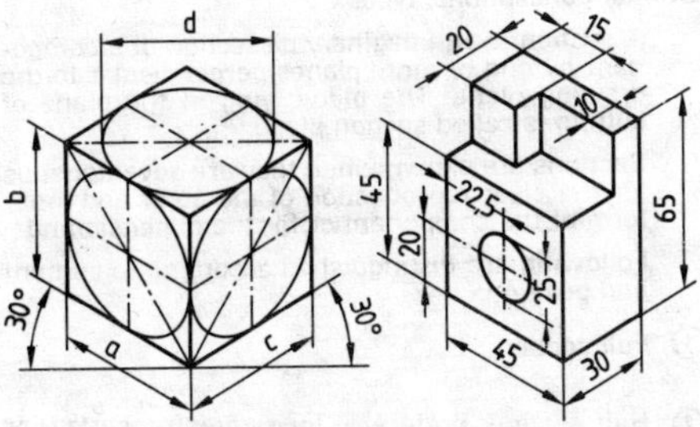

Isometric projection

The isometric projection is used for representations in which details should be shown clearly in all three views.

Representation
Side a at –30° to the horizontal
Side b vertical
Side c at +30° to the horizontal

Ratio of sides a:b:c=1:1:1

Ratio of axes in case of all ellipses=1:1:7

Ratio of edge of side to diameter=1:1.22

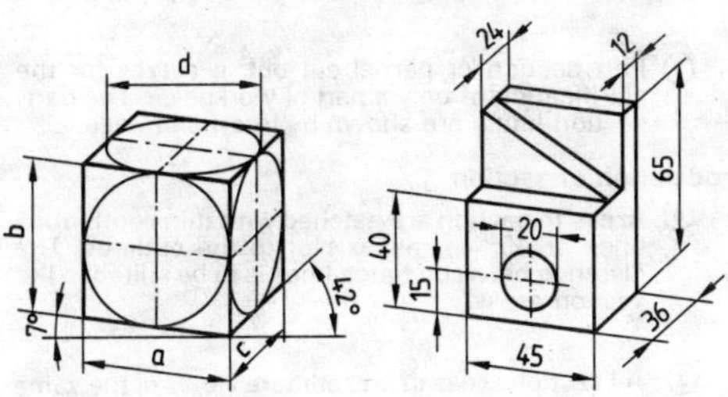

Dimetric projection

The dimetric projection is used for representations in which details should be shown in one view.

Representation
Side a at –7°
Side b vertical
Side c at + 42° to the horizontal

Ratio of sides a:b:c = 1:1:1/2

Ratio of axes
Ellipse 1 = 1:3 (major axis horizontal)
Ellipse 2 = 1:3 (minor axis perpendicular to 7°)
Ellipse 3 = 1:3 (it is simplified to the circle)

Major axis = 1.06 times the edge of side a or b

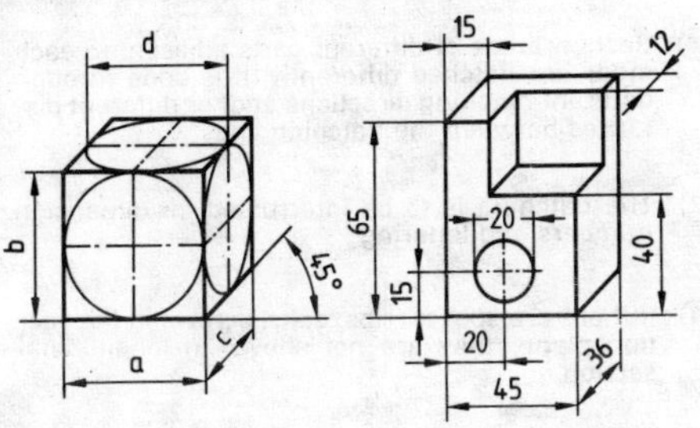

Cavalier perspective

The cavalier perspective belongs to the dimetric representations. It is not standardized and is used when details should be shown in one view.

Representation
Side a horizontal in the plane of the figure
Side b vertical
Side c at 45° to the horizontal

Ratio of sides a:b:c = 1:1:1/2
Ratio of axes
Ellipse 1 = 1:3.2 (major axis 7° to the horizontal)
Ellipse 2 = 1:3.2 (major axis 7° to the vertical)
Circles parallel to the plane of the figure remain unchanged
Major axis=1.07 times edge of side a or b

Representations in drawings

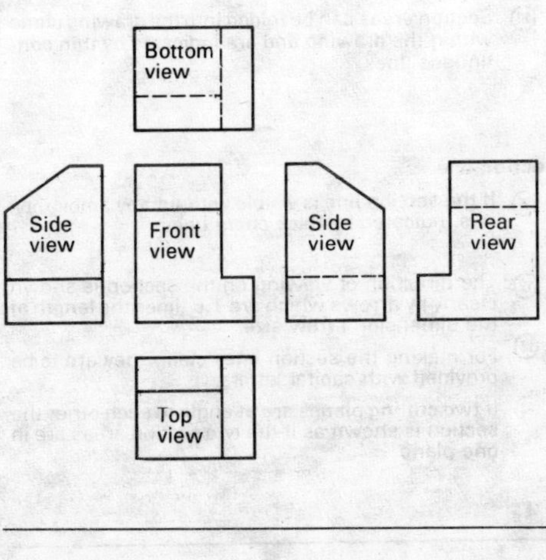

Views

In overall assembly drawings and group assembly drawings, the components are generally represented in the position of their use, i.e. components used in the standing position are drawn vertical and components used in the lying position are drawn horizontal.

In part drawings, the components are to be shown preferably in the manufacturing position.

The individual views must be arranged for all components as in the figure shown beside (ISO Method E). There is also the ISO Method A with opposite arrangement:
Top view above the Front view
Bottom view below the Front view
Side view from left on the left side
Side view from right on the right side.

The view which shows maximum possible detail in form and dimension is to be chosen as the Front view. Only such views that are necessary for the clear definition of the component are drawn.

The representation of hidden edges (dashed lines) is to be avoided to the extent possible.

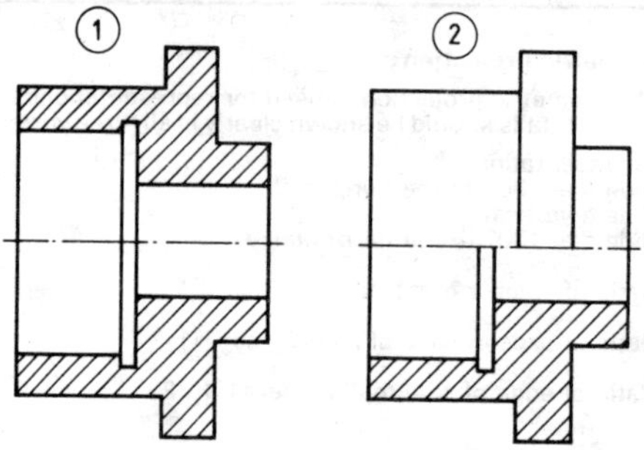

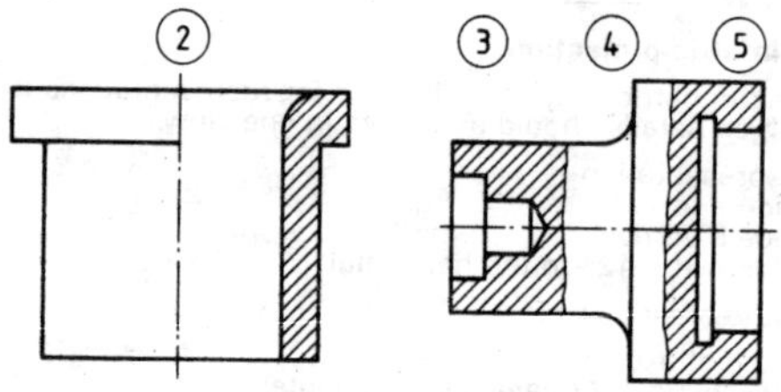

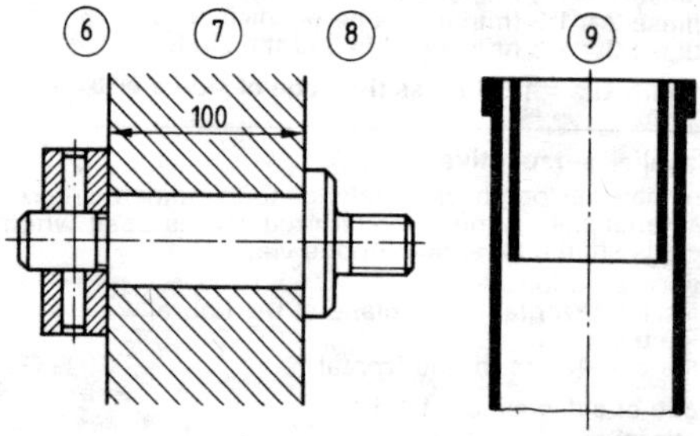

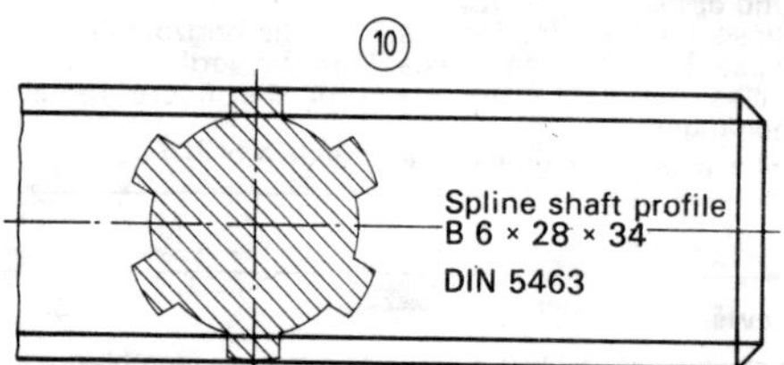

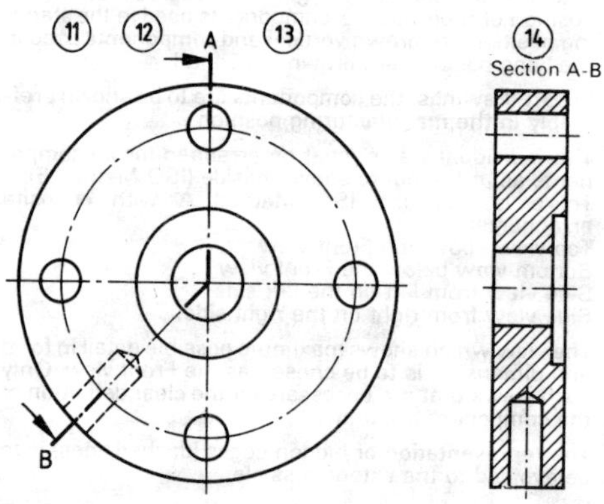

Sections, conceptions, types

A section is an imaginary dissection of a component by one or more planes perpendicular to the drawing plane. The plane lying in the plane of cutting is called section plane.

Sections are shown when they are advantageous for the clear reproduction of the outer and inner form of the component or for the dimensioning.

Following are distinguished according to content and position:

① Full section

② Half section, preferably for symmetric parts. For horizontal centre line, the half section is arranged at the bottom, for vertical centre line at the right.

③ Part section or partial cut out; it serves for the clarification of only a part of workpeice. The part-section limits are shown by free-hand lines.

Indication of section

④ Areas in section are hatched with thin continuous lines at 45° to the axis or to the outlines The distance between hatch lines is to be suited to the section areas.

⑤ All section areas in one or more views of the same part are hatched in the same manner.

⑥ Section areas of different parts adjacent to each other are hatched differently. It is done through different hatching directions and/or different distances between the hatching lines.

⑦ The hatching is to be interrupted for dimension numbers and lettering.

⑧ If there are spokes, ribs, bolts, screws in the section plane, they are not shown in longitudinal section.

⑨ Small section lines can be completely filled. If filled section areas overlap each other, they are shown with gaps.

⑩ Section areas can be folded into the drawing plane within the drawing and are indicated by thin continuous lines.

Section line

⑪ If the section line is visible without any ambiguity, it is indicated by thick chain lines.

⑫ The direction of viewing on the section is shown clearly by arrows which are 1.5 times the length of the dimension arrow size.

⑬ For making the section lines clear, they are to be provided with capital letters.

⑭ If two cutting planes are at angle to each other the section is shown as if the two section areas are in one plane.

(15) (16)

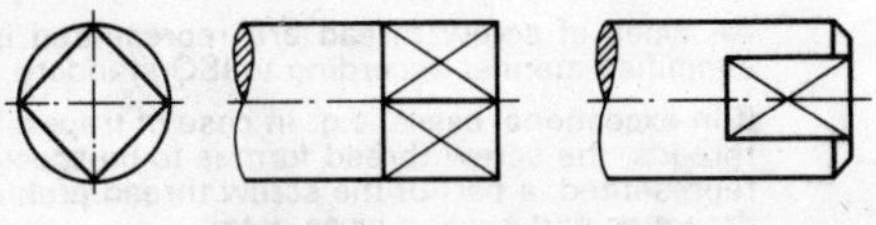

(17) (18)

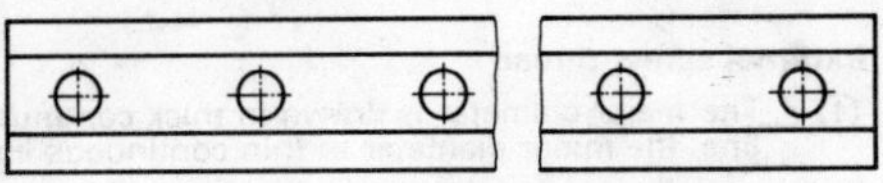

(19) (20)

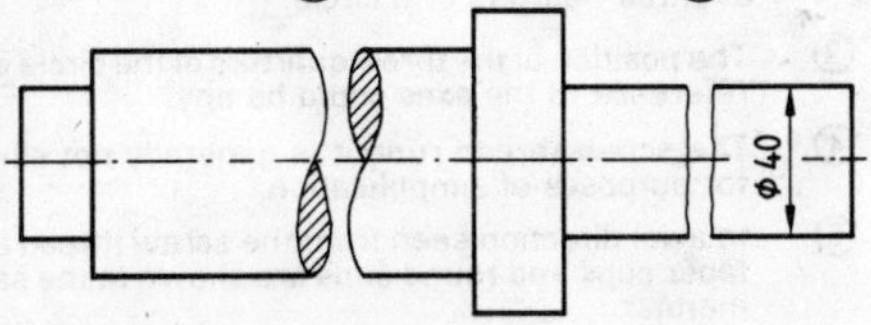

(21) (22)

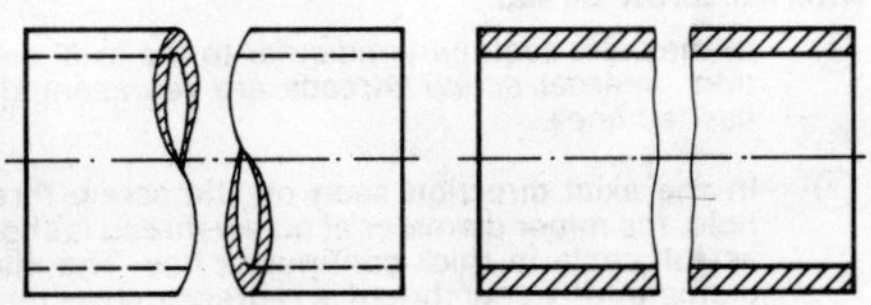

(23) (24) (23)

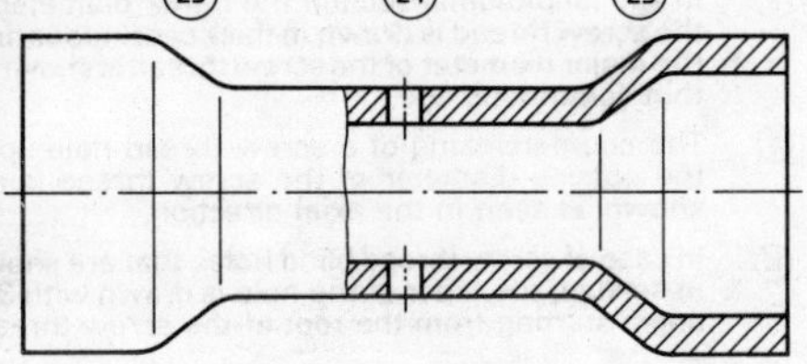

(25) (26)

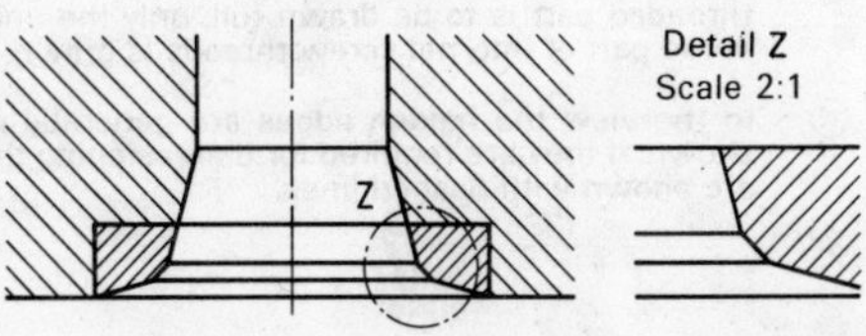

(27) (28) (29)

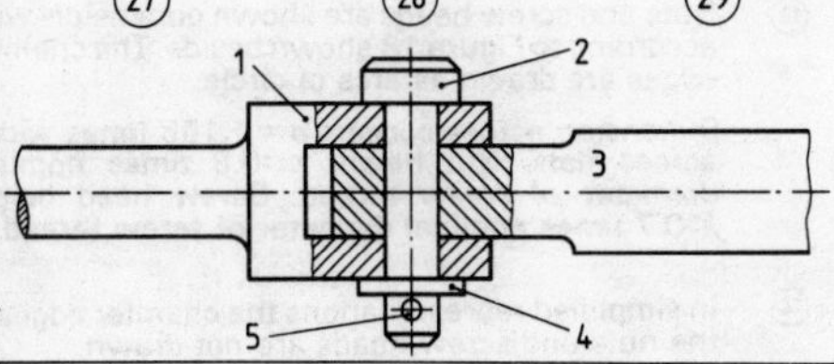

Diagonal intersection

(15) The diagonal intersection (thin continuous lines) indicates plane surfaces.

(16) The diagonal intersection must be used when side view or top view is not present. It is also permissible even in the presence of two views.

Break lines

(17) Components can be shown broken, e.g. for saving drawing area.

(18) Break lines are to be drawn free hand, however not excessively irregular.

(19) The break of round parts is shown by break loops in thin continuous lines.

(20) When the round part is to be clearly known from the dimensioning or a side view, the break line may also be drawn as free-hand line.

(21) The break of hollow round parts is shown by two loops in thin continuous lines.

(22) The break of hollow round parts in section is shown by free-hand lines.

Transitions, Penetrations

(23) Rounded-off transitions of penetrations (rounding-off of edges also) can be shown by thin continuous lines that end before the edges of the part if the figure is thereby clear.

(24) In case of penetrations of cylinders with different diameters, flat penetration curves can be avoided.

Details

(25) If details should be shown magnified, a chain-line circle is drawn around the extracted place which is to be identified with the capital letters Z, X or W.

(26) The magnified scale is to be specified and the magnified representation is to be arranged near the chain-line circle.

Serial numbers for part list

(27) For the designation of the individual parts in assembly drawings, numbers are used.

(28) The numbers are written approximately double the size of dimension numbers and placed by the side of or in the figure.

(29) Reference lines are as far as possible drawn straight. The end lying in the part is provided with a point if misunderstandings are likely.

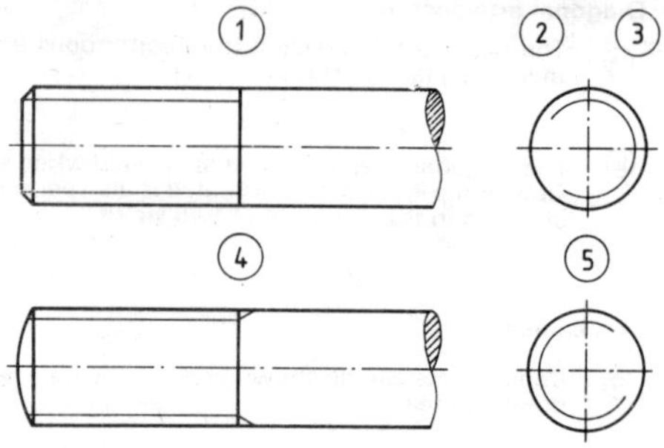

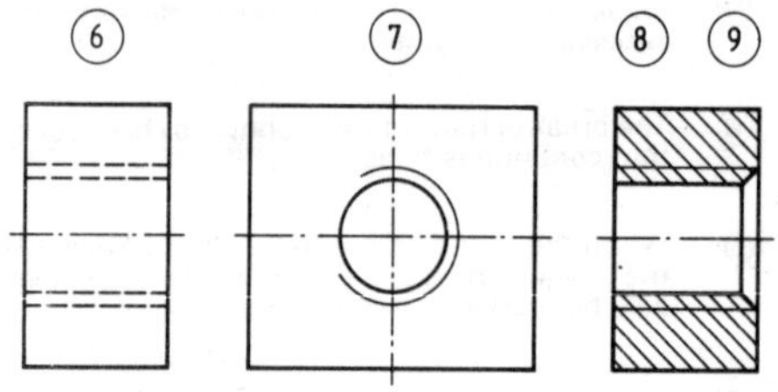

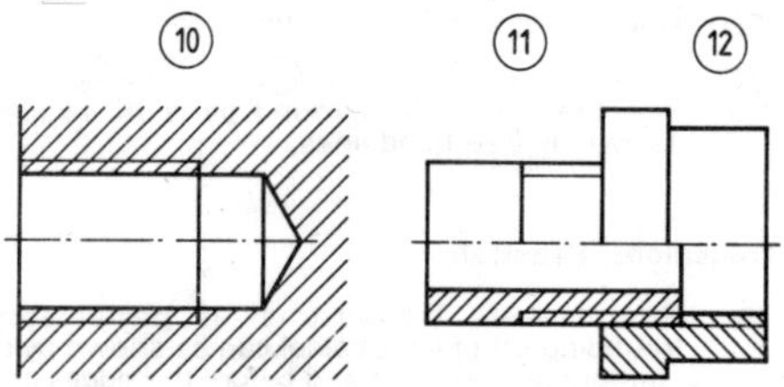

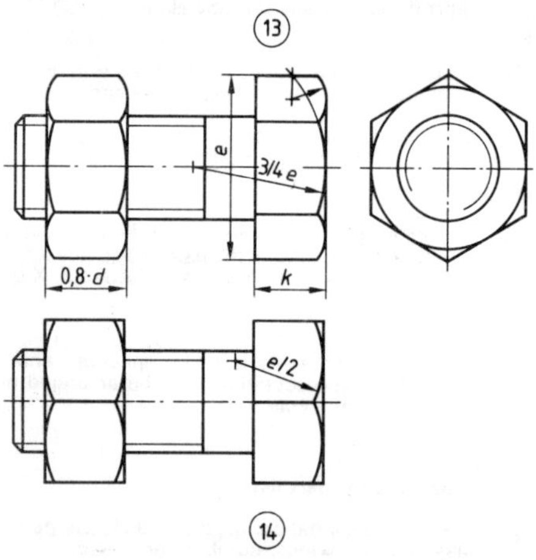

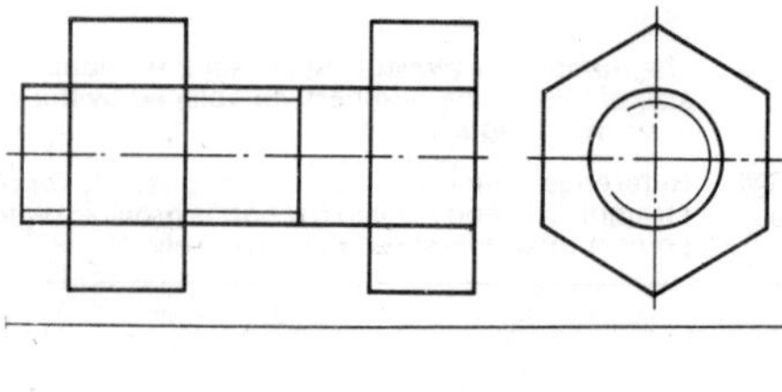

General

All types of screw thread are represented in simplified manner according to ISO standard.

If in exceptional cases, e.g. in case of trapezoid threads, the screw thread form is to be speciall represented, a part of the screw thread profile i drawn as part section or as detail.

External screw thread

① The major diameter is drawn in thick continuou line, the minor diameter in thin continuous line

② The minor diameter is shown in the axial directio as three-quarters of a circle.

③ The position of the three-quarters of the circle wi reference to the axes could be any.

④ The screw thread runout is generally not show for purposes of simplification.

⑤ In axial direction seen from the screw thread en taper cups and round cups are shown in the sam manner.

Internal screw thread

⑥ In the view seen perpendicular to the axial direc tion, internal screw threads are represented b dashed lines.

⑦ In the axial direction seen on the screw threa hole, the minor diameter of screw thread is show as full circle in thick continuous line. The majc diameter of screw thread is represented as three quarters of a circle in thin continuous line.

⑧ In the longitudinal section the minor diameter c the screw thread is drawn in thick continuous line the major diameter of the screw thread is shown i thin continuous line.

⑨ The countersinking of a screw thread hole up t the outside diameter of the screw thread is no shown as seen in the axial direction.

⑩ In case of screw thread blind holes that are show in section, the taper of the hole is drawn with 30 angle starting from the root of the screw thread

Screw joints

⑪ In section representation the screw thread of th threaded part is to be drawn full; only the unco vered part of internal screw threads is drawn.

⑫ In the view the hidden edges are generally no shown. If they are required for dimensioning the are shown with dashed lines.

Nuts and screw heads

⑬ Nuts and screw heads are shown dimension-wis according to Figure 13 shown beside. The chamfe edges are drawn as arcs of circle.

Dimension across corners e = 1.155 times width across flats. Nut height m=0.8 times nomina diameter of screw thread. Screw head heigh k=0.7 times nominal diameter of screw thread.

⑭ In simplified representations the chamfer edges o the nuts and screw heads are not drawn.

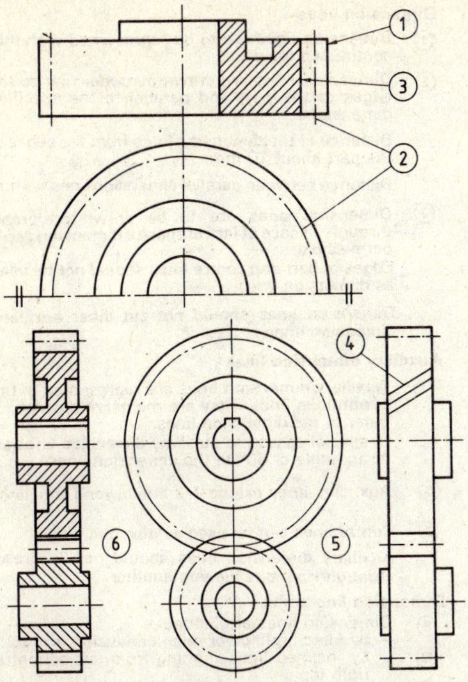

Gears and gear pairs (DIN ISO 2203)

① A gear is basically represented as a full part without individual teeth. The edges of the part are so drawn that they
— represent a full wheel limited by the tip surface in the unsectioned view.
— represent a gear with two opposite uncut teeth in the section.

② The reference surface of the teeth of a spur gear is to be identified with a thin chain line as follows:
— by the pitch circle in a representation perpendicular to the axis
— by the generating lines of the pitch cylinder in a representation parallel to the axis. These lines are to be drawn 1 to 2 mm extended beyond the edges of the part.

③ The tooth root area is represented only in sections. In case the inclusion appears meaningful in the view, the tooth root area is to be drawn with a thin continuous line.

④ In case it is required to specify the tooth direction in a projection parallel to the axis, three thin continuous lines of the appropriate form and direction are to be drawn.

⑤ In assembly drawings (gear pairs) the rules for the individual representation are suitably used.

⑥ If both gears are shown in axial section, the hidden edges of the part are to be represented if they are necessary for the clarity of the drawing.

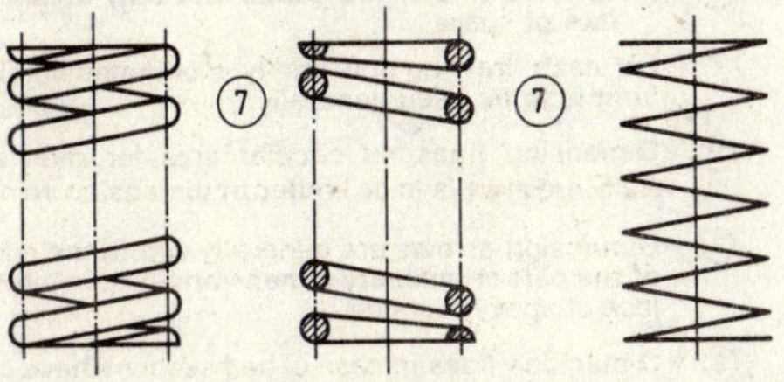

Springs (DIN ISO 2162)

⑦ Representation of a cylindrical helical compression spring made out of round wire
— in view
— in section and
— as name plate drawing

If required, the left or right direction of winding and the cross-section of the spring wire is to be specified with verbal note or symbol in the name plate drawing.

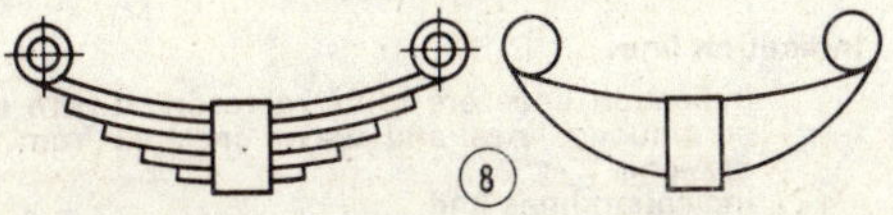

⑧ Representation of a semi-elliptic sheet spring with eyes and the middle joint
— in view and
— as name plate drawing.

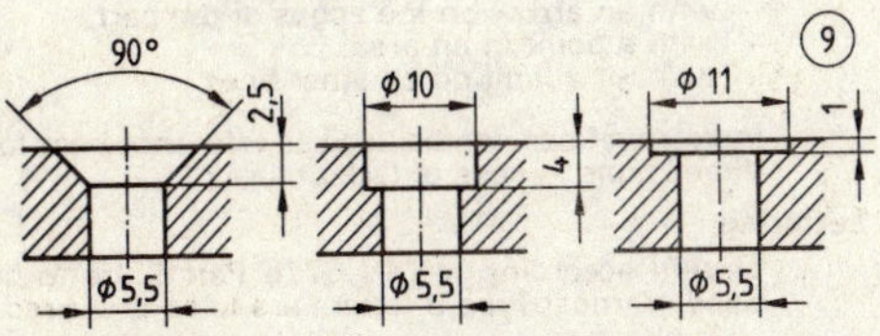

Counterbores and countersinks (DIN 74)

⑨ Representation of counterbores and countersinks in section
— for countersunk screws
— for cylindrical screws
— for hexagon head screws and hexagonal nuts

The dimension lines and anciliary dimension lines belonging to the representations can be replaced by reference lines that lead to the dimensions or to the designations from the representations. The DIN designations of the parts may be shortened.

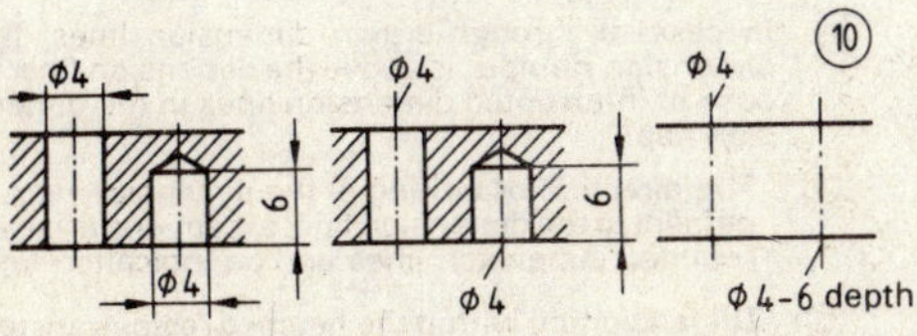

Holes (DIN 30)

⑩ Simplified representation of holes in section. This simplified representation applies appropriately even for counterbores, countersinks, screw threads, screw joints and rivet joints.

If the representations are too small and hence not clear they may be omitted. In their place, centre lines and centre line intersections appear from whom the reference lines are to be drawn up to the dimensions and designations.

Dimensioning in drawings

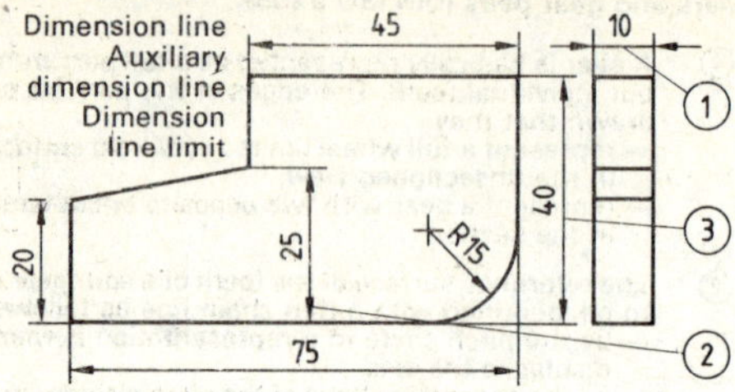

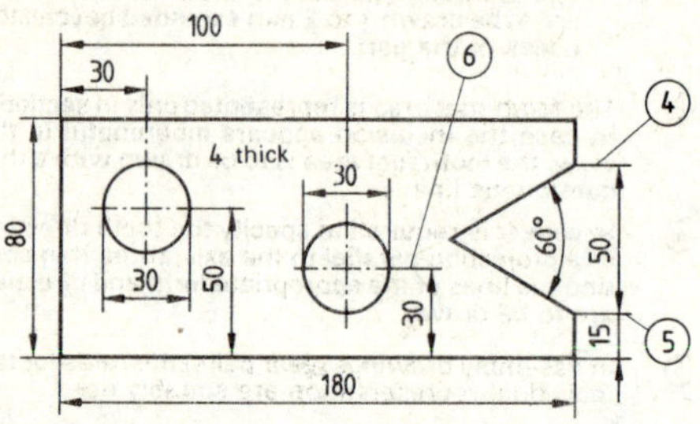

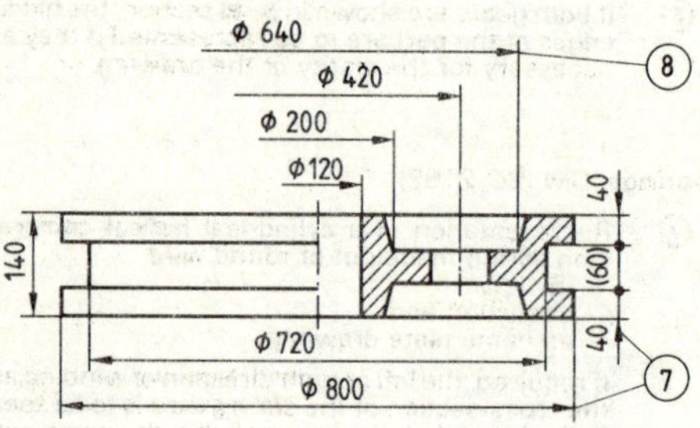

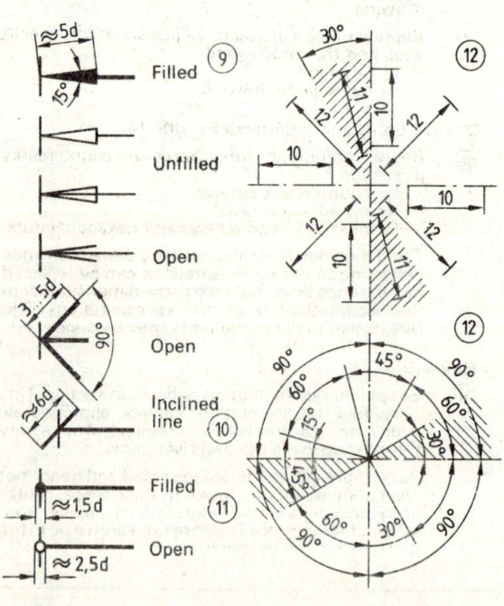

Dimension lines

(1) Dimension lines are to be represented with thin continuous lines.

(2) Dimension lines are drawn perpendicular to the edges of the part and parallel to the specified dimensions.

Distance of the dimension lines from the edges of the part about 10 mm.

Distance between parallel dimension lines ≈ 7mm.

(3) Dimension lines are to be drawn preferably through; in case of lack of space a dimension gap is permissible.
Edges of part and centre lines should not be used as dimension lines.

Dimension lines should not cut other auxiliary dimension lines.

Auxiliary dimension lines

Auxiliary dimension lines are represented in thin continuous lines. They are the general perpendicular to the dimension lines.
In special cases the auxiliary dimension lines are at an angle of 60° to the dimension lines.

(5) Aux. dim. lines extend 1-2 mm beyond dim. lines.

(6) Centre lines can be used as aux. dim. lines.

Auxiliary dimension lines should not be drawn through from one view to another.

Dimension line limits

(9) Dimension lines are limited
— by filled, unfilled or open dimension arrows
(10) — by inclined lines running from left bottom to right top
(11) — by filled or unfilled points, but only in case of lack of space.

For each drawing only one type of dimension line limit is to be used generally.

(23) Dimension lines on circular arcs for radii and diam. are always to be limited by dimension arrows.

(7) Dimension arrows are generally within the edges of the part or auxiliary dimension lines; in case of lack of space, outside.

(8) Dimension lines in case of half sections have only one dimension line limit.

Indication lines

Indication lines are to be represented with thin continuous lines and drawn inclined from the drawing.
Indication lines end
— with an arrow on the edges of the part
— with a point in an area
— without a limit on all other lines.

Indication lines can be used as reference lines for dimensions in case of lack of space.

Lettering

Letters according to DIN 6776 Part 1 are to be used; Vertical Type B letters are to be preferred.

Dimension letters should be in italics as symbols.

(3) In case of through-drawn dimension lines, the dimension number is above the dimension line; in case of interrupted dimension lines in the dimension gap.

(3) The direction of writing of the dimensioning runs parallel to the dimension line; exceptions in case of rounded dimension lines and on indication lines.

(12) Dimensioning within the hatched regions are to be avoided. However, if required, the dimension number must be readable from left. Interchangeable dimension numbers have a point.

(3) The lettering must be readable from bottom or from right.

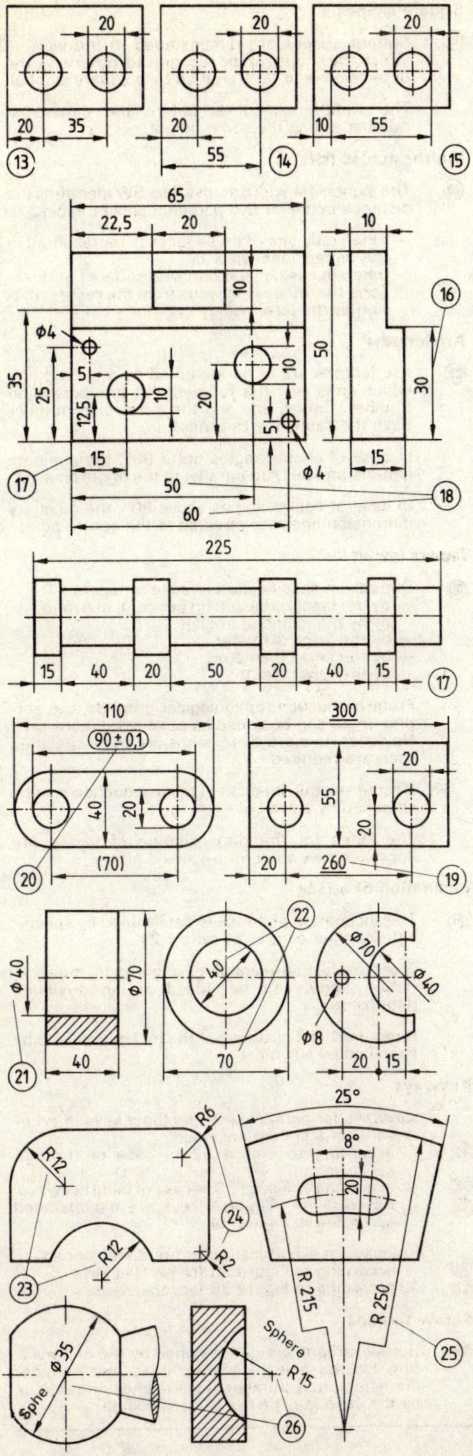

Scale

All dimensions in a drawing are to be specified in the same unit preferably in mm. Deviating units must be specially specified.

Scale for drawing is to be specified in the title block. In case of representations which are not to scale a horizontal line is to be put in the title block.

(19) In drawings with scale specification dimensions for edges of the part not drawn to scale are to be underlined.

Identification of dimension

(20) Following are to be specially identified:
- inspection dimensions by a round frame
- theoretical dimensions by a rectangular frame
- auxiliary dimensions by round brackets
- raw dimension by square brackets

Types of dimensioning

In engineering drawings, dimensioning can be carried out in three types:

(13) function oriented

(14) production oriented

(15) inspection oriented

Arrangement of dimensions, general

(16) Every dimension is to be entered only once in the drawing and at a place where the representation shows the dimension most clearly. No more dimensions than are necessary are to be specified. Dimensions should not make the representation unclear.

Dimension and auxiliary dimension lines are to be placed on continuous lines, not on dashed lines.

(17) Same type dimensioning is to be undertaken generally from a common reference element. Closed dimension chains are to be avoided. If this is not possible one length must remain undimensioned or must be identified as an auxiliary dimension. Related dimensions be as far as possible in one view.

(18) Dimensions lying one above the other are to be shown offset—Dimensions lying inside and dimensions lying outside are to be shown separate from one another.

Diameter

(21) The diameter symbol consists of a circle with an inclined line which is inclined at 15° to the right.

(22) The diameter symbol is placed before the dimension number and to the same height.

The diameter symbol is absent in case of diameter specifications which are in a circle or between the auxiliary dimension lines of a circle and whose dimension lines have two dimension limits.

Radius

(23) Radii are to be identified by the capital letter R which is to be placed before the dimension number

The centre point of the radius is to be identified in general by a centre line cross.

Dimension lines for radii have one dimension line limit either from outsider or from inside on circular arc.

(24) For dimension arrows from outside, the dimension line is to be drawn through up to centre point.

(25) In case of large radii the dimension line may be drawn, from space considerations, broken at right angles and shortened. The portion provided with the dimension arrow must be directed to the theoretical centre point.

Sphere

(26) The form is identified by the word specification "Sphere", it is to be placed before R or ϕ symbol

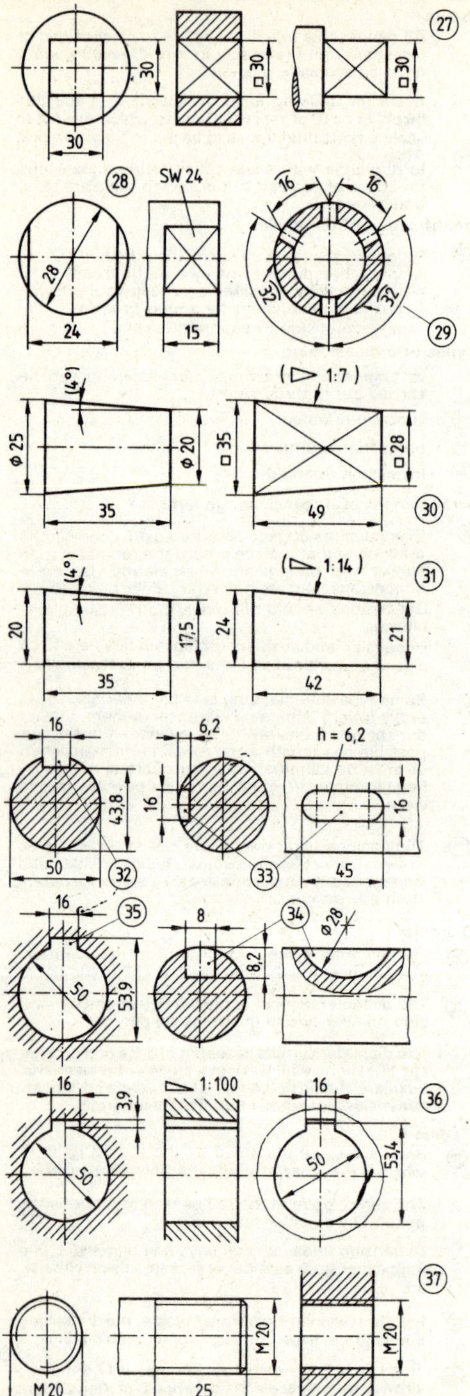

Square shapes

(27) Square shapes are dimensioned in the view in which the shape can be recognised, otherwise the dimension is to be identified by a square symbol.

The square symbol is before the dimension number and to the same height.

Widths across flats

(28) The symbol for width across flats SW identifies the distance between two parallel opposite faces.

— when only one of these faces is represented in the dimensional view or
— when in case of standardized squares or hexagons the shape is obvious from the representation or the terminology.

Arc lengths

(29) Arc lengths are to be identified by a curved line which runs over the full width of the dimension number. Curved line and the dimension number have the same line thickness.

In case of centre angles up to 90°, the auxiliary dimension lines run parallel to the angle bisector

In case of centre angles above 90°, the auxiliary dimension lines are directed to the centre point.

Tapers (pyramids)

(30) Dimension specification in case of tapers
— by the taper reduction (in per cent, in a ratio 1:*x*, or by the included angle)
— by the large diameter
— by the small diameter
— by the taper length.

From production-technological grounds, the setting angle can be added as auxiliary dimension. However no more dimensions are to be specified than are required.

If required, the direction of taper reduction is to be specified by a symbol.

The rules for the dimensioning of tapers are approximately valid for pyramids also.

Inclination of a face

(31) The inclination of a face is determined by specification in per cent or as ratio 1:2*x*.

From production-technological grounds, the inclination angle can be added as an auxilliary dimension.

If required, the direction of the inclination is to be specified by a symbol.

Keyways

Keyways for parallel keys and taper keys in cylindrical shafts are dimensioned:
(32) — according to Figure 32 in case of through keyways.
(33) — according to Figure 33 in case of blind keyways.
(34) — Keyways for Woodruff keys are dimensioned according to Figure 34.

Keyways in cylindrical holes are dimensioned:
(35) — according to Figure 35 for parallel keys.
(36) — according to Figure 36 for taper keys.

Screw threads

(37) Screw threads are dimensioned by the abbreviation for the type of screw thread and nominal diameter; pitch, number of starts and hand of helix as the case may be are to be specified.

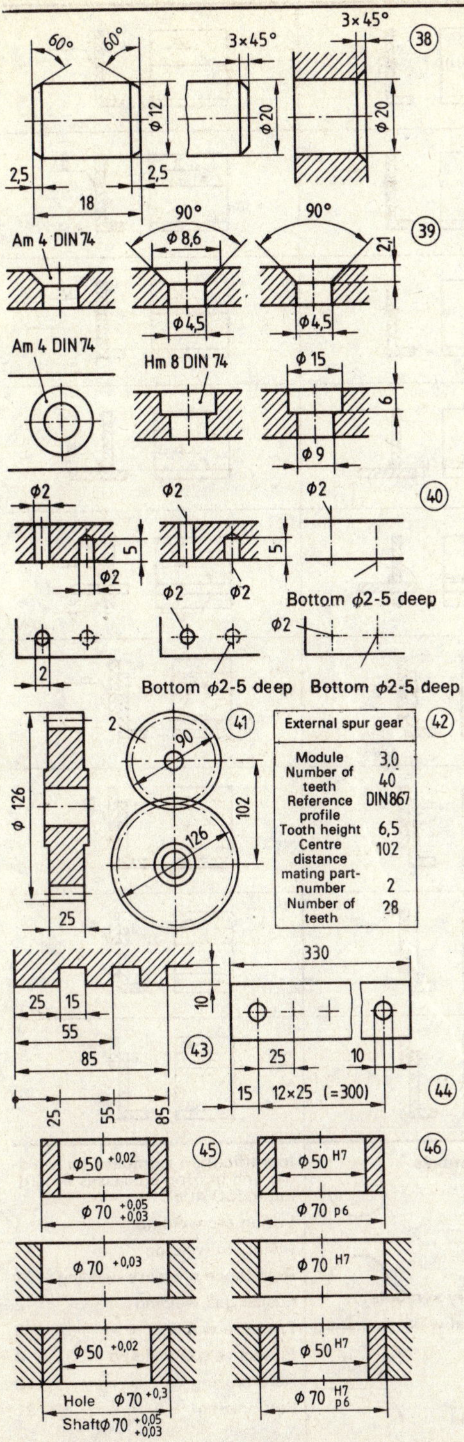

Chamfers, counterbores and countersinks

(38) In case of turned parts with chamfers, these are to be included in the length dimension.

Chamfers of 45° and countersinks of 90° can be dimensioned in simplified manner.

(39) Dimensioning of countersinks for countersunk-head screws and counterbores for cylindrical screws internal hexagon socket head screws and slotted cheese nead screws according to Fig. 39.

Screws and rivets

(40) The dimension lines and auxiliary dimension lines of holes, countersinks, counterbores and screw joints or rivet joints can be replaced by reference lines. In case of hidden holes or parts, the reference lines are shown as dashed lines over such width that they must remain hidden.

Gears (DIN 3966)

(41) In case of spur gears, the following are to be entered on the drawing: the tip circle diameter, the root circle diameter (if tooth height is not specified in the table) and the tooth width.

(42) In an accompanying table the quantities are to be specified which are required for the selection of the cutting tools, for the setting of the gear cutting machine and for the inspection of the gear teeth

For gear pairs, the centre distance is to be entered.

Pitches

(43) The dimensioning of several successive distances is undertaken individually from a common reference element. The beginning of the dimensioning is to be clearly identified by a point. Pitches for round holes are dimensioned in general from hole centre to hole centre, for square holes, notches or slots from edge to edge.

(44) In case of equal distances and equal geometric elements, a simplified dimensioning is permissible. Only the actual geometric element is to be fully dimensioned in this case, that is valid as reference element for all other specification. Total no. of same elements is specified by the entry, e.g. 5 *X*

In case of holes that are equally divided on the pitch circle, the specification of the pitch angle in the main view is not necessary . In the side view , the pitch circle diameter, the number of holes and the hole diameter may be entered together.

Tolerance specification

Tolerances can be specified
— by deviation, e.g. 100±0.5 or
— by ISO tolerance field abbreviation, e.g. 16H₇.

(45) Both deviations with the corresponding sign are to be added to the nominal dimension. The upper deviation is to be written higher than the nominal dimension and the lower deviation deeper than the nominal dimension. The deviation zero is in general not entered.

If upper and lower deviations have the same magnitude, the deviation with the pertinent signs appears only once after the nominal dimension.

In case of fits, the dimension with the deviations for the inside dimension is to be placed above the dimension with the deviations for the outside dimension. The coordination is to be identified by word specification "hole" or "shaft"

(46) If the deviations correspond to the ISO tolerance system, the ISO tolerance field abbreviations are to be used for the outside and inside dimensions

If ISO tolerance system is used, the common nominal dimension is to be entered only once in which the ISO tolerance field abbreviation for the inside dimension is above that for outside dimension.

Drawing representation of welding and brazing

Weld	Symbol	Number
Flange weld	⌒⌒	1
I-weld	‖	2
V-weld	V	3
Fillet weld	◺	10
Hole weld	⊓	11
Spot weld	○	12
Line weld	⊖	13
Surface weld	=	17

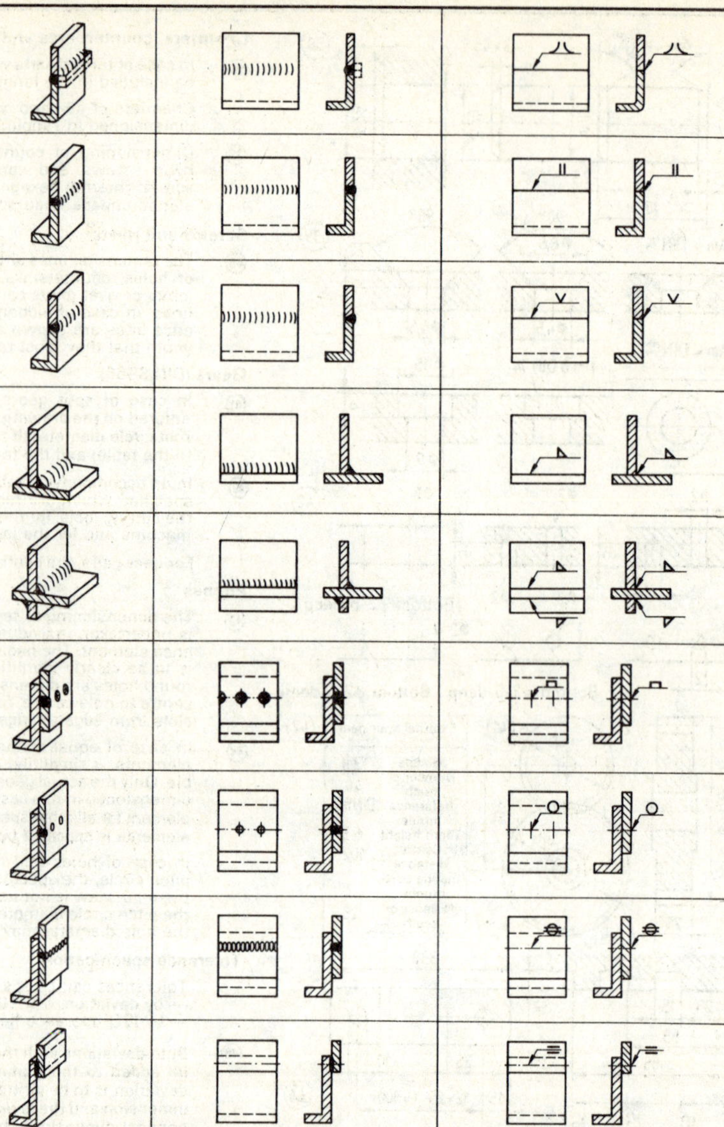

Reference symbols, position of the symbol

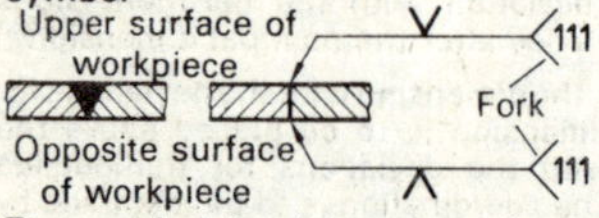

The symbol is above the reference line when the arrow line points to the upper surface of workpiece.
In case of arrow direction pointing to the opposite surface of workpiece, the symbol is below the reference line.
If required, the arrow line contains a fork for the identification number, e.g. 111.

Additional symbols

Upper surface concave

Flat (plane)

Convex

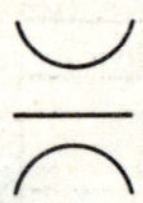

Supplementary symbols

Circumferential weld (fillet weld)

Assembly weld

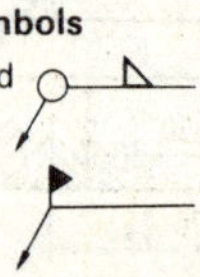

Identification numbers for welding and brazing processes according to ISO 4063

Fusion arc welding	1
Metal arc welding	11
Resistance pressure welding	111
Fusion gas welding	2
Pressure welding	3
Protective gas welding	4
Hard brazing	13
Soft brazing	91

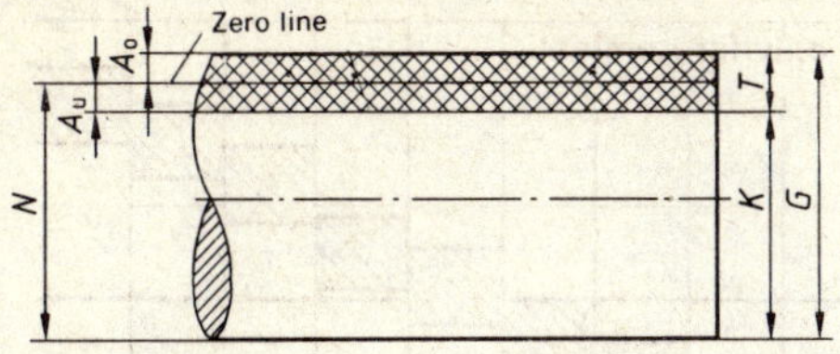

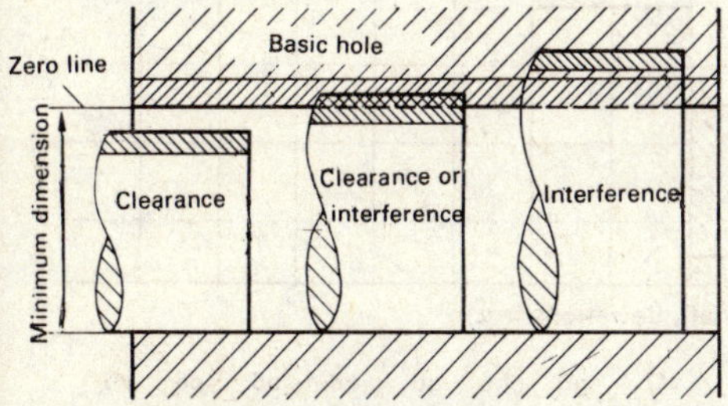

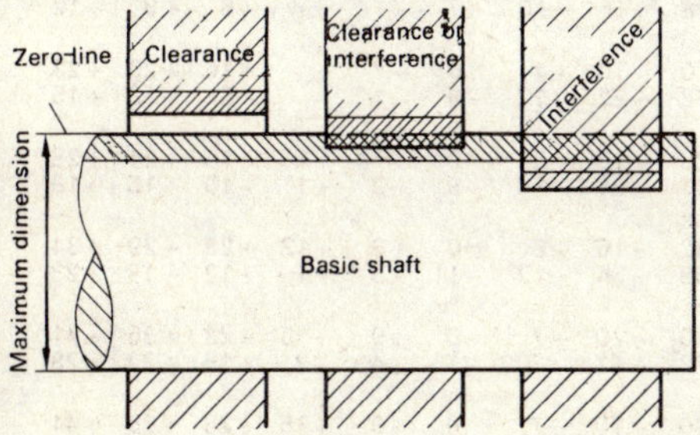

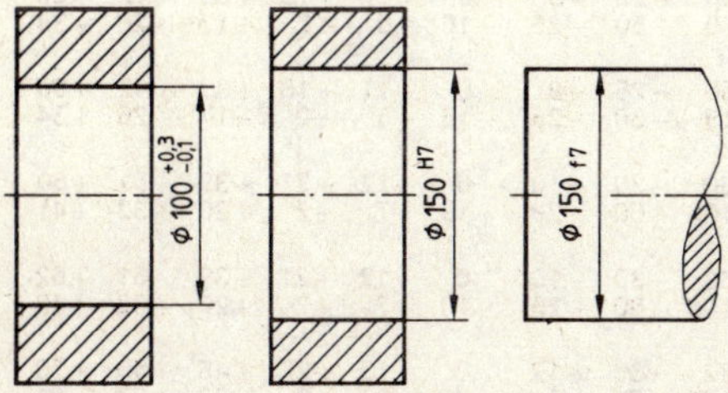

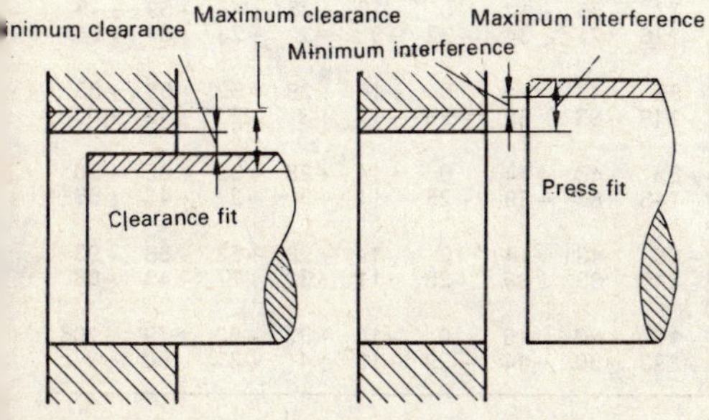

Basic conceptions

Nominal dimension (*N*) is the dimension mentioned in the drawing to which the deviations refer (zero line)

Actual dimension (*I*) is the dimension determined by measurement

Maximum dimension (*G*) is the maximum permissible dimension or the algebraic sum of nominal dimension and upper deviation.

Minimum dimension (*K*) is the minimum permissible dimension or the algebraic sum of nominal dimension and lower deviation.

Upper deviation (A_o) is the algebraic difference between the maximum dimension and the nominal dimension.

Lower deviation (A_u) is the algebraic difference between the minimum dimension and the nominal dimension.

Tolerance (*T*) is the upper deviation minus lower deviation or maximum dimension minus minimum dimension

Basic hole is the system of fits for holes with fundamental deviation zero, i.e., the minimum dimension goes up to the zero line. All holes have a H-tolerance.

Basic shaft is the system of fits for shafts with fundamental deviation O, i.e., the maximum dimension goes up to the zero line. All shafts have a H-tolerance

Fit is the relation between the mating surfaces of mating parts to be assembled or assemble mating parts.

Tolerance specification

The tolerances are specified by both the deviations and their signs. The algebraic difference of the deviations specifies the magnitude of tolerance field, the sign the position of the tolerance field. Deviations with the + sign lie above the zero line, those with the sign below the zero line. The deviation 0 is not specified.

Example: Dimension specification 100 + 0.3 —0.1

Upper deviation	0.3 mm	Max dimension	100.3 mm
Lower deviation	-(—0.1)mm	Min dimension	99.9 mm
Tolerance	0.4 mm	Tolerance	0.4 mm

According to ISO, the tolerances are specified by letters and numbers. The position of the tolerance field is specified with capital letters A to Z for internal dimensions (holes), with small letters a to z for external dimensions (shafts).

Tolerance position: for internal dimensions	for external dimensions
A to H above the zero line (nominal dimension)	a to h below the zero line (nominal dimension)
J to K either side of the zero line (nominal dimension)	j to k either side of zero line (nominal dimension)
M to Z below the zero line (nominal dimension)	m to z above the zero line (nominal dimension)

The magnitude of the tolerance fields is specified with numbers 1 to 18. Smaller the number, smaller is the tolerance. The numerical values for the deviations can be taken from the tables beside.

Example: Dimension specification Hole 150^{H7}

Upper deviation	+ 40 μm	Max dimension	150.040 mm
Lower deviation	— 0 μm	Min dimension	150.000 mm
Tolerance	40 μm	Tolerance	0.040 mm

Fits

Following can occur when 2 mating parts are assembled: Clearance, fit: Fit with clearance, i.e. the dimension of the hole is greater than the dimension of the shaft.

Transition fit: Fit with clearance or interference

Press fit: Fit with interference, i.e. the dimension of the shaft is greater than the dimension of the hole.

In case of basic hole of ISO system of fits, the desired fit is obtained by the selection of a suitable shaft tolerance.

In general, shaft tolerance a to h give clearance fits, j to n transition fits and p to z press fits.

Basic hole

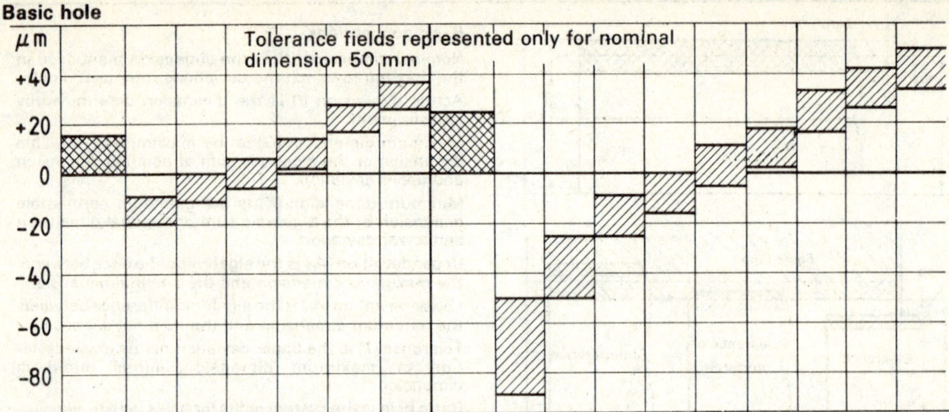

Nom. dimen. up to mm	Hole H 6	Shaft, deviations in μm g5	h5	j5	k5	n5	p5	Hole H 7	Shaft, deviations in μm e8	f7	g6	h6	j6	k6	n6	p6	r6
3	+7 0	−3 −8	0 −5	+4 −1	– –	+11 +6	+14 +9	+9 0	−14 −28	−7 −16	−3 −10	0 −7	+6 −1	– –	+13 +6	+16 +9	+19 +12
6	+8 0	−4 −9	0 −5	+4 −1	– –	+13 +8	+17 +12	+12 0	−20 −38	−10 −22	−4 −12	0 −8	+7 −1	– –	+16 +8	+20 +12	+23 +15
10	+9 0	−5 −11	0 −6	+4 −2	+7 +1	+16 +10	+21 +15	+15 0	−25 −47	−13 −28	−5 −14	0 −9	+7 −2	+10 +1	+19 +10	+24 +15	+28 +19
18	+11 0	−6 −14	0 −8	+5 −3	+9 +1	+20 +12	+26 +18	+18 0	−32 −59	−16 −34	−6 −17	0 −11	+8 −3	+12 +1	+23 +12	+29 +18	+34 +23
24	+13 0	−7 −16	0 −9	+5 −4	+11 +2	+24 +15	+31 +22	+21 0	−40 −73	−20 −41	−7 −20	0 −13	+9 −4	+15 +2	+28 +15	+35 +22	+41 +28
30	+13 0	−7 −16	0 −9	+5 −4	+11 +2	+24 +15	+31 +22	+21 0	−40 −73	−20 −41	−7 −20	0 −13	+9 −4	+15 +2	+28 +15	+35 +22	+41 +28
40	+16 0	−9 −20	0 −11	+6 −5	+13 +2	+28 +17	+37 +26	+25 0	−50 −89	−25 −50	−9 −25	0 −16	+11 −5	+18 +2	+33 +17	+42 +26	+50 +34
50	+16 0	−9 −20	0 −11	+6 −5	+13 +2	+28 +17	+37 +26	+25 0	−50 −89	−25 −50	−9 −25	0 −16	+11 −5	+18 +2	+33 +17	+42 +26	+50 +34
65	+19 0	−10 −23	0 −13	+6 −7	+15 +2	+33 +20	+45 +32	+30 0	−60 −106	−30 −60	−10 −29	0 −19	+12 −7	+21 +2	+39 +20	+51 +32	+60 +41
80	+19 0	−10 −23	0 13	+6 −7	+15 +2	+33 +20	+45 +32	+30 0	−60 −106	−30 −60	−10 −29	0 −19	+12 −7	+21 +2	+39 +20	+51 +32	+62 +43
100	+22 0	−12 −27	0 −15	+6 −9	+18 +3	+38 +23	+52 +37	+35 0	−72 −126	−36 −71	−12 −34	0 −22	+13 −9	+25 +3	+45 +23	+59 +37	+73 +51
120	+22 0	−12 −27	0 −15	+6 −9	+18 +3	+38 +23	+52 +37	+35 0	−72 −126	−36 −71	−12 −34	0 −22	+13 −9	+25 +3	+45 +23	+59 +37	+76 +54
140	+25 0	−14 −32	0 −18	+7 −11	+21 +3	+45 +27	+61 +43	+40 0	−85 −148	−43 −83	−14 −39	0 −25	+14 −11	+28 +3	+52 +27	+68 +43	+88 +63
160	+25 0	−14 −32	0 −18	+7 −11	+21 +3	+45 +27	+61 +43	+40 0	−85 −148	−43 −83	−14 −39	0 −25	+14 −11	+28 +3	+52 +27	+68 +43	+90 +65
180	+25 0	−14 −32	0 −18	+7 −11	+21 +3	+45 +27	+61 +43	+40 0	−85 −148	−43 −83	−14 −39	0 −25	+14 −11	+28 +3	+52 +27	+68 +43	+93 +68
200	+29 0	−15 −35	0 −20	+7 −13	+24 +4	+51 +31	+70 +50	+46 0	−100 −172	−50 −96	−15 −44	0 −29	+16 −13	+33 +4	+60 +31	+79 +50	+106 +77

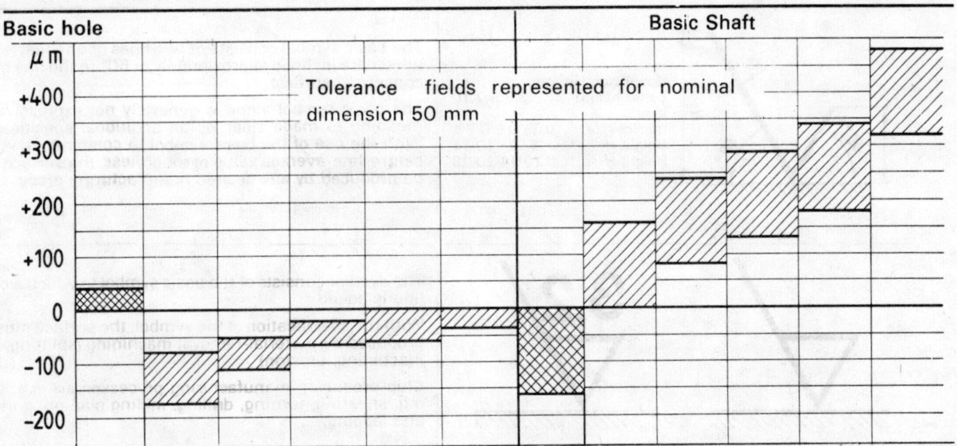

Nom. dimen. up to mm	Hole	Shaft, deviations in μm					Shaft	Hole, deviations in μm				
	H 8	d 10	e 9	f 8	h 9	h 8	**h 11**	H 11	D 11	C 11	B 11	A 11
3	+14 0	−20 −60	−14 −39	−7 −21	0 −25	0 −14	0 −60	+60 0	+80 +20	+120 +60	+200 +140	+330 +270
6	+18 0	−30 −78	−20 −50	−10 −28	0 −30	0 −18	0 −75	+75 0	+105 +30	+145 +70	+215 +140	+345 +270
10	+22 0	−40 −98	−25 −61	−13 −35	0 −36	0 −22	0 −90	+90 0	+130 +40	+170 +80	+240 +150	+370 +280
18	+27 0	−50 −120	−32 −75	−16 −43	0 −43	0 −27	0 −110	+110 0	+160 +50	+205 +95	+260 +150	+400 +290
30	+33 0	−65 149	−40 −92	−20 −53	0 −52	0 −33	0 −130	+130 0	+195 +65	+240 +110	+290 +160	+430 +300
40	+39 0	−80 −180	−50 −112	−25 −64	0 −62	0 −39	0 −160	+160 0	+240 +80	+280 +120	+330 +170	+470 +310
50	+39 0	−80 −180	−50 −112	−25 −64	0 −62	0 −39	0 −160	+160 0	+240 +80	+290 +130	+340 +180	+480 +320
65	+46 0	−100 −220	−60 −134	−30 −76	0 −74	0 −46	0 −190	+190 0	+290 +100	+330 +140	+380 +190	+530 +340
80	+46 0	−100 −220	−60 −134	−30 −76	0 −74	0 −46	0 −190	+190 0	+290 +100	+340 +150	+390 +200	+550 +360
100	+54 0	−120 −260	−72 −159	−36 −90	0 −87	0 −54	0 −220	+220 0	+340 +120	+390 +170	+440 +220	+600 +380
120	+54 0	−120 −260	−72 −159	−36 −90	0 −87	0 −54	0 −220	+220 0	+340 +120	+400 +180	+460 +240	+630 +410
140	+63 0	−145 −305	−85 −185	−43 −106	0 −100	0 −63	0 −250	+250 0	+395 +145	+450 +200	+510 +260	+710 +460
160	+63 0	−145 −305	−85 −185	−43 −106	0 −100	0 −63	0 −250	+250 0	+395 +145	+460 +210	+530 +280	+770 +520
180	+63 0	−145 −305	−85 −185	−43 −106	0 −100	0 −63	0 −250	+250 0	+395 +145	+480 +230	+560 +310	+830 +580
200	+72 0	−170 −355	−100 −225	−50 −122	0 −115	0 −72	0 −290	+290 0	+460 +170	+530 +240	+630 +340	+950 +660

Specification of surface quality

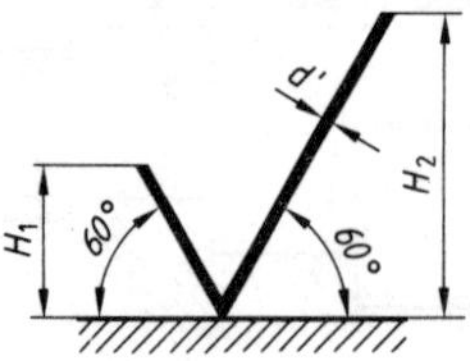

Dimensions in mm				
Letter height	3,5	5	7	10
Line thickness *d'*	0.35	0.5	0.7	1
Height H_1	5	7	10	14
Height H_2	10	14	20	28

The basic symbol consists of two lines of unequal length which are inclined approximately at 60° to the line of the concerned surface.

The basic symbol alone is generally not expressive. Its meaning is made clear by an additional specification. With the use of the basic symbol in combination with a centre-line-average value of roughness, the surface may be produced by any desired manufacturing process.

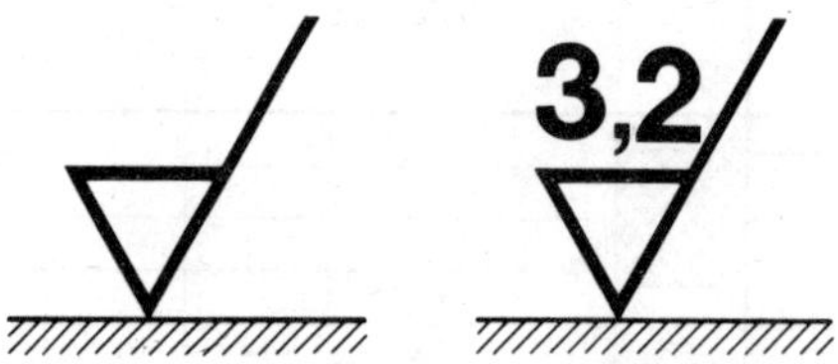

The symbol consists of the basic symbol to which a cross line is added.

With the specification of the symbol, the surface must be produced by material-removal machining (splitting, chip machining, erosion)

Chip-producing manufacturing processes are e.g. sawing, shearing, turning, drilling, milling planing, grinding and lapping.

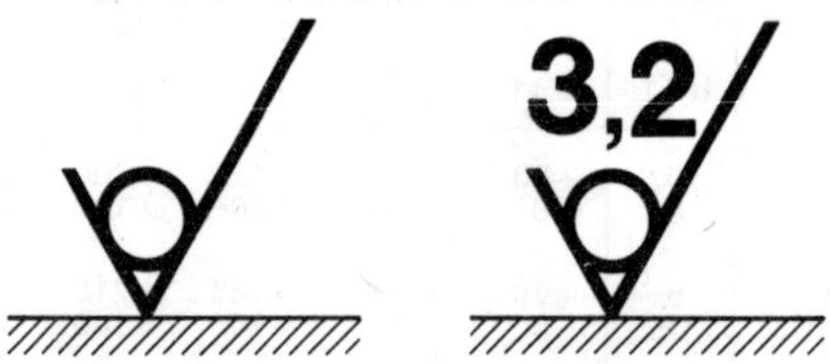

The symbol consists of the basic symbol to which a circle is added.
With the specification of the symbol
a) Without additional specifications, the surface must remain in the condition of the preceding manufacturing process.
b) With additional specifications, the surface must be produced without using material-removal machining (chipless).

Chipless manufacturing processes are e.g. casting, drawing, forging, bending, brazing, welding or galvanizing.

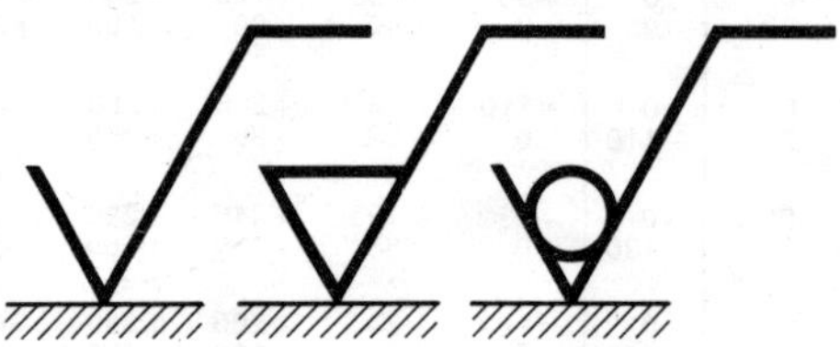

For special verbal specifications of the surface, the longer side of the above symbols contains an additional horizontal line. Its length depends upon the additional specifications.

With the specification of the symbol, the final condition of the surface must be produced by the manufacturing process specified on the horizontal line. The verbal specification can refer to the treatment or plating of the surface.

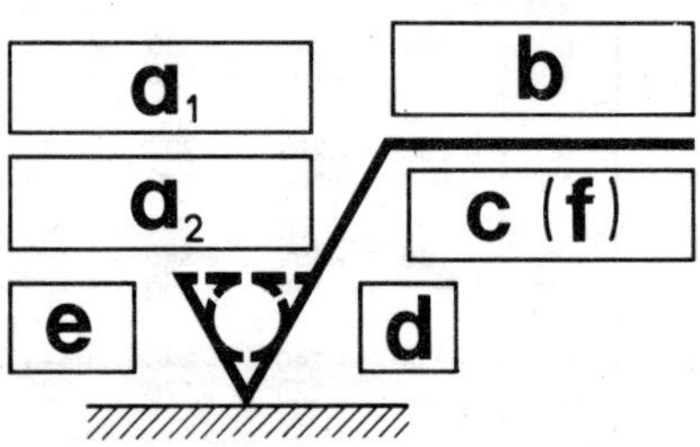

Roughness value R_a Class	in µm
N 1	0,025
N 2	0,050
N 3	0,100
N 4	0,200
N 5	0,500
N 6	0,800
N 7	1,600
N 8	3,200
N 9	6,300
N 10	12,50
N 11	25,00
N 12	50,00

Direction of grooves of the surface Symbol	Clarification
=	Parallel to the projection plane
⊥	Perpendicular to the projection plane
×	interesting in two inclined directions.
M	In several directions
C	Approximately circular
R	Approximately radial to the centre point.

Additional specifications of the surface quality that are necessary for the functional suitability of the part are coordinated to the symbols according to the figure beside:

a Centre-line-average roughness value R_a in µm or roughness classes from N_1 to N_{12}.

With the specification of the upper and lower values, the maximum value a_1 is to be placed above the minimum value a_2.

b Manufacturing process, treatment or plating of the surface.

c Reference length or limiting wave length in mm

d Direction of grooves of the surface

e Machining allowance in mm

f Other roughness measurement quantities R_z or R_{max} in µm.

The centre-line-average roughness value R_a is the arithmetic mean value of the distances of the actual profile from the mean line.

The average roughness depth R_z is the artihmetic mean value of the distances between the highest and the deepest point of the profile from five successive individual measuring lengths.

The maximum roughness depth R_{max} is the maximum distance between the highest and the deepest point of a profile inside of the measuring length.

Specification of surface quality

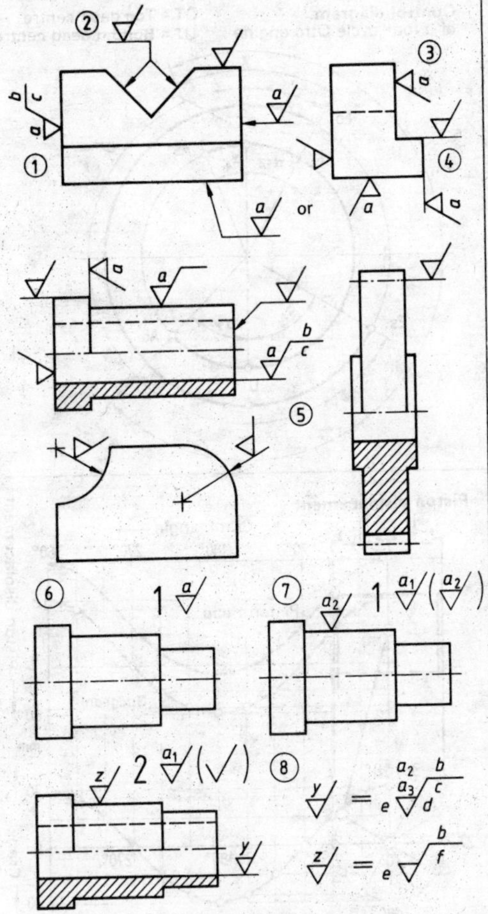

Entry in drawings

The symbols for surface quality are entered at all places for which they are required for functional suitability of the parts. They can be absent if the usual manufacturing process guarantees a sufficient end condition of the surface.

Rules for the entry

① The surface specifications are so arranged that they are to be read from bottom or from the right side (basic rule) and point from outside to the edge of the part or to the corresponding auxiliary dimension line.

② The surface specifications may be connected with the edge of the part by a reference line which is provided with a dimension arrow at one end.

③ If the basic rule cannot be adhered to, the symbols may be drawn without detailed specifications in any other position. The specification of the centre-line-average roughness value R_a is allowed when it can be read from bottom or from right.

④ For a particular surface the symbol corresponding to the dimensioning rules may be drawn only once possibly at the representation which also contains the dimensioning of the surface.

⑤ The surface specifications are to be drawn
a) on one of the two generators in case of turned parts
b) on the dimension lines of the radius in case of rounding
c) on the pitch circle lines in case of tooth flanks

⑥ In case of equal surface quality for all sides, one surface specification is sufficient which is placed beside the part number or the drawing.

⑦ In case of overwhelmingly same quality of the surface, the corresponding surface specification can be placed beside the part number and the exceptions in brackets beside that. Latter is to be placed on the particular edge of the part also.

⑧ In case of multiple repetition of complicated specifications, a single entry may be placed on the surface. The meaning of this symbol must be explained at another place of the drawing.

Symbol	Explanation	Symbol	Explanation
Without surface specification	Surfaces on which no specific requirements are placed.	R_z 4,0	Material-removal machining, average roughness $R_z = 4.0$ µm.
√	Basic symbol, not expressive, additional specifications required.	⊥	Material-removal machining, direction of grooves perpendicular
z	Simplified drawing entry, an additional explanation specifies the meaning of the symbol.	2,0	Material-removal machining, Machining allowance 2.0 mm.
3,2	Surface produced in any manner, maximum centre-line-average roughness value $R_a = 3.2$ µm.	Milled	Material-removal machining, surface produced by milling.
(R_z = 4,0)	Surface produced in any manner, average roughness depth $R_z = 4.0$ µm.	Milled 3,2 2,5 2,0 ⊥	Material-removal machining, $R_a = 3.2$ µm, milled surface, reference length 2.5 mm, direction of grooves perpendicular, allowance 2.0 mm.
3,2 2,5	Surface produced in any manner, centre-line-average roughness value $R_a = 3.2$ µm. reference length 2.5 mm.		Material removal not permissible, surface retained in the condition of preceding processing.
3,2	Material-removal machining, centre-line-average roughness value $R_a = 3.2$ µm.	3,2	Chipless or no machining, maximum centre-line-average roughness value $R_a = 3.2$ µm.

Automobile engineering drawings

Working sequence
of a four-cylinder in-line engine

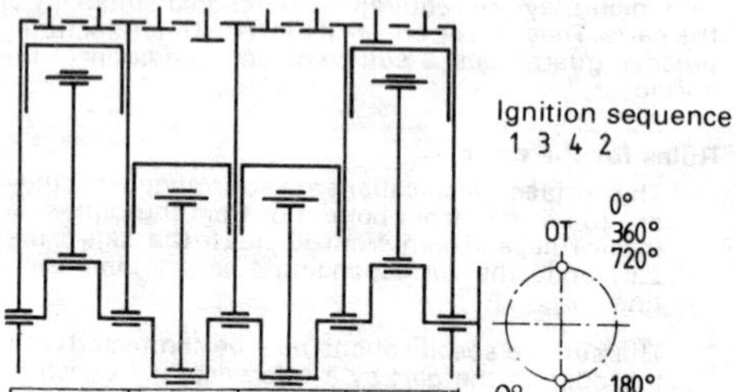

Power	Exhaust	Compr	Suction
Exhaust	Suction	Power	Compr
Suction	Compr	Exhaust	Power
Compr	Power	Suction	Exhaust
1. Cyl.	2. Cyl.	3. Cyl.	4. Cyl.

Control diagram
of a four-cycle Otto engine

OT = Top dead centre
UT = Bottom dead centre

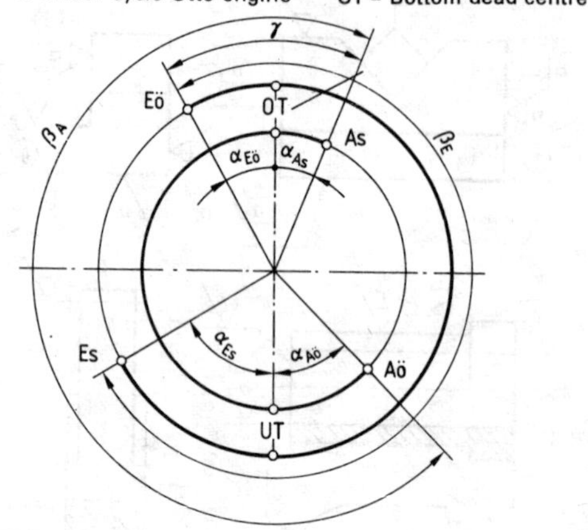

Piston speed

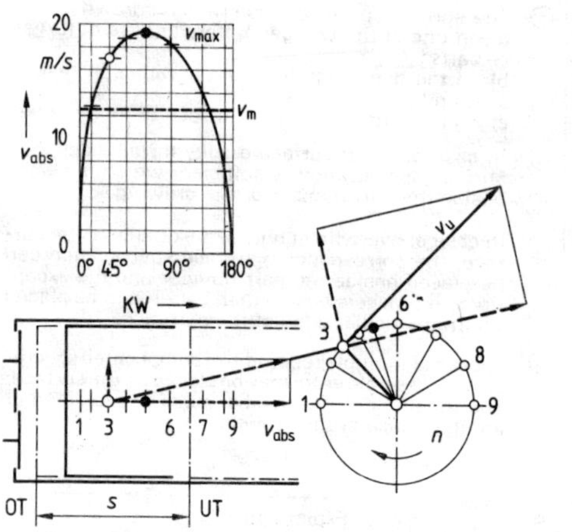

Piston displacement

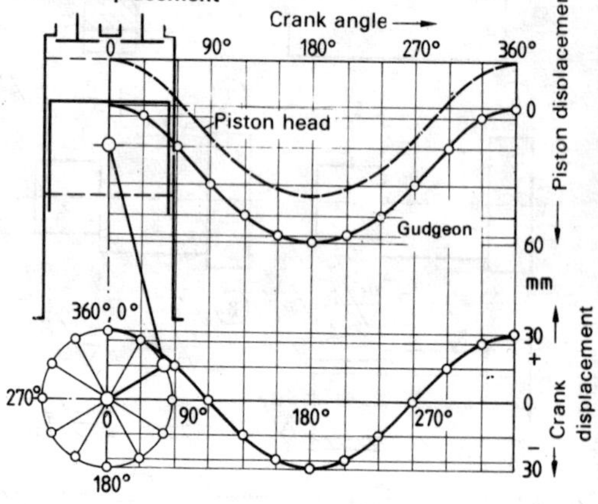

Cooling water circulation

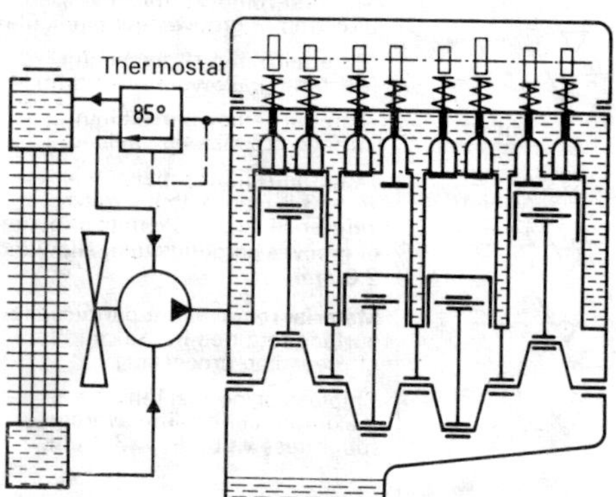

Four-speed change gear box

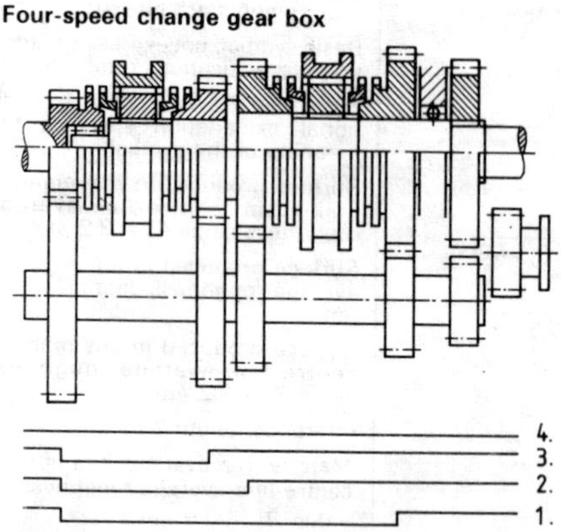

Track differential angles

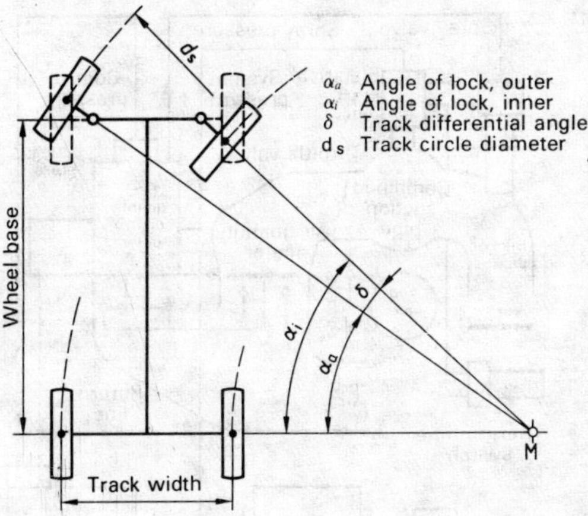

Steering trapezium
of a rigid axle and individual wheel suspension

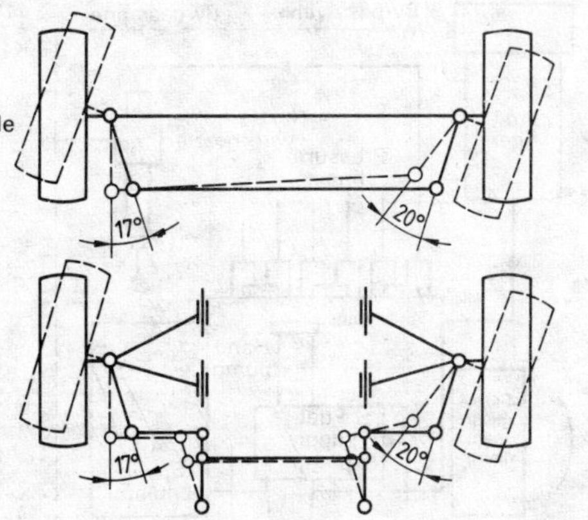

Rigid axle

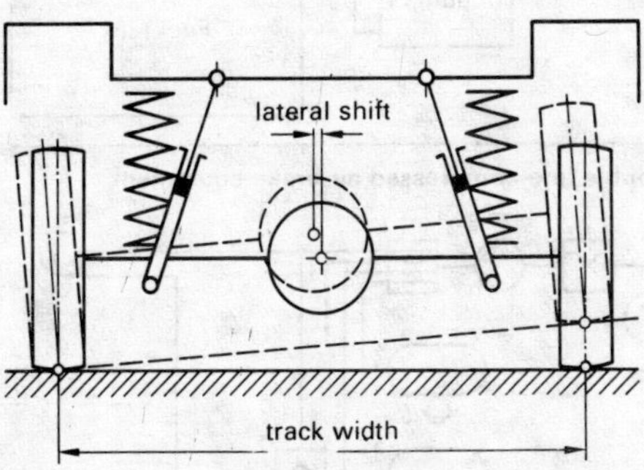

Double-hinge pendulum axle

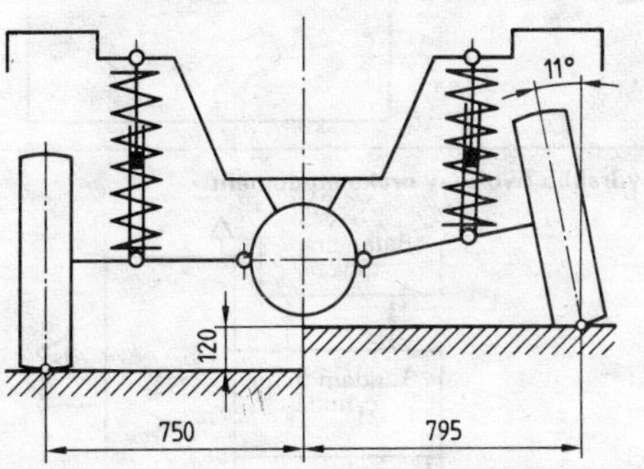

Double cross steer

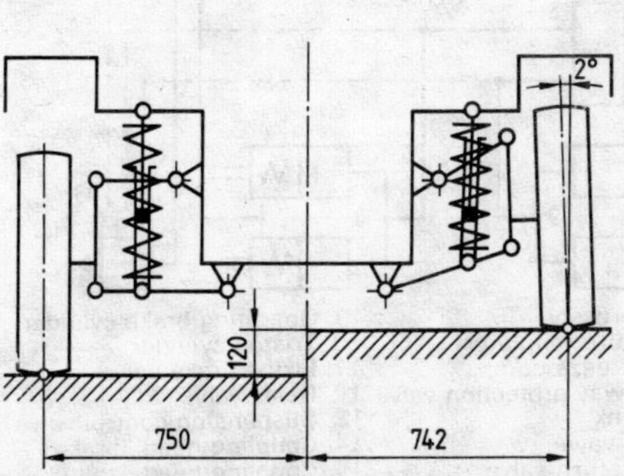

McPherson shock absorber leg

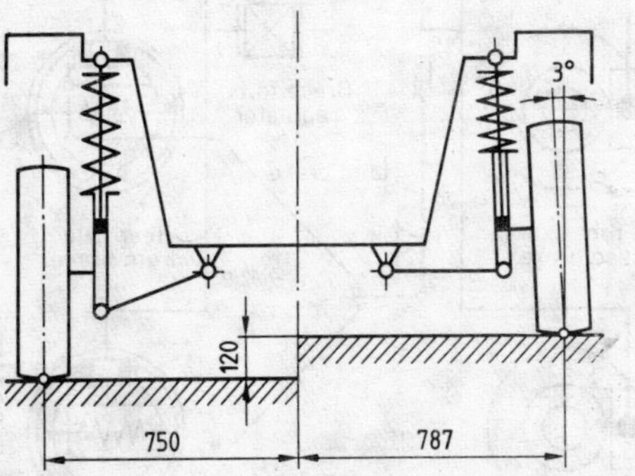

Automobile engineering drawings

Fuel equipment for diesel engines

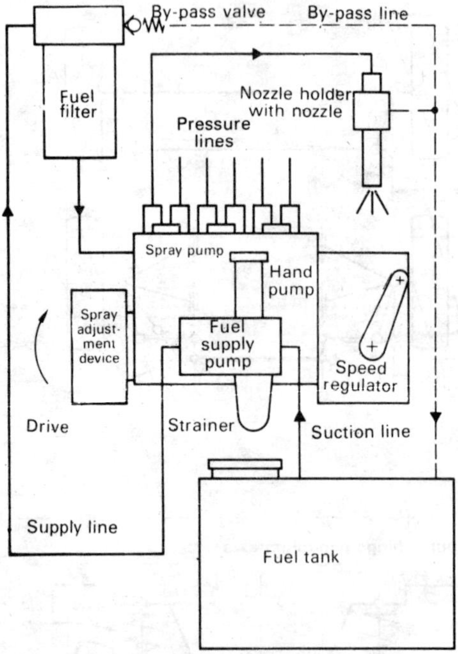

Spray system K-Jetronic

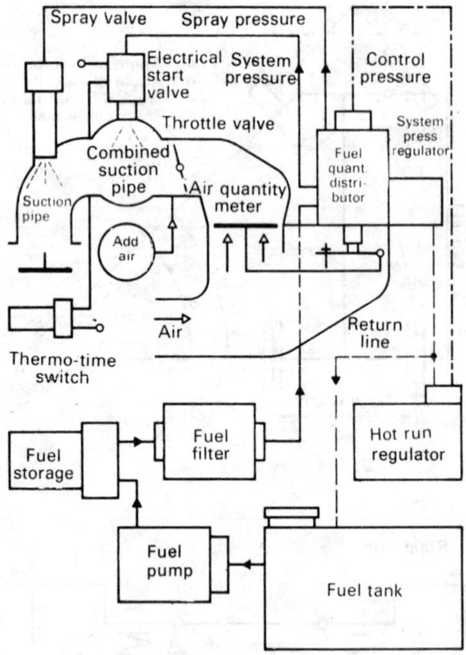

Hydraulic two-way brake equipment

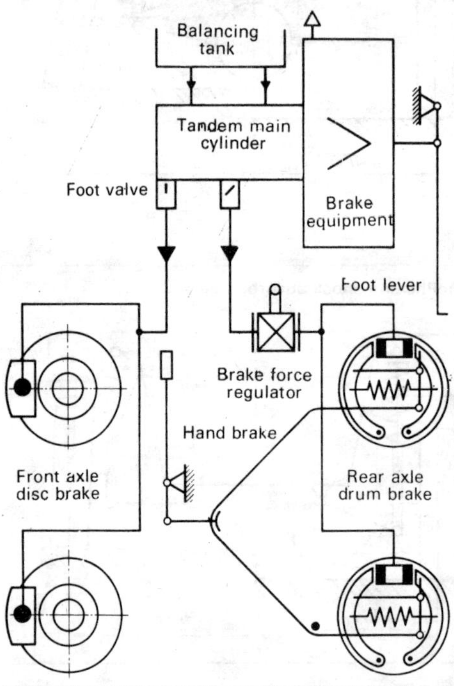

Double line compressed air brake equipment

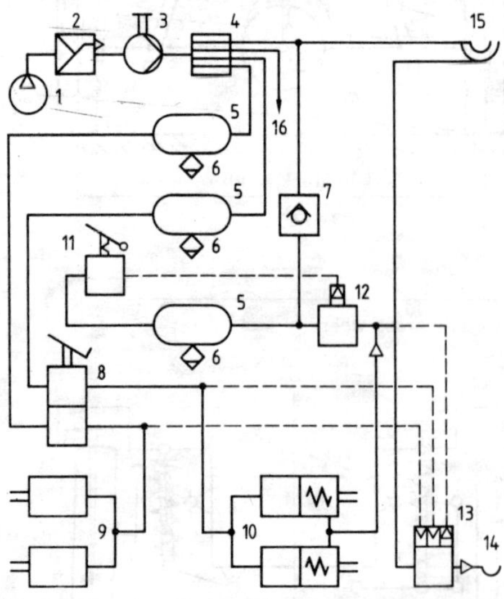

1. Compressor
2. Pressure regulator
3. Anti-freeze pump
4. Four-way protection valve
5. Air tank
6. Drain valve
7. Non-return valve
8. Operating brake valve
9. Operating brake cylinder
10. Tristop cylinder
11. Fixing brake valve
12. Relay valve
13. Suspension control valve
14. Coupling head "brake"
15. Coupling head "spare"
16. Standby use

Lighting equipment

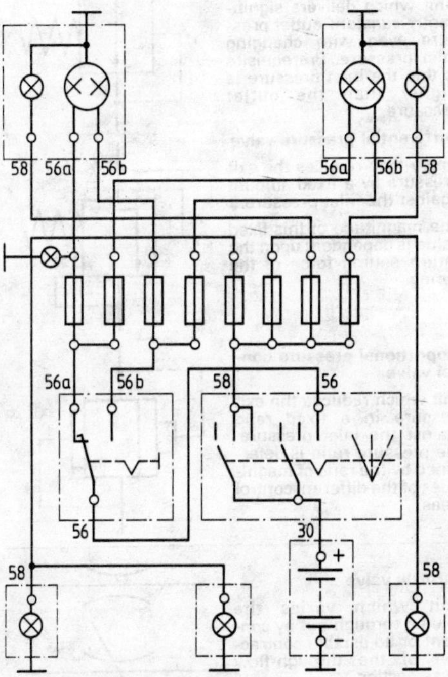

Coil ignition equipment, electronic

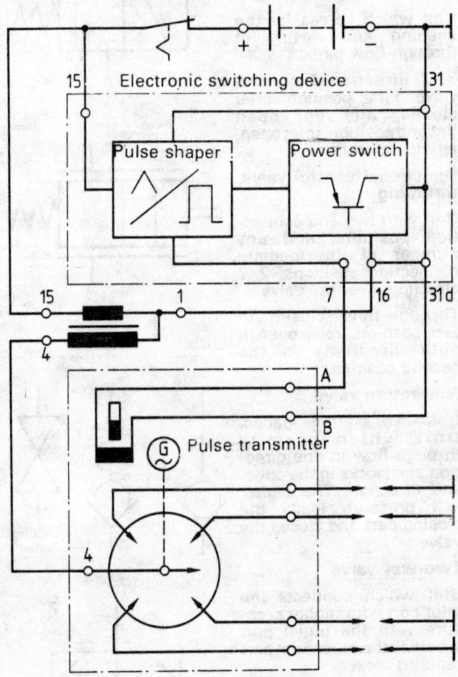

Preheating equipment

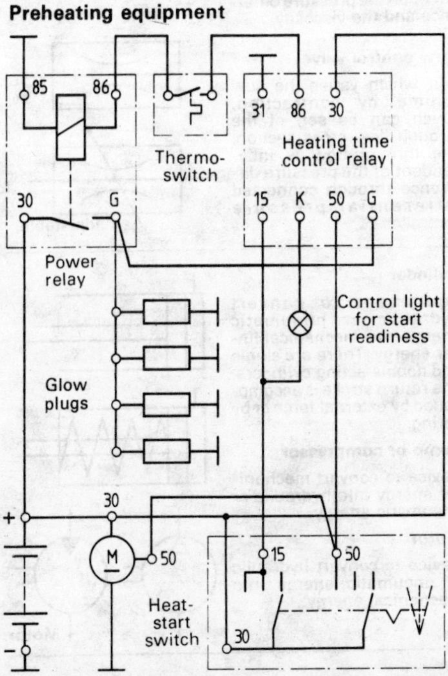

Blink equipment

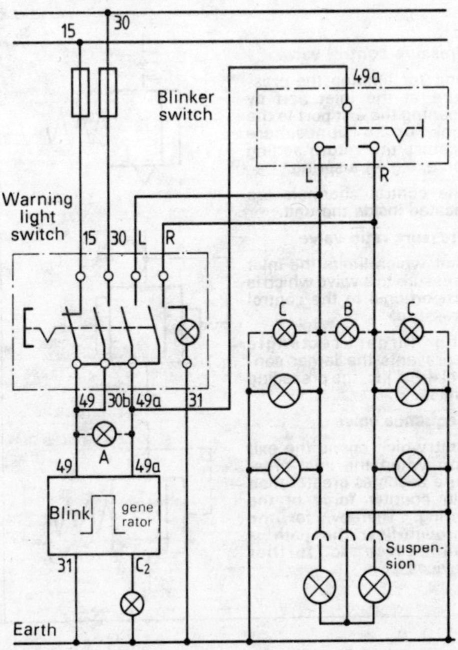

Directional control valve, non-throttling

Unit which serves for the opening and closing of through-flow paths.

3/2 directional control valve. First position: inlet closed, air vent open Second position: inlet open, air vent closed.

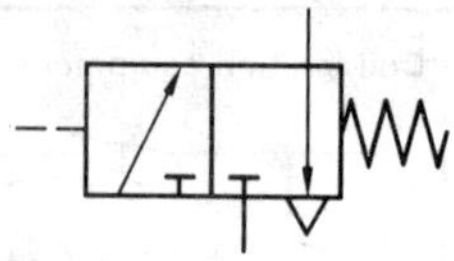

Directional control valve, throttling

Unit with two end connection positions and any number of intermediate connection positions. 2/2 directional control valve.

Through-flow closed in zero position. Valve open in both directions in the second position.

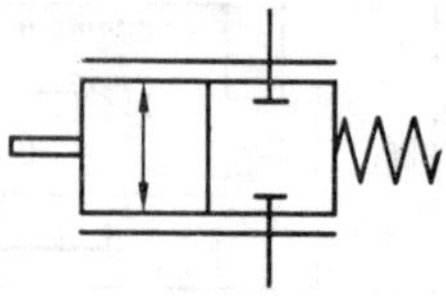

Non-return valve

Unloaded, spring loaded. Unit which permits through-flow in one direction and blocks in the opposite direction. The higher exit pressure loads the closing part and closes the valve.

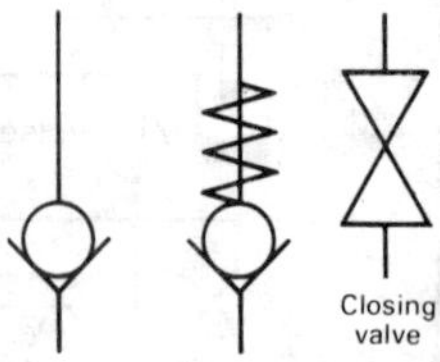

Closing valve

Two-way valve

Unit which connects the inlet port with higher pressure with the outlet port whereas the other inlet port remains closed.

The position remains in case of return flow.

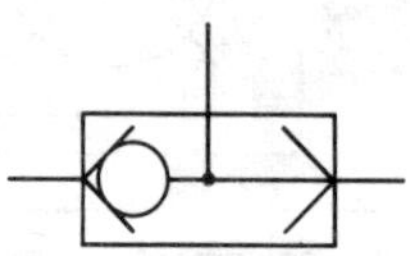

Pressure control valve

Unit for limiting the pressure at the inlet port by opening the exit port to the tank or to atmosphere against the return-setting force, e.g. of a spring.

The control channels are located inside the unit.

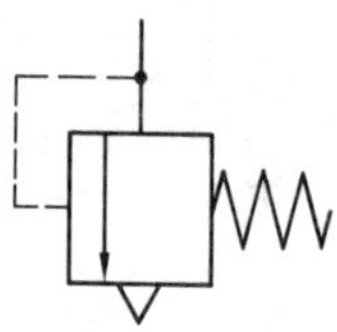

Pressure ratio valve

Unit which limits the inlet pressure to a valve which is proportional to the control pressure.

The larger rectangle represents the larger control area, i.e., the preceding phase.

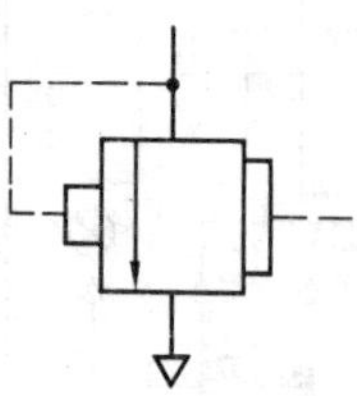

Sequence valve

Unit which opens the exit port when the inlet pressure becomes greater than the counter force of the spring. Thereby, for the through-flow the path is made free to further devices.

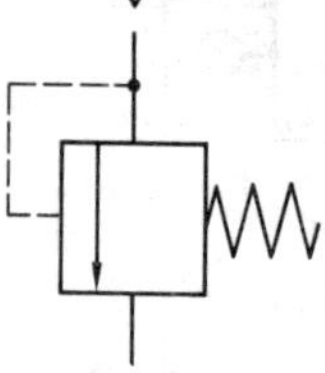

Pressure control valve

Unit which delivers significantly constant outlet pressure even with changing inlet pressure. Prerequisite is that the inlet pressure is higher than the outlet pressure.

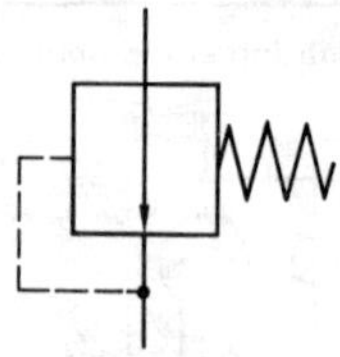

Differential pressure valve

Unit which reduces the exit pressure by a fixed amount against the inlet pressure.

The magnitude of this fixed value is dependent upon the return setting force of the spring.

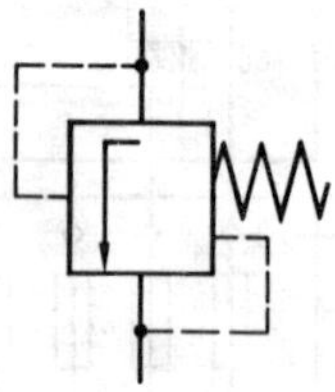

Proportional pressure control valve

Unit which reduces the exit pressure in a fixed ratio against the inlet pressure The pressure ratio is determined by the ratio of magnitudes of the different control areas.

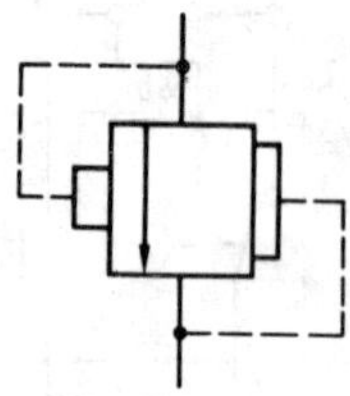

Throttle valve

Unit which varies the volume through put by constant or adjustable contraction of the through-flow cross-section

The through-flow is dependent upon the pressure difference and the viscosity.

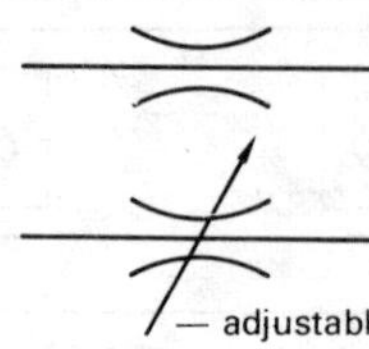

— adjustable

Flow control valve

Unit which varies the flow volume by contraction, which can be set, of the through-flow cross-section. The through-flow is independent of the pressure difference through connected differential pressure controller.

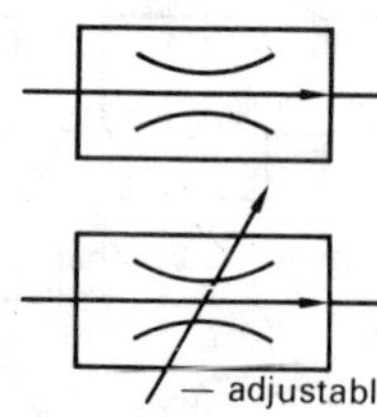

— adjustable.

Cylinder

Equipment to convert hydraulic or pneumatic energy into mechanical linear energy. There are single and double acting cylinders. The return stroke is accomplished by external force or by spring.

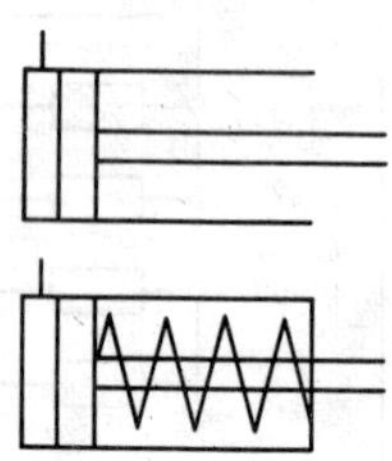

Pump or compressor

Device to convert mechanical energy into hydraulic or pneumatic energy.

Motor

Device to convert hydraulic or pneumatic energy into mechanical energy.

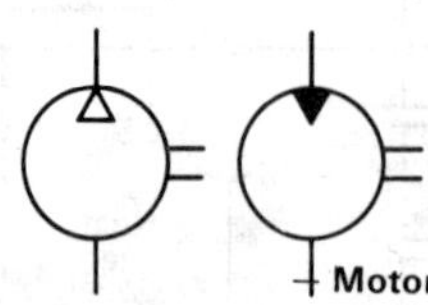

Motor

According to DIN ISO 1219

Circuit symbols for brake circuit schemes

Pressure regulator

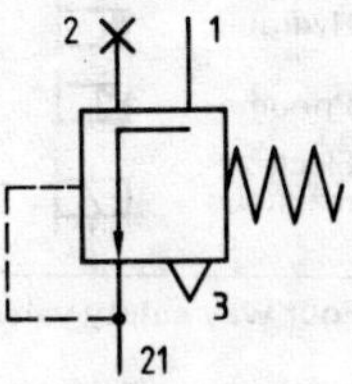

By-pass valve
without return flow

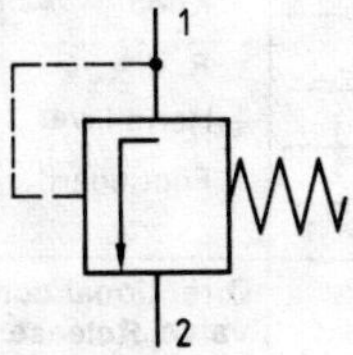

By-pass valve
with return flow

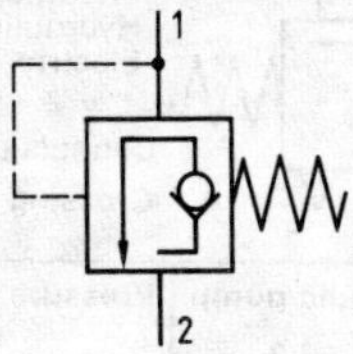

Operation brake valve

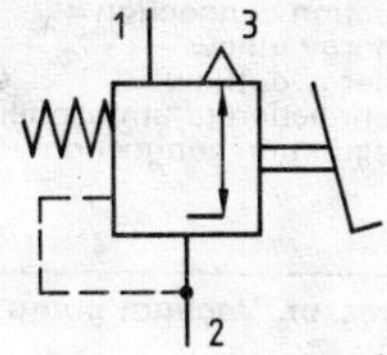

Fixed brake valve de-aerated

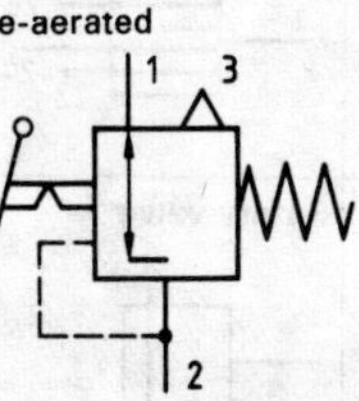

Relay valve
de-aerated

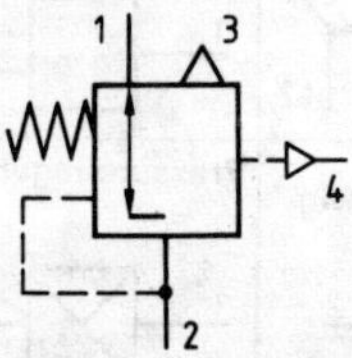

Relay valve aerated

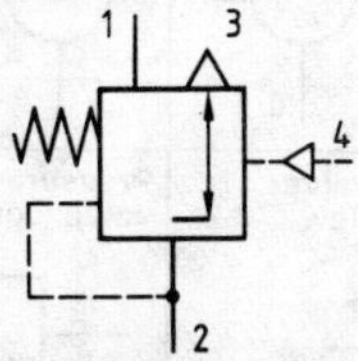

Solenoid relay valve aerated

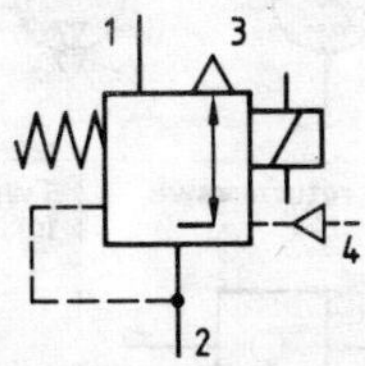

Trailor control valve
for single line braking device

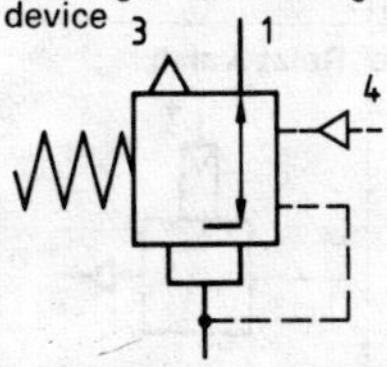

Trailor control valve
for double line braking device

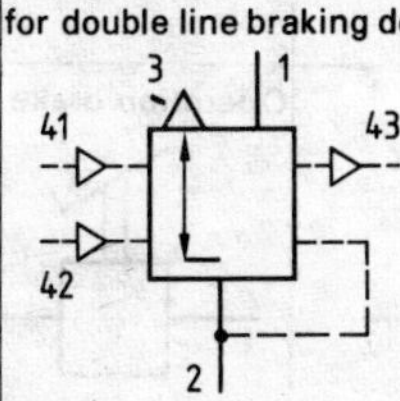

Trailor brake valve
for single line braking device

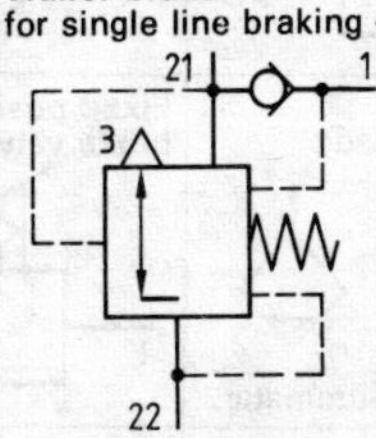

Trailor brake valve
for double line braking device

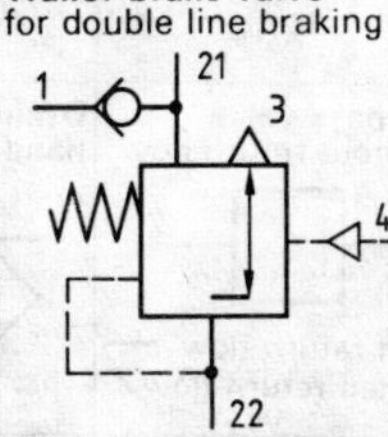

Brake force regulator

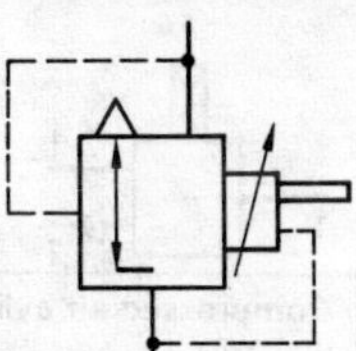

Compressed air motor

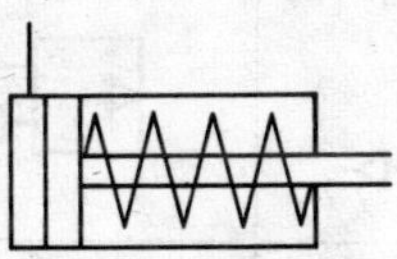

Spring loaded Cylinder

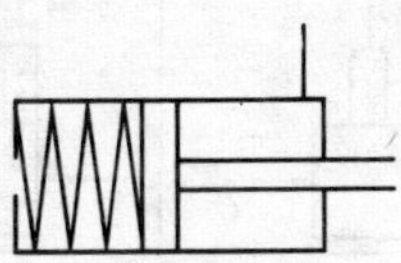

Coupling, without or with connecting element

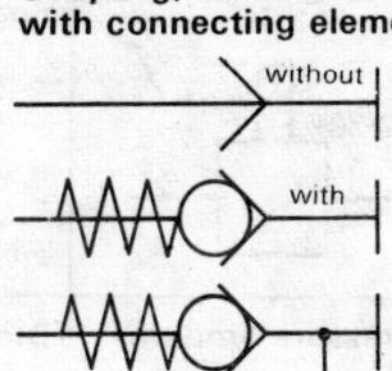

Brake amplifier for vacuum

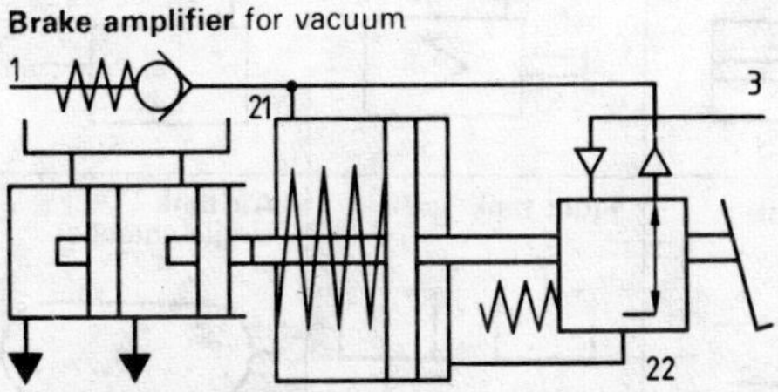

Four way safety valve

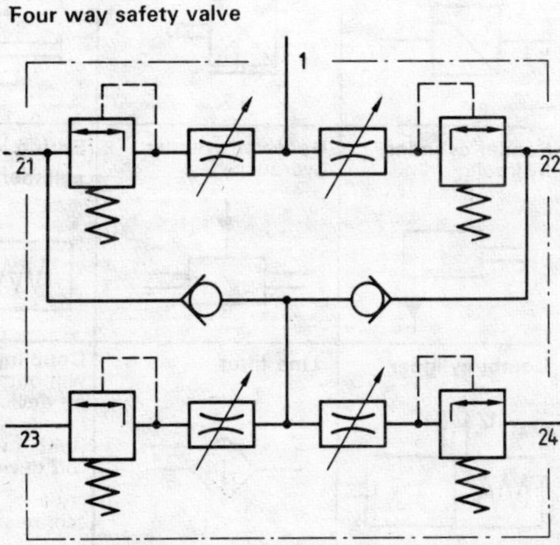

Brake amplifier for compressed air

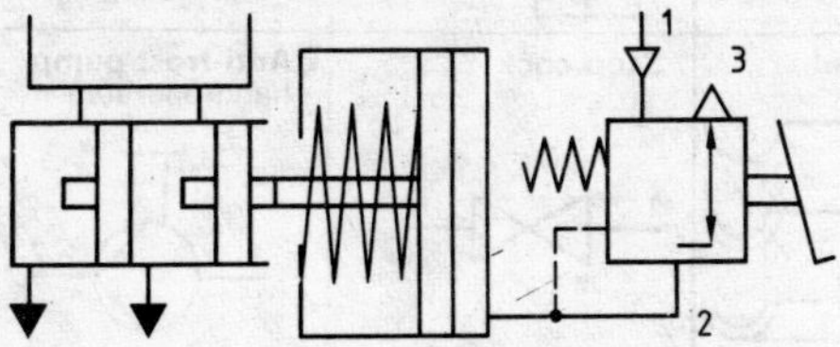

Graphic symbols for brake circuit schemes

DIN 74253, DIN 74254

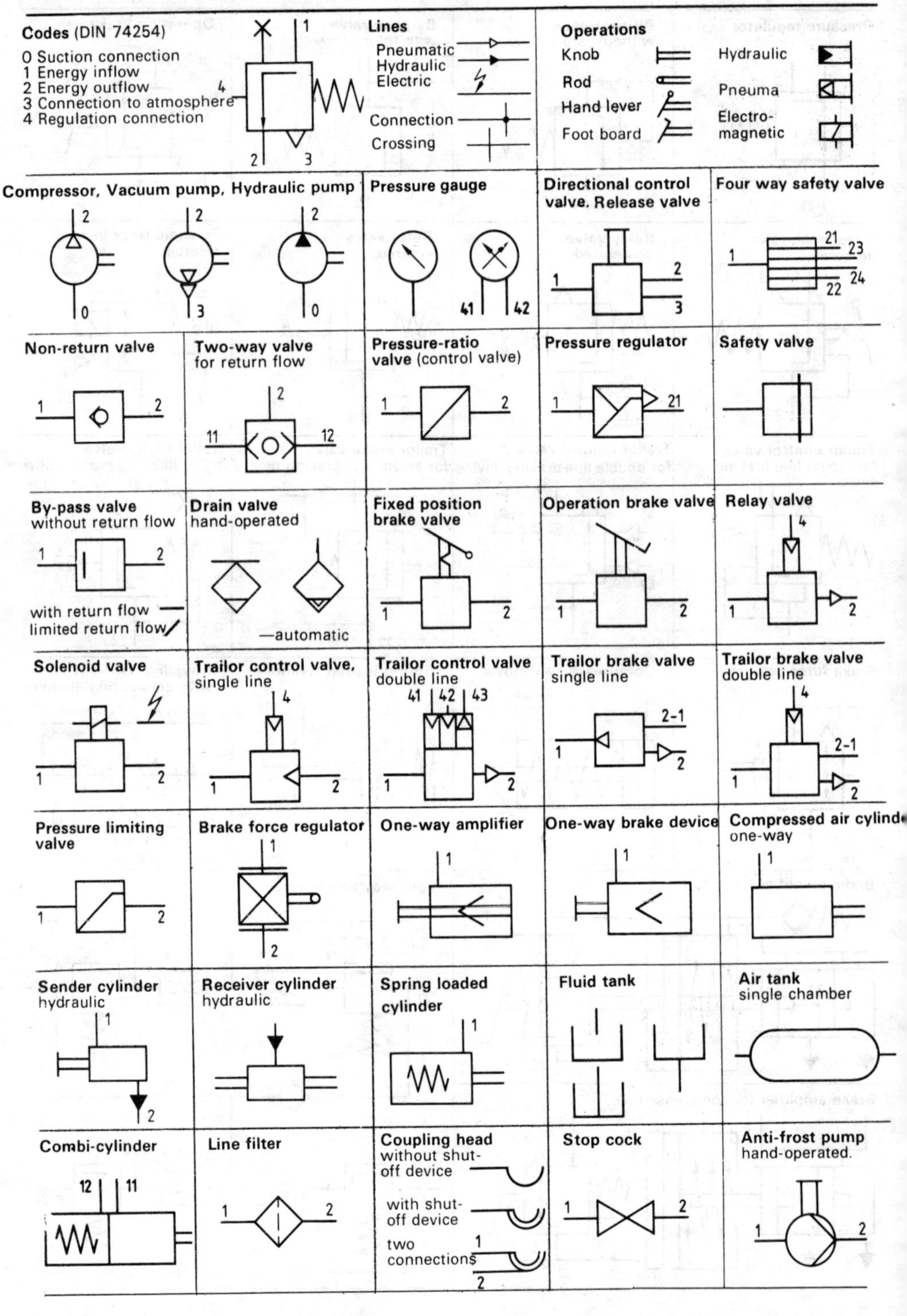

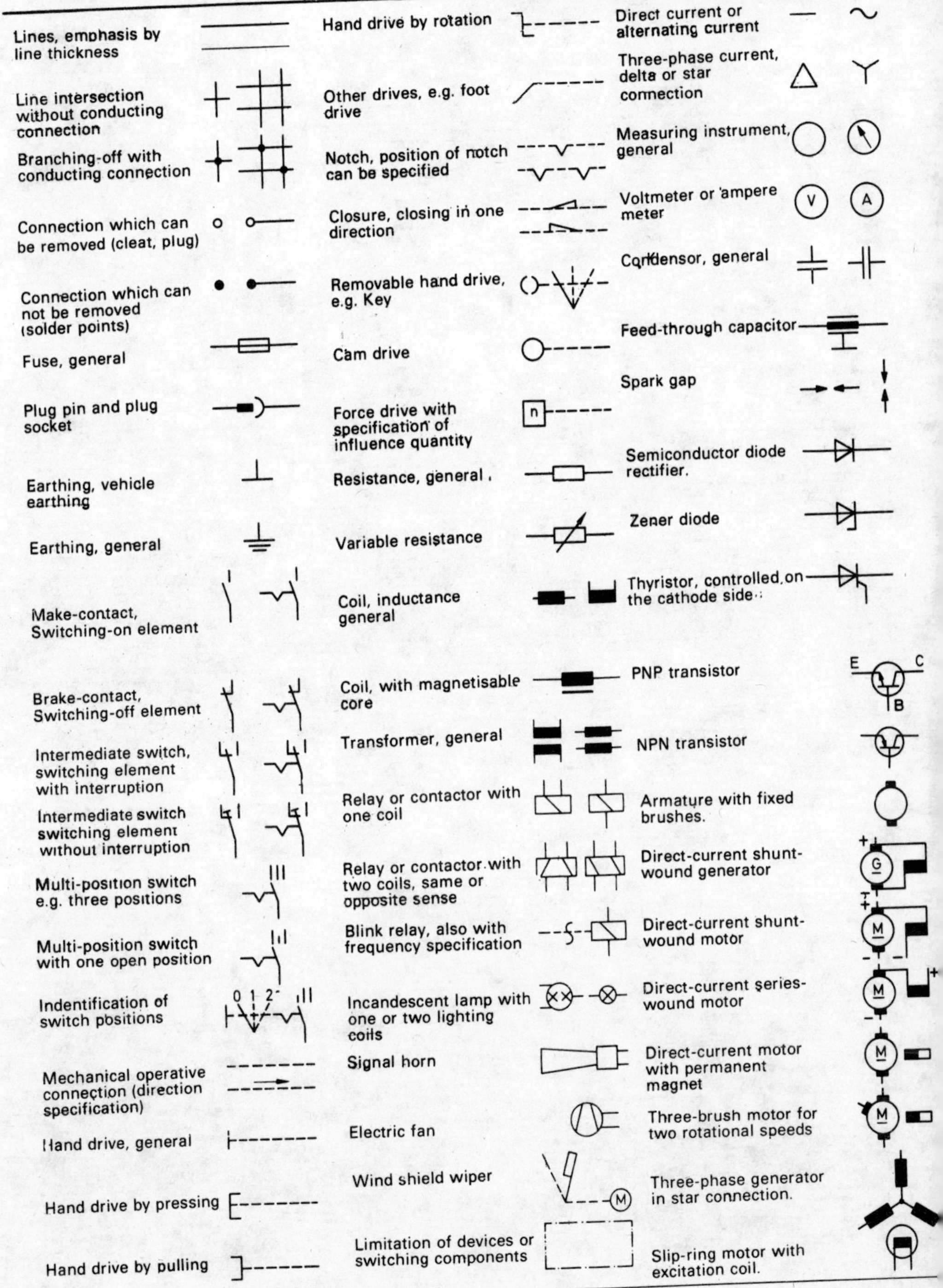
Lines, emphasis by line thickness
Line intersection without conducting connection
Branching-off with conducting connection
Connection which can be removed (cleat, plug)
Connection which can not be removed (solder points)
Fuse, general
Plug pin and plug socket
Earthing, vehicle earthing
Earthing, general
Make-contact, Switching-on element
Brake-contact, Switching-off element
Intermediate switch, switching element with interruption
Intermediate switch switching element without interruption
Multi-position switch e.g. three positions
Multi-position switch with one open position
Indentification of switch positions
0 1 2
Mechanical operative connection (direction specification)
Hand drive, general
Hand drive by pressing
Hand drive by pulling
Hand drive by rotation
Other drives, e.g. foot drive
Notch, position of notch can be specified
Closure, closing in one direction
Removable hand drive, e.g. Key
Cam drive
Force drive with specification of influence quantity
n
Resistance, general
Variable resistance
Coil, inductance general
Coil, with magnetisable core
Transformer, general
Relay or contactor with one coil
Relay or contactor with two coils, same or opposite sense
Blink relay, also with frequency specification
Incandescent lamp with one or two lighting coils
Signal horn
Electric fan
Wind shield wiper
M
Limitation of devices or switching components
Direct current or alternating current
Three-phase current, delta or star connection
Measuring instrument, general
Voltmeter or ampere meter
V
A
Condensor, general
Feed-through capacitor
Spark gap
Semiconductor diode rectifier.
Zener diode
Thyristor, controlled on the cathode side
PNP transistor
E
C
B
NPN transistor
Armature with fixed brushes.
Direct-current shunt-wound generator
G
Direct-current shunt-wound motor
M
Direct-current series-wound motor
M
Direct-current motor with permanent magnet
M
Three-brush motor for two rotational speeds
M
Three-phase generator in star connection.
Slip-ring motor with excitation coil.

INDEX